The Role of Bioenergy in the Emerging Bioeconomy

The Role of Bioenergy in the Emerging Bioeconomy

Resources, Technologies, Sustainability and Policy

Edited by

CARMEN LAGO
Energy System Analysis Unit, Energy Department, CIEMAT, Madrid, Spain

NATALIA CALDÉS
Energy System Analysis Unit, Energy Department, CIEMAT, Madrid, Spain

YOLANDA LECHÓN
Energy System Analysis Unit, Energy Department, CIEMAT, Madrid, Spain

ELSEVIER

ACADEMIC PRESS
An imprint of Elsevier

Academic Press is an imprint of Elsevier
125 London Wall, London EC2Y 5AS, United Kingdom
525 B Street, Suite 1650, San Diego, CA 92101, United States
50 Hampshire Street, 5th Floor, Cambridge, MA 02139, United States
The Boulevard, Langford Lane, Kidlington, Oxford OX5 1GB, United Kingdom

British Library Cataloguing-in-Publication Data
A catalogue record for this book is available from the British Library

Library of Congress Cataloging-in-Publication Data
A catalog record for this book is available from the Library of Congress

ISBN: 978-0-12-813056-8

For Information on all Academic Press publications
visit our website at https://www.elsevier.com/books-and-journals

Publisher: Jonathan Simpson
Acquisition Editor: Raquel Zanol
Editorial Project Manager: Charlotte Rowley/Gabriela Capille
Production Project Manager: Sruthi Satheesh
Cover Designer: Christian Bilbow

Typeset by MPS Limited, Chennai, India

CONTENTS

List of Contributors *xi*
Preface *xv*
Acknowledgments *xvii*

Part I Introduction, Resources and Technology

1. Nexus Bioenergy—Bioeconomy **3**

Carmen Lago, Israel Herrera, Natalia Caldés and Yolanda Lechón

1.1 Nexus Bioenergy—BE 4
1.2 BE and Sustainable Development 6
1.3 Bioeconomy—Circular Economy 12
1.4 Cascading Use 16
1.5 Bioenergy and Policy Context 17
References 20

2. Biomass Resources **25**

Javier Sánchez, María Dolores Curt, Nicolas Robert and Jesús Fernández

2.1 Biomass as a Renewable Source of Energy 25
2.2 Chemical Composition and Characterization of Biomass 30
2.3 Classification of Biomass Types 35
2.4 Biomass Resources 40
List of Abbreviations 97
References 98

3. Liquid Biofuels **113**

Mercedes Ballesteros and Paloma Manzanares

3.1 Introduction 113
3.2 First-Generation Technologies for Liquid Biofuels Production 114
3.3 Second-Generation Technologies for Liquid Biofuels Production 130
3.4 Prospective Technologies 138
3.5 Challenges, Opportunities, and Barriers 139
3.6 Concluding Remarks 141
References 141

4. Solid Biomass to Heat and Power **145**

Jorge Islas, Fabio Manzini, Omar Masera and Viridiana Vargas

4.1 Introduction 145
4.2 Biomass for Solid Biofuels 150
4.3 Solid Biomass Consumption 157
4.4 Solid Biomass Heat and Power Generation Technologies 158
4.5 Thermal Energy and/or Electricity Generation Costs With Biomass 171
4.6 Conclusions 172
Acknowledgments 173
References 173

5. Biorefineries **179**

Maria Hingsamer and Gerfried Jungmeier

5.1 Introduction 180
5.2 Description of Biorefineries 181
5.3 Concepts of Biorefineries 189
5.4 Integration in Existing Infrastructure 196
5.5 Biorefinery Complexity Index 201
5.6 Economic and Environmental Aspects of Biorefineries 206
5.7 Perspectives 217
Endnotes 220
References 221

Part II Sustainability

6. Sustainability of Bioenergy **225**

6.1 State of the Art 226
Uwe R. Fritsche
6.1.1 Sustainability: From Stockholm to Rio to New York—and Beyond 226
6.1.2 Sustainable Biomass: Principles, Criteria, and Indicators 229
6.1.3 Sustainability Challenges for Bioenergy: Open Issues 232
6.1.4 Governance of Sustainable Bioenergy 235
6.1.5 The Way Forward 235
References 236
Further Reading 241
6.2 Sustainability of Bioeconomy 241
Rocio Diaz-Chavez
6.2.1 Social Considerations for the Bioeconomy Sustainability Assessment 241

6.2.2 Social Methods in Bioeconomy 245
6.2.3 Job Creation, Working Conditions, and Rural Development 247
6.2.4 Food Production, GMOs, and Links to the Sustainable Development Goals 252
6.2.5 Stakeholders, Gender, Skills, and Training 254
6.2.6 Society Perception, Corporate Sustainability Reporting, and Monitoring 255
6.2.7 Conclusions 256
References 256
Further Reading 258
6.3 Socioeconomic Pillar. Methodology and Case Study 259
Cristina de la Rúa and Yolanda Lechón
6.3.1 Overview 259
6.3.2 Socioeconomic Implications of Bioenergy: A Review 259
6.3.3 Methodologies to Estimate the Socioeconomic Implications of Bioenergy 265
6.3.4 The IO Methodology 266
6.3.5 Discussion and Conclusions 276
References 277
Further Reading 280
6.4 Environmental Pillar: Methodology and Case Study 280
Benoît Gabriel and Aurelie Perrin
6.4.1 Methodological Background 280
6.4.2 Application to Bioenergy Chains 282
6.4.3 Bourgogne Pellets: A Case Study Based on Perennial Biomass Crops 284
6.4.4 Advanced Topics 292
References 294
Further Reading 296

7. Key Challenges and Opportunities 297

7.1 Contribution of Bioenergy to Rural Development 298
Francisco José Domínguez Pérez
7.1.1 The Status of World's Rural Population 298
7.1.2 Energy Availability and Supply in Rural Areas 300
7.1.3 Access to Modern Energy Services by Means of Locally Produced Bioenergy 303
7.1.4 Multifunctionality of Bioenergy in an Integrated Production System for Food, Feed, Energy, and Other Products 304
7.1.5 Multiplier Effect of Bioenergy in Local Economic Development 305
7.1.6 Environmental Benefits 309

7.1.7 The Experience of "Bioenergy Villages" 311
7.1.8 Specific Considerations for Developing Countries 312
7.1.9 Potential Risks and Necessary Requirements for a Sustainable and Equitable Development 316
7.1.10 Conclusions 317
References 319
7.2 Factors for Bioenergy Market Development 320
Paul O'Reilly
7.2.1 Introduction 320
7.2.2 Review of Bioenergy Consumer Studies 321
7.2.3 A Framework for Developing the Bioeconomy 325
7.2.4 Bioenergy: Policy Context in Europe 326
7.2.5 National Bioeconomy Strategies 328
7.2.6 European Commission Market Pull Recommendations 330
7.2.7 B2B Marketing of Bio-Based Products 333
7.2.8 Public Procurement as a Market Pull Measure for Bioenergy 333
7.2.9 Conclusions 336
References 338
Further Reading 340
7.3 Climate Change and Bioenergy Governance 341
Felipe García Cardona and Mateo Prada Quintero
7.3.1 Developing a Climate Change Agenda 342
7.3.2 Bioenergy Opportunities: Setting the Agenda 344
7.3.3 Why Creating an International Agenda on Bioenergy? 347
7.3.4 Bioenergy Challenges: Governance Coordination and NDCs Implementation 348
7.3.5 Case Study: Colombia's Biodiversity Strategy as a Multilevel Approach to Governance 351
References 352
Further Reading 354
7.4 Bioenergy and Food Security: Synergies and Trade-offs 355
Siwa Msangi
7.4.1 Introduction 355
7.4.2 The "Food Versus Fuel" Debate: A False Trade-off? 357
7.4.3 Rethinking Food Security and Bioenergy Linkages 361
7.4.4 Where Bioenergy Can Help (or Hurt) Food Security 366
7.4.5 Conclusions 373
References 375

8. Land—Water—Energy Nexus of Biofuels Development in Emerging Economies: A Case Study of Bioethanol Policy in Thailand **379**

Thapat Silalertruksa and Shabbir H. Gheewala

8.1 Emerging Economies and Biofuels Development 380
8.2 Land—Water—Energy Nexus and Sustainability of Food and Biofuels Production 381
8.3 Life Cycle—Based Indicators for Land—Water—Energy Nexus Assessment 384
8.4 A Case Study: Land—Water—Energy Nexus of Biofuels Policy Mandate in Thailand 388
8.5 Conclusions 398
References 399

Part III Future Trends and Policy

9. Innovation on Bioenergy **405**

Francisco Gírio

9.1 Introduction 405
9.2 Emerging Biofuels From Biochemical-Based Technologies 406
9.3 Emerging Biofuels From Thermochemical-Based Technologies 413
9.4 Advanced Renewable Fuels 418
9.5 Future Trends on Heat, Cooling, and Power Technologies 421
9.6 Conclusions 427
Acknowledgments 429
References 429

10. Future Role of Bioenergy **435**

Nicolae Scarlat and Jean-Francois Dallemand

10.1 Bioenergy Perspectives 435
10.2 Policies and Measures to Support Sustainable Bioenergy 489
Disclaimer 537
References 537
Further Reading 546

Index *549*

LIST OF CONTRIBUTORS

Mercedes Ballesteros
Biofuels Unit, Energy Department, CIEMAT, Madrid, Spain

Natalia Caldés
Energy System Analysis Unit, Energy Department, CIEMAT, Madrid, Spain

María Dolores Curt
Agro-Energy Group, College of Agricultural Engineering, Technical University of Madrid, Madrid, Spain

Jean-Francois Dallemand
Directorate for Energy, Transport and Climate, European Commission Joint Research Centre, Ispra (VA), Italy

Cristina de la Rúa
Technical University of Munich, Munich, Germany

Rocio Diaz-Chavez
Stockholm Environment Institute Africa Centre, Nairobi, Kenya

Jesús Fernández
Agro-Energy Group, College of Agricultural Engineering, Technical University of Madrid, Madrid, Spain

Uwe R. Fritsche
International Institute for Sustainability Analysis and Strategy (IINAS), Darmstadt, Germany

Benoît Gabriel
EcoSys Joint Research Unit, AgroParisTech, INRA, Paris, France

Felipe García Cardona
Dirección General, Colombia Bio. Bogotá, Colombia

Shabbir H. Gheewala
The Joint Graduate School of Energy and Environment, King Mongkut's University of Technology Thonburi, Bangkok, Thailand; Center for Energy Technology and Environment, Ministry of Education, Bangkok, Thailand

Francisco Gírio
LNEG, Bioenergy Unit, Lisboa, Portugal

Israel Herrera
Energy System Analysis Unit, Energy Department, CIEMAT, Madrid, Spain

Maria Hingsamer
Joanneum Research Forschungsgesellschaft mbH, LIFE—Centre for Climate, Energy and Society, Graz, Austria

Jorge Islas
Instituto de Energías Renovables, Universidad Nacional Autónoma de México, Temixco, Morelos, Mexico

Gerfried Jungmeier
Joanneum Research Forschungsgesellschaft mbH, LIFE—Centre for Climate, Energy and Society, Graz, Austria

Carmen Lago
Energy System Analysis Unit, Energy Department, CIEMAT, Madrid, Spain

Yolanda Lechón
Energy System Analysis Unit, Energy Department, CIEMAT, Madrid, Spain

Paloma Manzanares
Biofuels Unit, Energy Department, CIEMAT, Madrid, Spain

Fabio Manzini
Instituto de Energías Renovables, Universidad Nacional Autónoma de México, Temixco, Morelos, Mexico

Omar Masera
Centro de Investigaciones en Ecosistemas, Universidad Nacional Autónoma de México, Morelia, Michoacán, Mexico

Siwa Msangi
Environment & Production Technology Division, International Food Policy Research Institute, Washington DC, United States

Paul O'Reilly
College of Business, Dublin Institute of Technology, Dublin, Ireland

Francisco José Domínguez Pérez
Institute for Energy Diversification and Saving (IDAE), Madrid, Spain

Aurelie Perrin
EcoSys Joint Research Unit, AgroParisTech, INRA, Paris, France

Mateo Prada Quintero
Dirección General, Colombia Bio. Bogotá, Colombia

Nicolas Robert
European Commission—DG Joint Research Centre (JRC), Directorate D—Sustainable Resources, Unit D1—Bio-Economy, Ispra, Italy

Javier Sánchez
European Commission—DG Joint Research Centre (JRC), Directorate D—Sustainable Resources, Unit D1—Bio-Economy, Ispra, Italy

Nicolae Scarlat
Directorate for Energy, Transport and Climate, European Commission Joint Research Centre, Ispra (VA), Italy

Thapat Silalertruksa
The Joint Graduate School of Energy and Environment, King Mongkut's University of Technology Thonburi, Bangkok, Thailand; Center for Energy Technology and Environment, Ministry of Education, Bangkok, Thailand

Viridiana Vargas
Instituto de Energías Renovables, Universidad Nacional Autónoma de México, Temixco, Morelos, Mexico

PREFACE

This book provides the reader with a complete understanding of the role that bioenergy may play in the new bioeconomy paradigm. Given the authors' expertise and outstanding contributions to this book, this publication provides an excellent, highly readable, and comprehensive overview of the diverse issues of relevance to bioenergy in the emerging bioeconomy. Given the current relevance and even increasing importance that this topic is going to have in the future, it is a useful reference document, not just for the sector itself, but also more widely for policy and decision makers as well as researchers, analysts, and graduate students.

This book covers all key topics such as the nexus between bioenergy and bioeconomy, resources, transport biofuels, heat and power biomass, biorefineries, sustainability of bioenergy, challenges and opportunities, the water–energy–land nexus, innovation, and the future role of bioenergy. It is worth mentioning that for some of these topics, a more detailed analysis and information is provided. For example, within sustainability of bioenergy, a chapter on the state of the art is provided followed by three chapters that cover the three sustainability pillars—social, environmental, socioeconomic—not only from a conceptual point of view but also from an assessment methodological point of view through concrete bioenergy case studies. Similarly, within the section on key challenges and opportunities, four distinctive chapters cover address: the potential contribution of bioenergy to rural developments, bioenergy market developments, relation between bioenergy and food security and bioenergy governance.

The proposed book enlarges the existing body of literature by providing a complete overview of bioenergy main issues focusing on the challenges that bioenergy will have to overcome in order to play a key role in the future energy system.

Sustainable development of bioenergy into a circular economy will require the integration and consideration of the interests and concerns of policy makers, researchers and technology developers, project developers, as well as society. For sustainable bioenergy to have a key role in future energy systems, it will be important to make an optimal use of biomass resources with less environmental impact (particularly reducing greenhouse gas emissions) while fostering positive social and economic effects.

Additionally, novel issues such as land–energy–water nexus implications and governance aspects are treated in detail in the book, being these aspects usually missing in the existing literature in this area. The presentation of the future role of biomass in transition pathways scenarios in competition with other alternative technologies gives the reader with a more complete understanding of the challenges and opportunities of bioenergy.

We hope the book will help researches, analysts, policy makers, and students to update specific knowledge and expertise other aspects of bio-based economy not included into their day-by-day activities but needed in the development of their investigations. The book also offers knowledge of valuable case studies of the three pillars of sustainability as well as the competition for land, water, and food and the future trends and challenges.

ACKNOWLEDGMENTS

First, we would like to gratefully acknowledge all authors who have written the various chapters included in the *Bioenergy in the Emerging Bioeconomy* book. Our sincere gratitude for contributing with their time and expertise to make this a relevant, comprehensive, scientifically sound, up-to-date, and high quality publication. Without their dedication and outstanding contributions, this would not have been possible.

Second, we would like to thank the reviewers of this book for their support and guidance to improve the quality, coherence, and content presentation of chapters through valuable and outstanding comments and suggestions. Besides the initial revision by the editors, the review process has been a time-consuming task by a distinguished group of researchers in the field. For this reason, we would like to acknowledge the immense contribution of Ana Luisa Fernando (Universidade Nova de Lisboa, Lisboa); Barbara Bramble (The National Wildlife Federation, Washington); and Miguel Brandao (KTH Royal Institute of Technology, Stockholm).

Finally, we would also like to expand our deepest gratitude to all those who have directly and indirectly guided us in writing this assignment. We want also to make a special mention to Elsevier editorial personnel for their patient, professional support, and encouragement throughout the book development process.

PART I

Introduction, Resources and Technology

CHAPTER ONE

Nexus Bioenergy—Bioeconomy

Carmen Lago, Israel Herrera, Natalia Caldés and Yolanda Lechón
Energy System Analysis Unit, Energy Department, CIEMAT, Madrid, Spain

Contents

1.1 Nexus Bioenergy—BE 4
1.2 BE and Sustainable Development 6
1.3 Bioeconomy—Circular Economy 12
1.4 Cascading Use 16
1.5 Bioenergy and Policy Context 17
References 20

The role of bioenergy in the emerging bioeconomy (BE) aims to provide the reader with a complete understanding of the possible future role that bioenergy can play in the new BE paradigm. For this purpose, the book has gathered some of the most prominent authors to cover a wide range of issues that will determine the present and future role of bioenergy in the BE.

Besides describing the content of the book, this chapter also introduces some key concepts that are considered relevant and will be used throughout the following chapters as well as a brief summary about current status and policy context of bioenergy.

As for the content of the book, it starts with an introduction to the nexus between bioenergy and BE as well as the resources and technologies that could play an important role in future bioenergy production. Next, a whole section is also devoted to analyze how sustainability of bioenergy can be assessed from a methodological point of view. In this sense, the three sustainability pillars are addressed, and the applicability of the proposed methodological approaches is demonstrated through concrete bioenergy case studies. Additionally, some of the most relevant challenges and opportunities for bioenergy are also covered in this book. Examples of those include its potential impacts in rural development, food security, and other related markets. The role that institutions and governance currently play in the present and future of bioenergy is also

The Role of Bioenergy in the Emerging Bioeconomy.
DOI: https://doi.org/10.1016/B978-0-12-813056-8.00001-7

covered in this book together with the implications of bioenergy in the water–energy–land nexus. Furthermore, the expected evolution of biomass technologies as well as the role that bioenergy can play in the possible transition pathways to a low carbon economy are presented using the results from the energy modeling literature. Finally, the last section is dedicated to explore and propose a portfolio of policies and measures to support sustainable bioenergy in the future taking into consideration its challenges and opportunities. As a note of caution to the reader, it is important to take into consideration that many of the concepts covered in this book—such as BE, circular economy (CE), cascading use, and even sustainable development—do not yet have a universal and commonly agreed definition. In this sense, one can find different meanings with slightly different nuances of the same concept in the literature depending on the author and context.

The remaining part of this chapter will focus on introducing the two concepts that are pillars to this book: bioenergy and BE.

1.1 NEXUS BIOENERGY–BE

Bioenergy is defined as a renewable energy produced from natural sources capable of replacing fossil energy. There exist a wide range of biological resources that can be used to produce bioenergy such as energy crops, biomass residues from forest and crops, wastes and by-products from agro-industries and pulp and paper industries, wet organic wastes and the organic fraction of municipal solid wastes. Macro algae, microalgae, seaweeds, and aquatic plants are also considered very attractive future biological resources (OECD/IEA, 2017). Regardless of the resource, bioenergy can be used as a sustainable source of power providing heat, gas, and fuel to produce heat, electricity, and cogeneration and transportation fuels. More recently, biomass can also be used for the production of biomaterials and bioproducts.

Projections by the International Energy Agency (IEA) provide a clear vision of the importance of bioenergy expansion. According to the same source, bioenergy will provide nearly 17% of the final energy demand by 2060. At the same time, bioenergy can reach 20% of cumulative greenhouse gas (GHG) emission savings (OECD/IEA, 2017; Scatlat et al., 2015).

These remarkable figures must be obtained in a sustainable way contributing to decarbonize the energy systems. Modern bioenergy is obtained by mature technologies currently available in the biomethane market from wastes and residues, district heating networks, agricultural residues to generate electricity, and different transportation fuels, while other technologies, mostly related to transportation fuels are close to market (biomass gasification, pyrolysis and ethanol from lignocellulosic feedstock) and require additional support.

The BE concept is still under development. According to the OECD, BE refers to the set of economic activities relating to the invention, development, production, and use of biological products and processes. According to the same source, BE could bring major future benefits related to improvements in people's health, agriculture, and industrial productivity boosts while enhancing environmental sustainability (OECD, 2009). Other authors such as McCormick and Kautto understand BE as an economy where the basic components for the production of materials, chemicals, and energy are derived from renewable biological resources (McCormick and Kautto, 2013). The European Commission first defined BE as an economy where production of food, feed, fiber, bio-based products, and bioenergy is efficient and sustainably obtained from renewable resources from land, fisheries, and aquaculture environments including the related public goods (EC, 2012). Later, in 2014, the concept was modified to include that efficient and sustainable primary production and processing meet industry demand, consumer's needs, and at the same time address environmental challenges such as climate change (van de Pas, 2015). Other authors stated that BE uses renewable biological biomass (trees, shrubs, crops, plantations, algae and aquatic plants, waste and primary agricultural residues, waste and secondary agricultural residues, and successive generations of waste and residues) in its production processes obtaining different types of outputs (thermal energy, liquid fuels, chemicals, bioproducts, food and fodder, as well as cosmetics and medicines; Adamowicz, 2017). More recently, BE has gone from being perceived as a promise to become a tangible reality that will play a remarkable role when it comes to achieving global sustainability (Maciejczak, 2017). On the same line, the revised Bioeconomy Strategy and Action Plan broadens the definition of the BE, which must to make connections with other industrial sectors (construction, engineering, manufacturing, information and communication technology, and urban planning) to make human activity more environmentally friendly, more circular, more bio-based, more nature-inspired, more inclusive, and more competitive, and ultimately more sustainable (Rauschen and Esch, 2017).

Over the next few years, the concept of BE will probably continue to evolve until a more precise and consensual definition is achieved.

1.2 BE AND SUSTAINABLE DEVELOPMENT

A new concept of economy that takes into account the environmental variables and recovery times of "bio" raw materials, residues, and wastes must be promoted. For this purpose, sustainable assessments of bioenergy must cover the three sustainability pillars: social, environmental, and economic, so that their results can incline decision-makers towards those chains that are most beneficial for the environment, society, and the economy. In this context, the sustainable use of bioenergy must boost the social, environmental, and economic benefits gained from the better management of renewable resources in such emerging BE.

The most important roles that both bioenergy and BE are expected to play are to significantly reduce GHG emissions (in comparison to the fossil counterpart) and to help achieve the sustainable development goals (SDGs) (Bioenergy for Sustainable Development, 2017; GBEP Sustainability Indictors for Bioenergy (GSI)). One of the climate change mitigation measures stated in the Paris Agreement includes the use of renewable energies in substitution of fossil fuels as a way to meet the GHG emission reduction targets and to keep the increase in global temperature this century below 2°C. For these objectives to be achieved, the implication of all parties, the necessary support to developing countries, the appropriate finance funding, and a new technological framework development are required (UN, 2015a).

Bioenergy impacts on the three sustainability pillars must not be generalized nor taken for granted. Instead, it is important to keep in mind that bioenergy development could lead to both positive and negative impacts depending on how, where, and under what circumstances biomass is produced, harvested, transported, and processed. To shed some light to this issue and given the importance of sustainability in the current and future debate on bioenergy, several research studies have been carried out by the scientific community. While most of the studies have focused on the positive aspects related to climate change mitigation such as GHG emission savings, negative impacts have also received special attention such as biodiversity loss and indirect land use change (iLUC).

Among the latter, iLUC has been a predominant concern, especially in the case of biofuels (Ben Aoun et al., 2013; Yeh and Witcover, 2010; Finkbeiner, 2014; Kim et al., 2014; Searchinger et al., 2008; Malins et al., 2014). Several models have been developed addressing iLUC effects (Fehrenbach et al., 2009) such as the FSOM-FAPRI (US Environmental Protection Agency, 2010), various GTAP versions (Laborde, 2011; Darlington et al., 2013; Taheripour et al., 2011), the GCAM model (Plevin, 2016), and the GLOBIOM model (Valin et al., 2015). Other authors have developed new frameworks for a consistent modeling of iLUC within the Life Cycle Assessment (LCA) methodology (Finkbeiner, 2013; Schmidt et al., 2015). More recently, iLUC effects from lignocellulosic non-food biomass—*Miscanthus*, switchgrass, short rotation coppice, perennial grasses, etc.—have also been studied, identifying the strong influence of assumptions on yield, land conversion, and carbon stocks change on the final results (Pavlenko and Searle, 2018).

Biodiversity loss is another key concern related to bioenergy. The sixth extinction predicted by several authors (Tilman et al., 2017; Ceballos et al., 2015; Barnosky et al., 2011) is also relevant to bioenergy and BE development. Several studies stated land-use change as the main driver for biodiversity loss, whose main impacts are related to loss of valuable habitats and shifts in species richness and abundance. The Immerzeel and co-workers' review of 53 selected publications (Immerzeel et al., 2014) confirm both direct and indirect negative impacts with respect to the expansion of energy crops. Nevertheless, it also concludes the possibility of positive effects on biodiversity, particularly with the second-generation perennial crops. Mitigation measures have been proposed to reduce the potential negative impacts on biodiversity and boost the positive effects: (1) better management practices, (2) use of marginal, degraded, or abandoned agricultural land, (3) adequate land-use planning, defining conservation areas as well as landscape design maintaining landscape heterogeneity, (4) policy and regulatory measures to ensure protection to high-value natural ecosystems, avoiding deforestation, providing ecological corridors, ecological buffer zones, and (5) policy alignments between agriculture, environmental, and rural development sectors.

However, there is a consensus among the scientific community about the existence of knowledge gaps in biodiversity impacts from bioenergy.

To address such limitation, several lines of future research have been identified which include, among others, the need to (1) study all the vulnerable regions, (2) integrate multiple spatial scales, (3) combine o complementary biodiversity indicators, (4) assessment of species extinction risks, (5) include taxonomic groups in the analysis and quantification of pollution effects on biodiversity, especially in aquatic ecosystems, and (6) assess the effects of different mitigation measurements on biodiversity and long-term studies (Immerzeel et al., 2014).

Huge efforts have also been devoted to develop sustainability criteria and indicators as well as to explore synergies between different land uses, as well as to mitigate negative impacts on land, water, and biodiversity (Fritsche et al., 2017b; GBEP-IEA Bioenergy Examples of Positive Bioenergy and Water Relationships, 2016; Thrän et al., 2017; Thrän et al., 2016; Fritsche et al., 2011; FAO-GBEP, 2011). As to the search for sustainability criteria and indicators, one of the pioneer initiatives was the Global Bioenergy Partnership (http://www.globalbioenergy.org/). This effort has resulted in the proposal of 24 sustainability indicators for bioenergy (eight dedicated to environmental pillar, eight to social, and eight to economic one). The consideration of such indicators is expected to provide assistance to "the development of national bioenergy policies and programmes, monitor the impact of these policies and programmes, as well as interpret and respond to the environmental, social and economic impacts of their bioenergy production and use" (FAO-GBEP, 2011).

The future of bioenergy must necessarily be closely linked not only to BE but also to sustainable development. Back in September 2015, countries adopted 17 goals to end poverty, protect the planet, and ensure prosperity for all as part of a new sustainable development path. The next step was taken on January 2016, when the 17 SDGs of the 2030 Agenda for Sustainable Development—adopted by world leaders in September 2015 at a historic UN Summit—officially came into force. As a result, the Agenda is an action plan for people, the planet, and prosperity, which requires international cooperation from all countries and the stakeholders' involvement (UN, 2015b).

Although most SDG goals are somehow interconnected to the new BE concept, some of them are particularly related to bioenergy (Fritsche et al., 2017a). For example, bioenergy can clearly contribute to the achievement of SDG-7 "promote to ensure access to affordable, reliable, sustainable and modern energy for all people". In turn, such access to affordable and clean energy can stimulate productive activities and contribute to eradicate poverty, improve food security, and promote sustainable agriculture (IRENA, 2017b). Along the same lines, bioenergy can contribute to SGD-11 "Make cities inclusive, safe, resilient and sustainable" and SGD-12 "responsible consumption and production". Furthermore, bioenergy can also contribute to SGD-13 "Take urgent action to combat climate change and its impacts" and SGD-15 "Sustainably manage forests, combat desertification, halt and reverse land degradation, halt biodiversity loss" (OECD/IEA, 2017; Bioenergy for Sustainable Development, 2017; IRENA, 2017b). Such desirable pattern requires the collaboration of all citizens so that they make a rational use of natural resources throughout the entire life cycle both as consumers and producers.

Another crucial aspect resides in the interdependence between bioenergy and water consumption, which is likely to be exacerbated in the future. Such link is demonstrated, for example, by the fact that water is needed for energy production and water supply needs energy for its extraction and transportation. Under the expected future scenarios of water scarcity, increased bioenergy production and growing population will require a successful and smart management of the binomial energy—water. This challenge will also have to be addressed to fulfill SDG 6 related to "clean water and sanitation." Jointly addressing the problem of water scarcity and increasing energy demand will allow for smart and efficient solutions to supply clean water and energy (World Energy Outlook, 2016).

Another aspect to take into consideration is the expected future increase in the demand for bioenergy, biomass, and biomaterials which will be accompanied by increased competition for natural resources such as land and water. Unless considered and managed wisely, such competition will generate negative impacts on the environment, biodiversity, and land use. These impacts will be partly offset by expected increases in crop productivity and better breeding management. Similarly, there will be strong competition between traditional uses of biomass and its waste and the most innovative new uses (biomaterials and bioproducts production), a framework in which both cascade use and biorefineries will become very important. Achieving sustainability in the use of biomass is therefore one of the key points of the BE (Scatlat et al., 2015).

The work by Popp (Popp et al., 2014) provides a holistic vision of bioenergy expansion and its influence on food, energy, and environment. Results from this work indicate that increased demand of land as a result of biomass production will most likely affect food prices and water scarcity. Thus, foreseen conflicts for land must be taken into account by policy makers and planners by integrating land use and water scarcity. Furthermore, existing environmental impacts such as, among others, erosion, water stress, and increasing soil salinity could be amplified by climate change, which could also exacerbate land conflicts. Finally, social aspects must also be considered to avoid that communities are, as a result of bioenergy expansion, displaced and lose access to land and other natural resources (FAO-GBEP, 2011).

The more sustainable scenarios are those that avoid expanding land for bioenergy into sensitive areas such as primary forest, natural grasslands, and/or peatlands. In this sense, national policies that promote bioenergy and BE as a whole should take into account potential GHG emissions derived from changes in carbon stocks around the world (Thrän et al., 2017). Thus, to ensure sustainable development, most of the bioenergy expansion must be accomplished on existing agricultural land.

Finally, competition for bio-based resources is one of the challenges that bioenergy will have to overcome to play a key role in the future energy system. In this sense, bioenergy will continue to be a possible alternative within the BE as long as its use of bio-based raw materials, residues, and wastes competes with other production chains from other economic sectors such as agriculture, forestry, fisheries, food, and pulp and paper production, as well as chemical, biotechnological, and biorefineries capable of producing outputs for most of the above-mentioned sectors (biochemical, biomedicines, bioplastics, energy, etc.). The future will tell us which of the above-mentioned bio-based raw materials' use will persist when moving towards a more sustainable, fairer, and more habitable planet.

IRENA, IEA, and FAO, the three recognized institutions in favor of sustainable development (Bioenergy for Sustainable Development, 2017) consider that it is possible to combine food, feed, fiber, bioenergy, and bio-based products production in a sustainable way as long as several principles are followed: (1) identify areas best suited for bioenergy production (agro-ecological zoning) through inclusive multi-stakeholder processes, (2) sustainable intensification and landscape planning can make land available for additional production while enhancing ecosystem services, (3) develop good management practices at a local level for growing a mix of high-yielding food and fuel crops in different soils and climates, (4) secure land tenure can give farmers financial incentives to manage their land for higher yields while sustaining soil productivity, (5) promote a widespread use of modern farming techniques, (6) utilize organic residues within healthy soil conditions, (7) reduce losses in the food chain, which can also lower the pressure on land (Bioenergy for Sustainable Development, 2017).

As for the importance and role that BE is currently playing in today's world economy, in 2014, economic goods related to BE (agriculture, forestry, food, bioenergy, biotechnology, and green chemistry) were marketed for a value around US$ 2 trillion, which accounts for 13% of world trade (El-Chichakli et al., 2016). In addition, these sectors are essential to comply with those SDG related to food security, energy, and health. However, establishing a global BE is not an easy affair, which often clashes with divergent national policies. According to some authors (Arundel and Sawaya, 2009; Communiqué Global Bioeconomy Summit, 2015; El-Chichakli et al., 2016), the main challenges to achieve a global sustainable BE include (1) fostering regional and international coordination and cooperation activities which are essential to share knowledge, skills, and experience to optimize the use of the resource base, (2) achieving a smooth and continuous conversation between the interested parties to define the criteria for the evaluation of the BE as well as the national monitoring systems, (3) developing UN-sponsored policy initiatives that can coordinate and interconnect efforts to achieve the international treaties' goals related to sustainability such as the Paris Accord on Climate Change, 2030 Sustainable Agenda or the Aichi agreement on biodiversity, (4) educating and providing capacity building on the topic of sustainable BE from an interdisciplinary point of view. Such education must consider local specificities but also include global environmental, social, and economic assessment as well as planning policy and new ways of thinking. Additionally, there is a need to (5) fostering innovation, research, and cooperation about strategic topics such as biorefineries, new food systems, sustainable cities, aquaculture, and so on.

As for the stakeholder involvement and governance required to achieve sustainable development, all actors must be involved and participate in the process. Among others, citizens, politicians, business, consumers, policy makers, local, regional, and national authorities, scientific and technological community, retailers, media, non-governmental organizations, as well as development cooperation agencies will have a role to play.

In this sense, the private sector is expected to be a corner stone in the process towards achieving sustainability of bioenergy and the BE. Partly in response to the growing concerns and pressure from consumers and civil society in general, the private sector is expected to develop and join innovative initiatives aimed at promoting sustainable development. Independently of public measures which may take a longer time to be designed and implemented, private sector representatives, i.e., farmers, farmers associations, manufacturers, retailers, and agro-forestry industries, are expected to play a very important role in promoting sustainable bioenergy and BE. Proof of that is the current adoption of practices such as advanced control systems (i.e., low nutrient application, more efficient irrigation systems, use of renewable energies in agro-forestry sector, and zero deforestation actions). Similarly, the private sector has already been very active in developing and embracing numerous sustainability standards and certification voluntary schemes. Examples of those include the Roundtable for Sustainable Palm Oil—RSPO (https://rspo.org/); Forest Stewardship Council—FSC (https://ic.fsc.org/en); Sustainable Biomass Program—SBD (https://sbp-cert.org/); Fair Trade Labelling Organizations International (https://www.fairtrade.net/); Green Gold Label Program (http://www.greengoldcertified.org/); International Federation of Organic Agriculture Movements—FOAM (https://www.ifoam.bio/); Sustainable Agricultural Standards (https://www.rainforest-alliance.org/business/sas/); Ethical Trading Initiative (https://www.ethicaltrade.org/), etc.

Finally, education, innovation, as well as international, regional, and local cooperation and coordination will also be crucial to fulfill sustainable development into a low carbon BE.

1.3 BIOECONOMY–CIRCULAR ECONOMY

As introduced before, BE can be defined as the type of economy that uses natural renewable biological resources, from both land and ocean, to obtain food, materials, and energy in a sustainable way without compromising their availability for future generations. Crops, fisheries, forestry, microorganisms, algae, and animals as well as wastes or residues generated by them are included into the "renewable biological resources" category, while waste and residues become inputs for the production of food, feed, industrial bioproducts, and energy. Extending the concept from an economic point of view, BE can be described as those economic activities related to the invention, development, production, and use of biological products and processes (OECD, 2009).

Over the last few years, the inclusion of bioenergy into the CE concept has required the simultaneous consideration and conciliation of the sometime conflicting interests and concerns of policy makers, researchers, technology developers, project developers, and society. For sustainable bioenergy to play a key role in future energy systems, it will be important to make an optimal use of biomass resources with less environmental impact (particularly reducing greenhouse gas emissions) while fostering positive social and economic effects.

The traditional conceptualization of an economy is defined by its linear character: "take, make, and dispose." As a result, current economy is conceived as basically extractive. However, CE is based on the fundamental "closing the loop" that implies collecting waste from the different processes, recycling, and using it to make new products. The CE strategies to close the loop are related to reuse, remanufacturing and recycling. In other words, CE is based on reducing/avoiding waste generation and pollution, keeping wastes and materials in use in the cycle, and protecting natural systems (Ellen MacArthur Foundation, 2015). However, the major challenge lies in the selection of supply chains and economic models that generate the lowest environmental impacts with the greatest economic benefits. For this reason, decision support models are being developed to allow adaptation to circular model (Lieder et al., 2017).

Circularity is not a novel concept. Over the last decade several authors have been actively working analyzing different dimensions, production patterns, features, and perspectives at different scales (Ghisellini et al., 2016; Geisendorf and Pietrulla, 2017; Moreau et al., 2017). Even the concept of CE itself has been analyzed from different angles, and it could be said that we are still in the phase of having to redefine the concept and develop guidelines that clearly indicate the way forward to the CE [50].

The benefits associated to CE cover economic, social, and environmental dimensions. CE has the potential to decouple economic growth from resource consumption and environmental impacts (Lieder et al., 2017). Decoupling can be defined as reducing the amount of resources used to generate economic growth while decreasing environmental deterioration and ecological scarcity. In today's globalized world, strategies are also interconnected; therefore, CE packages should be aligned with the

Paris Agreement on Climate Change and Sustainable Development Goals ensuring that future systematic approaches are developed in close cooperation to gain a better future for all.

According to European Commission and Ellen MacArthur Foundation, the CE concept is defined by several principles: It is restorative and regenerative by design, aims to maintain the economic value of products, materials, and resources as much as possible in the life cycle minimizing waste generation and contributing to sustainable and low-carbon development, encouraging the use of renewable energies, while generating competitive advantages (Ellen MacArthur Foundation, 2015; Communication From the Commission to the European Parliament, 2014). In other words, the pillars of the CE can be summarized in "keeping products and materials in use while reducing and/or eliminating waste and pollution and contributing to the regeneration of natural ecosystems" (Ellen MacArthur Foundation, 2015). Strengthened by a transition to renewable energy sources, the circular model builds on the economic, natural, and social capital (Ellen MacArthur Foundation, 2015). As such, CE is an opportunity to reinvent the classic economy to get to a more sustainable and competitive development. Summarizing, CE helps to save energy and avoids overexploitation of natural resources without exceeding the earth's regenerative capacity in terms of climate, biodiversity, air, and soil and water pollution (Communication From the Commission to the European Parliament, 2015).

As we are moving forward in the CE paradigm, innovation will be the key in the search for new business opportunities to achieve new ways of producing and consuming without depleting natural resources. As such, the development of the CE will be linked to the creation of new jobs at all levels, generating social cohesion and integration (Lieder et al., 2017). Furthermore, moving to a circular economic model will lead society to rethink production and consumption chains toward increased resource efficiency. All the sectors will be affected as well as all the policy domains. The CE transition has just started (Moreau et al., 2017) and, as such, the next steps are related to the involvement of all the stakeholders and actors in society, collaboration at governmental organizations, NGOs, associations, companies and citizen levels, and exchange of successful experiences and knowledge gained. CE is moving through interdisciplinary frameworks to generate stable tools, guidelines, and policies. Societal challenges on circular business model innovation need to improve collaborative consumption, higher levels of collaboration, environmental protection, and even leading to lifestyle changes (Mentink, 2014). In this sense, CE could be considered as a new way to advance for new business based on innovation under the umbrella of appropriate use, reuse, and recycle, developing at the same time models and regenerative strategies making more with less. The "new way" has to be properly developed through international collaboration based on research and new models' approach for a successful implementation worldwide (Ghisellini et al., 2016).

The transition towards "CE" is unstoppable and is underway. Countries across the world are developing strategies for moving to CE. Europe, The United States, China, Japan, Australia, Canada, Korea, Vietnam, Argentina, Brazil, México, Colombia, Rwanda, Nigeria, South Africa, among others, have been implementing programs, strategies, and plans related to CE ("Zero waste," "Zero emissions," "Closing the loop," "Effective utilization of recyclables," "Solid waste management," "Action plans," and so on) (Ghisellini et al., 2016). CE will benefit not only developed economies but also to developing countries and emerging economies bringing new opportunities because better natural resources and waste management is a key element to eradicate poverty. This is why CE can help to end extreme poverty worldwide too.

BE and CE are concepts whose boundaries overlap in many cases and are closely linked. Both of them are intended to obtain a steady economic development for the future without causing environmental damage and providing social benefits. As shown by D'Amato et al. (2017) and Gobbo Fernandes (2016), while CE is dedicated to decoupling the resources used in the urban industrial development to the economic production, BE is mainly based on the use of biological resources based on innovation framework and the land-use practices in rural development. The circular bioeconomy concept merges these two trends and links urban and rural communities, something which is currently lacking (D'Amato et al., 2017; Gobbo Fernandes, 2016). Furthermore, both BE and CE are intrinsically linked, and the positive synergies between both must be fostered to achieve a more sustainable development. There exist many cross-cutting policies affecting both types of economies (BE and CE). Examples of those include, among others, climate change, environmental protection, waste management, industrial strategies, regional guidelines, research and innovation policies, which should drive forward together to achieve the societal challenges.

1.4 CASCADING USE

Another key topic for both bioenergy and bioeconomy is the so called "Cascading use," which can be described as the efficient utilization of resources by using residues and recycled materials for material use to extend total biomass availability within a given system (Vis et al., 2016). Again, as it is the case with other terms used in this book, there exist other conceptions with complementary meanings as demonstrated by the study conducted by World Wildlife Fund that found close to 40 definitions of this concept (Dammer et al., 2016). Its comprehensive and wide definition includes all types of biomass sources, taken into account the life cycle of the products and by-products aiming to achieve the highest resource efficiency.

The application of the cascading concept to bioenergy chains consists in considering the CE to the renewable biological resources. The key question is: *How to make the best use of our natural resources?* To address this issue, the environmental benefits and impacts should be previously assessed for the different possible chains through several procedures such as the internationally standardized LCA methodology (Vis et al., 2016).

Competition for biomass resources and its increase demand in the future will require prioritization in the cascading use of biomass maintaining the optimal value creation and resource efficiency while decreasing at the same time the GHG emissions (Scatlat et al., 2015; Immerzeel et al., 2014). In this way, cascading use of biomass could minimize depletion of biomass resource and could reduce the competition between different uses: food and feed, chemicals, materials, fuel, and energy (Scatlat et al., 2015). The cascade could be single or multiple. In a single cascade, biomass is processed into a product and this product is used once more for energy purposes. In a multistage cascade, the biomass is processed into a product and this product is used at least once more in material form before disposal or recovery for energy purposes (Vis et al., 2016). Some studies have suggested a hierarchical sequence prioritizing first the higher value uses, after that reuse and recycling of the products, raw materials, waste, and streams, and finally the energy use (Vis et al., 2016; Keegan et al., 2013; Risse et al., 2017). Among the proposed actions to support the development of a bio-based economy in the European Commission's Public Online Consultation, the preferential use of raw materials is (1) low meat diet, (2) bio-based material production, (3) recycling, and finally (4) bioenergy and biofuel production (EC, 2011). Recent studies show higher efficiency, more employment, higher added-value, and greater carbon sequestration benefits when wood is using materially (Vis et al., 2016; Keegan et al., 2013; Risse et al., 2017). However, biomass production, logistics, and uses are very dependent on local conditions. Therefore, carefully conducting a "case-by-case" analysis is needed to define which are the best chains to select them, including environmental, socioeconomic, and social aspects. Probably, the most analyzed cascade is related to wood sector, and even in this case, the cascading uses are very varied and complex. When all possible types of biomass, waste, and streams are included, the number of possible cascades rises and many more sectors are involved, increasing the complexity considerably. It can be concluded that the cascading use concept has a long way to run until getting maturity (Risse et al., 2017).

Governmental policies are still giving priority to energy use as a common strategy to decarbonize energy sector. Most policies encourage the energy use of biomass as part of the strategy to combat climate change promoting renewable energies. In the near future, sustainability biomass analysis will involve all marketable products including energy. These marketable products encompass intermediates and final products of diverse nature: food, feed, chemicals, biomaterials, and energy.

1.5 BIOENERGY AND POLICY CONTEXT

The present and future of bioenergy and its role in the bioeconomy greatly depends on the effectiveness of the policies and support measures that are put in place. Among other reasons, such support measures are justified because of its capacity to reduce fossil fuel dependence as well as its contribution to mitigate climate change.

Over the last decade, the fast growth in the production and the use of bioenergy has raised interest from both legislators and investors around the world. For example, within the European Union (EU) context, the Renewable Energy Directive (European Commission, 2009), the 2020 Energy Strategy (European Commission, 2010), the Energy Roadmap 2050 (European Commission, 2011a), and the Roadmap for a Low-Carbon Economy in 2050 (IRENA, 2017b) have defined ambitious targets for the future deployment of renewable energy technologies and energy efficiency up to 2020 and beyond. The EU is committed to reducing greenhouse gas emissions to 80%–95% below 1990 levels by 2050 in the context of necessary reductions by developed countries as a group (European Commission, 2011b). In this context, bioenergy will be of crucial importance for achieving both short- and long-term climate and energy policy targets up to 2050. The large-scale production of crop-based biofuels has been one of the fastest and most controversial global changes of recent years. Global biofuel outputs increased sixfold between 2000 and 2010, and a growing number of countries are adopting bioenergy promotion policies. Meanwhile, multilateral bodies have been created and a wide patchwork regarding bioenergy policies is emerging (Bastos and Gupta, 2013). All of these efforts are geared towards setting a long-term framework dealing with the issue of energy sustainability and the cross-border effects of crop-based biofuels expansion that cannot be dealt with at the national level alone. In this sense, international efforts must continue to find ways in which bioenergy can contribute to decarbonize the worlds' economy while ensuring the security of energy supply and sector's competitiveness in a sustainable way.

In terms of bioenergy's role in the current energy system, biomass and waste already represent a significant global energy source, accounting for over 70% of all renewable energy production in 2015 (OECD/IEA, 2017). To continue this positive trend, most support measures are oriented towards promoting higher investments in relevant research areas. In this sense, more support is needed to ensure that the acquired knowledge reaches the commercialization stage. Similarly, the various sectors of bioeconomy must become attractive career options through secondary and tertiary education.

Nowadays, it is unquestionable that bioenergy is the most important renewable energy source and will continue to play a major role in the future as demonstrated by its inclusion in the future energy and climate policies. Far more than any other type of renewable energy, bioenergy is strongly related to the whole system of land use and agricultural and forestry production that make up the global bioeconomy. Modern bioenergy refers to biomass that can be generated directly, processed in dry and densified solid fuels, or converted into liquid or gaseous fuels, using the so-called first- or second-generation technologies, depending on the level of development. However, the current rate of bioenergy deployment is well below the levels required within IEA long-term climate models. Acceleration is urgently needed to ramp up the contribution of bioenergy across all sectors notably in the transport sector where consumption is expected to triple by 2030 (OECD/IEA, 2017).

The increase in the demand for the key products of bioeconomy can be covered by increasing not only the cultivation of crops used for biofuels production but also profiting co-products of these chains. In this sense, biotechnology is used to increase the productivity of biofuel production and is related to many policies and regulations affecting bioeconomy. There are many established bio-based industries of agriculture, food, and forest-based industries dealing with biotechnology.

The analysis of the current status and future predictions of bioeconomy worldwide could help identify some aspects that appear key to strengthen this sector. Among others, these aspects are related to international collaborations on bioenergy, bioeconomy compared to fossil fuel economy, the growth of industrial biotechnology joint in bioeconomy, and finally, the bio-based industries for bioeconomy.

World bioenergy sector (liquid biofuels, solid biomass, and biogas) is the largest employer among renewable energies, with a total of 2.7 million jobs in 2016. The main jobs have been concentrated in feedstock supply in Brazil, China, the United States, and India (IRENA, 2017a). By 2015, the EU had 1.6 million people working on renewables and energy efficiency. This represents a growth of 13% since 2010, which is more than seven times greater when compared to the 1.7% employment increase for the whole EU economy (SET Plan, 2017).

To foster the synergies between bioenergy with bioeconomy, some barriers should be tackled. Among others, these barriers include current regulatory instability, tax levy, lack of coordination among regulators, and how bioprocesses are being involved in the bioeconomy context (ERRIN-ERIAFF-BIC, 2015).

Regarding the regulatory instability, the most fundamental barrier is the frequent change in quotas and support level for renewable energies. To decrease the regulatory uncertainty, it is important to set the rates in the early stages of the project (Held et al., 2013).

As for the issue of taxation, some of the taxes levied on entrepreneurs are not yet recognized as such. For this matter it is necessary to map the costs imposed on the employer and include them in the final tariff.

Finally, in terms of the coordination between regulators, renewable energy entrepreneurs need to work with regulators to enhance the coordination among them. One way to overcome this obstacle is to create a regulatory mechanism aimed at fostering cooperation among the various regulators.

REFERENCES

Adamowicz, M., 2017. Bio-economy as a concept of development strategies in the European Union. JIBRM 2 (4), 7–12.

Arundel A., Sawaya, D., 2009. The Bioeconomy to 2030: Designing a Policy Agenda. ISBN:978-92-64-03853-0.

Barnosky, A.D., Matzke, N., Tomiya, S., Wogan, G.O.U., Swartz, B., Quental, T.B., et al., 2011. Has the Earth's sixth mass extinction already arrived? Nature 471, 51–57.

Bastos, M.G., Gupta, J., 2013. The policy context of biofuels: a case of non-governance at the global level? Global Environ. Polit. 13 (2), 46–64.

Ben Aoun, W., Gabrielle, B., Gagnepain, B., 2013. The importance of land use change in the environmental balance of biofuels. Oilseeds Fat Crops Lipids (OCL) 20 (5), D505.

Bioenergy for Sustainable Development, 2017. IEA Bioenergy. IRENA and FAO. Available from: http://www.ieabioenergy.com/wp-content/uploads/2017/01/BIOENERGY-AND-SUSTAINABLE-DEVELOPMENT_v20170105-cob-x.pdf.

Ceballos, G., Ehrlich, P.R., Barnosky, A.D., García, A., Pringle, R.M., Palme, T.M., 2015. Accelerated modern human-induced species losses: entering the sixth mass extinction. Sci. Adv. 1, e1400253.

Communication from the Commission to the European Parliament, the Council, the European Economic and Social Committee and the Committee of the Regions, 2014. Towards a circular economy: a zero waste programme for Europe. COM. 398 final/2.

Communication from the Commission to the European Parliament, the Council, the European Economic and Social Committee and the Committee of the Regions, 2015. Closing the loop—an EU action plan for the circular economy. COM. 614 final.

Communiqué Global Bioeconomy Summit, 2015. Making bioeconomy work for sustainable development. Berlin, Germany.

Dammer L., Bowyer C., Breitmayer E., Eder A., Nanni S., Allen B., et al., 2016. Mapping study on cascading use of wood products. WWF & Mondi Group. Technical Report.

Darlington, T., Kahlbaum, Dennis, O'Connor, Don, Mueller, Steffen, 2013. Land Use Change Greenhouse Gas Emissions of European Biofuel Policies Utilizing the Global Trade Analysis Project (GTAP) Model. Air Improvement Resource, Inc., (S&T)2 Consultants Inc., University of Illinois, Chicago, IL. Available from: http://www.ebb-eu.org/studiesreports/GTAP%20Report%20ILUC%20Aug%2030%202013%20Final.pdf.

D'Amato, D., Droste, N., Allen, B., Kettunen, M., Lähtinen, K., Korhonen, J., et al., 2017. Green, circular, bio economy: a comparative analysis of sustainability concepts, J. Clean. Prod., 168. pp. 716–734.

European Commission, 2011. Bio-based Economy for Europe: State of Play and Future Potential—Part 2 Summary of Position Papers Received in Response to the European Commission's Public Online Consultation. Publications Office of the European Union. Available from: https://doi.org/10.2777/67596.

European Commission, 2012. Communication from the Commission to the European Parliament, the Council, the European Economic and Social Committee and the Committee of the Regions: A Bioeconomy for Europe, COM, 60.

ERRIN-ERIAFF-BIC, 2015. How to build effective regional strategies? Guideline for Regions, National Governments and Bioeconomy sectors. Consensus document. European Regions for Research and Innovation Network (ERRIN). European

Regions for Innovation in Agriculture, Food and Forestry Network (ERIAFF). Biobased Industries Consortium (BIC).
El-Chichakli, B., von Braun, J., Lang, C., Barben, D., Philp, J., 2016. Five cornerstones of a global bioeconomy. Nature. 35, 221–223.
Ellen MacArthur Foundation, 2015. The circular economy concept—regenerative economy. www.ellenmacarthurfoundation.org/circular-economy/overview/concept. Accessed December 2017.
European Commission, 2009. Renewable Energy Directive.
European Commission, 2010. 2020 Energy Strategy.
European Commission, 2011a. The Energy Roadmap 2050.
European Commission, 2011b. The Roadmap for a Low-Carbon Economy in 2050.
FAO-GBEP, 2011. The Global Bioenergy Partnership Sustainability Indicators for Bioenergy, first ed. http://www.globalbioenergy.org/fileadmin/user_upload/gbep/docs/Indicators/The_GBEP_Sustainability_Indicators_for_Bioenergy_FINAL.pdf.
Fehrenbach, H., Giegrich, J., Reinhardt, G., Rettenmaier, N., 2009. Synopsis of current models and methods applicable to indirect land use change (ILUC). Report Commissioned by Bundesverband der deutschen Bioethanolwirtschaft e.V, ifeu—Institut für Energie—und Umweltforschung. Heidelberg, Germany: Heidelberg GmbH.
Finkbeiner M., 2013. Indirect land use change (ILUC) within life cycle assessment (LCA) e-scientific robustness and consistency with international standards. VDB & OVID Report, May 2013. Available in: http://www.ufop.de/files/5213/.
Finkbeiner, M., 2014. Indirect land use change e help beyond the hype? Biomass Bioenerg. 218–e221. Available from: http://www.sciencedirect.com/science/article/pii/S0961953414000257.
Fritsche, U., et al., 2017a. Linkages between the Sustainable Development Goals (SDGs) and the GBEP Sustainability Indictors for Bioenergy (GSI). Technical Paper for the GBEP Task Force on Sustainability. International Institute for Sustainability Analysis and Strategy (IINSAS), Darmstadt, and Institut für Energie—und Umweltforschung [Institute for Energy and Environmental Research] (IFEU), Heidelberg.
Fritsche U., Berndes G., Cowie A.L., Dale V.H., Kline K.L., Johnson F.X., et al., 2017b. Energy and land use. In: Working Paper for the UNCCD Global Land Outlook, United Nations Convention to Combat Desertification (UNCCD) and International Renewable Energy Agency (IRENA). https://static1.squarespace.com/static/5694c48bd82d5e9597570999/t/593a4294b8a79b4be75f6078/1496990366441/Energy + and + Land + Use__U_Fritsche.pdf.
Fritsche U.R., Hennenberg K.J., Hünecke K., Herrera R., Wiegmann K., 2011. Sustainable bioenergy: a tool for biodiversity, rural development and food security. Contribution to the Panel Discussion: Renewable Energies, green economies, and carbon offsets. UNESCO Programme Man & Biosphere Conference. "For life, for the future. Biosphere reserves and climate change". Öko-Institut.
GBEP Sustainability Indictors for Bioenergy (GSI). http://www.globalbioenergy.org/fileadmin/user_upload/gbep/docs/2017_events/TFS_workshop_SDGs_and_GBEp_GSIs_-_3-4_July_2017_Bonn/IINAS_IFEU_SDG-GSI-linkages.pdf.
GBEP-IEA, 2016. Bioenergy examples of Positive Bioenergy and Water Relationships, a report from the GBEP-IEA Bioenergy collaboration within the GBEP Activity Group 6, Bioenergy and Water. http://www.globalbioenergy.org/fileadmin/user_upload/gbep/docs/2015_events/AG6_workshop_25-26_August_2015/AG6_Examples_of_Positive_Bioenergy_and_Water_Relationships_Final.pdf.
Geisendorf, S., Pietrulla, F., 2017. The circular economy and circular economic concepts—a literature analysis and redefinition. Thunderbird Int. Bus. Rev. 1–12.

Ghisellini, P., Cialani, C., Ulgiati, S., 2016. A review on circular economy: the expected transition to a balanced interplay of environmental and economic systems. J. Clean. Prod. 114, 11–32.

Gobbo Fernandes, A., 2016. Closing the Loop. The Benefits of the Circular Economy for Developing Countries and Emerging Economies. EPEA, NuReS, Tearfund. Available from: http://www.sustainablebrands.com/digital_learning/research_report/next_economy/closing_loop_benefits_circular_economy_developing_coun.

Held A., Ragwitz M., Gephart M., de Visser E., Klessmann C. 2013. Design features of support schemes for renewable electricity. Task 2 Report. Ecofys.

IRENA, 2017a. Renewable Energy and Jobs—Annual Review 2017. International Renewable Energy Agency, Abu Dhabi.

IRENA, 2017b. REthinking Energy 2017: Accelerating the Global Energy Transformation. International Renewable Energy Agency, Abu Dhabi, ISBN 978-92-95111-05-9 (print) | ISBN 978-92-95111-06-6 (PDF). https://www.irena.org/DocumentDownloads/Publications/IRENA_REthinking_Energy_2017.pdf.

Immerzeel, D.J., Verweij, P.A., Vanderhilst, F., Faaij, A.P.C., 2014. Biodiversity impacts of bioenergy crop production: a state-of-the-art review. GCB Bioenergy 6, 183–209. Available from: https://doi.org/10.1111/gcbb.12067.

Keegan, D., Kretschmer, M., Elbersen, B.S., Panoutsou, C., 2013. Cascading use: a systematic approach to biomass beyond the energy sector. Biofuel Bioprod Bior 7 (2), 193–206. Available from: https://doi.org/10.1002/bbb.1351. ISSN 1932-104X.

Kim, S., Dale, B.E., Heijungs, R., Azapagic, A., Darlingtom, T., Kalhbaum, D., 2014. Indirect land use change and biofuels: mathematical analysis reveals a fundamental flaw in the regulatory approach. Biomass Bioenerg. 71, 408e412. Available from: http://www.sciencedirect.com/science/article/pii/S0961953414004255.

Laborde, D., 2011. Assessing the land use change. Consequences of European biofuel policies. Final Report. ATLASS Consortium. Specific Contract No SI2. 580403 implementing Framework Contract No TRADE/07/A2. http://ebrary.ifpri.org/cdm/ref/collection/p15738coll5/id/197.

Lieder, M., Asif, F.M.A., Rashid, A., Mihelič, A., Kotnik, S., 2017. Towards circular economy implementation in manufacturing systems using a multi-method simulation approach to link design and business strategy. Int. J. Adv. Manuf. Technol. 93, 1953–1970. Available from: https://doi.org/10.1007/s00170-017-0610-9.

Maciejczak, M., 2017. Bioeconomy as a complex adaptive system of sustainable development. JIBRM 2 (2), 7–10.

Malins, C., Searle, S., Baral, A., 2014. A guide for the perplexed to the indirect effects of [biofuels production. Int. Counc. Clean Transp 166. Available from: http://www.theicct.org/sites/default/files/publications/ICCT_A-Guide-for-the-Perplexed_Sept2014.pdf.

McCormick, K., Kautto, N., 2013. The bioeconomy in Europe: an overview. Sustainability 5, 2589–2608. Available from: https://doi.org/10.3390/su5062589.

Mentink B., 2014. Circular business model innovation. A process framework and a tool for business model innovation in a circular economy.

Moreau, V., Sahakian, M., van Griethuysen, P., Vuille, F., 2017. Coming full circle. Why social and institutional dimensions matter for the circular economy. J. Ind. Ecol. 21, 497–506.

OECD, 2009. The Bioeconomy to 2030: Designing a Policy Agenda. Main Findings and Policy Conclusions. OECD International Futures Project.

OECD/IEA, 2017. Technology roadmap. Delivering sustainable bioenergy.

Pavlenko N., Searle S., 2018. A comparison of induced land-use change emissions estimates from energy crops. White paper. The International Council on Clean Transportation (ICCT).

Plevin, R.J., 2017. Assessing the climate effects of biofuels using integrated assessment models, part I methodological considerations. J. Ind. Ecol. 21 (6), 1478–1487. Available from: https://doi.org/10.1111/jiec.12507.

Popp, J., Lakner, Z., Harangi-Rákos, M., Fári, M., 2014. The effect of bioenergy expansion: food, energy, and environment. Renew. Sust. Energ. Rev. 32, 559–578. Available from: https://doi.org/10.1016/j.rser.2014.01.056. ISSN 1364-0321.

Rauschen S., Esch J.V., 2017. Policy Brief on the Future of the European Bioeconomy Strategy. Standing Committee of Agricultural Research (SCAR). https://www.scar-swg-sbgb.eu/lw_resource/datapool/_items/item_28/policy-brief-23082017_final_template.pdf.

Risse, M., Weber-Blaschke, G., Richter, K.A., 2017. Resource efficiency of multifunctional wood cascade chains using LCA and exergy analysis, exemplified by a case study for Germany. Resources, Conservation and Recycling 126, 141–152. Available from: https://doi.org/10.1016/j.resconrec.2017.07.045.

Scatlat, N., Dallemand, J.-F., Monforti-Ferrario, F., Nita, V., 2015. The role of biomass and bioenergy in a future bioeconomy: policies and facts. Environ. Dev. 15, 3–34.

Schmidt, J.H., Weidema, B.P., Brandao, M., 2015. A framework for modelling indirect land-use change in life cycle assessment. J. Clean. Prod. 99, 230e238. Available from: http://www.sciencedirect.com/science/article/pii/S0959652615002309.

Searchinger, T., Heimlich, R., Houghton, R.A., Dong, F.X., Elobeid, A., Fabiosa, J., 2008. Use of US croplands for biofuels increases greenhouse gases through emissions from land-use change. Science 319, 1238–e1240.

SET Plan, 2017. The Strategic Energy (SET) Plan at the Heart of Energy Research and Innovation in Europe.

Taheripour, F., Tyner, W.E., Wang, M.Q., 2011. Global land use changes due to the U.S. cellulosic biofuel program simulated with the GTAP model. Final Version. https://greet.es.anl.gov/publication-luc_ethanol.

Thrän, D., Schaldach, R., Millinger, M., Wolf, V., Arendt, O., Ponitka, J., et al., 2016. The milestones modelling framework: An integrated analysis of national bioenergy strategies and their global environmental impacts. Environ. Model. Software. 86, 14–29.

Thrän, D., Arendt, O., Banse, M., Braun, J., Fritsche, U., Gärtner, S., et al., 2017. Strategy elements for a sustainable bioenergy policy based on scenarios and systems modeling: Germany as example. Chem. Eng. Technol. 40 (2), 211–226.

Tilman, D., Clark, M., Williams, D.R., Kimmel, K., Polasky, S., Packer, C., 2017. Future threats to biodiversity and pathways to their prevention. Nature . Available from: https://doi.org/10.1038/nature22900. Available from: www.nature.com/nature/journal/. . .ull/nature22900.html.

US Environmental Protection Agency, Assessment and Standards Division, Office of Transportation and Air Quality, 2010. Renewable fuel standard program (RFS2) regulatory impact analysis. (EPA-420-R-10-006). Retrieved from https://nepis.epa.gov/Exe/ZyPURL.cgi?Dockey = P1006DXP.txt.

UN, 2015a. Climate Change. The Paris Agreement (accessed 15.12.17).

UN, 2015b. Transforming our world: the 2030 agenda for sustainable development. http://www.un.org/ga/search/view_doc.asp?symbol = A/RES/70/1&refer-er = http://www.un.org/sustainabledevelopment/development-agenda/&Lang = E.

Valin, H., Peters, D., van den Berg, M., Frank, S., Havlik, P., Forsell, N., et al., 2015. The Land Use Change Impact of Biofuels Consumed in the EU: Quantification of Area and Greenhouse Gas Impacts. ECOFYS Netherlands BV, Utrecht, Netherlands, BIENL13120. https://ec.europa.eu/energy/sites/ener/files/documents/Final%20Report_GLOBIOM_publication.pdf.

van de Pas J., 2015. The bio-economy: definitions and measurement. http://edepot.wur.nl/338454.

Vis M., Mantau U., Allen B. (Eds)., 2016. Study on the Optimised Cascading Use of Wood. No 394/PP/RCH/14/7689. Final Report. Brussels, Germany. 337 pages.
World Energy Outlook. 2016. Executive summary. IEA. https://www.iea.org/Textbase/npsum/WEO2016SUM.pdf.
Yeh, S., Witcover, J., 2010. Policy brief: indirect land-use change from biofuels: recent developments in modelling and policy landscapes. Carbon Standards in Agricultural Production and Trade. International Food & Agricultural Trade Policy Council (IPC), Sao Paulo. Available from: http://www.its.ucdavis.edu/research/publications/publication-detail/?pub_id¼1434.

CHAPTER TWO

Biomass Resources

Javier Sánchez[1], María Dolores Curt[2], Nicolas Robert[1] and Jesús Fernández[2]
[1]European Commission—DG Joint Research Centre (JRC), Directorate D—Sustainable Resources, Unit D1—Bio-Economy, Ispra, Italy
[2]Agro-Energy Group, College of Agricultural Engineering, Technical University of Madrid, Madrid, Spain

Contents

2.1 Biomass as a Renewable Source of Energy 25
  2.1.1 Nature of biomass and its terminology 25
  2.1.2 Process of photosynthesis 27
2.2 Chemical Composition and Characterization of Biomass 30
  2.2.1 Elemental composition 30
  2.2.2 Organic matter 31
  2.2.3 Water content and the heating value of biomass 34
2.3 Classification of Biomass Types 35
  2.3.1 According to chemical composition 35
  2.3.2 According to origin 36
  2.3.3 According to biomass end use 39
2.4 Biomass Resources 40
  2.4.1 Agriculture 42
  2.4.2 Forestry 50
  2.4.3 By-products, residues, and wastes 58
  2.4.4 Algae for bioenergy 81
List of Abbreviations 97
References 98

2.1 BIOMASS AS A RENEWABLE SOURCE OF ENERGY

2.1.1 Nature of biomass and its terminology

Biomass is broad concept and currently there is no agreed definition of the term. It can be generally defined as the "organic matter derived from living, or recently living organisms" (Lewandowski et al., 2018). In the context of bioenergy, biomass is defined in the Directive 2009/28/EC

The Role of Bioenergy in the Emerging Bioeconomy.
DOI: https://doi.org/10.1016/B978-0-12-813056-8.00002-9

(European Parliament, 2009) as the "biodegradable fraction of products, waste and residues from biological origin from agriculture (including vegetal and animal substances), forestry and related industries including fisheries and aquaculture, as well as the biodegradable fraction of industrial and municipal waste." Consequently, the material embedded in geological formations and transformed to fossil is excluded (FAO, 2004).

In the bioeconomy context, biomass comprises the renewable biological resources for the conversion of these resources and waste streams into value-added products such as food, feed, bio-based products, and bioenergy (European Commission, 2012). Thus, it includes edible (food) and non-edible (non-food) biomass from plants, animals, and waste origin. Therefore, it is important to avoid partial definitions of the term, especially purely energy-focused definitions or, on the other hand, clearly state the framework and boundaries of the term used, e.g., for energy statistics, (Eurostat, 2017a).

Biomass is a renewable resource as its energy content is ultimately coming from the sun. But it is also limited, because for its production finite resources such as land, water, and nutrients are required. Moreover, there is currently a growing concern on the competitive uses of biomass and its limited availability which requires an accurate analysis of the biomass demand in relation to the existing potential for the development of a sustainable bioeconomy (Scarlat et al., 2015).

According to an ecological point of view, it can be distinguished: (1) *Primary biomass*, the organic matter directly produced by photosynthetic organisms (algae, plants, and other autotrophic organisms), which comprises all plant biomass; (2) *Secondary biomass* produced by the heterotrophic organisms that consume primary biomass, such as meat and waste from herbivores, and (3) *Tertiary biomass*, which encompasses the biomass produced from organisms fed on secondary biomass. Along this ecological chain, there is a transfer and a loss of energy. Bioenergy chains are supplied with primary biomass and the waste streams coming from the production of secondary and tertiary biomass.

According to FAO (2004), fuel produced directly or indirectly from biomass is named biofuel. Biofuel is defined as "liquid, solid, or gaseous fuel produced by conversion of biomass such as bioethanol from sugar cane or corn, charcoal or woodchips, and biogas from anaerobic decomposition of wastes" (OECD, 2002). In parallel, the term biofuels is generally used to exclusively refer to liquid or gaseous biofuels for the transport sector. Hence, whenever biofuels are mentioned, it is convenient to specify the type of biofuels being addressed.

Within the literature, it is also distinguished as primary (unprocessed) from secondary (processed) biofuels, being the former used in their natural form (as harvested) while the latter have undergone a conversion process to better serve for their final use (FAO, 2004).

Additionally, to set a sustainability framework for the production of transport biofuels as well as dealing with the food versus energy issue, a distinction between conventional and advanced biofuels has been established. This distinction is addressed in Chapter 3, Liquid Biofuels.

2.1.2 Process of photosynthesis

Biomass is a renewable source of energy since its chemical energy content ultimately comes from the solar energy through the photosynthetic process. The photosynthetic process occurs in the chloroplasts and consists of two series of reactions: light reactions (absorption of solar energy) and dark reactions (CO_2 assimilation reactions). In light reactions the energy of the visible portion of solar radiation (from 400 to 700 nm approximately) is absorbed by Chlorophyll of the photosynthetic units located in the thylakoids (Fig. 2.1).

$$\begin{aligned}2H_2O + CO_2 + 8\ \text{photons} &\rightarrow (\text{H-C-OH}) + H_2O \\ &+ O_2\ (\Delta G = 114\ \text{kcal mole}^{-1})\end{aligned}$$

As overall result of the Calvin cycle, in order to transform three molecules of CO_2 into sugars (among others glucose, fructose, sucrose) and starch, six molecules of NADPH and nine molecules of ATP are required in total (Fig. 2.1). From the sugars initially produced in the photosynthesis, the main components of organic matter (carbohydrates, lipids, proteins, and nucleic acids) are formed (see also Section 2.2.2).

Considering the average energy of 49.7 $\text{kcal}\cdot\text{mole}^{-1}$ for the photosynthetic active radiation that corresponds to the solar spectrum between 400 and 700 nm approximately, the maximum theoretical energy efficiency of the photosynthetic process is 28.6%:

$$\begin{aligned}&e_{Max}\ (\text{maximum photosynthesis efficiency}) \\ &\quad = 114\ \text{kcal mole}^{-1} \times 100/(8\ \text{photons} \times 49.7\ \text{kcal mole}^{-1}) = 28.6\%\end{aligned}$$

However, there are energy losses in the process of photosynthesis, e.g., in the form of heat during the metabolism and the energy consumption

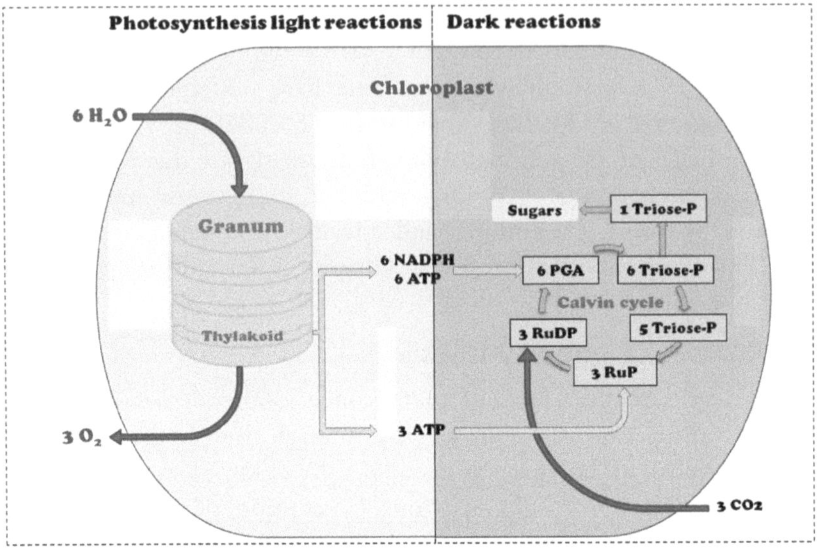

Figure 2.1 Diagram of the two stages of photosynthesis with the Calvin cycle. For each three turns of the cycle, three molecules of CO_2 are absorbed so that three atoms of carbon are converted into sugar (triose-P). In this process, six molecules of NADPH and nine of ATP are consumed. Out of the total ATP, six molecules are used to reduce PGA and three for the phosphorylation of ribulose 5-phosphate (RuP). NADPH, nicotinamide adenine dinucleotide phosphate; ATP, adenosine triphosphate; RuDP, ribulose diphosphate; PGA, phosphoglyceric acid; GAP, glyceraldehyde phosphate.

through photorespiration. Additionally, many environmental factors also influence in the actual efficiency of plant ecosystems, such as disease, predation, inadequate inorganic nutrient supply, and suboptimal water supply. Thus, efficiencies of agricultural systems range between 0% and 2% (Monteith, 1972).

The efficiency of different ecosystems to produce biomass is also measured by the net primary production (NPP), i.e., the total amount of carbon assimilated by plants within a given area over a specified time frame minus the flux of autotrophic respiration of assimilate used for the plant's own metabolism (Roxburgh et al., 2005). Fig. 2.2 shows the NPP of the main continental and marine ecosystems. Among continental ecosystems, swamps and marshes followed by tropical forest are the most productive (2,500 and 2,000 $g \cdot m^{-2} \cdot year^{-1}$, respectively), while cultivated

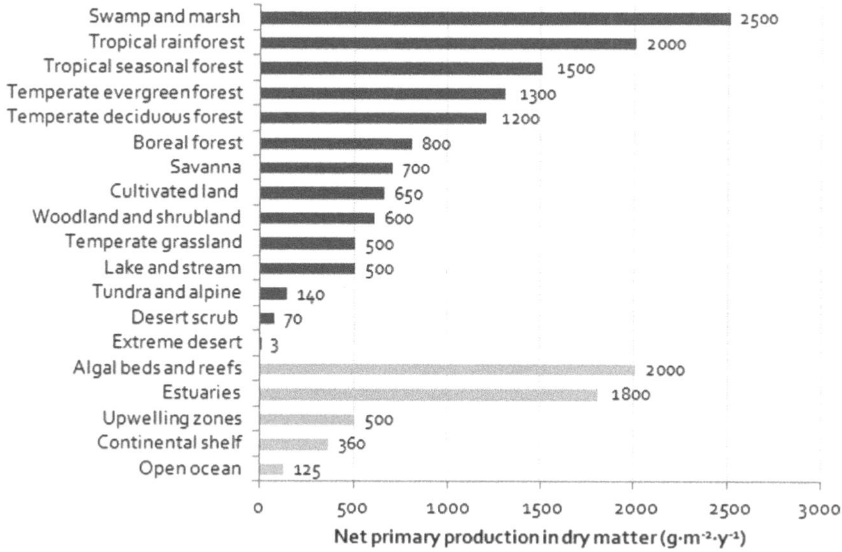

Figure 2.2 NPP of different continental and marine ecosystems. *Data from Whittaker, R.H., Likens, G.E., 1973. Primary productivity: the biosphere and man. Human Ecol. 1 (4), 357–369.*

land shows a NPP of 650 $g \cdot m^{-2} \cdot year^{-1}$. "Algal beds and reefs" is the most productive marine ecosystem with 2,000 $g \cdot m^{-2} \cdot year^{-1}$ (Whittaker and Likens, 1973) (Box 2.1).

BOX 2.1 Efficiency of annual biomass production in the biosphere

Biomass in the biosphere grows at a yearly rate of 170,000 Mt, which in energy terms represents around 2.85×10^{15} MJ. Solar global radiation reaching the earth surface is on average 5.36×10^{18} $MJ \cdot year^{-1}$. Thus, the global average efficiency of photosynthesis results in 0.05%. At a smaller scale, one hectare of maize grown in the Iberian Peninsula receives yearly around 6.15×10^{7} MJ of solar radiation. Assuming a yield of 20 tons of dry matter per hectare and year ($t \cdot ha^{-1} \cdot year^{-1}$) ($3.35 \times 10^{5}$ MJ considering an energy content of the whole biomass of 16.74 $kJ \cdot g^{-1}$ of dry matter), the photosynthesis efficiency amounts to 0.54%. The same calculation for a corn leaf results in a direct photosynthesis efficiency of 8% (Fig. 2.3).

(Continued)

BOX 2.1 (Continued)

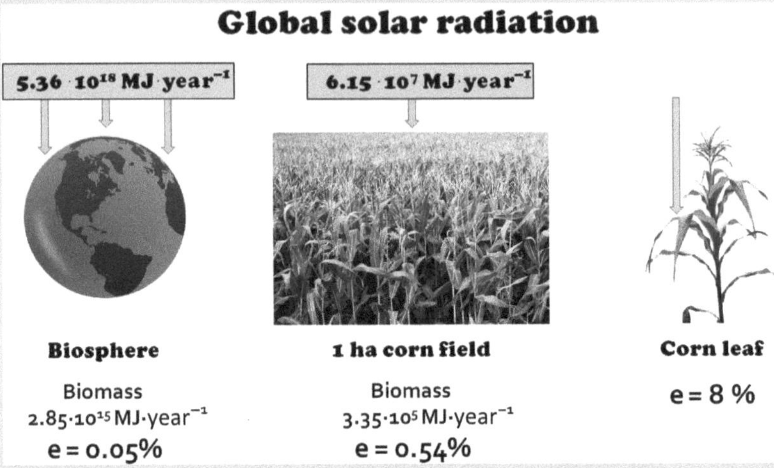

Figure 2.3 Photosynthesis efficiency at different ecosystem scales. *Based on Fernández, J., 2015. Tecnologiјas para el uso y transformacioјn de biomasa energeјtica. Capitulo XII. In: Energia de la biomasa. Madrid, Spain: Mundi-Prensa, pp. 397–445.*

2.2 CHEMICAL COMPOSITION AND CHARACTERIZATION OF BIOMASS

2.2.1 Elemental composition

Plant biomass is mostly composed of three elements: 42%–47% of carbon (C), 40%–44% of oxygen (O), and 6% of hydrogen (H), all percentages in dry matter. This elemental composition of biomass is followed by the so-called macronutrients, which are essential for biomass production: nitrogen (N), phosphorus (P), potassium (K), calcium (Ca), magnesium (Mg), and sulfur (S). Moreover, plants also need some additional elements in lower quantities, micronutrients, and trace elements, such as sodium (Na), chlorine (Cl), iron (Fe), manganese (Mn), copper (Cu), zinc (Zn), molybdenum (Mo), nickel (Ni), selenium (Se), and silicon (Si), summing all together up to 4%. Biomass also contains, namely in the ashes, some different elements like aluminum (Al), arsenic (As), barium (Ba),

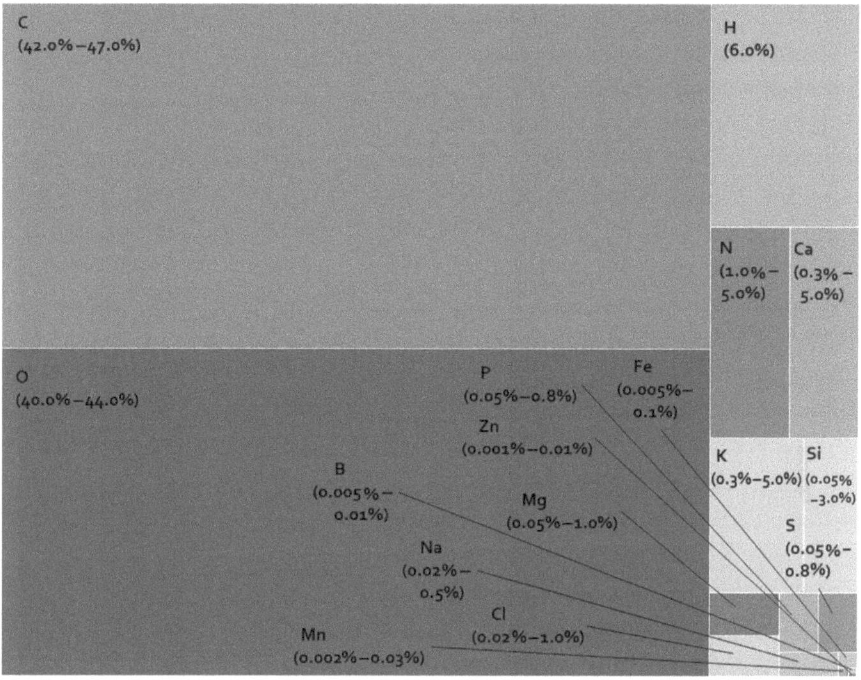

Figure 2.4 Elemental composition of plant biomass. The size of the boxes corresponds to the amount of each element in the composition of biomass (%). *Based on Lewandowski, I., Gaudet, N., Lask, J., Maier, J., Tchouga, B., Vargas-Carpintero, R., 2018. Bioeconomy. In: Lewandowski (Ed.), Shaping the Transition to a Sustainable, Biobased Economy I. Cham, Switzerland: Springer International Publishing AG.*

cadmium (Cd), chromium (Cr), mercury (Hg), lead (Pb), antimony (Sb), titanium (Ti), thallium (Tl), vanadium (V), and tungsten (W).

Fig. 2.4 shows the average values of the elemental composition of plant biomass.

2.2.2 Organic matter

Sugars synthesized in the Calvin cycle are the basis for the formation of the organic components which give then rise to the vegetal tissues. These organic components can be classified into four main groups: carbohydrates, proteins (polypeptides), lipids, and nucleic acids.

2.2.2.1 Carbohydrates

These are compounds from the combination of carbon, hydrogen, and oxygen to form soluble sugars (monosaccharide and disaccharides) and polymeric carbohydrates (polysaccharides). Among the most important monosaccharides, glucose and fructose should be mentioned, which combined constitute sucrose (disaccharide).

Polysaccharides are formed through the aggregation of different monosaccharides, which are then used for either reserve or structural functions. Reserve polysaccharides are accumulated during periods of intense photosynthetic activity, and are later used to deliver carbohydrate monomers for plant metabolism. Starch and inulin are the most important reserve polysaccharides from an energy point of view. The former is a glucose polymer present in many seeds (cereal grains), tubers (e.g., potato (*Solanum tuberosum* L.)), and roots (e.g., parsnip (*Pastinaca sativa*)), while inulin is composed of fructose and glucose, and it can be typically found in roots (e.g., chicory (*Cichorium intybus* L.)) and tubers (e.g., Jerusalem artichoke (*Helianthus tuberosus* L.)). In the bioenergy context, both carbohydrates can be hydrolyzed into monomers and then fermented to produce ethanol or even directly fermented with specific microorganisms.

Structural polysaccharides are used to build the cell walls and consist mainly of four organic compounds: cellulose, hemicellulose, pectins, and lignin. These compounds are also major components of natural lignocellulosic materials in different ratios which determine the hardness of the cell wall and thus the type of biomass: woody (hardwood and softwood) and herbaceous biomass.

The cellulose is a polysaccharide made of 200–5000 molecules of glucose, aggregated in linear chains or bundles $[(C_6H_{10}O_5)_n]$ to build microfibers (Ø = 3 nm) and fibers (Ø = 10 nm–0.20 μm). The hemicellulose consists of polymers of pentoses and hexoses entangle among the cellulose fibers. Both polymers, cellulose and hemicelluloses, are relatively easy to hydrolyze and represent two-thirds of the lignocellulosic biomass. Meanwhile, lignins $[C_n(H_2O)_m]$ are high-molecular weight, insoluble plant polymers, which have complex and variable structures, made from phenylpropanoid alcohols. It is the second most abundant polymer of plant biomass, after cellulose (the lignin content is about 27%–32% in woody plants and about 14%–25% in herbaceous plants), and it provides the plant with stiffness and rigidity, also avoiding tissue degradation. Thus, it requires strong acid or bases or other hydrothermal treatments to be hydrolyzed and make cellulose and hemicelluloses accessible. Pectin is a minor component of lignocellulosic biomass and consists of polymers of galacturonic acid which allow cell wall extension and plant growth. The feasibility and energy demand for hydrolyzing the structural

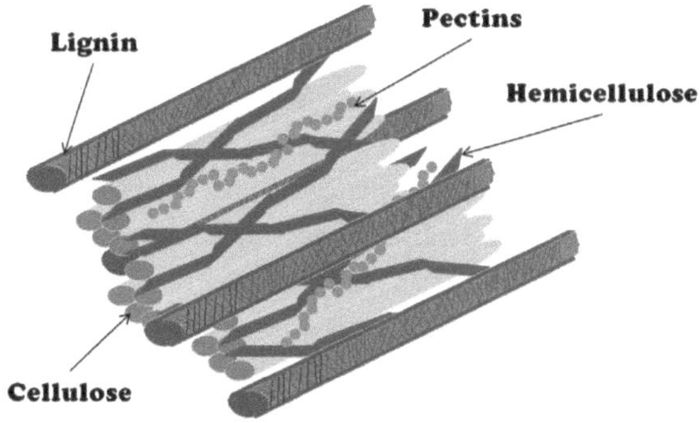

Figure 2.5 Structural polysaccharides of lignocellulosic biomass.

polysaccharides are essential parameters for the development of second-generation biofuels which are addressed in Chapter 3, Liquid Biofuels.

A representation of the structural polysaccharides of lignocellulosic biomass is shown in Fig. 2.5.

2.2.2.2 Proteins

Proteins are made of chains of amino acids, organic compounds containing amine (-NH_2) and carboxyl (-COOH) functional groups, which provide plants with enzymatic and structural functions. Thus, their nitrogen content is relatively high (16%) entailing potential emissions of nitrogen oxides (NO_x) during the combustion of protein-rich biomass feedstocks and, hence, contributing to air pollution. Moreover, the production of proteins by plants requires high quantities of energy ($40.33\ kJ \cdot g^{-1}$) in comparison with other organic compounds (e.g., cellulose, $19.73\ kJ \cdot g^{-1}$). Considering the higher heating value (HHV) of proteins ($22.19\ kJ \cdot g^{-1}$) and cellulose ($19.05\ kJ \cdot g^{-1}$), the energy yield is 52.5% and 96.5%, respectively (Fernández, 2015). Therefore, protein-rich biomasses are more interesting for food or feed production rather than for energy uses.

2.2.2.3 Lipids

Lipids are heterogeneous and hydrophobic organic compounds that make up the building blocks of the structure and function of living cells. The main lipids contained in biomass feedstocks are fats and oils, phospholipids, and waxes. Major components of fats or oils are tri-esters of fatty acids and glycerol (triglycerides or triacylglycerols). According to the saturation level of the fatty acids, whether the containing carbon is saturated by hydrogen atoms, double or triple bonds, they are classified in saturated or unsaturated fats, respectively. Saturated fatty acids, as contained in animal fats, have a higher melting point, and thus, they are solids at room temperature. On the other hand, vegetal lipids usually have lower melting point because they contain fatty acids of longer chains and higher proportion of unsaturated fatty acids, and hence they are also called oils. Waxes are esters made from the union of long chains of alcohols and acids with the aim of acting as waterproof layers and avoiding water loss in certain parts of the plants. Regarding phospholipids, they are composed of glycerol, fatty acids, and a phosphate molecule to provide structure and protection to cells.

From an energy point of view, the production of fats entails an energy demand of $50.5\ kJ \cdot g^{-1}$ with an energy yield of 77.2% (considering a HHV of $38.93\ kJ \cdot g^{-1}$).

2.2.2.4 *Nucleic acids*

The nucleic acids are composed of nucleotides, which are monomers made of three components: a pentose, a phosphate group, and a nitrogenous base. According to the containing sugar, there are two types of nucleic acids: DNA and RNA. They are responsible for the encoding and transcription of proteins.

2.2.3 Water content and the heating value of biomass

The moisture content of biomass is the quantity of water existing within the biomass, expressed as a percentage of the total material's mass. Moisture content of biomass in natural conditions (without any further processing) varies enormously depending on the type of biomass, ranging from less than 15% in cereals straw to more than 90% as in algae biomass. This is a critical parameter when using biomass for energy purposes since it has a marked effect on the conversion efficiency and heating value. Moreover, high moisture content entails logistic issues since it increases the tendency to decompose (resulting in energy loss during storage) and reduces the energy and cost balances.

The heating value of a biomass feedstock represents the energy amount per unit mass or volume released on complete combustion (FAO, 2004). The heating value is referenced in two different ways, the higher (or gross) heating value (HHV) and low (or net) heating value (LHV). The HHV includes the latent heat contained in the water vapor that in practice cannot be used effectively, while the LHV excludes the heat of evaporation of the water formed from the hydrogen contained in the biomass feedstock and its moisture content. Thus, the LHV is the appropriate value to assess the energy available for subsequent use.

However, for some bioenergy pathways or processing, e.g., fermentation or anaerobic digestion, high moisture content is not counterproductive. Hence, based primarily on the biomass moisture content, the type of biomass selected subsequently dictates the most likely form of energy conversion process (McKendry, 2002).

2.2.4 Inorganic compounds and ash composition

As stated at the beginning of this section, many elements in low quantities and varying concentrations are present in the biomass feedstocks, such as Si, Ca, Mg, K, Na, P, S, CI, AI, Fe, and Mn, as well as heavy metals such as Cu, Zn, Co, Mo, As, Ni, Cr, Pb, Cd, V, and Hg. The presence of these inorganic elements has a strong influence in the combustion process, by forming gaseous and solid emissions, as well as influencing the ash melting behavior, which may derive in corrosion processes (Obernberger et al., 1997). While Na and K could lead to ash vitrification (depending on their concentration and the combustion temperature), high content of Cl entails emission of dioxins and material corrosion. The oxidation of S produces sulfur oxides (mainly SO_2) which in combination with steam generate sulfuric acid contributing to acid rain formation. On the other hand, the presence of elements such as As, Ba, Cd, Co, Cr, Cu, Fe, Hg, K, Mn, Mg, Mo, Ni, P, Pb, Sb, Tl, V, and Zn allows the use of the generated ashes as fertilizers, which improves the environmental performance of the use of biomass for energy purposes.

Hence, knowing the biomass composition of inorganic elements is crucial for process control and for handling co-products and wastes resulting from energy and fuel utilization of biomass (Thy et al., 2013).

2.3 CLASSIFICATION OF BIOMASS TYPES

2.3.1 According to chemical composition

As already mentioned in this chapter, the main components of biomass feedstocks are carbohydrates (namely lignocellulosic, sugar, and starch—polysaccharides—carbohydrates), lipids, and proteins. The ratio in which these components are present in the biomass varies greatly. Thus, biomass types can be grouped according to the following classification:

- *Lignocellulosic biomass* with a predominance of plant fibers, i.e., cellulose, hemicelluloses, and lignin, such as straw, wood, and energy grasses. This biomass type is intrinsically linked to the classification of biomass into *herbaceous biomass* (biomass from plants that have a non-woody stem and which dies back at the end of the growing season) and *woody biomass* (biomass from trees, bushes, and shrubs) (FAO, 2004).
- *Sugar-rich biomasses* with carbohydrates in the form of monosaccharides (mainly glucose or fructose) and disaccharides (sucrose), such as sugar beet and sugar cane.
- *Starch-rich biomasses* with a high proportion of reserve polysaccharides, basically starch and inulin, such as grain cereals (wheat, corn, etc.) and tubers (potato, Jerusalem artichoke, etc.).
- *Oil-rich biomasses* with a high lipid content, especially in some specific parts, such as rapeseed and some micro- and macro-algae.
- *Protein-rich biomasses* from plant biomass such as oilseed (e.g., soybean, sunflower) and legumes (e.g., peas) and also from animal biomass (e.g., pig meat and fish). Due to its nitrogen content, high price, and the food-first principle, this kind of biomass is not feasible, in the long term, for its use as bioenergy feedstock.

Finally, biomass can also be characterized according to its physical conditions into wet and dry biomass. The water content of the former varies from 15% to more than 90% (as in the case of algae biomass), while the latter presents moisture content below 13%. This factor will influence the requirements for its harvest, transport, storage, and processing since wet biomass requires a higher transport effort (because more water is transported) and additional processing before storage, such as drying or ensiling (e.g., maize is ensiled for feed and biogas applications).

2.3.2 According to origin

Biomass for energy uses can also be classified according to its origin or the sector where it is produced. Thus, biomass feedstocks can be grouped as listed below:

- *Agricultural biomass*: biomass grown in agricultural land (Agricultural land defined as the land area that is either arable, under permanent crops, or under permanent pastures (OECD, 2018)) which includes all kind of agricultural produce, regardless of the chemical composition (i.e., lignocellulosic, starch, oil crops, etc.) and whether or not it is edible (i.e., food and energy crops).

- *Forest biomass*: wood from forest (Forest is defined by FAO as Land spanning more than 0.5 hectares with trees higher than 5 meters and a canopy cover of more than 10 percent, or trees able to reach these thresholds in situ. It does not include land that is predominantly under agricultural or urban land use (FAO, 2012)) and other wooded land (Land not defined as "Forest," spanning more than 0.5 hectares; with trees higher than 5 meters and a canopy cover of 5–10 percent, or trees able to reach these thresholds; or with a combined cover of shrubs, bushes and trees above 10 percent (FAO, 2012)) including tree plantations in forest land for energy and woody biomass from forest management (pruning, thinning, etc.).
- *By-products, residues, and waste*: can be defined as biomass from well-defined side-streams from agricultural, forestry, and related industrial operations (FAO, 2004). It also includes organic residues from municipal solid wastes (MSWs).
- *Aquatic biomass*: refers to any plant or animal material that has formed in water, such as microalgae, seaweed, and aquatic plants.

It is worth clarifying that biomass classifications often mix both criteria, the origin and nature of biomass. Thus, agricultural biomass comprises both herbaceous and woody biomass. At the same time, short rotation coppice (SRC) (woody biomass grown as a raw material and/or for its fuel value harvested on a 2- to 5-year cycle (FAO, 2004, DEFRA, Department for Environment Food & Rural Affairs, 2004)) does not refer to the type of land where it is produced and can be produced in both forest and agricultural land (e.g., poplar grove) (Box 2.2).

BOX 2.2 The concept of residual biomass vs biomass by-product vs waste

The consideration of biomass resources as by-product or residue is basically linked to how the terminology is used since the concept behind is similar: by-product can be defined as "an incidental product deriving from a manufacturing process or chemical reaction, and not the primary product or service being produced. A by-product can be useful and marketable, or it can have negative ecological consequences." In the context of sustainability scheme for biofuels and bioliquids, the review of the European Renewables Directive (European

(Continued)

BOX 2.2 (Continued)

Commission, 2017) defines "residue" (Article 2) as "a substance that is not the end product(s) that a production process directly seeks to produce; it is not a primary aim of the production process and the process has not been deliberately modified to produce it." As it can be seen in both, residue and by-product, there is a primary commodity that drives the production chain.

On the other hand, waste is considered as defined in Article 3(1) of Directive 2008/98/EC (European Commission, 2008), i.e., "any substance or object which the holder discards or intends or is required to discard."

Thus, regardless of the competitive uses (e.g., fodder and animal bedding vs bioenergy), feedstocks like cereal straw are considered crop residues or by-product but not waste and they count as "non-food cellulosic material" for the production of advanced biofuels.

Within the mentioned category of by-products residues and waste, the following classes can be distinguished:

- Agricultural by-products or crop residues: originated from production, harvesting, and processing in farm areas, such as cereal straw and fruit stones, shells, and husks, and also in animal by-products from livestock keeping such as solid excreta of animals.
- Forestry by-products/residues: residues from harvesting of forest and forestry plantations such as small branches, leaves and needles. Products of non-commercial thinnings could be also included in this category.
- Agro-industrial by-product/residues: including biomass materials produced chiefly in food and fiber processing industries such as sugarcane bagasse, rice/paddy husks and hulls, coconut shells, husks, fiber and pith, olive pressing, glycerin, oil press-cake, and manure.
- By-products and residues from the wood industry: originated from the wood processing as well as the pulp and paper industry, such as bark, cork by-products, cross-cut ends, edgings, fiber board by-products, grinding dust, particle board by-products, plywood by-products, saw dust, slabs, and wood shavings.
- Other processing residues: substance that is not the end product(s) that a production process directly seeks to produce. It is not a primary aim of the production process and the process has not been deliberately modified to produce it (European Commission, 2010) such as tall oil pitch and manure.
- Waste biomass: includes biowaste (biodegradable garden and park waste, food and kitchen waste from households, restaurants, caterers and retail premises, and comparable waste from the food processing industry), renewable MSW, etc.

The use of residual biomass and wastes as bioenergy feedstock entails several benefits but also drawbacks. As benefits it reduces the associated pollution and fire risks; the production costs and environmental burdens are

(*Continued*)

BOX 2.2 (Continued)

associated to the main produce; it minimizes waste accumulation and it entails lower transport costs since they are usually gathered in a reduced area. Drawbacks are the price fluctuation; insecurity of long-term supply when considered as by-products; market competition with other raw materials, and potential soil impoverishment (due to exportation of minerals and carbon matter) and other environmental impacts.

2.3.3 According to biomass end use

An additional and usual classification of biomass resources is based on the end use of the bioenergy feedstock. According to the type of energy produced, biomass can be classified as follows:

- Transport biofuels: biomass used in the transport sector (road, rail, air, and maritime transport) regardless of the physical state, i.e., including (1) bioliquids for internal combustion engines (e.g., bioethanol, biobutanol, bio-ethyl tertiary butyl ether, and bio-methanol); (2) bioliquids for compression ignition engines (biodiesel, bio-dimethyl ether, and hydrotreated and pure vegetable oils) and jet engines (biokerosene); and (3) gaseous biofuels for internal combustion engines (e.g., biogas, biosynthetic gas, and bio-hydrogen).
- Biomass for heat and power: refers to biomass used for the production of heat, electricity, or both simultaneously (cogeneration or combined heat and power—CHP). Heat and power are usually produced from solid biomass. However, for these purposes the above-mentioned gaseous bioenergy carriers (biogas, bio-syngas, etc.) and bioliquids (e.g., ethanol, biodiesel, bio-crude oil, or vegetable oil) can also be used. The latter are used in a lower proportion for stationary decentralized energy generation since they are high energy-intensive and the product costs are high compared to other resources. Regarding solid biomass, it should be mentioned: (1) chipped and baled herbaceous biomass mainly made from straw and dedicated energy crops; (2) pellets and briquettes, cylinders made of biomass, mainly from woody but also herbaceous biomass, with different sizes. Pellets usually have 6 mm diameter and 5–30 mm length (FAO, 2004), while briquettes are of 50–130 mm diameter and 50–30 cm of length; (3) sawdust and wood chips from woody biomass, and (4) charcoal produced by a slow pyrolysis.
- Biomass for biorefineries: Biorefining could be defined as the sustainable processing of biomass into a spectrum of marketable products and

energy. Biorefineries can be fed by all kind of biomass, mainly sugar, starch, and oil-rich biomass, but lignocellulosic biomass seems to be more promising since it would avoid the competition with food and feed uses and the availability of raw material would be greater (Cherubini, 2010). The concept of biorefinery is addressed more in depth in Chapter 5, Biorefineries.

- Intermediates or energy carriers: refers to processed biomass containing energy that can be later converted to other forms of energy. In addition to the bioenergy carriers already mentioned, it can be highlighted within this category the torrefied pellets (roasted, ground, and pelletized biomass) and the bio-oil (pyrolysis oil).

2.4 BIOMASS RESOURCES

Bioenergy is the main source of renewable energy in the world when referred as primary energy production (1,319 Mtoe), accounting for 9.6% of total energy sources and 71.1% of total renewable energies (Fig. 2.6) (IEA, 2017a,b).

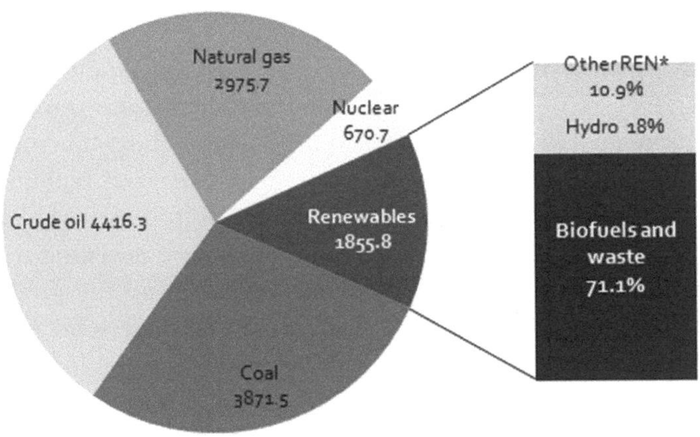

Figure 2.6 Production of energy (primary energy) in 2015 in the world (Mtoe). *Data for "Other renewables" (REN) includes geothermal, solar, wind, and heat from heat pumps. "Biofuels and waste" comprises solid biomass, liquid biofuels, biogases, industrial waste, and municipal waste. *From IEA, 2017a. Renewables Information: Overview. Available at: https://www.iea.org/publications/freepublications/publication/renewables-information---2017-edition---overview.html; IEA, 2017b. Key World Energy Statistics, p. 96. doi:10.1017/CBO9781107415324.004. ISBN: 9788578110796.*

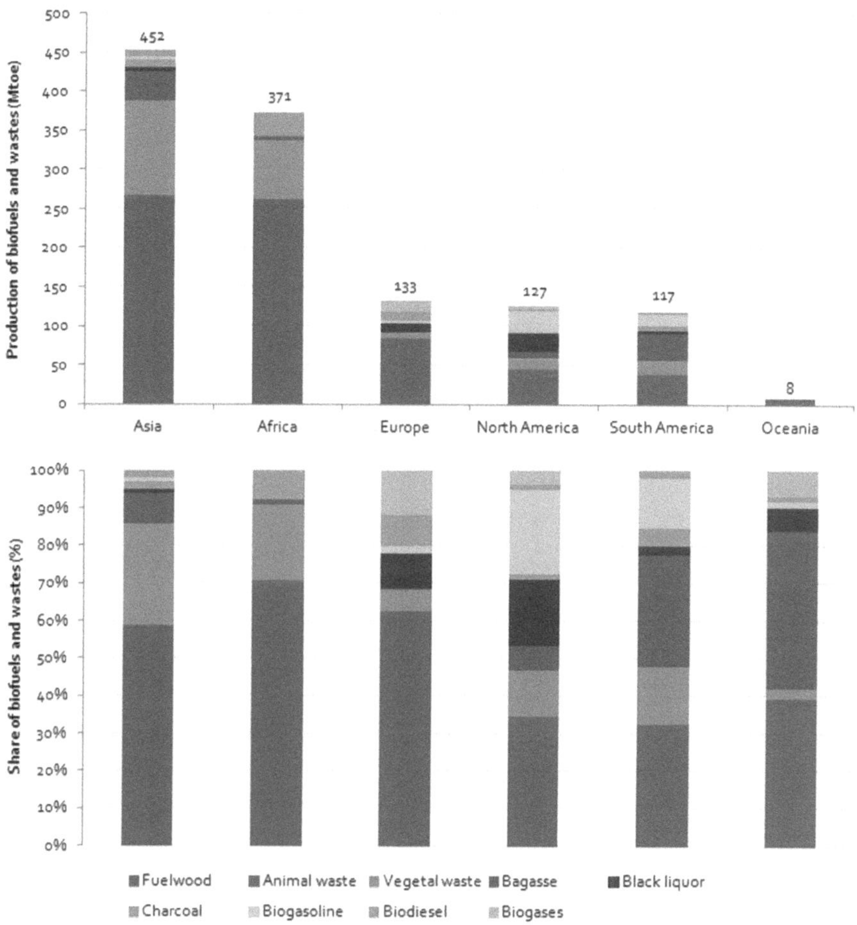

Figure 2.7 Total production and share of main bioenergy carriers and wastes in 2015 by region. Data for municipal and industrial waste were not included since the share of renewable fraction for these categories were not specified. *From United Nations, 2018. Energy Statistics Yearbook 2015. Available at: https://unstats.un.org/unsd/energy/yearbook/.*

According to the United Nations (2018), the bioenergy production by regions in 2015 was led by Asia, which was the continent with the highest production of biofuels and waste, accounting for 452 Mtoe, followed by Africa (371 Mtoe) (see Fig. 2.7). The predominant type of bioenergy produced in each region is somehow representative of the prevailing type of biomass in each area and the availability of technologies. Thus, fuelwood is the main source of biomass in all continents except for Oceania where bagasse is more used. In total, fuelwood represents more than 56% of total bioenergy. Vegetal waste is the second source of bioenergy in Asia (122 Mtoe) and Africa (76 Mtoe), while the second sources are biogas in Europe (15.8 Mtoe) and biogasoline in North and Central America (28.6 Mtoe). In South America bagasse (34.9 Mtoe) and biogasoline (15.7 Mtoe) are, after fuelwood, the main bioenergy sources. Charcoal still has its relevance in developing areas, e.g., Africa (29 Mtoe), Asia (10 Mtoe), and South America (5.5 Mtoe). The use of black liquor for bioenergy is mainly occurring in North and Central America and Europe with 22.2 Mtoe and 12.0 Mtoe, respectively. Bioenergy from animal waste is mainly generated in North and Central America (8.6 Mtoe).

Nevertheless, these statistics could be misleading since they mix types of raw material (e.g., fuelwood, animal wastes, bagasse, etc.) with processed bioenergy carriers (biogas, biodiesel, etc.). Therefore, in the following sections, the description of the main feedstocks for the production of bioenergy is given according to a classification which tries to avoid overlaps and a jumble of biomass resources and bioenergy carriers.

2.4.1 Agriculture

Agriculture ranks second in the order of importance of biomass supply in the world, contributing with about 10% of all the biomass feedstock for bioenergy (Kummamuru, 2017). Agriculture provides biomass for bioenergy from three main sources: energy crops, by-products of other crops, and wastes of farming activities, which can be directed to the production of liquid, solid, and gaseous biofuels. Thus, agricultural biomass has been defined by OECD (OECD, 2004) as "a subset of biomass produced directly from agricultural activities, including cereal grains; sugar crops; oilseeds; other arable crops and crop by-products such as straw; vegetative grasses, farm forestry (e.g. willow and poplar); and livestock by-products, for example, manure and animal fats." Notwithstanding that definition, this section only deals with crops grown for bioenergy, on the grounds that issues concerning biomass from plant species (whether there are arable crops, perennial crops, or SRC) grown on agricultural land are very different from those of the agricultural wastes; the latter category is separately addressed in Section 2.4.3.

Two categories of agricultural energy crops have been considered by Brown and Le Feuvre (2017): (1) sugar, starch, and oilseed crops, which are grown for liquid biofuels and (2) lignocellulosic plants and SRC grown for solid biomass, whether it is for the thermal applications pathway or for the second-generation biofuels. On a global scale, the contribution of dedicated crops to biomass supply for bioenergy has not been well assessed except for the sector of commercial liquid biofuels that come from the first-generation feedstocks, i.e., conventional crops such as corn, oil palm, and soybean. In fact, the World Energy Council state that agricultural biomass is mostly a by-product, waste product, or residue, with the exception of some first-generation biofuels (World Energy Council, 2016b). Bioenergy categories provided by the data service of the IEA include renewable municipal waste, charcoal, solid bioenergy carriers (excluding charcoal), biogasoline, biogases, biodiesel, and other liquid biofuels, without distinction of agricultural or forest origin (IEA, 2018). Likewise, the World Bioenergy Association follows the bioenergy breakdown into municipal waste, industrial waste, solid biomass, biogas, and liquid biofuels (Kummamuru, 2017).

Global statistics indicate that agriculture is a significant contributor to the biomass supply. Thus, the cultivation of crops for biofuels production represented 4.3% total primary energy supply of biomass in the world in 2015. Global biofuels production attained 80,024 ktoe (final energy) in 2015; from 2015 to 2016 the global production rose by 2.6%; such increase varied among biofuels, it was only 0.7% for ethanol while for biodiesel, 6.5%. The allocation of the global biofuels was 61.5% OECD countries, 38.5% non-OECD countries, 17.2% European Union (EU) (BP, 2017). In OECD countries, a share of 10.3% of biomass primary energy supply has been reported (IEA, 2017b). Furthermore, the rise on liquid biofuels production has been policy-driven; more specifically, biofuels target or mandates in the EU (10% renewables in transport) and in countries such as the United States (9.02% renewable fuels), China (bioethanol blend: E10 in nine provinces, biodiesel blend: B5), or Brazil (E27, B7) have been of particular importance for that (OECD/IEA and FAO, 2017).

The choice of a particular crop for bioenergy is driven by a number of factors such as local conditions (soil, climate, and water availability), previous knowledge (technical knowledge, know-how), existing conversion facilities (first- or second-generation ethanol plants, biodiesel plant, and power plant), and policy measures (HLPE, 2013). Growing conventional crops has been the easiest way to accomplish renewable targets in bioenergy but has resulted in great controversy because of their potential impact or effect on food security, land-use change, and deforestation.

Sustainability has turned into a necessary requirement for bioenergy feedstocks (OECD/IEA and FAO, 2017). Although the range of plant species proposed as energy crops has been very wide (El Bassam, 2010), the development of non-food crops for bioenergy (dedicated energy crops) has been proved to be a challenge for researchers and extension agents—in particular as regards scaling-up—and for technologists. Nowadays, the focus has been placed on the development of lignocellulosic crops along with the second-generation technologies. However, in terms of land area and trading, conventional crops dominate the contribution of agriculture to bioenergy. Bioenergy crops are usually classified according to the component of economic interest (Zörb and Lewandowski, 2018) into sugar crops, starch crops, oil crops, and lignocellulosic crops. The first two categories gather crops for first-generation ethanol; the third one crops for biodiesel or HVO biodiesel (hydrotreated vegetable oil); and the last one crops for solid biomass and second-generation biofuels. Following is an overview of the main crops in each category.

2.4.1.1 Sugar-producing crops

Sugarcane has been the most important sugar crop for fuel ethanol in the world, due to intrinsic crop characteristics (high-yielding C4 plant, sucrose-rich feedstock) and historical implementation of ambitious policies on fuel ethanol in Brazil. Brazil is the largest producer of sugarcane in the world, with 10.2 Mha sugarcane, yielding 41% of the world sugarcane production (768.6 Mt in 2016); far from Brazil, India ranks second with 18% total production (348.4 Mt) (FAO, 2017a). As early as 1931, ethanol blending was compulsory in this country. As a result, sugarcane facilities for ethanol have become conventional agro-industries in certain regions of Brazil subjected to market fluctuations and fossil fuels competition (Horta Nogueira and Silva Capaz, 2013). Currently, sugar-based ethanol from Brazil ranks second in the world after the US corn ethanol. However, in terms of ethanol yield, greenhouse gas emissions reduction and production costs sugarcane is more competitive than corn (Manochio et al., 2017; Zabed et al., 2017). Bioethanol produced by Argentina is partly manufactured from sugarcane; in 2016 the Argentinean production amounted to 0.89 Mm^3 bioethanol, 45% from sugarcane (Ciani, 2017).

Very far from sugarcane, other sugar crop commercially used for bioethanol is sugar beet that, different from sugar cane, is a crop of temperate regions. Most sugar beet is grown in Europe, with 3 Mha and 185 Mt production in 2016, representing 66.8% of the world sugar beet production (FAO, 2017a). As compared to sugarcane, the energy efficiency and energy gain of the whole ethanol chain is about 10 times lower (Manochio et al., 2017), but their climate requirements are very different. Moreover, the use of sugar beet for ethanol responds to a concept of biorefinery because it is coupled with the production of crystallized sugar for food. The share of ethanol from sugar beet in Europe is about 45% (Flach et al., 2017). No ethanol plants in Europe use sugar beet directly; instead, they process sugar juice, namely molasses from sugar beet agro-industries (Eurobserv'Er, 2017b).

Sweet sorghum is a sugar crop botanically close to sugarcane. The same as sugarcane, it is a C4 plant and accumulates sucrose in the stem. This crop has been extensively studied for bioethanol production in temperate regions (i.e., out of tropical regions) (Zegada-Lizarazu and Monti, 2012; Regassa and Wortmann, 2014; Nghiem et al., 2016). Land area cultivated with sweet sorghum and values of bioethanol production from this crop are not available in global statistics but ICRISAT reported that sweet sorghum production was commercialized in Brazil and China and was taking off in India (ICRISAT, 2016); in addition it has been reported that China promoted ethanol production using sweet sorghum as well as cassava and other non-food grain feedstocks (Macke and Ward, 2017).

2.4.1.2 Starch rich crops

Cereal crops represent the major source of feedstock for ethanol production, based on the starch in grains; the United States leads the world biofuel production, using maize as main crop. It is well known and highly technified field crop, with C4 photosynthetic pathway that can be highly productive in non-limiting conditions of temperature, water, and solar radiation. The United States, with 35.1 Mha maize and nearly 385 Mt corn production, is the first producer of maize in the world (FAO, 2017a) and ranks first in ethanol production. Ethanol made from corn starch from kernels accounted for 94% of all biofuel production in 2012. Bioenergy statistics provided by this country indicated that the share of total corn used for ethanol attained 31.7% and that the ethanol production reached 60 Mm^3 in 2017 (USDA, 2018). Current technologies allow the whole exploitation of the crop produce in different ways, depending on the type of facilities; ethanol and animal feed ("distilled grains") from the kernel, oil from maize germ, and bioenergy (second-generation ethanol or solid biomass) or bio-based products from the crop residues (stover) (Rosentrater, 2015).

Corn is also a major feedstock for bioethanol in China, along with cassava, a staple crop grown for its starchy tuberous root in tropical and subtropical regions of the world. The top four producers of cassava in the world are Nigeria (57.1 Mt in 2016), Thailand (31.1 Mt), Brazil (21.1 Mt), and Indonesia (20.7 Mt). The production of cassava in China is far from those countries, but it is significant; it attained 4.8 Mt in 2016 (FAO, 2017a). In China, corn and cassava represented 70% and 25% of the feedstocks of fuel ethanol in 2016, respectively (Macke, 2017), while this country ranked fourth in fuel ethanol production (2 Mtoe) in the world after the United States, Brazil, and the EU (Eurobserv'Er, 2017b). The production of fuel ethanol for 2018 was forecast at 3.95 Mm^3 (Macke, 2017).

In the EU corn does not dominate the market of feedstocks for ethanol production as in United States; wheat (37% share) and other cereals (9%), besides corn (38%) are used for that purpose. In all 5.21×10^3 t corn, 5.07×10^3 t wheat, and 1.3×10^3 t other cereals and starch rich crops were used for ethanol production in 2015 (Calderón et al., 2017). Ethanol consumption for transport in 2016 amounted to 2.1 Mm^3 (2.6 Mtoe) in the EU (Eurobserv'Er, 2017b). Although cereal grains are commonly traded commodities, it is worth noting that the land area cultivated with wheat and barley in the EU was 39.4 Mha (68% wheat) with a total production of 200.9 Mt grain (71% wheat) in 2016. In contrast, maize crop occupied 8.8 Mha and yielded 62.7 Mt total corn production (FAO, 2017a).

Among the tuber crops, it is worth mentioning Jerusalem artichoke (JA), an American native plant that accumulates inulin—a biopolymer of fructose and glucose—instead of starch which provides some technological advantages in the process plant. JA is a hardy crop that has been extensively studied for ethanol and other bioproducts (Kays and Nottingham, 2012; Gunnarsson et al., 2014; Yang et al., 2015). Reports on the commercial use of Jerusalem artichoke for bioethanol have not been provided yet. In Europe JA is not a conventional crop but the European Directive 2009/28/EC placed this crop in the same category as cereals; in practice, this fact implies that the implementation of this crop for bioethanol would not be possible. Maybe in the future other countries such as China could undertake the challenge of growing Jerusalem artichoke for ethanol (Liu et al., 2011).

2.4.1.3 *Oil crops*

The contribution of oil crops to bioenergy is framed within the sector of biodiesel, which is currently produced by transesterification or hydrotreatment of vegetable oils. Vegetable oil is expected to continue as the feedstock of choice in biodiesel production (OECD-FAO, 2017). Vegetable oils are internationally traded commodities, particularly palm oil and soybean oil; these oils are the fourth and sixth most produced commodities in the world, respectively. In fact, in terms of area harvested, yield, and total production, the most important oil crop is oil palm, with 21.1 Mha, 1.4 t palm fruit per ha, and 300.2 Mt fruit harvested in 2016 (57.3 Mt palm oil and 6.6 Mt palm kernel oil in 2014). Next in importance are soybean oil (45.7 Mt oil in 2014), rapeseed oil (25.9 Mt), and sunflower oil (15.8 Mt) (FAO, 2017a).

Palm oil is primarily —but not exclusively—feedstock for biodiesel in major palm oil producing countries, which are located in Asia. The top three producer countries in the world are Indonesia (29.3 Mt oil in 2014), Malaysia (19.7 Mt oil), and Thailand (1.8 Mt oil) (FAO, 2017a). Availability and low cost of palm oil make that biodiesel produced by these countries is based on palm oil. Biodiesel production from palm oil in 2017 reached 2.90 Mm^3 in Indonesia (Wright and Rahmanulloh, 2017), 0.48 Mm^3 in Malaysia (Ghani Wahab, 2017), and 1.42 Mm^3 in Thailand (Preechajan and Prasertsri, 2017). Oil palm has been frequently subjected to considerable controversy on the grounds that the increase experimented in the demand for palm oil has often been met at the expense of rainforest, giving rise to impacts on biodiversity, land use, ecosystem services, and others (Khatun et al., 2017). In this regard, there are on-going initiatives to improve the sustainability of palm oil and certify its sustainable origin, particularly with a view to biodiesel sustainability (RSPO, 2018).

So far the sector of biodiesel has been led by the European Union, where the biodiesel consumption attained 11.2 Mtoe (approximately, 14.3 Mm^3) in 2016 (Eurostat, 2018b). The EU biodiesel is based on 70% own-produced feedstocks (specifically, rapeseed oil) as well as internationally traded oils (23% palm oil, mostly imported from Indonesia, and 6% soybean oil) (European Parliament, 2017). Rapeseed is an important crop in central Europe in terms of area harvested, yield, and production. Thus, France, Germany, Poland, and Czech Republic ranked 4th, 5th, 7th, and 10th in world rapeseed production in 2016; mean rapeseed yield in those countries (3,145 $kg \cdot ha^{-1}$) is above the mean yield of the largest producer of the world (Canada, 2,306 $kg \cdot ha^{-1}$) and above the world mean (2,043 $kg \cdot ha^{-1}$) (FAO, 2017a). Projections made by OECD-FAO show that the regional distribution of world biodiesel production in 2025 would be 28% in EU, 24% in the United States, 10% in Brazil, Indonesia in 9%, Thailand in 3%, and others 17%, with a 12% increase from the 2016 level (OECD-FAO, 2018).

Linked to the fact that the United States is the largest producer of soybean in the world (35% of the world production is allocated to the United States) (FAO, 2017a), soybean oil is the feedstock of choice for biodiesel in the United States. However, the current production of US biodiesel can be considered of little significance in relation to bioethanol, because of the composition of diesel and non-diesel fleet (96% gasoline, 4% diesel) (Chambers and Schmitt, 2015). The share of biodiesel was less than 6% of all biofuel production in the United States in 2016; with 99 facilities in the whole country, the total production of biodiesel in the United States amounted to 8.9 Mm^3 in 2017 (USDA, 2018). Biodiesel in Brazil is also based on soybean, despite the efforts made by Brazilian institutions to promote other oil crops, such as castor beans or oil palm for biodiesel that would benefit more small farmers (Horta Nogueira and Silva Capaz, 2013); nevertheless, Brazil ranks second in world soybean production (29%) (FAO, 2017a). After Brazil, Argentina is the third largest producer of soybean in the world; likewise, feedstock of Argentinean biodiesel is own-produced soybean oil; in 2016, Argentinean production attained 2.66 Mt biodiesel from 2.75 Mt soybean oil; biodiesel production in 2017 was expected to be higher, since the production in the period January–September was higher in 2017 than in 2016 (Ciani, 2017).

2.4.1.4 Lignocellulosic crops

In the last decades much research effort has been made to develop a new and challenging line of agricultural biomass production based on lignocellulosic crops. This type of crops is intended to be dedicated energy crops grown for solid biomass or advanced biofuels (see Chapters 3–5: Liquid Biofuels; Solid Biomass to Heat and Power; Biorefineries, respectively). Reasons for the promotion of lignocellulosic species are various. In nature, the range of lignocellulosic plant species is by far wider than sugar or oil plant species; hence, there is almost always the possibility of growing a lignocellulosic plant species everywhere; lignocellulosic species are usually hardier than conventional energy crops. Moreover, the energy cost of producing cellulose or hemicellulose for a plant is approximately half the lipid cost (Bentsen and Møller, 2017). The fact that cellulose is the most abundant biopolymer on the earth has pushed the development of advanced concepts for biofuels production (Carere et al., 2008).

In the late 1970s, the United States undertook a specific R&D program on energy crops conducive to the selection of herbaceous and woody crops for solid biomass that resulted in the identification of switchgrass as a model species in the category of the perennial herbs and poplar and willow as model crops for SRC, and the advent of new lines of research (Kszos et al., 2001). In parallel with the US research, a wide range of perennial herbs and woody plant species were investigated for lignocellulosic biomass production in subsequent decades in Europe (Venendaal et al., 1997). Progress made in the field of plant production has been impressive concerning the agronomy of perennial grasses (Barth et al., 2016) and short rotation woody species (Dimitriou and Rutz, 2015; Njakou Djomo et al., 2015), as well as in the field of lignocellulose conversion technologies (Aditiya et al., 2016). Current trends in lignocellulosic energy crops are focused on their suitability to marginal lands (Shortall, 2013; Fernando et al., 2015; Mehmood et al., 2017).

In the category of perennial grasses, probably the most studied lignocellulosic energy crops have been switchgrass (*Panicum virgatum*) and Miscanthus (*Miscanthus x giganteus*). Both species are C4 and exhibit great biomass potential in temperate/cold climates. Switchgrass is a species native to North America that has been studied for over 70 years in the United States (Wright, 2007). Miscanthus is a relatively novel crop in western countries; it was originated in East-Asia and was first introduced in Europe in 1930 (Lewandowski et al., 2003). There have been numerous R&D programs and consolidated international networks on these two crops, which have resulted in extensive literature (Jones and Walsh, 2000; Monti, 2012; Lee et al., 2018) and some commercial plantations.

As regards SRC, the potential of genotypes in genus *Salix* (willow), *Populus* (poplar), and *Eucalyptus*, among others, have been studied for biomass production in regions of temperate climates (Volk et al., 2006; Hinchee et al., 2011; Dimitriou and Rutz, 2015; Njakou Djomo et al., 2015). *Salix* and *Populus* are botanically related genera; they belong to the same family (*Salicaceae*), and both of them are deciduous woody species. In contrast, *Eucalyptus* spp. are evergreen. This fact brings advantages for willow and poplar over *Eucalyptus* in thermal biomass conversion, due to lower quality of leaves as feedstock for energy production (Senelwa and Sims, 1999). SRC of willow for biomass production was reported to be fully developed in 1990 in Northern Europe (Venendaal et al., 1997). Currently SRC plantations are based on willow and poplar in Europe (Calderón et al., 2017).

Statistics on lignocellulosic crops grown for biomass have not been provided yet at a global scale, probably because the cultivation of this type of crops for energy purposes started recently and it is not significant in the world. For instance, in Europe the land area devoted to lignocellulosic crops has been estimated at 50,764 ha in all for the European Union (EU28) (Calderón et al., 2017), 29% Miscanthus, 51% Willow (SRC), and 25% Poplar (SRC). The Department for Environment Food & Rural Affairs of the United Kingdom stated that miscanthus represented around 0.2% of the total arable area in England in 2016 (7057 ha), while for SRC it was reported less than 0.1% (2952 ha) (DEFRA: Department for Environment Food & Rural Affairs, 2017). A power plant of 2.6 MW_e capacity is currently in operation in Britain, fueled on miscanthus biomass (Bacovsky et al., 2016). From these figures, it could be said that lignocellulosic energy crops have just taken off. Nevertheless, there are great expectations about the contribution of these crops to bioenergy and greenhouse gas mitigation, particularly with a view to the scaling of second-generation technologies.

2.4.2 Forestry

2.4.2.1 Forest and other wooded lands

At global level, forests and other wooded land are the main supplier of biomass for energy. According to the World Bioenergy Association (Kummamuru, 2017), 88% (in quantity of energy) of the biomass used for energy in the world originated from forestry. The United Nations (2018) estimated that 728 Mtoe of energy were produced from wood (6.5% of which is in the form of charcoal) in 2015. As for energy in general, the use of woody biomass for energy is highly contrasted. On the one hand, wood is the principal source of energy for cooking and heating for 2.4 billion people—mainly in lower income countries (FAO, 2014). It is usually collected manually from local woodlands and burnt in open stoves and fireplaces (Richards et al., 2015). On the other hand, in higher income countries, wood fuels, which had been progressively replaced by fossil fuels during the 20th century, have regained attention in the last decades as a renewable source of energy. The new increase in uses goes in pair with the transformation of wood in new types of fuel (production of, among others, wood chips, pellets, and liquid biofuels) and the development of new high-efficiency burners. The contrasts between regional contexts are associated with important differences in the current intensity of wood mobilization and in the approach towards a more sustainable use of the wood fuels. This section details the different types of wood fuels, develops an overview of their supply and uses in the world, and describes the principal wood fuel value chains and their interactions with material uses of wood.

2.4.2.2 Different types of wood fuels

Wood fuel denominates all types of bioenergy carriers originating directly or indirectly from woody biomass (FAO, 2004). It includes *direct wood fuels*, i.e., wood extracted from forests, other wooded lands, and other lands with trees for energy use, *indirect wood fuels* (mainly solid bioenergy carriers) produced from wood processing activities, and *recovered wood fuels* which correspond to wood used directly or indirectly as fuel, derived from socioeconomic activities outside the forest sector. In this part, the analysis focuses on the direct wood fuels, while the two others are presented in Section 2.4.3.3. Wood fuels can be used in the form of fuelwood (where the original composition of the wood is preserved), charcoal, black liquor, and other wood fuels such as ethanol (see Table 2.1).

All forests, including the ones dedicated to industrial wood, can supply direct wood fuel, since products of the early thinnings and part of the residual biomass of final harvest cannot be used as industrial wood. Therefore, direct wood fuel can be either a primary product (forest managed for wood fuel production such as traditional coppices) or a byproduct. The wood fuel production is therefore directly linked to the

Table 2.1 Wood fuel sources and types

	Supply side (sources)		
Commodities (wood energy carriers)	**Direct wood fuels**	**Indirect wood fuels**	**Recovered wood fuels**
Fuelwood	×	×	×
Charcoal	×	×	×
Black liquor		×	
Other (methanol, ethanol, pyrolytic gas)	×	×	×

Source: From FAO, 2004. Unified Bioenergy Terminology (UBET). Available at: ftp://ftp.fao.org/docrep/fao/007/j4504e/j4504e00.pdf.

general wood production, as a synergetic by-product (the more industrial wood is produced, the more wood fuel is available) or as a competing product (some wood could be used as fuel, instead of being used for particle boards).

More than 88% of the direct wood fuel is made of non-coniferous species (FAO, 2018a). Fuelwood is the predominant type of direct wood fuel produced and used. If fuelwood is still mainly used as fire logs, the share of new products such as wood chips and pellets increases with the development of large industrial boilers, CHP plants, and pellet stoves. Wood chips are rectangular shaped pieces of wood with a typical length of 5–50 mm and low thickness, obtained by mechanical treatment. Pellets are principally made of wood, but can also be produced with other agricultural biomass (herbaceous, fruit) or biomass blends.

Globally, nearly 20% of the wood fuel is transformed into charcoal (FAO, 2018a). The transformation rate is higher in African countries (30% of the wood fuel in 2015) where wood is the main source of energy for a large part of the population. Charcoal is sometimes preferred over fuelwood for its greater heating value (almost twice as much energy per kilogram as fuelwood) which makes it easier and less energy intensive to transport (Keita, 1987), its higher versatility, and its lower emissions of small particles.

The production of liquid biofuels and biogas from direct fuelwood is still limited. Innovative treatment of lignocellulosic biomass for the production of liquid biofuels, and ethanol in particular, has been developed over the last years and might become economically viable in the future (Kallio et al., 2018). Data are currently too scarce to provide statistics.

2.4.2.3 Wood resources in forest and other wooded lands

Forests cover almost 4 thousand million ha in the world in 2015, which correspond to 31% of the terrestrial land of the earth (without inland water, FAO, 2017b). South America, with 48%, and Europe (including Russia), with 46%, are the continents with the highest share of forest cover, as opposed to Asia, with 19% forest cover. This variability can be explained by natural factors such as climate and soil conditions such as altitude and drought as well as human factors such as the pressure on the land for agriculture and urbanization. In Africa, other wooded lands and other lands, occupying 12% and 7% of the land, respectively, are important components of the wooded land that supply wood fuel in addition to the forest land (Fig. 2.8).

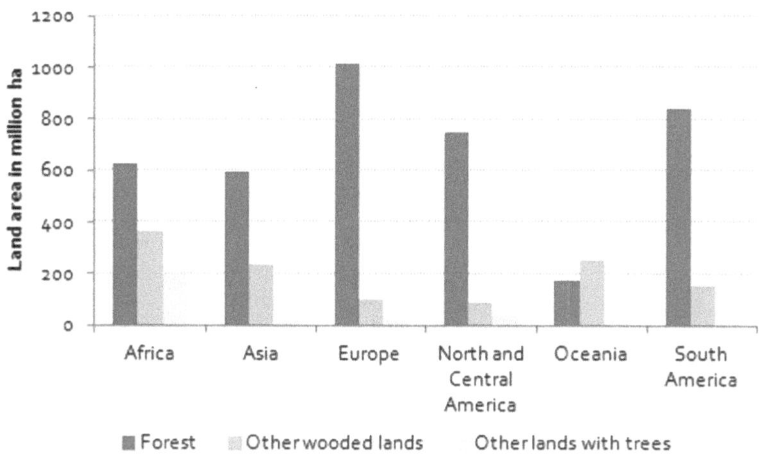

Figure 2.8 Area of wooded land area in 2015 by continent. *From FAO, 2017b. Flude—Global Forest Resources Assessments. Rome, Italy: Food and Agriculture Organization of the United Nations. http://www.fao.org/forest-resources-assessment/explore-data/flude/en/ (accessed February 20, 2018).*

Globally, the forest area has been decreasing at a rhythm of 0.13% per year from 1990 to 2015. This decrease mainly took place in Africa and South America, while in Europe the forest area has been slightly increasing over the last decades after the decline in the middle age and the industrial area.

There is on average slightly more than 0.5 ha of forest per capita in the world, 0.3 ha of other wooded land. Asia, with its high population density offers 0.1 ha of forest, 0.1 ha of other wooded land, whereas Oceania has 4.4 ha of forest and 6.5 ha of other wooded land per capita. However, only one part of this land is available for wood supply because of technical, economic, or regulatory reasons. Moreover, because of the global growth of the population and the decrease in forest areas, the area of forest per capita is decreasing except in Europe where it is almost stable. This causes new problems for the sustainable management of forest resources and the efficient use of these resources.

The volume of growing stock in the world's forests was close to 480 thousand million m^3 in 2015 (own calculation from FAO data, see Fig. 2.9). The countries having the largest resources are Brazil (20% of the world growing stock), the Russian Federation (17%), Canada (10%), and the United States (9%). These countries, together with India and China, are among the six first wood producers in the world. However, the resources in Brazil are decreasing while they are increasing in the

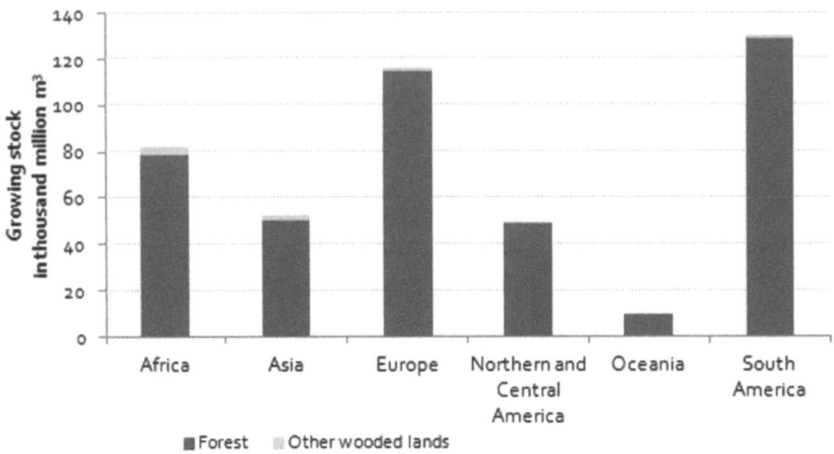

Figure 2.9 Growing stock in forests and other wooded land in 2015 by continent. Data corresponding to 2015 or most recent information available. *From FAO, 2017b. Flude—Global Forest Resources Assessments. Rome, Italy: Food and Agriculture Organization of the United Nations. http://www.fao.org/forest-resources-assessment/explore-data/flude/en/ (accessed February 20, 2018).*

United States and the Russian Federation. Globally, about 2% of the wood resources are located in other wooded lands. This share reaches slightly more than 4% in Africa and 3% in Asia. Other wooded lands are essential for the provision of wood in areas such as Sahel and the Sudan region, the south Mediterranean region.

2.4.2.4 Wood fuel production

Nearly 1.86 thousand million solid cubic meters of wood were produced in 2016 (FAO, 2018a). This corresponds to half of the annual production of wood with important differences between countries. Most of this production takes place in Asia and Africa where wood is traditionally used for cooking and heating. The most important players are India, China, and Brazil (see Fig. 2.10), which are large countries, with considerable resources. However, wood resources are under threat (i.e., the resources have been decreasing recently) in most of the top 10 providers except in China and in the United States. A similar threat can be seen in most tropical countries. When wood production exceeds the capacity of the forest ecosystems to supply wood, wood cannot be considered as a renewable source of energy under current practices. Many countries facing (or risking) overconsumption of wood for cooking and heating take measures to reduce wood fuel consumption by incentivizing alternatives sources of

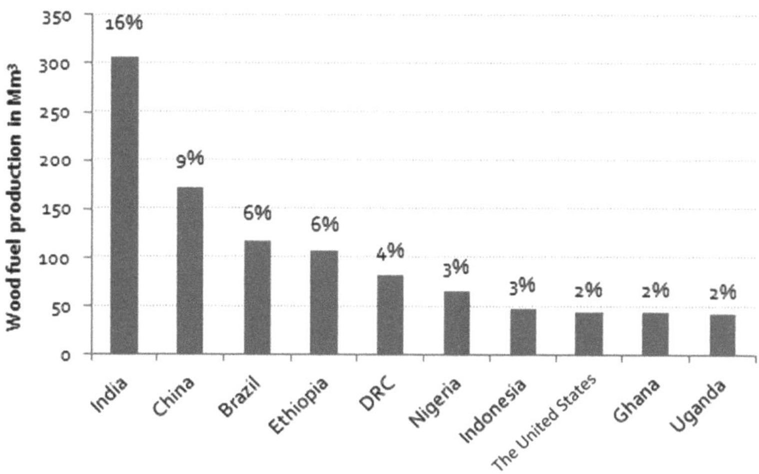

Figure 2.10 Production by the top 10 wood fuel producers in the world in 2016. Total production in the country (bar) and share of the world production (percentage). DRC, Democratic Republic of the Congo. *From FAO, 2018a. FAOSTAT—Forestry Production and Trade. Food and Agriculture Organization of the United Nations, Rome. http://www.fao.org/faostat/en/#data/FO (accessed March 05, 2018).*

energy such as natural gas or improving the energetic performance of the appliances (Sulaiman et al., 2017).

In Northern America and in Europe, the stock of standing wood is increasing. This gives opportunities to intensify the production of wood (including wood fuel) without compromising the forest ecosystems (FAO, 2017c, Camia et al., 2018). Policies are developed to take advantages of these resources to increase the sustainability of the economies and to reduce the net emissions of greenhouse gases (EC, 2012).

The share of wood production dedicated to wood fuel varies considerably between countries and regions. In the least developed countries, 92% of the wood is harvested for energy. Most of these countries are in Africa and Asia where the average share of wood production for fuel are 90% and 65%, respectively, Fig. 2.11). In America, there is a high contrast between the North, where more than 90% of the wood is harvested for industrial use, and Central America and the Caribbean where 90% and 82% of the wood is produced as fuel, respectively. These differences stem from the structure of the demand for energy which relies highly on traditional uses of wood and charcoal in the least developed countries, while a switch to other sources of energies (fossil fuels in particular) was performed over the last century in industrialized countries.

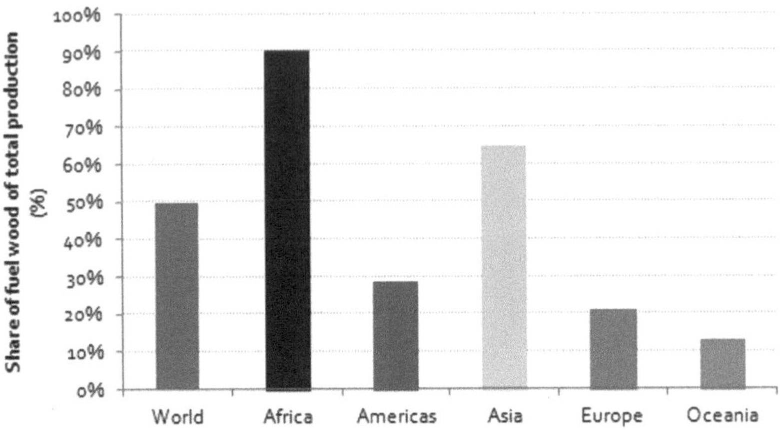

Figure 2.11 Share of fuelwood in the total production by continent in 2016. *From FAO, 2018a. FAOSTAT—Forestry Production and Trade. Food and Agriculture Organization of the United Nations, Rome. http://www.fao.org/faostat/en/#data/FO (accessed March 05, 2018).*

The return to biomass for energy in industrialized countries is accompanied by a switch to more efficient technologies and to the use of wood in the form of chips or pellets, which can be produced from wood extracted from the forest or from industrial residues (see section 2.4.3). Part of the wood fuel production is transformed into charcoal, in particular in Africa where about 30% of the wood fuel production is converted into charcoal. Brazil is the world leader in charcoal production and is also the main consumer.

2.4.2.5 Changes in wood fuel production

The production of wood fuel has been increasing globally, from 1.5 billion m^3 in 1961 to 1.86 billion m^3 in 1991. Since then, wood fuel production decreased slightly in the 1990s and early 2000 before returning to the 1.86 billion m^3 in 2016. The change has been dominated by the increase in the production in Africa, connected to the increase in population (Fig. 2.12), while the production decreased in Asia since the 1990, resulting from a switch to fossil fuels. In Europe, the production went down in the 1970s compared to the 1960s. The production is growing since the year 2000, which is partially a result of the policies to support a switch to renewable resources and to reduce greenhouse gases emissions. In Northern America, the production went up after the oil crisis in 1973,

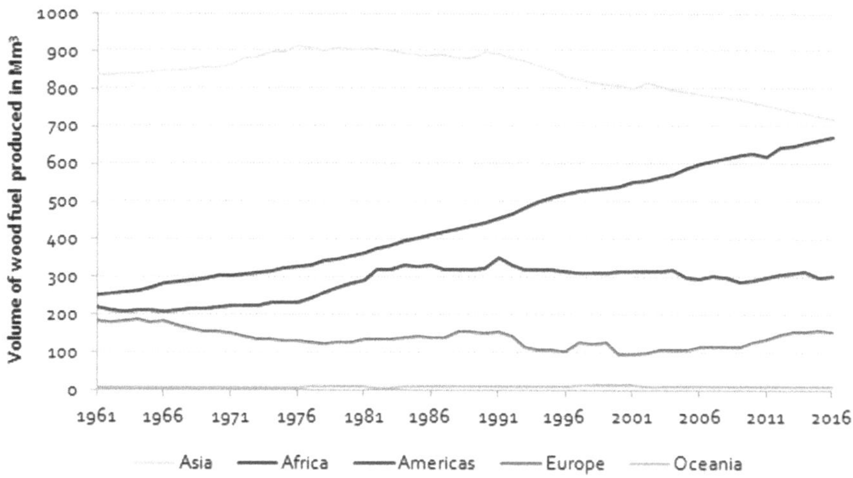

Figure 2.12 Direct wood fuel production by continent from 1961 to 2016. *From FAO, 2018a. FAOSTAT—Forestry Production and Trade. Food and Agriculture Organization of the United Nations, Rome. http://www.fao.org/faostat/en/#data/FO (accessed March 05, 2018).*

when wood was seen as an alternative to fossil fuels (Mittlefehldt, 2016). It went down after the end of the gulf war in 1991.

According to FAO (2018a), the apparent consumption of fuelwood per capita has been decreasing from 0.48 m^3/capita in 1961 to 0.25 m^3/capita in 2016 due to the switch and other sources of energies and the use of more efficient technologies. This decrease is mainly visible in Asia, where it went down from 0.48 to 0.16 m^3/capita over the period 1961–2016. The consumption per capita in Africa has decreased as well over the same period but to a smaller extent (from 0.86 to 0.55 m^3/capita).

Many energy policies (e.g., in the European Union, Switzerland, Colombia, etc.) target the replacement of fossil fuels by biomass to increase the share of renewable in the energy mix and to reduce greenhouse gasses emissions (OECD, 2015). This would lead to higher consumption of wood fuel. In this context, the sustainability of the forest management practices must be ascertained. In Europe, an additional quantity of wood corresponding to approximately one-third of the current production could be harvested without compromising the forest ecosystems (Camia et al., 2018). Part of the increase in the production could be allocated to energy, either as direct or indirect sources.

This analysis makes use of internationally available information on wood fuel. Wood fuel production is often underestimated because of the lack of statistics on wood fuel harvested by households for their own consumption from their own land or from public land (Camia et al., 2018). Therefore, results shall be considered with caution.

2.4.3 By-products, residues, and wastes

2.4.3.1 Crop residues

Crop residues is meant as the primary biomass fraction of the agricultural residues; crop residues include wastes associated with harvest (e.g., straw, stubble, or stover) and residues from crop operations (e.g., orchard pruning) (Jölli and Giljum, 2005; Torén et al., 2011; Bentsen et al., 2017). Most agricultural crops are grown only for a fraction of their biomass, which means that, in addition to the economic product, a certain amount of residual biomass is generated from agricultural systems, whether they are based on annual crop systems (field crops such as cereals, vegetables, sugar crops, or fiber crops) or permanent crops (such as olive groves, vineyards, or fruit trees).

Residues from agricultural crops have been used as traditional biomass (OECD/IEA and FAO, 2017) for bioenergy in rural areas; that has been the case of fruit tree and olive pruning residues used as wood fuel for heat in the Mediterranean region. Nowadays the use of crop residues for modern bioenergy (solid bioenergy carriers, advanced liquid biofuels, and power generation) has been promoted as a way of avoiding some of the problems caused by the cultivation of conventional crops for bioenergy, essentially derived from competence for land, water, and other resources, land-use change, deforestation, and food security.

In context of great concern on the availability of lands for biomass production and the effects of land-use change associated with conventional crops used for the first- generation biofuels, crop residues have been envisaged as a part of the solution to sustainable biofuels.

Assessing the size of the resource is key to bioenergy projects based on crop residues. In this respect, values reported in literature for residues' assessments should be taken with caution. A recent report published by the EIA pointed out the importance of the approach used and the assumptions made for assessing residual biomass; differences among estimations and actual field data can be great (Bentsen et al., 2017). Thus, the resource assessment can be made at the level of theoretical potential (maximum amount of biomass that can be considered theoretically available for bioenergy production), technical potential (the fraction of the theoretical potential that is available under current framework conditions, such as infrastructure, harvesting techniques, accessibility, and others), or sustainable potential (the share of the technical potential that meets criteria of economic profitability, while being environmentally and socially sustainable) (OECD/IEA and FAO, 2017) and at different geographical scales (field, municipality, region, and country). Bentsen et al. (2014) showed that literature data varied greatly; estimates ranged from 10 to 69 $EJ \cdot year^{-1}$. Results from a study conducted in Europe showed that the maximum sustainable residue potential accounted for 24%–48% of the theoretical potential (competing uses of residues were not considered) (Haase et al., 2016).

Similar to agricultural economic products, yields in residual biomass vary among plant species, crop varieties, and abiotic and biotic and crop conditions. In Agronomy, an efficiency index of a particular annual crop (a variety grown in a location) is the Harvest Index (HI), which is calculated as the ratio between the economic production and the above ground biomass production. In biomass assessments, if the values of HI and crop yield (Y) are known and the above ground biomass production is unused, the value of HI can be used to estimate the theoretical residue production or residue yield (RY).

On the other hand, if the value of HI is unknown, an assumed value of biomass production on a land area basis (amount of residues per hectare of crop) could be used as an alternative although the study by Jölli and Giljum (2005) showed that the calculation with the HI was more exact than the calculation by land area. In line with HI definition, the amount of residual biomass that a crop system may generate can be assessed by the so-called residue to product ratio (RPR), which refers to the production of agricultural residues to the production of the economic product (Bentsen et al., 2014). The amount of residue is assumed to be directly proportional to crop yield (economic product), especially when assessing non-constrained crops (see, for instance, the study by Guerrero et al., 2016). Thus, a widely known approach based on RPR is the FAO's BEFS (Bioenergy and Food Security) approach, which comprises two sets of methodologies and tools to conduct a bioenergy assessment, including rapid appraisal based on assumed RPR values (FAO, 2014). Tools for the rapid appraisal have been made available on the Internet (FAO, 2018b). However, it is widely known that biomass partitioning is influenced by plant growth which, in turn, depends on crop conditions (soil, climate, pest, and diseases). In practice it means that the proportion of residues varies with the crop yield, assuming that a linear relationship between Y and RY could be simplistic in some cases. Other studies were based on plant growth models (Camia et al., 2018). Bentsen et al. (2014) assumed for their assessment of the agricultural residue production in the world that the relation between crop yield and RY was exponential. These authors highlighted the high level of uncertainty entailed in this type of assessments and the need for statistics on residue production based on measurements in the field.

Sustainability of crop residues for bioenergy has been supposed to be higher than for energy crops but it has become a subject of concern, especially in the environmental and economic pillars. It has been claimed that the use of crop residues for bioenergy may have an impact on soil organic carbon (SOC), soil quality, and ecosystem services (Lal, 2005; Bentsen et al., 2017). According to Lal (2005), impacts can be short term and long term, and the long-term impact on SOC pool and soil quality cannot be ignored. In the opinion of this author: "Production of biomass for biofuel ... must be undertaken on specifically dedicated land to grow species with a potential to produce high biomass. The economics and environmental consequences of competing uses of crop residue must be assessed objectively with a holistic approach and long-term perspective." Indeed the IPCC has stressed the importance of agricultural activities to soil C; nevertheless, the global carbon stock in soils represents about 57% of the C in terrestrial ecosystems (atmospheric C represents about 30%) (IPCC, 2000). Thus, restrictions used for the assessment of crop residues potential for bioenergy have included the criteria of soil carbon preservation (Monforti et al., 2013; Haase et al., 2016). As a reference, Haase et al. (2016) excluded areas with <1% SOC content in topsoil (Corg) and assumed that the proportion of residues that was left on the field was (1) 80% if Corg <2% and (2) 40% if Corg >2%.

Studies conducted so far have shown that each case should be separately analyzed in relation to the applicable context (Monforti et al., 2013; Haase et al., 2016; Bentsen et al., 2017). The IEA (Bentsen et al., 2017) analyzed three national cases of supply chains based on crop residues: Denmark, Canada, and the United States; final applications and critical issues observed for each country were different: risk of depleting soil organic matter due to removal of crop residues for heat and electricity in Denmark; high economic risk of the lignocellulosic ethanol industry in the United States; and baseline of SOC and crop yield, along with concerns on the profitability of the corn stover chain (feedstock and technologies) in Canada. The overall conclusion of that work was that trends cannot be extrapolated from one case study to another. These authors stated that there are opportunities for sustainable use of agricultural residues in the world, but they are country and site specific.

Statistics on the contribution of crop residues to world bioenergy have not been made available yet. The World Bioenergy Association reported that the agricultural sector (including energy crops and animal and agricultural by-products) represented 10% of all biomass feedstocks for bioenergy in the world and that the share of agricultural by-products was 4% (on all biomass feedstocks); with a total biomass supply of 59.2 EJ, agricultural by-products represent 2.4 EJ (57.3 Mtoe). The share of each crop residue was not specified but in that study it was stated that crop residues could be a major source of energy generation based on theoretical estimates, which is in the range of 3.6−17.2 billion tons (347−2,938 Mtoe) (Kummamuru, 2017). The absence of statistical data could be interpreted in the following terms:

- Lack of convenient biomass supply categorization. For instance, crop residues are not accounted separately from other agricultural residues (manure, fruit biomass, and others) in global bioenergy statistics (Kummamuru, 2017).
- Low contribution of crop residues to modern bioenergy. Energy applications compete with traditional applications of crop residues, such as animal bedding, feed, and mushroom substrate (Muth et al., 2013). In addition, the development of advanced biofuels is still low (16.4 of 126 billion liters of liquid biofuels in the world, according to Kummamuru, 2017). However, it is expected that lignocellulosic feedstocks (such as straw) will be increasingly used for advanced biofuels.
- Energy applications of crop residues are not restricted to this type of biomass resource. For instance, solid bioenergy carriers can be produced from mixtures of agriculture and wood residues.

In order to ensure sufficient biomass supply from crop residues, the focus has been placed on residues from extensive crops, specifically residues from cereals, which are the most extensive category of crops in the world with 718 Mha harvested area in 2016 (grain production: 2,849 Mt grain). In 2016 maize ranked first with 37.2% of world cereal production; main producers in the world were the United States (36.3% production) and China (26.1%), while the share of Europe was 11.1%. Wheat was the second most important cereal in the world with 26.3% of production; main producers were Europe (33.4% wheat production) and China

(17.6%); the share of the United States was 8.4% (FAO, 2017a). In Mediterranean regions, large areas of land are devoted to olive groves, from which large amounts of pruning residues (woody biomass) can be made available for bioenergy (Velázquez-Martí et al., 2011). Spain is the first producer of the world with 2.5 Mha olive groves (FAO, 2017a); in this country olive tree pruning has been marketed as solid bioenergy carrier (pellets and firewood) for thermal energy (García-Maraver et al., 2012).

The EU has supported straw energy applications for a long time; an important milestone was the EU Expert Consultation on "Cereals straw resources for bioenergy in the European Union" held in 2006 (Dallemand et al., 2006). Denmark has a long tradition in using this biomass resource for modern bioenergy. The first target on this regard dated back to 1993, when it was agreed a mandate of 1.2 Mt of straw for 2000. Straw energy applications, ranked according to their importance, were individual heating, district heating, and CHP until 2000; afterwards the main application has been CHP. Straw has also been used for power generation in direct combustion in Spain since 2007 (López González et al., 2007); currently there are three power plants (a total of 61 MWe) running on straw in Spain (Acciona, 2018). Biomass supply in Denmark has been described in the National Renewable Energy Action Plan (NREAP) of this country; in 2006 straw represented 82% (18,538 $TJ \cdot year^{-1} \leftrightarrow$ 1.3 Mt of straw assuming 14.5 $GJ \cdot t^{-1}$ calorific value) of the biomass supply from agricultural by-products/processed residues and fisheries by-products for energy generation (B2 grouping in NREAP). Straw biomass supply in Denmark for 2015 and 2010 was estimated at +500 ktoe and +1,000 ktoe, respectively (Denmark, 2010).

Current policies in Europe and the United States have boosted advanced biofuels (second-generation biofuels) with a view to replacing conventional liquid biofuels (first-generation biofuels); thus, most advanced biofuels facilities in the world in 2012 were based in the United States and Europe (Bacovsky et al., 2013). However, the fact that technologies are expensive and that conversion efficiencies are still low represents a barrier to their development (Bentsen et al., 2017). Approximate ratios of lignocellulose conversion are 214 liters of gasoline equivalent (lge) $\cdot t^{-1}$ (dry mass of lignocellulosic biomass) for cellulosic ethanol, 217 lge $\cdot t^{-1}$ for BTL diesel, and 307 lge $\cdot t^{-1}$ for biosynthetic natural gas (Eisentraut, 2010). For most advanced biofuels, plant capital costs have been estimated at 5–10 times as much as corn ethanol plants, and productions costs, two to five times more expensive than corn ethanol (Kim and Kim, 2014). Production of cellulosic ethanol from straw has been recorded for Spain and Denmark since 2008 (Janssen et al., 2013; Bentsen et al., 2017).

Crop residues and other lignocellulosic feedstocks for bioenergy have not been separately tracked in the United States so far, but policies have promoted the use of these biomass resources by the Renewable Fuel Act since 2007. The production target of advanced biofuels in the United States was set at 80 billion liters of advanced biofuels (58% cellulosic ethanol) by 2022. The second-generation biofuel plant volume projection for 2018 was 310 $ML \cdot year^{-1}$ (USDA, 2018). The status of production facilities for the second-generation biofuels in the United States and in the EU was reviewed by Janssen et al. (2013); the authors stated that feedstocks were various and included corn stover, corn cobs, and wheat straw. Corn stover was reported as feedstock of two advanced biofuel plants in China (Eisentraut, 2010). It should be noted that, in spite of the projections made, the second-generation biofuel sector has experimented a recession in the United States and in Europe in the last years due to unfavorable economic circumstances, among other reasons. In the United States cellulosic ethanol plants owned by Abengoa Bioenergy (Hugoton) and DowDuPont (Nevada) were put up for sale, and in Italy, the Crescentino cellulosic refinery shut down (BiofuelsDigest, 2016, 2017). The study by Golecha and Gan (2016) showed that the true potential of corn stover for cellulosic biofuels ("derisked" stover) in the United States was 63% of the collectable stover, while the amount of collectable stover was about 40% of the produced stover.

There is general consensus that agricultural residues represent a large biomass resource that can be used as feedstock for bioenergy in different ways but the fact that are residual does not ensure the economic and environmental sustainability of this resource; controversy on the assessment of the actual biomass resource and the impacts of residues removal suggest that each case should be independently analyzed and that statistics on crop residues should be developed. Literature data show that the contribution of crop residues to modern bioenergy is modest, and it has been essentially driven by policies; among the energy applications, thermal applications, and power generation have been more successful than advanced biofuels, coming to light that business and economics go hand by hand.

2.4.3.2 *Agro-industrial residues*

Agro-industrial residues consist of residues from the processing of a particular crop or animal product. The variability of residues along the value chain of agro-industries is very broad in terms of origin, composition, quantities, and temporality. With the increasing interest on the circular bioeconomy and biorefinery concepts, the use of residues and waste for the production of bio-based value-added products is gaining importance, and their use for energy generation is normally placed at the end of the cascading process. In any event, residues and waste-to-energy (WtE) technologies are, by the time being, in a more advanced stage in terms of technology readiness. In the work of De Corato et al. (2018), the reader can find a comprehensive review of the main biorefining opportunities for the conversion of agro-industrial residues into high-value-added products and bioenergy carriers.

Table 2.2 provides a classification of agro-industries producing residues potentially used for bioenergy purposes. Other industries also producing renewable residues with bioenergy potential, but not agro-industries in a strict sense, such as fish farming (producing, e.g., brown, red, and green seaweeds; fish scales; viscera; and scraps), are not included. Algae biomass, the main source of aquatic biomass used for bioenergy, is addressed in Section 2.4.4.

Specific information and statistics on bioenergy production from agro-industrial residues and wastes are scarce, especially at a global scale. The main agro-industrial residues in terms of production and bioenergy importance are detailed next.

Bagasse (fuel obtained from the fiber which remains after juice extraction in sugar cane processing) is one of the main agro-industrial residues used for bioenergy production. Its composition in wet basis is basically cellulose (36%), hemicelluloses (27%), lignin (23%), and ash (4%) (Pippo and Luengo, 2013). For each ton of sugarcane processed, it is possible to obtain approximately 38 kg molasses and 250 kg of bagasse (Mena, 1985). Bagasse has more than 150 potential final uses, being the main one its use as fuel to generate steam in the sugarcane factory and a small fraction to produce pulp and board. It can also be used for animal feed and bedding and other bio-based products such as furfural or xylitol (Paturau, 1986). In 2015, it is estimated that more than 474.8 thousand of metric tons (dry matter) of bagasse were produced globally, with Brazil and India being the countries with the largest production (34% and 20%, respectively). More than 66% of the total production was used for energy

Table 2.2 Agro-industries classification and their residues as bioenergy feedstocks and potential bioenergy carriers

Type of agro-industry		Residues as feedstock for bioenergy	Main potential bioenergy carrier
Vegetal biomass industries	Grain industry (cereals processing)	Husks, cobs	Biochar, syngas, biogas
	Sugar industry	Bagasse, molasses, sugar beet tops	Solid biomass for heating and electricity, biogas, syngas, second-generation liquid biofuel, biochar
	Nuts industry	Nutshells (almonds, macadamia, walnuts...) and empty fruit bunches	Solid biomass for heating and electricity
	Oil-seeds processing and oil mills	Press cakes from rape and soybean seeds, olive marc, lees and pomace, palm kernel meals, palm oil mill effluents	Biogas, syngas, second-generation liquid biofuel, biochar
	Winery industry	Grape stalks, pomace and seeds, wine lees	Bioethanol and hydrogen
	Other food processing	Vegetable processing (tomato, potato, orange, mushrooms, etc.) and other food-processing wastes (peels, seeds, skins, etc.)	Biogas, syngas, second-generation liquid biofuel, biochar
Animal industries	Livestock farms	Manure, slurry, poultry litter, whey	Biogas, hydrogen, biomethanol, biochar, bio-oil, syngas
	Slaughterhouse and other meat-processing industries	Animal fats, blood, paunch, soft offal (intestinal residues, fat and meat trimmings, and some blood), dissolved air flotation sludge	

consumption (The United Nations Statistics Division (UNSD), 2018), which would be equivalent to 118 Mtoe of primary energy (considering a calorific value of 15.62 $MJ \cdot kg^{-1}$) (Energy research Centre of the Netherlands, 2018).

Molasses (sweet sirup obtained during the manufacture of beet or cane sugar containing unextractable sugars, vitamins, and minerals, such as calcium, sodium, potassium, and magnesium) represent a key feedstock for the production of bioethanol. Molasses are also used for feed as well as for the production of yeast, citric acid, vitamins, amino acids, and lactic acid by the yeast and fermentation industries. It is estimated that more than 57.7 M tons of molasses in 2016 are produced worldwide and 38.1% is used for bioethanol production, 25.5% for feed, and 36.4% for other uses purposes (OECD-FAO, 2018).

The olive oil industry generates several residues or by-products, namely olive pomace (olive skin, pulp, seed and fragments of stones, and residual oil), extracted olive pomace, olive stones, and olive leaves from olive cleaning operations at olive mills (Manzanares et al., 2017). In two-phase extracting systems, pomace has a high content on moisture, organic matter, and phenolic compounds which makes it highly pollutant and difficult to dispose. Exhausted olive pomace (after pomace oil extraction) presents lower moisture content ($\sim$10%) and higher calorific value (low heating value = 20.6 $MJ \cdot kg^{-1}$) and thus, it is usually burned in domestic heating systems or in industrial boilers of agro-industries or livestock farms. Actually, the use of olive mill residues for energy is well established. For instance, in Andalusia (Spain), which produces 50% of the olive oil of EU28, the energy generation from olive mill residues accounts for 80% (47% for electricity generation and 33% for thermal energy), while composting or direct field application represents 14.3% and landfill 0.7% (Berbel and Posadillo, 2018). Fokaides (2017) provides a state-of-the-art study in the field of energy recovery alternatives of the residues and the by-products of the olive oil industry, also including solid olive kernel for the production of solid bioenergy carriers (by pelletizing and torrefaction). However, as for other agro-industrial residues, the revalorization of oil mill residues for high value-added products in a biorefinery approach is getting more economic and research attention (Manzanares et al., 2017; Berbel and Posadillo, 2018; Ruiz et al., 2017; Negro et al., 2017).

Similar to olive mills, wineries also produce high amounts of grape pomace. Approximately, 17% of the grape weight results eventually in pomace. It is composed by 8% of seeds, 10% of foot-stalks and fractions of the stem, 25% of skins of pressed grapes, and 57% of marrow berries (Burg et al., 2016). It can be used as soil amendment and animal fodder due to its high fiber content. From the bioenergy point of view, grape pomace can be revalorized as bioethanol (5 L of ethanol per 100 kg of pomace) and solid bioenergy carrier for thermal and power production (LHV between 14.6 and 17.75 $MJ \cdot kg^{-1}$). According to Burg et al. (2016), 8 million tons of pomace are yearly produced in Europe.

In olive mills and vineries, a considerable amount of wastewater (e.g., 0.5–14 L per liter of wine) with a high content on organic matter (chemical oxygen demand—COD) is produced. Wastewater has been usually treated by anaerobic digestion aiming at the reduction of COD rather than the production of biogas. However, efforts are being made to use it as feedstock to produce platform chemicals such as lactic acid, biofuels including ethanol, enzymes, and chemical intermediates (Zacharof, 2017).

Residual biomass from livestock farms has been traditionally used as organic soil fertilizer in extensive breeding systems. However, with the intensification of the production systems, residues from animal industries have increased enormously raising a concern on how to treat and dispose them. Especially problematic are animal wastes with high moisture content, production ratio, and volume of production, such as cattle, pork, and poultry farms. Table 2.3 details the quantity of manure produced daily and the moisture content of different livestock residues.

Most of the residual biomass from livestock farms is revalorized via anaerobic fermentation for the production of biogas. According to the US Environmental Protection Agency (2017), in the United States there

Table 2.3 Average values of yearly manure production and water content by type of feedstock

Livestock type	Manure/slurry (kg/kg animal weight per year)	Water content (%)
Dairy	24.9	88.8
Beef	29.4	91.2
Swine	24.9	89.9
Poultry	24.3	74.4
Sheep	15.0	75.0
Horse	20.1	86.0

Source: Based on data from Lorimor, J., Powers, W., Sutton, A., 2004. Manure Characteristics. MWPS-18 Section 1. Second Edition. Available at: http://msue.anr.msu.edu/uploads/files/ManureCharacteristicsMWPS-18_1.pdf.

are 250 operational anaerobic digesters on livestock farms and 15 under construction, with a total production capacity of 895,136 $m^3 \cdot d^{-1}$ and 993 $GWh \cdot year^{-1}$ of electricity production. Other biological and thermochemical conversion technologies such as pyrolysis, direct liquefaction, and gasification to convert animal waste into gaseous fuels, combustible oils, and charcoal are moving forward. Cantrell et al. (2008) provide a review on available biological and thermochemical conversion technologies that could be applied for the valorization of these waste streams.

Dry manure (dung) is extensively used in developing countries, especially in dry and cold areas where fuelwood is scarce (e.g., Tibet, Mongolia, countries in high Andes). Sometimes it is mixed with other types of fuels such as coal dust and crop residues to increase the energy density and enhance burning performance. It is mainly used as biomass for heating and cooking purposes. Even though there is no consistent and comprehensive data available for this type of biomass, it is estimated that more than 2 million people burn dry animal dung for cooking (Helbig and Roth, 2017). More efficient ways to produce energy from dry manure should be spread to reduce the high emissions of carbon monoxide, hydrocarbons, and particulate matter because of the inefficient burn of dung in households.

Regarding wastes from slaughterhouses and meat processing industries, the main residues with bioenergy potential are waste animal fats. Thanks to their high content in fatty acids (40%–60%) (Banković-Ilić et al., 2014), they can be used as feedstocks for biodiesel production, being basically tallow (beef tallow from domestic cattle and mutton tallow from sheep), pork lard (rendered pork fat), chicken fat, and grease. According to OECD-FAO (2018), it is estimated that nearly 110,000 M liters of biodiesel were produced in 2016 from waste. In 2015, animal fats accounted for 8% of the total biodiesel produced at both, global and European scale (UFOP, 2017).

In 2015, the production of bioenergy from animal waste in the world is estimated at 16 Mtoe, being North America, Africa, and Asia the continents with the highest bioenergy production from animal waste with 53.3%, 24.1%, and 21.0%, respectively (United Nations, 2018).

Food waste is produced along the food supply chain, mostly in the consumption stage (56%) and in the food processing phase (38%) (Monier et al., 2010). The former is normally included in the MSW category, either in the household or the industry (services, e.g., restaurants) fractions (see Section 2.4.3.4 on MSW). The compositional heterogeneity in this stage is high, which hinders the optimization of its valorization processes. On the other hand, waste streams from the agro-food processing industries present a greater homogeneity which facilitates their conversion into value-added products and bioenergy. Food wastes include peelings and scraps from fruit and vegetables, food that does not meet quality control standards, pulp and fiber from sugar and starch extraction, and coffee grounds.

With a view of the progress and optimization of the biorefinery processes, the use of food waste for its conversion into added-value bio-based products and chemicals is currently under intensive study. As for other residues and wastes, the production of energy from food waste is technologically ahead, some technologies being applied at a commercial scale, such as the production of biogas through anaerobic digestion (Ren et al., 2018). It can also be fermented to produce bioethanol. Other strategies for the treatment and valorization of food waste are composting (humification of organic matter for organic fertilization and soil conditioning) and their use as animal feed (as a source of fiber-rich nutrition). Kiran et al. (2014) provides a state of the art of food waste fermentation technologies for renewable energy generation.

Due to its relatively high moisture content, the anaerobic digestion of food waste seems to be more suitable than thermochemical conversion technologies, such as combustion or gasification. The methane yield from the anaerobic digestion of food waste can reach 340 m^3 CH_4 per ton of dry mass (Jouhara et al., 2017; Zhang et al., 2007). On the other hand, thanks to its organic- and nutrient-rich composition, food waste can, theoretically, be utilized as a useful resource for the production of bioenergy through various fermentation processes providing biogas, hydrogen, ethanol, and biodiesel as final products.

A particular case represents the cooking oil waste from households, restaurants, and food-processing industries, which can be easily transformed into biodiesel with an efficiency of 85%–96% (Patil et al., 2010). In 2015, used cooking oil accounted for 9% of the total biodiesel produced worldwide, while in EU28, it accounted for 17% (UFOP, 2017).

Statistics on the production of food waste are difficult to gather and thus not reported. Efforts are being made for the quantification of this biomass resource (Corrado et al., 2017). Cristóbal et al. (2018) estimated the food waste production in Europe along the food supply chain (processing, distribution, and consumption) of four representative food processing industries (tomato, potato, orange, and olive processing). As a result, it was estimated that the tomato industrial losses range between 5% and 19.5% (referred to the input and excluding processing water) with a total industrial waste production of 1.48 $Mt \cdot year^{-1}$. Wastes from potato industry of specific waste streams range from 8% (e.g., peels) to 82% (e.g., starch) with a total European production of potato peels estimated at 2.34 $Mt \cdot year^{-1}$. Regarding food wastes from the orange processing industry, loss percentage is estimated to vary from 6% (distribution losses of fresh oranges) to 50% (e.g., peels), which leads to an estimated total European production of waste form the orange industry of 3.16 $Mt \cdot year^{-1}$.

2.4.3.3 Wood processing

Throughout the industrial wood value chains, residues are produced. Wood residues exist in different forms such as bark, sawdust, trimmings, split wood, planer shavings, sanderdust, and black liquor, each of which might be used for different purposes. The share of residues produced depends on the industry, its size, and its modernity. For example, according to JRC (Camia et al., 2018), in the European Union, about 85% of the wood entering the panel industry and 50% of the wood entering the sawmill and pulp industries are contained in the final products. In Honduras, sawmill residues, including bark, vary between 47% and 52% of the incoming logs, depending of the efficiency of the sawmill and the extent of the recovery of secondary products from the solid residues. In Ghana the average residue rate for sawmills is 55% (Mahin, 1991). With the higher economic valuation of the residues, in particular for energy, some actors of the wood-based panel and the wood pulp industries considered their residues as a valuable co-product that became part of their business models.

These residues can be used by the industry that produced them. For example, many paper industries use their residues as a source of energy for their processing, in particular for drying. Residues can also be used by other material industries and energy industries. In the European case, 36% of the residues are used by other industries, while 64% are used for energy. Altogether, industrial residues constitute 30% of the wood fuels consumed in the EU in 2013 (Camia et al., 2018). Progress can be made to improve the share of residues used (Bais-Moleman et al., 2018).

2.4.3.4 Municipal solid waste

The production of MSW is an inevitable fact in modern human society, and its reduction and sustainable management is one of the greatest challenges for future generations. However, while a big effort is being made to minimize the amount of waste produced and to increase the recycling of larger fractions of waste materials, the "resource use is still largely unsustainable and inefficient, and waste is not yet properly managed" (EU, 2013). From the circular bioeconomy perspective, wastes (namely renewable or organic wastes) are not undesired products any longer but a valuable resource for energy and bio-based products.

The definition and classification of MSW vary depending on the country and the management practice employed. Eurostat defines MSW as "waste collected by or on behalf of municipal authorities and disposed of through the waste management system." This definition excludes waste from municipal sewage network and treatment as well as municipal construction and demolition waste (Eurostat, 2017b). The lack of an agreed definition creates uncertainty and inconsistencies for assessing waste management and data collection. A classification of types and sources of the different fractions of the MSW is shown in Table 2.4. The bioenergy sector, and generally the bioeconomy, is focused on the organic and renewable fraction of the MSW, regardless of the source or generating activity, i.e., covering wastes from the three sources of MSW, residential, industrial, and commercial and institutional.

Worldwide production of MSW is currently estimated to be up to 1.3 billion tons per year. However, this estimation varies considerably by region, country, and city (Hoornweg and Bhada-Tata, 2012). In general terms, the higher the economic development and rate of urbanization, the greater the amount of MSW produced. Thus, OECD countries generate around 44% of world's waste, while Africa and eastern and central Asia are the regions that produce less waste (Fig. 2.13). Concurrently, the composition of the MSW also varies significantly with the economic development and thus, with the geographical location, but it is also influenced by other factors such as cultural norms, energy sources, and climate. Fig. 2.13 shows how low- and middle-income countries have a high percentage of organic matter in the urban waste stream, while paper, plastic, glass, and metal fractions increase in the waste stream of middle- and high-income countries. The geographical location also influences waste composition by determining building materials (e.g., wood versus steel), ash content (often from household heating), and horticultural waste. Likewise, the energy source has its impact on the composition of MSW especially in low-income countries, where energy for cooking,

Table 2.4 Types and sources of MSW fractions

Source/type of residue		Generating activity	Composition
MSW	Residential	Single and multifamily dwellings	Food wastes, paper, cardboard, plastics, textiles, leather, yard wastes, wood, glass, metals, ashes, special wastes (e.g., bulky items, consumer electronics, white goods, batteries, oil, and tires), household hazardous wastes (e.g., paints, aerosols, gas tanks, waste containing mercury, motor oil, and cleaning agents), and e-wastes (e.g., computers, phones, TVs)
	Industrial	Light and heavy manufacturing, fabrication, construction sites, power and chemical plants	Housekeeping wastes, packaging, food wastes, wood, steel, concrete, bricks, ashes, hazardous wastes, ashes, and special wastes
	Commercial and institutional	Stores, hotels, restaurants, markets, office buildings	Paper, cardboard, plastics, wood, food wastes, glass, metals, special wastes, hazardous wastes, and e-wastes
	Construction and demolition	New construction sites, road repair, renovation sites, and demolition of buildings	Wood, steel, concrete, soil, bricks, tiles, glass, plastics, insulation, and hazardous waste
	Municipal services	Street cleaning, landscaping, parks, beaches, and other recreational areas	Street sweepings, landscape and tree trimmings, sludge, and wastes from recreational areas

Source: From World Energy Council, 2016b. World Energy Resources Bioenergy. Available at: https://books.google.com/books?id = WLC7CdLOZosC&pgis = 1; Hoornweg, D., Bhada-Tata, P., 2012. What a waste: a global review of solid waste management. In: Urban Development Series Knowledge Papers, Washington, DC: World Bank Group, p. 98. Available at: www.worldbank.org/urban.

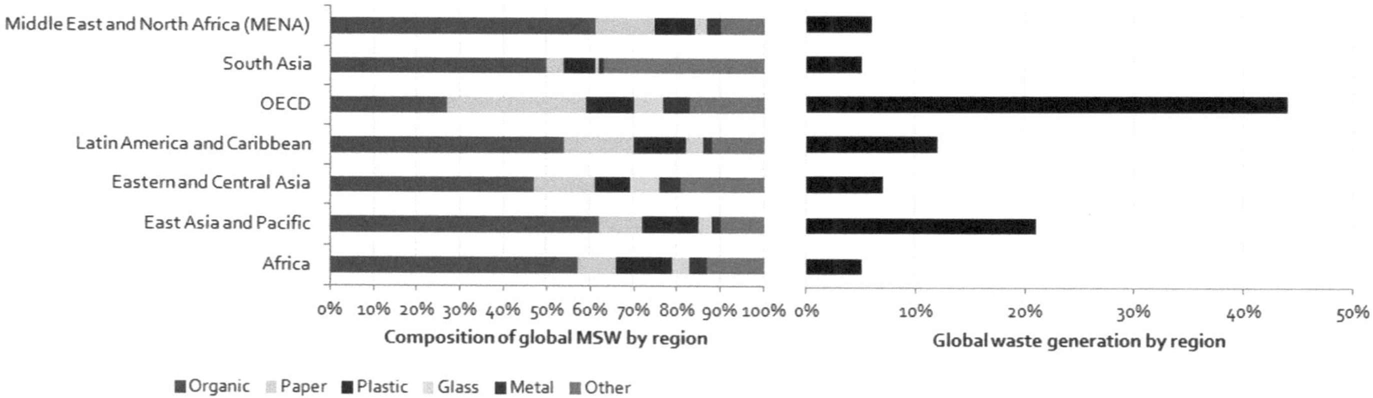

Figure 2.13 Composition of global MSW (left panel) and global waste generation (right panel) by region. Other residues include, e.g., textiles, leather, and rubber. *From Nizami, A.S., Rehan, M., Waqas, M., Naqvi, M., Ouda, O.K.M., Shahzad, K., et al., 2017. Waste biorefineries: enabling circular economies in developing countries. Bioresour. Technol. 241, 1101–1117. Available at: http://dx.doi.org/10.1016/j.biortech.2017.05.097; Hoornweg, D., Bhada-Tata, P., 2012. What a waste: a global review of solid waste management. In: Urban Development Series Knowledge Papers, Washington, DC: World Bank Group, p. 98. Available at: www.worldbank.org/urban; Cardoen, D. Joshi, P., Diels, L., Sarma, P.M., Pant, D., 2015. Agriculture biomass in India: part 1. Estimation and characterization. Resour. Conserv. Recycl. 102, 39–48. Available at: https://www.sciencedirect.com/science/article/pii/S0921344915300227?via%3Dihub (accessed March 10, 2018).*

heating, and lighting are not sourced from district heating systems or the electricity grid (Hoornweg and Bhada-Tata, 2012).

The strategy for waste management and energy production from waste can vary significantly, depending on multiple factors, mainly social, economic, and environmental criteria and constraints. The production of energy from waste, also worded as WtE, can contribute to both, a correct waste treatment strategy and an environment-friendly energy production, if sustainability principles are in place (World Energy Council, 2016a).

In a broader context, also including waste biorefineries approach, three pathways can be identified:

- Thermochemical technologies which transform waste into fuels, electricity, and heat via pyrolysis, gasification, incineration, and carbonization.
- Physicochemical technologies converting residues into liquid biofuels through rendering and transesterification with chemical agents.
- Biochemical technologies such as anaerobic digestion and fermentation to convert wastes into liquid and gaseous biofuels and bio-based products.

In Table 2.5, the main characteristics, also including advantages and limitations of some waste management strategies, are shown.

The WtE technology will depend on the nature and volume of the waste stream, the technology available, as well as the wastes' net calorific value, which determines how much energy can be extracted from it. Table 2.6 shows the approximate net calorific values of common MSW fractions. Other factors influencing the selected system for the energy conversion from waste are the desired energy form, the thermodynamic and chemical conditions in which the plant can operate, and the energy efficiency.

In general terms, waste incineration (with energy recovery) should only be considered if the incoming waste stream has an average net calorific value of at least $7\ \text{MJ} \cdot \text{kg}^{-1}$, so that the combustion has, at least, a positive energy balance (World Energy Council, 2016b). Incineration without energy recovery is, however, the last preferred option in waste management after controlled dump and/or landfill. The energy recovery prior to final disposal is preferable to direct landfilling as long as pollution control requirements and costs are addressed (Hoornweg and Bhada-Tata, 2012). On the other hand, it requires a lower complexity of technology and labor skills.

Table 2.5 Main characteristics of different waste management strategies for bioenergy production

Type of technology	Suitable waste	Advantages	Limitations	Daily power generation (MW per ton waste)	Technology complexity	Labor skill level	Process efficiency
Incineration	General waste stream	Reduces waste volume by up to 80%	Causes pollutions (air and waterborne)	0.01–0.02[a]	Low	Low	25%
		Reduces waste mass by up to 70%	Production and release of carcinogenic chemicals (dioxins)				
		Most streams of MSW can be treated	Require larger capital cost				
		Easy set up and fast treatment	Public refusal Produce harmful solid waste (lag)				
Pyrolysis	Organic and inorganic waste	Up to 80% energy can be recovered from waste	Impurities in products	0.01–0.014[a]	High	Intermediate	17%
		Land acquisition reduced	Low liquid products yields				
		High energy products	Coke production				
		Liquid and gas product easily separated	Products required cleaning				

(Continued)

Table 2.5 (Continued)

Type of technology	Suitable waste	Advantages	Limitations	Daily power generation (MW per ton waste)	Technology complexity	Labor skill level	Process efficiency
		MSW volume can be reduced by 50%–90%	Metal tubes and system columns corroded over time High energy input Expensive operational and maintenance cost				
Gasification	Organic and inorganic waste	No greenhouse gas emissions All waste types can be treated Easily expandable technology	High energy input High capital and operational cost	0.04–0.045[a]	Very high	Very high	32%
Anaerobic digestion	Organic waste	Less solid produced High energy carrier biogas Organic fertilizer from nutrient-rich digestate Cheaper technology	Impurities Larger scale plants are not attractive Attractive Low system stability Require extra space	0.015–0.02[b]	Low	Low	25%–30%

[a]Power generated in a daily basis; [b]Power generation spread over the life span of biomethanation plant.

Source: From Nizami, A.S., Rehan, M., Waqas, M., Naqvi, M., Ouda, O.K.M., Shahzad, K., et al., 2017. Waste biorefineries: enabling circular economies in developing countries. Bioresour. Technol. 241, 1101–1117. Available at: http://dx.doi.org/10.1016/j.biortech.2017.05.097.

Table 2.6 Approximate net calorific values of common MSW fractions

Fraction	Net calorific value ($MJ \cdot kg^{-1}$)
Organic	4
Paper[a]	16
Plastic	35
Glass	0
Metal	0
Other	11

[a]Paper wastes could be also classified as organic wastes.
Source: From World Energy Council, 2016b. World Energy Resources Bioenergy. Available at: https://books.google.com/books?id = WLC7 CdLOZosC&pgis = 1; Kamuk, B., Haukohl, J., 2013. Guidelines: Waste to Energy in Low and Middle Income Countries. International Solid Waste Association.

Anaerobic digestion provides an economical solution for problems of waste disposal with the lower technology complexity and skilled labor requirements. On average, 120–150 m^3 of biogas (2,500 $MJ \cdot t^{-1}$) can be produced from one ton of dry MSW (Tan et al., 2015).

The products of the anaerobic decomposition are compost or biogas and digestate. The biogas produced from anaerobic digestion can be further upgraded to a virtually pure stream of biomethane, which can then be injected into the local gas grid or used locally for heating or fueling purposes (biomethane can also be produced from cleaned syngas from gasification of biomass). The methane yield from the anaerobic digestion of MSW can range from 201 to 470 m^3 CH_4 per ton of dry mass (Jouhara et al., 2017).

Landfill is still the most spread practice at global scale despite being the least preferred option. According to Hoornweg and Bhada-Tata (2012), the amount of MSW treated with WtE technologies amounts up to 122 million tons per year (Fig. 2.14). High-income countries lead all waste management strategies except for dumping, which is the most utilized in middle-income countries.

According to the International Energy Agency (IEA, 2017a,b), 16 Mtoe of energy were recovered from renewable MSW worldwide (see Fig. 2.15). Data presented in this report indicate that the production of energy from renewable MSW has increased at an annual rate of 5.8% during 2000–14. While Europe leads greatly the production of energy from MSW (62%), North and South America (23%) and Asia (15%) are quite behind in this aspect. Africa and Oceania, on the contrary, do not have

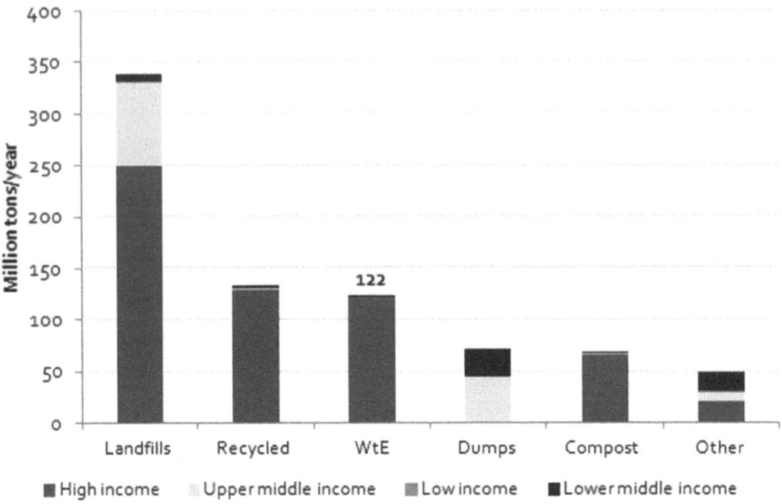

Figure 2.14 Amount of waste disposed by technique and income level in 2012. *From Hoornweg, D., Bhada-Tata, P., 2012. What a waste: a global review of solid waste management. In: Urban Development Series Knowledge Papers, Washington, DC: World Bank Group, p. 98. Available at: www.worldbank.org/urban.*

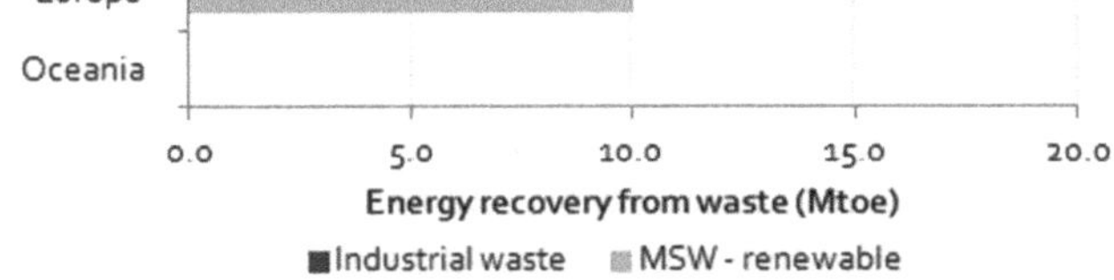

Figure 2.15 Energy recovery from MSW and industrial waste by continent in 2014. Industrial waste included here is waste consisting of solid and liquid products (e.g., tires) combusted directly in specialized plants. *From IEA, 2017a. Renewables Information: Overview. Available at: https://www.iea.org/publications/freepublications/publication/renewables-information---2017-edition---overview.html; IEA, 2017b. Key World Energy Statistics, p. 96. doi:10.1017/CBO9781107415324.004. ISBN: 9788578110796; Kummamuru, B., 2017. WBA Global Bioenergy Statistics 2017, Available at: http://www.worldbioenergy.org/uploads/WBA.GBS 2017_hq.pdf. doi: 10.1016/0165-232X(80)90063-4.*

facilities in the same scale as the former continents, and thus, they barely contribute to the energy recovery from MSW.

Either by direct incineration or through an energy carrier as biogas, the energy recovery form MSW is mainly conducted via electricity generation. Worldwide, it is estimated that the production of electricity from MSW in 2014 was 68.7 TWh (IEA, 2017a,b; Kummamuru, 2017).

In the United States, the amount of biogenic MSW used for electricity production in 2016 attained 16,994 thousand tons (US Energy Information Administration (EIA), 2018). In EU28, the amount of MSW in 2016 incinerated with energy recovery was 62,312 thousand tons (25% of total MSW generated), while the quantity composted and digested amounted to 40,726 thousand tons (16.5% of total) (Eurostat, 2018a).

The production of biogas worldwide increased from 6.7 Mtoe in 2000 to 30.3 Mtoe in 2014 (Statista, 2018). In EU28, the production of landfill biogas in 2016 is estimated at 2.77 Mtoe, which represents the 17.2% of total biogas produced, while sewage sludge biogas and other biogas (e.g., decentralized agricultural plants) represent 8.7% and 74.1%, respectively. For landfill biogas, the United Kingdom is the leading country with 50.5% of total, while Germany is the European country with the greatest total biogas production with 7.96 Mtoe or 49.4% of total biogas (EurObserv'Er, 2017a). Nevertheless, the share of landfill biogas is decreasing because of the reduction in deposits of organic waste in landfills in favor of using more efficient recovery modes. On the other hand, biogas from organic waste (including form MSW recovery) is expected to grow from 2 to more than 8.4 Mtoe in 2020 (Kampman et al., 2017).

In 2016 in Europe, there were registered 17,662 biogas plants from which 497 upgraded their production to biomethane. The number of biomethane plants in Europe increased from 187 to 497 from 2011 to 2017 reaching a capacity of 283,300 $m^3 \cdot h^{-1}$ and a yearly production of 1.5 bm^3 (GIE and EBA, 2018). According to this source, in 2017 Germany and United Kingdom are the countries with the highest biomethane installed capacity, with 58% and 22% of the total of Europe, respectively. Out of the total number of plants reporting on the main substrate used (334 plants), 19% use mainly agricultural residues, manure, and plant residues (10% of the installed capacity, i.e., 21,100 $m^3 \cdot h^{-1}$), while 14% use MSW (9% of capacity, i.e., 19,100 $m^3 \cdot h^{-1}$). It should be noted that the United Kingdom does not specify the main substrate used.

2.4.3.5 Wastewater

Municipal wastewater and the sludge produced during its treatment could provide with a valuable source of water, organic matter, and nutrients to be revalorized, not only as energy but also as fertilizer and other compounds. However, due to its potential content on heavy metals, different uses to bioenergy are, in those cases, limited.

In wastewater treatment, three processes are distinguished with different targets: (1) primary treatment to remove suspended solids, both organic and inorganic through a mechanical process; (2) secondary treatment to degrade and remove soluble biodegradable organics through aerobic or anaerobic biological processes, thermochemical or mechanical processes; (3) tertiary treatment to remove nutrients (mainly nitrogen and phosphorus), microorganisms, and the remaining suspended matter by different technologies (e.g., membrane filtration, infiltration/percolation, activated carbon, and disinfection).

Anaerobic digestion is the most common strategy for sludge stabilization resulting in the production of biogas for electricity and heat or to be upgraded and be exploited as transport biofuel. Fuel cells generate electricity directly from the chemical energy contained in the biogas, with an electric efficiency of 50%–55% (Bachmann, 2015). Sludge can also be directly incinerated for the production of heat and electricity. The processes of gasification and pyrolysis of waste sludge are also possible although not so commonly used. The wet-air (chemical) oxidation of sludge at high temperatures and pressure is well known and can be used to produce heat (Tyagi and Lo, 2016). The hydrothermal treatment involves heating sludge in the water phase at temperatures between 150°C and 450°C in the absence of oxygen or another oxidant, generating dissolved organic compounds that are used as a source of carbon to increase C/N ratio in the production of biogas.

The methane yield obtained through anaerobic digestion is very dependent on the sludge composition. Theoretically, it reaches $0.590\ m^3 \cdot kg^{-1}$ organic dry solids (Owens and Chynoweth, 1993; Appels et al., 2011). By physical, chemical, thermal, mechanical, or biological pretreatment steps (e.g., microwave, heating, ultrasonication, zonation, wet oxidation, etc.), the yield of biogas production can be increased up to 50% (Tyagi and Lo, 2016) as well as with the addition of different cosubstrates such as glycerol (Nartker et al., 2014) and other substances (Maragkaki et al., 2018).

Information on the global levels of wastewater generation and treatment nowadays is crucial but also scarce since it is not systematically monitored or reported in many countries (Mateo-Sagasta et al., 2015). Data on global bioenergy or biogas produced specifically from waste sludge were not found. However, according to Bachmann (2015), Germany is the country with the highest production of electricity through biogas from wastewater treatment plants with 3050 $GWh \cdot year^{-1}$. Data for South Korea (969 $GWh \cdot year^{-1}$), Sweden (672 $GWh \cdot year^{-1}$), and Switzerland (550 $GWh \cdot year^{-1}$), among others, are reported.

According to GIE and EBA (2018), 34 biomethane plants in Europe are based on sewage sludge, gas, and waste as main substrate, reaching nearly 6,600 $m^3 \cdot h^{-1}$ of production capacity (excluding UK plants).

On the other hand, more than 2,700 million people still need sludge management (Koncagül et al., 2017) due to the absence of treatment facilities. In contrast, countries where wastewater treatment plants are available (e.g., the United States), this resource is as well underutilized (Seiple et al., 2017). Thus, the potential for improvement in this field is still high.

2.4.4 Algae for bioenergy

The term "algae" includes a wide range of polyphyletic taxa of prokaryotic and eukaryotic organisms originated from different evolution lines. Similar to higher plants, algae exhibit oxygenic photosynthesis, but their level of organization is rather different: algae are protophytes or thallophytes, not cormophytes. According to Barsanti and Gualtieri (2014), the number of species of algae is estimated at around 72,500 but new algae species are reported every year. Most algae are microalgae (about 80%); among the microalgae, a major group is the diatoms, with about 20,000 species.

2.4.4.1 Categories of algae

Two major categories of algae can be distinguished according to their cell structure: prokaryotes and eukaryotes. Prokaryotic cells do not have a differentiated nucleus (cells do not have membrane-bound organelles), and the genetic material is dispersed in the cytoplasm; on the contrary, eukaryotic cells do have a full nucleus that contains the genetic material; eukaryotic cells also have other membrane-bound organelles. Most algae species are unicellular and are commonly called microalgae. Macroalgae are pluricellular but, regardless of their size (species in genus *Macrocystis* can attain several tens of meters in length), algae do not have pure tissues.

The main two categories of algae organisms are described below.

2.4.4.1.1 Prokaryotic algae

Among the prokaryotic organisms—in the Archaea and Bacteria Kingdoms—a subgroup in bacteria that gathers Gram-negative eubacteria with oxygenic photosynthesis is considered as algae; this is an only phylum (division), which receives the name Cyanophyta. Within this phylum, two categories (classes) are recognized: Cyanophyceae and Prochlorophyceae.

The ecological distribution of the Cyanophyceae—also known as blue-green algae—is very wide. Actually, they can grow in very different habitats. Among the characteristics of Cyanophyceae, it is worth mentioning that they have chlorophyll a as main photosynthetic pigment and accessory pigments such as phycobilins (phycocyanin, allophycocyanin, and phycoerythrin) and carotenoids. Common genera of Cyanophyceae, used in microalgae cultures for different products, are *Anabaena*, *Arthrospira (Spirulina)*, *Microcystis*, *Nostoc*, *Oscillatoria*, and *Phormidium*, among others.

The Prochlorophyceae were recently discovered and include a limited number of species found in symbiotic associations with marine worms (genus *Prochloron*) and other free marine phytoplankton and freshwater filamentous organisms (genera *Prochlorococcus* and *Prochlorothrix*, respectively). The Prochlorophyceae have chlorophyll a and b, β-carotene, and zeaxanthin as main pigments but, unlike the Cyanophyceae, they lack phycobilins. For that reason, the Prochlorophyceae is considered an additional category to Cyanophyceae. Moreover, the Prochlorophyceae are related to chloroplasts of higher plants and to other categories of eukaryotic algae such as chlorophytes and euglenophytes.

2.4.4.1.2 Eukaryotic algae

The group of the eukaryotic algae is very heterogeneous. Even though their taxonomy classification is still in progress as it has not been fully accomplished, 10 Phyla have been already identified: 4 in Kingdom Plantae (Glaucophyta, Rhodophyta, Chlorophyta, and Charophyta), 4 in Kingdom Chromista (Haptophyta, Criptophyta, Ochrophyta, and Cercozoa), and 2 in Kingdom Protozoa (Mizozoa and Euglenozoa). A detailed description of eukaryotic algae Phyla and Classes can be found in the work by Barsanti and Gualtieri (2014). A summary of Phyla, Classes, and representative Genera, including macroalgae, is given in Table 2.7.

Table 2.7 Summary of Phyla, Classes, and representative Genera of eukaryotic algae

Phyla	Classes	Representative Genera	Main photosynthetic pigment	Accessory pigments	Storage compound
Glaucophyta	**Glaucophyceae**	*Cyanophora*	Chlorophyll a	Phycoerythrocyanin C-phycocyanin Allophycocyanin β-carotene Zeaxanthin	Starch
Rhodophyta	Seven classes, among them:				
	Bangiophyceae	*Porphyra*	Chlorophyll a	B-phycoerythrin R-phycoerythrin R-phycocyanin Allophycocyanin Carotenes (α and β) Lutein	Floridean starch
	Florideophyceae	*Kappaphycus* *Gelidium* *Gracilaria* *Palmaria* *Phyllophora*			
	Porphyridiophyceae	*Porphyridium*			
Chlorophyta	Ten Classes, the major ones:				
	Chlorodendrophyceae	*Tetraselmis*	Chlorophylls a and b	Carotenes (α, β, and γ) Lutein Prasinoxanthin	Starch
	Chlorophyceae	*Botryococcus* *Chlamydomonas* *Dunaliella* *Haematococcus* *Neochloris* *Scenedesmus*			
	Dasycladophyceae	*Acetabularia*			
	Trebouxiophyceae	*Chlorella*			
	Ulvophyceae	*Caulerpa* *Ulva*			

(Continued)

Table 2.7 (Continued)

Phyla	Classes	Representative Genera	Main photosynthetic pigment	Accessory pigments	Storage compound
<u>**Charophyta**</u>	Six classes, among them:				
	Charophyceae	*Nitella* *Chara*	Chlorophylls a and b	Carotenes (α, β, and γ) Lutein Prasinoxanthin	Starch
<u>**Haptophyta**</u>	Several classes, among them:				
	Isochrysidaceae **Pavlovophyceae**	*Isochrysis* *Pavlova*	Chlorophyll a, c_1 and c_2	Carotenes (α and β) Fucoxanthin	Chrysolaminarin
<u>**Cryptophyta**</u>	**Cryptophyceae**	*Chilomonas* *Cryptomonas*	Chlorophylls a and c_2	B-phycoerythrin R-phycocyanin Allophycocyanin, Carotenes (α, β, and ε) Alloxanthin	Starch
<u>**Ochrophyta**</u> (Brown and golden algae)	Eighteen classes, the major ones:				
	Bacillariophyceae or **Diatomeae**	*Achnanthes* *Cyclotella* *Navicula* *Phaeodactylum*	Chlorophylls a, c_1, c_2, and c_3	Carotenes (α, β, and ε) Fucoxanthin Violaxanthin	Chrysolaminarin
	Chrysophyceae	<u>*Dinobryon*</u> *Ochromonas*			

	Dictyochophyceae	*Dictyocha*			
	Eustigmatophyceae	*Chlorobotrys*			
		Nannochloropsis			
	Phaeophyceae	*Fucus*			
		Laminaria			
		Macrocystis			
		Sargassum			
	Raphidophyceae	*Coleochaete*			
		Gonyostomum			
	Synurophyceae	*Mallomonas*			
		Synura			
	Xanthophyceae	*Botrydium*			
		Vaucheria			
Cercozoa	**Chlorarachniophyceae**	*Gymnochlora*	Chlorophylls a and b	Lutein	Paramylon
				Neoxanthin	
				Violaxanthin	
Myzozoa	**Dinophyceae**	*Ceratium*	Chlorophylls a, c_1, and c_2	β-carotene	Starch
		Crypthecodinium		Peridinin	
		Dinophysis		Fucoxanthin	
		Dinothrix		Diadinoxanthin	
		Gonyaulax		Dinoxanthin	
		Noctiluca		Gyroxanthin	
		Peridinium			
Euglenozoa	**Euglenophyceae**	*Euglena*	Chlorophylls a and b	Carotenes (β and γ)	Paramylon
		Phacus		Diadinoxanthin	
		Trachelomonas			

Usage of fonts: Bold and underlined for Phylum (plural Phyla); bold for Class; italics for genera of microalgae; underlined and italic for genera of macroalgae.
Source: Adapted from Barsanti L., Gualtieri P., 2014. Algae: Anatomy, Biochemistry and Biotechnology, second ed. Boca Raton, FL: CRC Press. 345 pp.

2.4.4.2 Microalgae cultures

2.4.4.2.1 History of algae utilization

Microalgae were used as food in ancient civilizations such as the Aztecs in Mexico. Aztecs used cultures of *Arthrospira (Spirulina) maxima*, a Cyanophyceae, to prepare a type of cake named "Tecuitlatl." Likewise, natives to Chad used the same microalga species to prepare a foodstuff named "dihe" (Abalde et al., 1995). *Nostoc flagelliforme* has been used for food for more than 2000 years in China; such nourishment has been named "fa cai," which means "vegetal hair" (Gao, 1998).

The first scientific studies on microalgae started at the end of the 19th century, when the microbiologist Beijerinck (1890) succeeded in growing pure cultures of *Chlorella vulgaris*. This achievement allowed using microalgae for research purposes such as for photosynthesis research. Otto Warburg (Warburg, 1919) tested *Chlorella* to study the release of oxygen in the course of photosynthetic reactions. Several years later, Melvin Calvin and his team in Berkeley identified the cycle of photosynthetic CO_2 assimilation—the so-called Calvin cycle—using cultures of *Chlorella pyrenoidosa* and applying ^{14}C-labeled CO_2 in very short pulses of photoassimilation (Bassham and Calvin, 1960), for which Professor Melvin Calvin was awarded with the 1961 Nobel prize in Chemistry (Calvin, 1961).

Research on lipids production from microalgae as feedstock for biofuels started in Göttingen (Germany) in the course of World War II (Harder and Von Witsch, 1942), soon after other countries showed interest in growing microalgae for a number of applications. In the work by Burlew (1953) a compilation of microalgae mass production experiments, laboratory photobioreactor (PBR) developments, and prototypes of pilot plants in the United States, Germany, Israel, and Japan can be found.

Experiments on the potential of microalgae for wastewater treatment and energy production started in the decade of 1950 (Golucke et al., 1957, Golucke and Oswald, 1959, Oswald and Golueke, 1960). The development of closed production systems of microalgae (PBRs) was initiated in the decade of 1960 aiming at their use in space missions (Nichiporovich et al., 1962).

Following the 1973 oil crisis, numerous research programs on renewable energies for fossil fuel replacement were launched, including programs on liquid and gaseous biofuels production from microalgae. Regarding scope and duration, the US Aquatic Species Program (ASP) at the Solar Energy Research Institute (SERI) (currently the National

Renewable Energy Laboratory, NREL), which was developed in the period 1978–96 with a budget of 25 M$ should be highlighted; a main conclusion was that the production of biofuels from microalgae was technically possible but not economically feasible (Sheehan et al., 1998).

Research programs launched in the following years focused on the advantages that would entail producing microalgae for CO_2 capture, in an attempt to mitigate the greenhouse effect. In this line of research, a program amounting to 250 M$ was carried out in Japan in the decade of 1990 to develop high technology PBRs, which were proved not to be economically feasible (Lundquist et al., 2010). As a section of the program Greenhouse Gas R&D of the International Energy Agency (IEA), with the support of the US DOE-NETL and ENI (Italian oil company) an "International Network for Biofixation of CO2 and Greenhouse Gas Abatement with Microalgae" was promoted. However, interests of companies to protect intellectual property made the Program failed. As a result, the network was abandoned in 2008 (Lundquist et al., 2010).

The policy for biofuels promotion in the EU—which arose from the Directive Scrivener in 1992—has been gradually developed with various regulations and the adoption of the EU Directive on the promotion of renewable energies (Directive 20-20-20 or RES Directive) in 2009. The RES Directive established a target of 10% share of renewable energies in transport by 2020 to tackle the increase in greenhouse gas emissions. The target was proved to be difficult to achieve using feedstocks from agriculture for biofuels in a sustainable way. The use of arable lands for biofuels raised controversies and concerns on the impact of the land-use change but gave a boost to the development of new projects for biofuel production from microalgae; microalgae yields were expected to be very high based on values in scientific literature (Chisti, 2007). In practice, actual yields were much lower, and culture and harvesting costs resulted very high. These findings, together with low oil prices, increased the difficulty of achieving economic feasibility of biofuels from microalgae. Research projects on microalgae have been reorientated towards the production of high-value products following the biorefinery concept so that the production of biofuels from microalgae could attain economic feasibility (IEA Bioenergy, 2017).

2.4.4.2.2 Microalgae production systems

2.4.4.2.2.1 Open systems In open systems of culture, microalgae can move freely in the growth medium, whose surface is in direct contact

Figure 2.16 Examples of microalgae cultivation systems: (A) Open pond raceway type (ITC, Canarias, Spain); (B) horizontal and vertical arrangements of a tubular PBR (Estación Experimental Cajamar, Almeria, Spain); (C) flat panel in polyethylene material (CICYTEX, Badajoz, Spain); (D) vertical biofilm PBR (UPM, Madrid, Spain) *Photographs by J. Fernández.*

with the atmosphere. As compared to other strategies, open systems are cheaper and assure affordable microalgae mass production. They are designed as ponds or round channels (raceways) in which the growth medium is constantly moved by mechanical devices (generally, paddle-wheels) (see Fig. 2.16A). Solar radiation strikes on the water surface and a fraction of that, penetrates the culture, which leads to a negative correlation between the microalgae concentration and the light penetration. Microalgae concentration in open systems can attain a maximum of about $1\ g \cdot L^{-1}$. In these conditions, the solar radiation penetrates only few millimeters in depth. In case of continuous microalgae cultures, concentrations of about $0.5\ g \cdot L^{-1}$ are adequate when the pond is 0.2–0.5 m deep. Algal mass production is dependent on irradiance, temperature of the culture medium (both factors are related to the latitude of the site, if they are not artificially modified), and microalgae species, if the growth medium is not limiting. Values of the potential microalgae mass production in open systems vary greatly; in autotrophic conditions, the values provided by IEA Bioenergy (2017), about $13\ g \cdot m^{-2} \cdot day^{-1}$ representing $47\ t \cdot ha^{-1} \cdot year^{-1}$, can be assumed. A drawback of open systems is the need of removing a large amount of water (growth medium) to concentrate the microalgae mass; the removal rate is estimated at $2\ m^3$ water per kg dry matter of algal biomass.

2.4.4.2.2.2 Suspended microalgae PBRs Similar to open systems, microalgae can move freely in the growth medium but in this case, the growth medium is inside a transparent container or closed glass vessel, specially designed to facilitate light penetration and gas exchange between the culture and the atmosphere. Artificial lighting is also possible, particularly with LED technology. Suspended PBRs have been designed in different ways, with different shapes and arrangements, such as vertical or inclined flat panels, tubular PBRs horizontally or vertically arranged, or vertical cylinders (Zittelli et al., 2013).

In this type of PBRs, if the cultivation conditions (particularly, the lighting conditions) are optimized, it is possible to achieve algal concentrations higher than in open systems (up to $5\ g \cdot L^{-1}$). Yields in dry algal biomass can be either referred to the volume of the growth medium ($g \cdot L^{-1} \cdot day^{-1}$) or to the footprint or ground area occupied by PBR ($g \cdot m^{-2} \cdot day^{-1}$). The latter measure is more practical to compare the performance of different production systems as long as the photosynthetic photon flux density received by the culture is taken into account. De Vree et al. (2015) studied the performance of three types of PBR (horizontal tubular, vertical tubular, and flat panel) against an open raceway pond using several microalgae species. The highest production was obtained for the flat panel with a culture of *Nannochloropsis* sp. ($20.5-27.5\ g \cdot m^{-2} \cdot day^{-1}$) in the Netherlands. The average production of microalgae in PBR cultures was estimated at $20.7\ g \cdot m^{-2} \cdot day^{-1}$ by the IEA Bioenergy (2017), equivalent to $75\ t \cdot ha^{-1} \cdot year^{-1}$ (Fig. 2.16B and C).

2.4.4.2.2.3 Algae biofilm PBRs In biofilm PBRs the microalgae are attached or immobilized on the surface of a support and grow in layers associated with other microorganisms—generally bacteria—although one species can dominate others. When the growth medium is rich in mineral nutrients and poor in organic matter, microalgae surpass other organisms. Biofilm cultures can be immersed in growth medium—whether there is continuously or intermittently immersion—or in contact with the atmosphere, i.e., only wetted by a thin water layer (Berner et al., 2016).

Advantages of biofilm PBRs over suspended microalgae PBRs are (1) better lighting conditions for algal population, (2) higher algal concentration, on a volume basis, (3) easier harvesting of algal mass with higher algal concentration (10%–20%) without centrifugation, (4) lower energy consumption, and (5) higher radiation efficiency per ground area.

According to support mobility, there are two types of biofilm PBRs: (1) static support (horizontal or vertical) or (2) mobile support.

Horizontal static biofilm systems: These systems include the reactors called Algal Turf Scrubber based on microalgae growth (like a turf) on a horizontal support—generally such support is a screen on which water flows. The first patent dates back to 1982 (Adey, 1982); later on, it was modified/improved, mainly for wastewater treatment (Mulbry and Wilkie, 2001, Kibede-Westhead et al., 2003, Mulbry et al., 2008). There have been developments with horizontal or even slightly inclined supports, such as the one described by Ozkan et al. (2012) which consisted of a sloping concrete panel (0.2% slope) to facilitate water flow and biofilm growth. In static biofilm systems the maximum value of the ratio biofilm area to ground area is 1. Hence, the photosynthetic efficiency is limited. Average yields are usually $<5\ g \cdot m^{-2} \cdot day^{-1}$ (Gross, 2015).

Vertical static biofilm systems: In these systems, microalgae biofilms grow on a vertical support exposed to the air while the support is continuously wetted with growth medium. This concept has been developed in two ways so that these systems can be scaled up for mass production. The development by the University of Cologne (Germany) is a porous twin layer PBR where the growth medium is applied by means of a glass fiber panel that acts as a porous substrate; in addition, the glass fiber panel is covered on both sides by porous films of different materials (paper, plastic films, metal films, etc.) to which microalgae grow attached, giving rise to biofilms (Shi et al., 2005, Naumann et al., 2013, Podola et al., 2017). The Technical University of Madrid (Spain) has developed a laminar PBR made up of hollow panels covered on both faces by fabric sheets (polypropylene geotextile); growth medium flows down from the upper part, laminated over both faces of the panel; CO_2-enriched air can be added in between both sides to enhance microalgae growth (Fernández, 2010, Borges et al., 2016, Martin-Girela et al., 2017) (Fig. 2.16D). On average, yields obtained with these biofilm systems in large facilities are in the range of 7–15 g algal mass (dry weight) $m^{-2} \cdot day^{-1}$, referred to footprint area (including corridors), a value equivalent to $21-45\ t \cdot ha^{-1} \cdot year^{-1}$ assuming 300 day production cycle.

Biofilm PBRs with mobile supports: This is the case of the system developed by Johnson (2009). In essence, it consists of a panel that pivots on a central shaft and a device for artificial lighting over the panel; the faces of the panel are alternatively immersed in the growth medium or exposed to the air by panel movement.

Another original system is the one described in the master thesis by Christenson (2011) –director R.C. Sims at the University of Utah (the United States)-, which was developed for wastewater treatment. The system is a rotating cylinder that has its surface covered by a cotton cord, which has been rolled up all around, like a coil, so that microalgae grow attached to the cord. The system by Gross and Wen (2013) is also a biofilm PBR with a mobile support but in this case, the mobile support is like a treadmill that has been arranged as a triangular prism; one of the lateral edges of the prism remains immersed in growth medium, while the others are exposed to the air. According to the authors, such system yields an average of $11.36\ g \cdot m^{-2} \cdot day^{-1}$ over 1 year.

The category of biofilm systems with mobile support also includes other systems derived from rotational biological contactors (bio-discs) for wastewater treatment, currently under development (Fernández, 2013, Sebestyén et al., 2016).

2.4.4.2.2.4 Fermenters Some microalgae species have lost photosynthetic capacity and become heterotrophic, i.e., they utilize organic matter to grow as saprobiontic or parasitic organisms. Other species like *Chlorella* are able to utilize both pathways and grow either as autotrophic or heterotrophic organisms; this ability is called mixotrophy. Based on this strategy, microalgae cultures can be carried out in conditions of heterotrophy or mixotrophy in fermenters filled with a liquid medium containing organic compounds. Different from PBRs, a continuous supply of oxygen—besides organic compounds—is needed to favor cell metabolism. Fermenters may be exposed to light in case of mixotrophy; light is not needed in case of pure heterotrophy. Algal concentration may attain over $100\ g \cdot L^{-1}$, in contrast to the low concentrations that are achieved in PBRs ($5\ g \cdot L^{-1}$ at the most). Under normal N-nutritional conditions, algae growth rate is about 0.5–0.6 g biomass per gram of sugar but in case of N deficiency, the algal metabolism shifts to lipids synthesis. When this is the case, lipids may represent around 60% of the algal biomass weight (Barclay et al., 2013). Detailed description of fermenters for microalgae culture can be found in Behrens (2005). This type of bioreactors (fermenters) enables aseptic conditions as well as metabolic engineering by adjusting the culture conditions. Thanks to that, fermenters can be used to produce axenic cultures of species selected for high commercial value products; for instance, docosahexanoic acid, an omega-3 fatty acid, from *Crypthecodinium* (a dinoflagellate) cultures, or feed for rotifers for use in aquaculture, from *Tetraselmis* cultures (IEA Bioenergy, 2017, Barclay et al., 2013, Behrens, 2005).

2.4.4.3 Macroalgae and bioenergy

Marine macroalgae (seaweeds) have been traditionally used by coastal populations for food, feed, disease remedy, and organic fertilizer since ancient times. There is evidence that species in Ochrophyta, Rhodophyta, and Chlorophyta Phyla have been used for food and medicine in south Chile (Monte Verde) about 13,000 years ago (Barsanti and Gualtieri, 2014). Likewise, they have been used for food in China, Japan, Korea, Indonesia, Malaysia, and other countries that have exported the practice of using macroalgae as a food to other parts of the world. Traditional practice has been based on harvesting macroalgae directly from the sea. However, in the last 50 years methodologies for macroalgae cultivation have been developed, boosting macroalgae cultures, whether there are sea-based cultures or cultures in seawater ponds (Sahoo and Yarish, 2005, Kim et al., 2017). Due to that, the production of macroalgae has increased dramatically in the last two decades. World production rose from 10.51 Mt in 2000 to 30.45 t in 2015; macroalgae cultivation represented over 95% of algae production (Araujo and Hoepffner, 2018).

According to FAO (2016), the world production of macroalgae amounted to 27.22 Mt (fresh matter) in 2014. Main genera (91.6% of the macroalgae production) were *Kappaphycus* and *Eucheuma* (10.99 Mt), *Gracilaria* (3.75 Mt), and *Porphyra* (2.36 Mt) in Phylum Rhodophyta; second in rank was the Phylum Ochrophyta, Class Phaeophyceae, with the species *Saccharina (Laminaria) japonica* (7.65 Mt) and *Sargassum fusiforme* (0.175 Mt). China (50.1%), Indonesia (34.6%), Philippines (5.8%), and South Korea (4.2%) represented 94.7% of macroalgae production. European production is very low; in 2015, when the world macroalgae production attained 30.45 Mt, Europe contributed with only 0.75% (0.23 Mt). Most of the European production came from traditional sea harvesting; the main producers were Norway (65%), France, and Ireland (Araujo and Hoepffner, 2018).

Over 90% macroalgae production is marketed for food and feed, while the rest is processed for phycocolloids (alginates, agar, and carrageenan), fertilizers, nutraceuticals, and cosmetics. Species in genera *Saccharina (Laminaria), Undaria, Porphyra,* and *Sargassum* are used for food; *Kappaphycus alvarezii* and *Eucheuma denticulatum*, for carrageenan; species in genera *Gracilaria* and *Gelidium* for agar; and species in genera *Saccharina (Laminaria)* and *Lessonia* for alginates.

Macroalgae have been envisaged as feedstocks for bioenergy (mainly for bioethanol and biogas). The basic rationale for their bioenergy use is the following: macroalgae are rich in polysaccharides (15%–20% FW) (Kraan, 2013); yields can be much higher than terrestrial plants (maximum value reported in literature: $131\ t \cdot d \cdot m \cdot ha^{-1} \cdot year^{-1}$) (Kraan, 2013; Milledge and Harvey, 2016); the range of high value-added products that can be produced from macroalgae is very wide; and arable land is not needed to grow macroalgae. Bioenergy from macroalgae is technically feasible but the economic feasibility is only possible under the biorefinery concept, i.e., the production of high value-added products along with bioenergy (Balina et al., 2017). The chemical composition of macroalgae determines the bioenergy pathway; macroalgae rich in polysaccharides are candidates for bioethanol, those rich in lipids could be used for biodiesel and macroalgae with adequate C/N ratio, may be fermented to biogas as long as they do not contain methanation inhibitors. A brief state-of-the-art is presented next.

Concerning the bioethanol pathway, macroalgae usually contain about 20%–65% (dry weight basis) carbohydrates, but this content varies among species. It should be noted that not every polysaccharide is fermentable to ethanol. In principle, the production of cellulosic ethanol from microalgae is possible; accessibility of hydrolytic enzymes to cellulose in macroalgae cell walls is higher than in plants because their cell walls are poor in lignin. Along with cellulose, algae contain certain polysaccharides that are specific of their taxa. Brown algae contain alginates (polymers of uronic acids) and fucoidans (heteropolymers of sulfated fucose, xylose, galactose, mannose, and glucuronic acid) associated to cell walls, and laminarin (polymer of glucose) as storage compound. Red algae contain agar (sulfated polymers of galactose and anhydrogalactose), carrageenans (similar to agar), xylans, and mannans. Green algae contain xylans, sulfated galactans, and ulvanes (sulfated polymers of rhamnose, xylose, and uronic acids). Most monomers from the hydrolysis of algae polysaccharides are fermentable by different microorganisms, but the ethanol yield is not the same as fermenting sugars with yeasts; methodologies and procedures are still under investigation. In addition, some products from the needed pre-treatments, such as furfural or 5-hydroxymethylfurfural, act as fermenting inhibitors. Literature data show that the yields in ethanol vary among algae species and authors ranging from 0.08 to $0.29\ g \cdot g^{-1}$ (dry algae mass). A more conservative value is $0.073\ g \cdot g^{-1}$, equivalent to about 90 L per ton of dry algal mass (Milledge et al., 2014, Rajkumar et al., 2014).

Chemical composition of macroalgae determines their conversion into biogas. In the previous hydrolytic phase, compounds resilient to hydrolysis (cellulose or agar) or even methanation inhibitors may appear. Such is the case of sulfur compounds from the hydrolysis of sulfated polymers (agar, carragenates, fucoidans, ulvanes, and others), when *Ulva lactuca* is intended for biogas. Polyphenols can also act as inhibitors, as in the case of *Himanthalia elongata* or *Sargassum muticum* (Jard et al., 2013). It is important that the ratio of C to N in the feedstock is between 20 and 30 for biogas production; some macroalgae such as *Ulva lactuca* have a ratio of 10, whereas in other species such as *Saccharina (Laminaria) lattissima*, this ratio varies with the season. Ratios are usually higher in spring-summer because the content in protein decreases at the beginning of the development cycle; when protein is low, the ratio can be higher than 20 which has an effect on biogas production (IEA Bioenergy, 2017). Another limiting factor is the content in sodium. Sodium contents over $10\ g \cdot L^{-1}$ affect methanation; if the content reaches $14\ g \cdot L^{-1}$ (similar to sea water) the production of biogas halves, although it may be resumed after a period of acclimation (Milledge et al., 2014). Yields in biogas vary among substrates; yields can be from 0.120 to 0.279 L of CH_4 per gram of volatile solids (Jard et al., 2013).

Literature on biodiesel produced from macroalgae is still scarce, a fact attributable to the low lipid content of cultivated macroalgae. Literature review by Wielgosz-Collin et al. (2016) reports 2% lipids (dry weight basis) as an average; average values and its variation found for total lipids in different algae Phyla are also reported: Ochrophyta 3.0% (0.1%–20%), Rhodophyta 1.4% (0.02%–6.2%), and Chlorophyta 3.9% (0.5%–17.3%). Despite the lipids content, fatty acids are relatively low (30%–50% of total lipids) (Gosch et al., 2012) and vary with the species and growth conditions. Further research is needed to reach actual applications of macroalgae biodiesel technologies.

2.4.4.4 Algae biorefinery

Biofuels production from microalgae has been widely studied at the laboratory and pilot plant levels, giving rise to extensive literature on the fields of biodiesel, bioethanol, biogas, bio-hydrogen (Rajkumar et al., 2014, Wang and Lan, 2010, Brennan and Owende, 2010, Barbosa and Wijffels, 2013, Show et al., 2018), and bioelectricity from microalgae-based hydrogen fuel cell (Baicha et al., 2016). As commented in a previous section, the IEA-Bioenergy Report (IEA Bioenergy, 2017) concluded that the production of bioenergy from microalgae is technically possible but not economically feasible under the current circumstances, i.e., if bioenergy is the only pursued product. Latest research trends are focused on microalgae biorefinering; this way, high value-added products would make microalgae bioenergy economically feasible (Vanthoor-Koopmans et al., 2013, Zhu, 2015, Chew et al., 2017).

A wide spectrum of high value-added products can be obtained from microalgae. Koller et al. (2014) reviewed this subject and pointed out the following products: pigments (chlorophylls, carotenoids, and phycobilins); lipids (hydrocarbons and polyunsaturated fatty acids); proteins (for food or feed); polysaccharides (starch, agar, carrageenan, alginates, and cellulose); bio-polyesters (poly-hydroxyalkanoates) for bioplastics; bioactive compounds with bactericidal, antifungal, anti-protozoan, or anti-algal activity (antibiotics in broad sense). Such products could be obtained in a cascading process aiming at the whole valorization of algal biomass, pursuant to the biorefinery concept.

Research on the integral exploitation of microalgae biomass has been addressed in several ways; two of them appeared more feasible than others to achieve this objective, and are summarized next.

1. Pretreatment of the whole biomass by mechanical (cell disruption) and chemical procedures for biomass fractionation into carbohydrates, proteins, lipids, and high-value products. Fractions are subsequently processed for marketed products. Biofuels production is envisaged as a further stage in the exploitation of microalgae mass; for instance, biodiesel could be produced from the lipids fraction, bioethanol from the fermentable carbohydrate fraction or biogas from the remains of the process.

2. High pressure and high temperature treatment for bio-oil production. Bio-oil would be subsequently refined to biofuel. Hydrothermal liquefaction is a promising thermochemical process for microalgal biomass since it can be applied to wet biomass (75%–85% moisture) in contrast to pyrolysis, which requires dry biomass (<10%). Drying algal biomass would imply unaffordable costs.

As compared to microalgae, macroalgae seem to have better prospects for biorefinery (Balina et al., 2017), among other reasons because the world production of macroalgae is 100 times the microalgae production and high value-added products are obtained such as nutraceuticals, phycocolloids, and pigments, leaving aside residual biomass suitable for biofuel production. The fact that the production of macroalgae is seasonal in some latitudes could be taken as an opportunity for the implementation of complementary microalgae facilities for biomass production.

Economic analysis of the biorefinery approach should be made on a case-by-case basis. Revenues from high value-added products might change if the supply exceeds the demand or if the consumers' perception moves to other products. Sensitive analysis would be highly recommended to assess the effect of an increase in the production of both high value-added products and bioenergy.

An extensive research has been done on the field of algal biomass for energy purposes. There is consensus on the technical feasibility of the production of algae for biofuels from both, macro and microalgae. However, such activity alone is not economically feasible, particularly under current circumstances of low oil prices. In order to achieve a positive economic balance, the value chain of algae biofuels should be incorporated into the production of other products, following the biorefinery concept. Further research is needed in this line. Concerns have been raised on heterogeneous ways in which results have been presented, as well as on data extrapolation from laboratory to large-scale production plants conducive to great but not reliable results. Such type of studies along with more cautious studies has added confusion on the sustainability of this activity. As the IEA states (IEA Bioenergy, 2017), the collaboration among research teams needs to be re-enforced as well as the harmonization of results presentation to optimize the resources devoted to the research on the energy applications of algal biomass. The use of algal biomass for bioenergy should be integrated into production chains of high added-value products, pursuant the biorefinery concept. Algae biorefining represents the most updated trend for exploiting the whole potential of both macro- and micro-algae biomass.

LIST OF ABBREVIATIONS

ATP	Adenosine Triphosphate
ATS	Algal Turf Scrubber
CO_2	Carbon dioxide
COD	Chemical oxygen demand
CHP	Combined heat and power
DNA	Deoxyribonucleic acid
DAF	Dissolved air flotation sludge
DHA	Docosahexanoic acid
EU28	European Union of 28 countries
ETBE	Ethyl tertiary butyl ether
GAP	Glyceraldehyde phosphate
HI	Harvest Index
Ha	Hectare
HHV	Higher Heating Value
HTL	Hydrothermal liquefaction
LGE	Litres of gasoline equivalent
LHV	Low Heating Value
MSW	Municipal Solid Waste
Mtoe	Million ton oil equivalent
NREAP	National Renewable Energy Action Plan
NPP	Net Primary Production
NADPH	Nicotinamide Adenine Dinucleotide Phosphate
PGA	Phosphoglyceric acid
PBR	Photobioreactor
PAR	Photosynthetic active radiation
PPFD	Photosynthetic photon flux density
REN	Renewable energy(ies)
RFA	Renewable Fuel Act
RPR	Residue to product ratio
RNA	Ribonucleic acid
RuP.	Ribulose 5-phosphate
RuDP	Ribulose diphosphate
SRC	Short Rotation Coppice
t.d.m	Tons of dry matter
WtE	Waste-to-energy

REFERENCES

Abalde, J., Cid, A., Fidalgo, P., Torres, E., Herrero, C., 1995. Microalgas: Cultivo y Aplicaciones. Universidad de Coruña, A Coruña, Spain, ISBN: 84-88301-84-7, 210 pp.

Acciona, 2018. Activities of Energy: Biomass. Available at: https://www.acciona.com/business-divisions/energy/ (accessed March 22, 2018).

Adey, W.H., 1982. Algal turf scrubber. United States Patent & Trademark Office #4333263.

Aditiya, H.B., Mahlia, T.M.I., Chong, W.T., Nur, H., Sebayang, A.H., 2016. Second generation bioethanol production: a critical review. Renew. Sust. Energy Rev. 66, 631–653. Available from: https://doi.org/10.1016/j.rser.2016.07.015.

Appels, L., Lauwers, J., Degrève, J., Helsen, L., Lievens, B., Willems, K., et al., 2011. Anaerobic digestion in global bio-energy production: potential and research challenges. Renew. Sust. Energy Rev. 15 (9), 4295–4301. Available at: https://www.sciencedirect.com/science/article/pii/S1364032111003686 (accessed March 31, 2018).

Araujo R., Hoepffner N., 2018. Biomass from algae. In: Biomass Production, Supply, Uses and Flows in the European Union (Ed. CE 2018) EUR 28993 EN, pp. 69–77.

Bachmann, N., 2015. Sustainable biogas production in municipal wastewater treatment plants, IEA Bioenergy Report. ISBN: 978-1-910154-22-9. Available at: http://www.iea-biogas.net/files/daten-redaktion/download/Technical Brochures/Wastewater_biogas_grey_web-1.pdf.

Bacovsky, D., Ludwiczek, N., Ognissanto, M., Wörgetter, M., 2013. Status of advanced biofuels demonstration facilities in 2012. IEA Bioenergy Task 39 Report, pp. 1–209.

Bacovsky, D., Ludwiczek, N., Pointner, C. and Verma, V.K., 2016. IEA bioenergy countries' report: bioenergy policies and status of implementation. Available at: http://www.ieabioenergy.com/wp-content/uploads/2016/09/IEA-Bioenergy-Countries-Report-23.09.2016.pdf.

Baicha, Z., Salar-Garcia, M.J., Ortiz-Martinez, V.M., Hernández-Fernández, F.J., de los Rios, A.P., Labjaar, N., et al., 2016. A critical review on microalgae as an alternative source for bioenergy production: a promising low cost substrate for microbial fuel cells. Fuel Process. Technol. 154, 104–116. Available from: https://doi.org/10.1016/j.fuproc.2016.08.017.

Bais-Moleman, A.L., Sikkema, R., Vis, M., Reumerman, P., Theurl, M.C., Erb, K., 2018. Assessing wood use efficiency and greenhouse gas emissions of wood product cascading in the European Union. J. Clean. Prod. 172, 3942–3954. Available from: https://doi.org/10.1016/j.jclepro.2017.04.153.

Balina, K., Romagnoli, F., Blumberga, D., 2017. Seaweed biorefinery concept for sustainable use of marine resources. Energy Procedia 128, 504–511. Available from: https://doi.org/10.1016/j.egypro.2017.09.067.

Banković-Ilić, I.B., Stojković, I.J., Stamenković, O.S., Veljkovic, V.B., Hung, Y.T., 2014. Waste animal fats as feedstocks for biodiesel production. Civil Environ. Eng. Faculty Pub. 104, 1–17. Available at: http://engagedscholarship.csuohio.edu/encee_facpub/104.

Barbosa, M.J., Wijffels, R.H., 2013. Biofuels from microalgae. In: Richmond, A., Hu, Q. (Eds.), Handbook of Microalgal Culture. Wiley Blackwell, Chichester, pp. 566–577.

Barclay, W., Apt, K., Dong, X.D., 2013. Commercial production of microalgae via fermentation. In: Richmond, A., Hu, Q. (Eds.), Handbook of Microalgal Culture. Wiley Blackwell, Chichester, pp. 134–145.

Barsanti, L., Gualtieri, P., 2014. Algae: Anatomy, Biochemistry and Biotechnology, second ed. CRC Press, Boca Raton, FL, 345 pp.

Barth, S., Murphy-Bokern, D., Kalinina, O., Taylor, G., Jones, M. (Eds.), 2016. Perennial biomass crops for resource-constrained world. Springer International Publishing. Available from: https://doi.org/10.1007/978-3-319-44530-4.

Bassham, J.A., and Calvin, M., 1960. The Path of Carbon in Photosynthesis, Ernest Orlando Lawrence Berkeley National Laboratory, University of California Lawrence Radiation Laboratory -Berkeley, United States Department of Energy (through predecessor agency the Atomic Energy Commission), (October 1960). Ref. UCRL-9583. 71 pp. https://www.osti.gov/accomplishments/documents/fullText/ACC0326.pdf.

Behrens, P.W., 2005. Photobioreactors and fermentors: the light and dark sides of growing algae. In: Andersen), R.A. (Ed.), Algal Culturing Tecniques. Elsevier, Oxford, pp. 189–204.

Beijerinck, M.W., 1890. Kulturversuche mit Zoochlorenllen, Lichenengoniden und anderen niederen. Algen. Bot. Zeitung. 48, 725. Reproduced in "algae base.org." http://img.algaebase.org/pdf/AC100CF003338161AAoHt43C4207/35049.pdf.

Bentsen, N.S., Møller, I.M., 2017. Solar energy conserved in biomass: Sustainable bioenergy use and reduction of land use change. Renew. Sust. Energy Rev. 71, 954–958. Available from: https://doi.org/10.1016/j.rser.2016.12.124.

Bentsen, N.S., Felby, C., Thorsen, B.J., 2014. Agricultural residue production and potentials for energy and materials services. Progress in Energy and Combustion Science. 40 (1), 59–73. Available from: https://doi.org/10.1016/j.pecs.2013.09.003. Elsevier Ltd.

Bentsen, N.S., Lamers, P., Lalonde, C., Wellisch, M., Dale, V.H., Bonner, I., et al., 2017. Mobilisation of agricultural residues for bioenergy and higher value bio-products: Resources, barriers and sustainability. IEA Bioenergy Report. Available at: http://www.ieabioenergy.com/wp-content/uploads/2018/01/TR2017-01-F.pdf.

Berbel, J., Posadillo, A., 2018. Review and analysis of alternatives for the valorisation of agro-industrial olive oil by-products. Sustainability (Switzerland) 10 (1), 1–9. Available from: https://doi.org/10.3390/su10010237.

Berner, F., Heimann, K., Sheehan, M., 2016. Microalgal biofilms for biomass production. J. Appl. Phycol. 27, 1793–1804.

BiofuelsDigest, 2016. Abengoa's Hugoton cellulosic ethanol project goes on the Block. Available at: http://www.biofuelsdigest.com/bdigest/2016/07/18/abengoas-hugoton-cellulosic-ethanol-project-goes-on-the-block/ (accessed March 21, 2018).

BiofuelsDigest, 2017. Top 17 of2017. Available at: http://www.biofuelsdigest.com/bdigest/2017/12/23/top-17-for-2017/ (accessed March 21, 2018).

Borges, T., Fernandez, J., Teixeira, A.L., 2016. Effect of carbon dioxide and sunlight on algae biomass produced in laminar photobioreactor. Ambiente & Agua 11 (3), 602–611. Available from: http://www.scielo.br/pdf/ambiagua/v11n3/1980-993X-ambiagua-11-03-00601.pdf.

BP, 2017. BP Statistical Review of World Energy, June 2017. BP, London. Available at: https://www.bp.com/content/dam/bp/en/corporate/pdf/energy-economics/statistical-review-2017/bp-statistical-review-of-world-energy-2017-full-report.pdf.

Brennan, L., Owende, P., 2010. Biofuels from microalgae- A review of technologies for production processing and extraction of biofuels and co-products. Renew. Sust. Energy Rev. 14, 557–577.

Brown, A., Le Feuvre, P., 2017. Technology Roadmap: Delivering Sustainable Bioenergy. International Energy Agency (IEA). Renewable Energy Division (RED). Available at: http://www.iea.org/publications/freepublications/publication/Technology_Roadmap_Delivering_Sustainable_Bioenergy.pdf.

Burg, P., Ludín, D., Rutkowski, K., Krakowiak-Bal, A., Trávníček, P., Zemánek, P., et al., 2016. Calorific evaluation and energy potential of grape pomace. Int. Agrophys. 30 (2), 261–265.

Burlew, J.S. (Ed.), 1953. Algal Culture From Laboratory to Pilot Plant, 600. Carnegie Institute of Washington, Washington, DC, p. 357. ISBN: 0-87279-611-6 http://shelf2.library.cmu.edu/Tech/01847533.pdf.

Calderón, C., Gauthier, G., Jossart, J.M., 2017. AEBIOM Statistical Report. European Bioenergy Outlook. Available at: http://www.aebiom.org/statistical-report-2017/statistical-report-2017-17-10-17/.

Calvin, M., 1961. The Path of Carbon in Photosynthesis (Nobel Prize Lecture). Ernest Orlando Lawrence Berkeley National Laboratory, University of California Radiation Laboratory-Berkeley, United States Department of Energy (through predecessor agency the Atomic Energy Commission), (December 11, 1961). Ref. UCRL-9966. 48 pp. https://www.osti.gov/accomplishments/documents/fullText/ACC0327.pdf.

Camia, A., Robert, N., Jonsson, R., Pilli, R., García-Condado, S., López-Lozano, R., et al., 2018. Biomass production, supply, uses and flows in the European Union. European Union, Luxembourg. Available from: http://dx.doi.org/10.2760/181536.

Cantrell, K.B., Ducey, T., Ro, K.S., Hunt, P.G., 2008. Livestock waste-to-bioenergy generation opportunities. Bioresour. Technol. 99 (17), 7941−7953.

Cardoen, D., Joshi, P., Diels, L., Sarma, P.M., Pant, D., 2015. Agriculture biomass in India: part 1. Estimation and characterization. Resour. Conserv. Recycl. 102, 39−48. Available at: https://www.sciencedirect.com/science/article/pii/S0921344915300227?via%3Dihub (accessed March 10, 2018).

Carere, C.R., Sparling, R., Cicek, N., Levin, D.B., 2008. Third generation biofuels via direct cellulose fermentation. Int. J. Mol. Sci. 9 (7), 1342−1360. Available from: https://doi.org/10.3390/ijms9071342.

Chambers, M. and Schmitt, R., 2015. Diesel powered passenger cars and light trucks. Available at: https://www.rita.dot.gov/bts/sites/rita.dot.gov.bts/files/DieselFactSheet.pdf.

Cherubini, F., 2010. The biorefinery concept: using biomass instead of oil for producing energy and chemicals. Energy Convers. Manag. 51 (7), 1412−1421. Available at: https://doi.org/10.1016/j.enconman.2010.01.015.

Chew, K.W., Yap, J.Y., Show, P.L., Suan, N.H., Juan, J.C., Ling, T.C., et al., 2017. Microalgae biorefinery: high value products perspectives. Bioresour. Technol. 229, 53−62.

Chisti, Y., 2007. Biodiesel from microalgae. Biotechnol. Adv. 25, 294−306.

Christenson L., 2011. Algal biofilm production and harvesting system for wastewater treatment with biofuels by-products. All Graduate Theses and Dissertations. Paper 994, Utah State University. http://digitalcommons.usu.edu/etd/994.

Ciani, M., 2017. Biocombustibles 11/17. Dirección de Agroenergía. Ministerio de Agrindustria de la Presidencia de la Nación. Argentina. Available at: http://www.minagri.gob.ar/sitio/areas/bioenergia/informes/_archivos//000001_Informes%20Biocombustibles%202017/000011_Informe%20biocombustibles%20(noviembre%202017).pdf.

Corrado, S., Ardente, F., Sala, S., Saouter, E., 2017. Modelling of food loss within life cycle assessment: from current practice towards a systematisation. J. Clean. Prod. 140, 847−859. Available at: https://doi.org/10.1016/j.jclepro.2016.06.050.

Cristóbal, J., Caldeira, C., Corrado, S., Sala, S., 2018. Techno-economic and profitability analysis of food waste biorefineries at European level. Bioresour. Technol. 259 (January), 244−252. Available at: http://linkinghub.elsevier.com/retrieve/pii/S0960852418303547.

Dallemand, J.F., Echeverria, I. and Sánchez, D., 2006. Cereals Straw Resources for bioenergy in the European Union. In: Proceedings of the Expert Consultation. Available at: http://publications.jrc.ec.europa.eu/repository/bitstream/JRC35679/PUBSY%205679%20-%20EUR%2022626%20-%20docnum.pdf.

De Corato, U., De Bari, I., Viola, E., Pugliese, M., 2018. Assessing the main opportunities of integrated biorefining from agro-bioenergy co/by-products and agroindustrial residues into high-value added products associated to some emerging markets: a review. Renew. Sust. Energy Rev. 88 (November 2017), 326–346. Available at: https://doi.org/10.1016/j.rser.2018.02.041.

DEFRA, Department for Environment Food & Rural Affairs, 2004. Growing short rotation coppice. Best Practice Guidelines for Applicants to Defra's Energy Crops Scheme. Defra Publications. Available at: https://www.forestry.gov.uk/pdf/Growing_Short_Rotation_Coppice_tcm6_2004.pdf/$FILE/Growing_Short_Rotation_Coppice_tcm6_2004.pdf.

DEFRA, Department for Environment Food & Rural Affairs, 2017. Crops Grown For Bioenergy in England and the UK: 2016. Defra Publications. Available at: https://www.gov.uk/government/uploads/system/uploads/attachment_data/file/664991/nonfood-statsnotice2016-6dec17b.pdf.

Denmark, 2010. National Action Plan for Renewable Energy in Denmark. Available at: https://ec.europa.eu/energy/en/topics/renewable-energy/national-action-plans.

Dimitriou, I. and Rutz, D., 2015. Sustainable Short Rotation Coppice. A Handbook. Available at: http://www.srcplus.eu/images/Handbook_SRCplus.pdf.

Eisentraut, A., 2010. Sustainable production of second-generation biofules, IEA Energy Papers. IEA, Paris, France. Available from: http://dx.doi.org/10.1787/9789264084247-en.

El Bassam, N., 2010. Handbook of Bioenergy Crops: A Complete Reference to Species, Development and Applications. Earthscan, London978-1-84407-854-7.

Energy Research Centre of the Netherlands, 2018. Phyllis2, Database for Biomass and Waste. Available at: https://www.ecn.nl/phyllis2 9 (accessed February 11, 2018).

EU, 2013. Decision No 1386/2013/EU of the European Parliament and of the Council of 20 November 2013 on a General Union Environment Action Programme to 2020 "Living well, within the limits of our planet," pp. 171–200.

EurObserv'Er, 2017a. Biogas Barometer 2017. p. 14 Available at: https://www.eurobserv-er.org/biogas-barometer-2017/.

Eurobserv'Er, 2017b. Biofuels Barometer 2017. p. 14. Available at: https://www.eurobserv-er.org/category/all-biofuels-barometers/.

European Commission, 2008. Directive 2008/98/EC of the European Parliament and of the Council of 19 November 2008 on Waste and Repealing Certain Directives. Available at: http://ec.europa.eu/environment/waste/framework/.

European Commission, 2010. Communication from the commission on the practical implementation of the EU biofuels and bioliquids sustainability scheme and on counting rules for biofuels. Off. J. Eur. Union 17 (443), 8–16. Available at: http://eur-lex.europa.eu/legal-content/EN/ALL/?uri = CELEX:52010XC0619(02).

European Commission, 2012. Innovating for sustainable growth: a bioeconomy for Europe. Off. J. Eur. Union 8 (2), 60.

European Commission, 2017. Proposal for a directive of the European Parliament and of the council on the promotion of the use of energy from renewable sources (recast). Off. J. Eur. Union 382 (2016), Available at: http://eur-lex.europa.eu/resource.html?uri=cellar:3eb9ae57-faa6-11e6-8a35-01aa75ed71a1.0007.02/DOC_1&format = PDF%0Ahttp://eur-lex.europa.eu/legal-content/EN/TXT/?uri=CELEX:52016PC0767R%2801%29.

European Parliament, 2009. Directive 2009/28/EC of the European Parliament and of the Council of 23 April 2009. Off. J. Eur. Union 140 (16), 16–62. Available at: https://eur-lex.europa.eu/legal-content/EN/ALL/?uri = celex%3A32009L0028.

European Parliament, 2017. Palm oil and deforestation of rainforest. European Parliament resolution of 4 April 2017, p. 15.

Eurostat, 2017a. Eurostat: Statistics Explained. Available at: http://ec.europa.eu/eurostat/statistics-explained/index.php/Glossary:Biomass (accessed January 28, 2018).
Eurostat, 2017b. Guidance on municipal waste data collection. Available at: http://ec.europa.eu/eurostat/documents/342366/351811/Municipal + Waste + guidance/bd38a449-7d30-44b6-a39f-8a20a9e67af2.
Eurostat, 2018a. Municipal waste by waste operations. Available at: http://ec.europa.eu/eurostat/web/environment/waste/database (accessed March 22, 2018).
Eurostat, 2018b. Supply, transformation and consumption of renewable energies—annual data. Available at: http://appsso.eurostat.ec.europa.eu/nui/submitViewTableAction.do (accessed February 20, 2018).
FAO, 2004. Unified Bioenergy Terminology (UBET), Available at: ftp://ftp.fao.org/docrep/fao/007/j4504e/j4504e00.pdf.
FAO, 2012. Forest Resources Assessment 2015: Terms and Definitions. FAO Report, p. 36. Available at: http://www.fao.org/docrep/017/ap862e/ap862e00.pdf.
FAO, 2014. FAO Bioenergy & Food Security. Approach: Implementation Guide. FAO Report, p. 25. ISBN 978-92-5-108222-5. Available at: http://www.fao.org/docrep/019/i3672e/i3672e.pdf.
FAO, 2016. The State of World Fisheries and Aquaculture 2016. Contributing to Food Security and Nutrition for All. FAO, Rome, 200 pp. ISBN 978-92-5-109185-2. Available at: http://www.fao.org/3/a-i5555e.pdf.
FAO, 2017a. FAOSTAT. Available at: http://www.fao.org/faostat/en/#data (accessed February 20, 2018).
FAO, 2017b. Flude—Global Forest Resources Assessments. Food and Agriculture Organization of the United Nations, Rome, http://www.fao.org/forest-resources-assessment/explore-data/flude/en/ (accessed February 20, 2018).
FAO, 2017c. Yearbook of Forest Products 2015. Food and Agriculture Organization of the United Nations, Rome978-92-5-009766-4.
FAO, 2018a. FAOSTAT—Forestry Production and Trade. Food and Agriculture Organization of the United Nations, Rome. Available from: http://www.fao.org/faostat/en/#data/FO (accessed March 05, 2018).
FAO, 2018b. Natural Resources—Biomass Potential Assessment. FAO Report. Available at: http://www.fao.org/energy/bioenergy/befs/assessment/befs-ra/natural-resources/en/ (accessed March 20, 2018).
Fernández, J., 2010. Laminar Photobioreactor for the Production of Microalgae, Spain patent ES 2 347 515 B2. PCT/ES2011/000104, May 3, 2010. WO 2011/138477 A1, November 10, 2011. Also Published as EP2568038A1, EP2568038A4, US20130115688.
Fernández J., 2013. Photobioreactor with a floating support WO 2015092093 A1 PCT WO 2015092093 A1. From "Fotobiorreactor de soporte flotante." Spanish Patent ES 1 099 707 U, priority 20-12-2013.
Fernández, J., 2015. Tecnologiјas para el uso y transformacioјn de biomasa energeјtica. Capitulo XII. In: Energia de la biomasa. Madrid, Spain: Mundi-Prensa, pp. 397–445.
Fernando, A.L., Boléo, S., Barbosa, B., Costa, J., Duarte, M.P., Monti, A., 2015. Perennial grass production opportunities on marginal Mediterranean land. Bioenergy Res. 8, 1523–1537. Available from: https://doi.org/10.1007/s12155-015-9692-0.
Flach, B., Lieberz, S. and Rossetti, A., 2017. EU-28 Biofuels Annual. EU Annual Biofuels Annual 2017, Global Agricultural Information Network (GAIN) Report. The Hague. Available at: https://gain.fas.usda.gov/Recent%20GAIN%20Publications/Biofuels%20Annual_The%20Hague_EU-28_6-19-2017.pdf.
Fokaides, P.A., 2017. Energy recovery alternatives for the sustainable management of olive oil industry. Olive Mill Waste. Elsevier, pp. 79–96. Available at: http://linkinghub.elsevier.com/retrieve/pii/B9780128053140000042 (accessed March 29, 2018).

Gao, K., 1998. Chinese studies on the edible blue-green alga, Nostoc flagelliforme: a review. J. Appl. Phycol. 10, 37–49.
García-Maraver, A., Zamorano, M., Ramos-Ridao, A., Díaz, L.F., 2012. Analysis of olive grove residual biomass potential for electric and thermal energy generation in Andalusia (Spain). Renew. Sust. Energy Rev. Available from: https://doi.org/10.1016/j.rser.2011.08.040.
Ghani Wahab, A., 2017. Malaysia Biofuels Annual 2017. Global Agricultural Information Network (GAIN) Report, Kuala Lumpur. Available at: https://gain.fas.usda.gov/Recent.GAIN Publications/Biofuels Annual_The Hague_EU-28_6-19-2017.pdf.
GIE and EBA, 2018. Gas Infrastructure Europe & European Biogas Association. European Biomethane Map. Available at: http://www.gie.eu/index.php/maps-data/bio-map (accessed March 10, 2018).
Golecha, R., Gan, J., 2016. Effects of corn stover year-to-year supply variability and market structure on biomass utilization and cost. Renew. Sust. Energy Rev. 57, 34–44. Available from: https://doi.org/10.1016/j.rser.2015.12.075.
Golucke, C.G., Oswald, W.J., 1959. Biological conversion of light energy to the chemical energy of methane. Appl. Microbiol. 7 (4), 219–227.
Golucke, C.G., Oswald, W.J., Gotaas, H.B., 1957. Anaerobic digestion of algae. Appl. Microbiol. 5 (1), 47–55.
Gosch, B.J., Magnusson, M., Paul, N.A., de Nys, R., 2012. Total lipid and fatty acid composition of seaweeds for the selection of species for oil-based biofuel and bioproducts. GCB Bioenergy 4, 919–930.
Gross M.A., 2015. Development and optimization of biofilm based algal cultivation. Graduate Theses and Dissertations 14850. http://lib.dr.iastate.edu/etd/14850
Gross M.A., Wen Z., 2013. Revolving algal biofilm photobioreactor systems and methods. US patent: US 20140273171 A1 priority, March 13, 2013.
Guerrero, A.B., Aguado, P.L., Sánchez, J., Curt, M.D., 2016. GIS-based assessment of banana residual biomass potential for ethanol production and power generation: a case study. Waste Biomass Valorization 7 (2). Available from: https://doi.org/10.1007/s12649-015-9455-3.
Gunnarsson, I.B., Svensson, S.E., Johansson, E., Karakashev, D., Angelidaki, I., 2014. Potential of Jerusalem artichoke (Helianthus tuberosus L.) as a biorefinery crop. Ind. Crops Prod. 56, 231–240. Available from: https://doi.org/10.1016/j.indcrop.2014.03.010.
Haase, M., Rösch, C., Ketzer, D., 2016. GIS-based assessment of sustainable crop residue potentials in European regions. Biomass Bioenergy 86, 156–171. Available from: https://doi.org/10.1016/j.biombioe.2016.01.020.
Harder, R., Von Witsch, H., 1942. Uber Massenkultur von diatomeen. Ber. Bot. Ges 60, 142. Available from: https://es.scribd.com/document/360174362/Harder-Von-Witsch-1942-Uber-Massenkultur-Von-Diatomeen.
Helbig, F., Roth, C., 2017. Solid Biomass Fuels for Cooking—Beyond Firewood and Charcoal. Bonn and Eschborn, Germany. Available at: http://www.ccacoalition.org/sites/default/files/resources/2017_solid-biomass-fuels-for-cooking_GIZ.pdf.
Hinchee, M., Rottmann, W., Mullinax, L., Zhang, C., Chang, S., Cunningham, M., et al., 2011. Short-rotation woody crops for bioenergy and biofuels applications. Biofuels: Global Impact on Renewable Energy, Production Agriculture, and Technological Advancements. Springer-Verlag, New York, pp. 139–156. Available from: http://dx.doi.org/10.1007/978-1-4419-7145-6_8.
HLPE, 2013. Biofuels and Food Security. A Report by the High Level Panel of Experts on Food Security and Nutrition of the Committee on World Food Security. FAO, Rome. Available at: http://www.fao.org/fileadmin/user_upload/hlpe/hlpe_documents/HLPE_Reports/HLPE-Report-5_Biofuels_and_food_security.pdf.

Hoornweg, D., Bhada-Tata, P., 2012. What a waste: a global review of solid waste management. Urban Development Series Knowledge Papers. World Bank Group, Washington, DC, p. 98. Available at: www.worldbank.org/urban.
Horta Nogueira, L.A., Silva Capaz, R., 2013. Biofuels in Brazil: evolution, achievements and perspectives on food security. Global Food Security 2, 117–125. Available from: https://doi.org/10.1016/j.gfs.2013.04.001.
ICRISAT, 2016. Increasing Ethanol Production in India Using Sweet Sorghum. Available at: http://www.icrisat.org/increasing-ethanol-production-in-india-using-sweet-sorghum/ (accessed February 20, 2018).
IEA, 2017a. Renewables Information: Overview. Available at: https://www.iea.org/publications/freepublications/publication/renewables-information---2017-edition---overview.html.
IEA, 2017b. Key World Energy Statistics, p. 96. doi:10.1017/CBO9781107415324.004. ISBN: 9788578110796.
IEA, 2018. International Energy Agency Data Services. Available at: http://wds.iea.org/WDS/ReportFolders/ReportFolders.aspx (accessed February 13, 2017).
IEA Bioenergy, 2017. State of Technology Review—Algae Bioenergy. http://www.iea-bioenergy.com/wp-content/uploads/2017/01/IEA-Bioenergy-Algae-report-update-20170114.pdf.
IPCC, 2000. Land use, land-use change, and forestry. In: Watson, R.T., Noble, I.R., Bolin, B., Ravindranath, N.H., Verardo, D.J., Dokken, D.J. (Eds.), Summary for Policymakers. IPCC, Geneva, Switzerland.
Janssen, R., Turhollow, A.F., Rutz, D., Mergner, R., 2013. Production facilities for second-generation biofuels in the U.S. and the EU - current status and future perspectives. Biofuels Bioprod. Biorefining 7 (6), 647–665.
Jard, G., Marfaing, H., Carrère, H., Delgenes, J.P., Steyer, J.P., Dumas, C., 2013. French Brittany macroalgae screening: composition and methane potential for potential alternative sources of energy and products. Bioresour. Technol. 144, 492–498.
Johnson M.B., 2009. Microalgal biodiesel production through a novel attached cultured system and conversion parameters. Master Science Thesis. Faculty of Virginia Polytechnic Institute and State University. http://scholar.lib.vt.edu/theses/available/etd-04292009-160208/unrestricted/ETD2.pdf
Jölli, D. and Giljum, S., 2005. Unused biomass extraction in agriculture, forestry and fishery, SERI Studies. Available at: http://web205.vbox-01.inode.at/Data/seri/publications/documents/SERI_Studies_3.pdf.
Jones, M., Walsh, M. (Eds.), 2000. Miscanthus for Energy and Fibre. James & James (Science Publishers) Ltd., London, United Kingdom.
Jouhara, H., Czajczyńska, D., Ghazal, H., Krzyżyńska, R., Anguilano, L., Reynolds, A.J., et al., 2017. Municipal waste management systems for domestic use. Energy 139, 485–506.
Kallio, A., Chudy, R., Solberg, B., 2018. Prospects for producing liquid wood-based biofuels and impacts in the wood using sectors in Europe. Biomass Bioenergy. 108, 415–425. Available from: https://doi.org/10.1016/J.BIOMBIOE.2017.11.022.
Kampman, B., Leguijt, C., Scholten, T., Tallat-Kelpsaite, J., Brückmann, R., Maroulis, G., et al., 2017. Optimal use of biogas from waste streams—an assessment of the potential of biogas from digestion in the EU beyond 2020, Available at: https://ec.europa.eu/energy/sites/ener/files/documents/ce_delft_3g84_biogas_beyond_2020_final_report.pdf.
Kamuk, B., Haukohl, J., 2013. Guidelines: Waste to Energy in Low and Middle Income Countries. International Solid Waste Association.
Kays, S.J., Nottingham, S.F. (Eds.), 2012. Biology and chemistry of Jerusalem artichoke *Helianthus tuberosus* L. CRC Press, Taylor & Francis Group, Boca Raton, FL.

Keita, J.D., 1987. Wood or charcoal—which is better? FAO Report. vol. 39, pp. 157–158. http://www.fao.org/docrep/s4550e/s4550e09.htm#wood%20or%20charcoal%20%20%20which%20is%20better (accessed March 21, 2018).

Khatun, R., Reza, M.I.H., Moniruzzaman, M., Yaakob, Z., 2017. Sustainable oil palm industry: the possibilities. Renew. Sust. Energy Rev. 76 (August 2016), 608–619. Available from: https://doi.org/10.1016/j.rser.2017.03.077.

Kibede-Westhead, E., Pizarro, C., Mulbry, W., Wilkie, A., 2003. Production and nutrient removal by periphyton grown under different loading rates of anaerobically digested flushed dairy manure. J Phycol 39, 1275–1282.

Kim, J.K., Yarish, C., Hwang, E.K., Park, M., Kim, Y., 2017. Seaweed culture: cultivation technologies, challenges and its ecosystem services. Algae 32 (1), 1–13.

Kim, T.H., Kim, T.H., 2014. Overview of Technical Barriers and Implementation of Cellulosic Ethanol in the U.S. Energy., 66. Elsevier Ltd., Oxford, United Kingdom, pp. 13–19. Available from: https://doi.org/10.1016/j.energy.2013.08.008.

Kiran, E., Trzcinski, A.P., Ng, W.J., Liu, Y., 2014. Bioconversion of food waste to energy: a review. Fuel 134, 389–399. Available at: https://www.sciencedirect.com/science/article/pii/S0016236114005365 (accessed March 30, 2018).

Koller, M., Muhr, A., Braunegg, G., 2014. Microalgae as versatile cellular factories for valued products. Algal Res. 6, 52–63.

Koncagül, E. Tran, M., Connor, R., Uhlenbrook, S., Renata, A., Ortigara, C., 2017. Wastewater the untapped resource. Facts and figures. The United Nations World Water Development Report 2017. Available at: http://unesdoc.unesco.org/images/0024/002475/247553e.pdf.

Kraan, S., 2013. Mass cultivation of carbohydrate rich macroalgae, a possible solution for sustainable biofuel production. Mitig. Adapt. Strateg. Glob. Change. 18, 27–46.

Kszos, L.A., Downing, M.E., Wright, L.L., Cushman, J.H., McLaughlin, S.B., Tolbert, V. R., et al., 2001. Bioenergy Feedstock Development Program Status Report (5049), pp. 1–68. Available at: https://info.ornl.gov/sites/publications/Files/Pub57004.pdf.

Kummamuru, B., 2017. WBA Global Bioenergy Statistics 2017, Available at: http://www.worldbioenergy.org/uploads/WBA.GBS 2017_hq.pdf. doi: 10.1016/0165-232X(80)90063-4.

Lal, R., 2005. World crop residues production and implications of its use as a biofuel. Environ. Int. 31 (4), 575–584. Available from: https://doi.org/10.1016/j.envint.2004.09.005.

Lee, D.K., Aberle, E., Anderson, E.K., Anderson, W., Baldwin, B.S., Baltensperger, D., et al., 2018. Biomass production of herbaceous energy crops in the United States: field trial results and yield potential maps from the multiyear regional feedstock partnership. GCB Bioenergy 1–19. Available from: https://doi.org/10.1111/gcbb.12493.

Lewandowski, I., Scurlock, J.M.O., Lindvall, E., Christou, M., 2003. The development and current status of perennial rhizomatous grasses as energy crops in the US and Europe. Biomass Bioenergy 25, 335–361. Available from: https://doi.org/10.1016/S0961-9534(03)00030-8.

Lewandowski, I., Gaudet, N., Lask, J., Maier, J., Tchouga, B., Vargas-Carpintero, R., 2018. Bioeconomy. In: Lewandowski, I. (Ed.), Shaping the Transition to a Sustainable, Biobased Economy. Springer International Publishing AG, Cham, Switzerland.

Liu, Z.X., Han, L. pu, Yosef, S., Xie, G. Hui, 2011. Genetic variation and yield performance of Jerusalem Artichoke germplasm collected in China. Agr. Sci. China 10 (5), 668–678. Available from: https://doi.org/10.1016/S1671-2927(11)60049-7.

López González, L.M., Sala Lizarraga, J.M., Míguez Tabarés, J.L., López Ochoa, L.M., 2007. Contribution of renewable energy sources to electricity production in the autonomous community of Navarre (Spain): a review. Renew. Sust. Energy Rev. 11 (8), 1776–1793. Available from: https://doi.org/10.1016/j.rser.2006.01.002.

Lorimor, J., Powers, W., Sutton, A., 2004. Manure Characteristics. MWPS-18 Section 1. Second Edition. Available at: http://msue.anr.msu.edu/uploads/files/ManureCharacteristicsMWPS-18_1.pdf.

Lundquist T.J., Woertz I.C., N.W.T., Quinn, Benemann, J.R., 2010. A Realistic Technology and Engineering Assessment of Algae Biofuel Production. Report of Energy Biosciences Institute of UC Berkeley (California). 153 pp.

Macke, Y., 2017. Peoples Republic of China. Biofuels Annual. GAIN Report Number CH17048, Global Agricultural Information Network (GAIN) Report. Available at: https://www.fas.usda.gov/data/china-biofuels-annual-3.

Macke, Y. and Ward, M., 2017. Growing Interest for Ethanol Brightens Prospects Biofuels Annual Peoples Republic of China. Beijing. Available at: https://gain.fas.usda.gov/Recent%20GAIN%20Publications/Biofuels%20Annual_Beijing_China%20-%20Peoples%20Republic%20of_10-20-2017.pdf.

Mahin, B., 1991. Industrial energy and electric power from wood residues. Bioenergy System Report. Winrock International, Arlington, Virginia. Available from: https://pdf.usaid.gov/pdf_docs/PNABI678.pdf.

Manochio, C., Andrade, B.R., Rodriguez, R.P., Morae, B.S., 2017. Ethanol from biomass: a comparative overview. Renew. Sust. Energy Rev. 80, 743–755.

Manzanares, P., Ruiz, E., Ballesteros, M., Negro, M.J., Gallego, F.J., López-Linares, J.C., et al., 2017. Residual biomass potential in olive tree cultivation and olive oil industry in Spain: valorization proposal in a biorefinery context. Span. J. Agric. Res. 15 (3), 1–12.

Maragkaki, A.E., Fountoulakis, M., Kyriakou, A., Lasaridi, K., Manios, T., 2018. Boosting biogas production from sewage sludge by adding small amount of agro-industrial by-products and food waste residues. Waste Manag. 71, 605–611. Available at: http://www.ncbi.nlm.nih.gov/pubmed/28427739 (accessed March 31, 2018).

Martin-Girela, I., Curt, M.D., Fernández, J., 2017. Flashing light effects on CO2 absorption by microalgae grown on a biofilm photobioreactor. Algal Res. 25, 421–430.

Mateo-Sagasta, J., Raschid-Sally, L., Thebo, A., 2015. Global wastewater and sludge production, treatment and use. Wastewater: Economic Asset in an Urbanizing World. Springer Netherlands, The Netherlands, pp. 1–282.

McKendry, P., 2002. Energy production from biomass (part 1): overview of biomass. Bioresour. Technol. 83 (1), 37–46. Available at: https://doi.org/10.1016/S0960-8524(01)00118-3.

Mehmood, M.A., Ibrahim, M., Rashid, U., Nawaz, M., Ali, S., Hussain, A., et al., 2017. Biomass production for bioenergy using marginal lands. Sustain. Prod. Consumption 9, 3–21. Available from: https://doi.org/10.1016/j.spc.2016.08.003.

Mena, A., 1985. The utilization of sugar cane by-products as substitutes for cereal in animal feed. Proceedings of the FAO Expert Consultation on the substitution of imported concentrate feeds in animal production systems in developing countries. Food and Agriculture Organization of the United Nations, Rome, Italy. Available at: http://www.fao.org/docrep/003/x6930e/X6930E00.htm#TOC.

Milledge, J.J., Harvey, P.J., 2016. Potential process "hurdles" in the use of macroalgae as feedstock for biofuel production in the British Isles. J. Chem. Technol. Biotechnol. 91, 2221–2234.

Milledge, J.J., Smith, B., Dyer, P.W., Harvey, P., 2014. Macroalgae-derived biofuels: a review of methods of energy extraction from seaweed biomass. Energies 2014 (7), 7194–7222.

Mittlefehldt, S., 2016. Seeing forests as fuel: how conflicting narratives have shaped woody biomass energy development in the United States since the 1970s. Energy Res. Soc. Sci. 14, 13–21. Available from: https://doi.org/10.1016/j.erss.2015.12.023.

Monforti, F., Bódis, K., Scarlat, N., Dallemand, J.F., 2013. The possible contribution of agricultural crop residues to renewable energy targets in Europe: a spatially explicit study. Renew. Sust. Energy Rev. 19, 666–677. Available from: https://doi.org/10.1016/j.rser.2012.11.060.

Monier, V., Mudgal, S., Escalon, V., O'Connor, C., Gibon, T., Anderson, G., et al., 2010. Preparatory Study on Food Waste Across EU 27. European Commission [DG ENV—Directorate C]. Available at: http://ec.europa.eu/environment/eussd/pdf/bio_foodwaste_report.pdf.

Monteith, J.L.L., 1972. Solar radiation and productivity in tropical ecosystems. J. Appl. Ecol. 9 (3), 747–766.

Monti, A. (Ed.), 2012. Switchgrass: A Valuable Biomass Crop for Energy. Springer-Verlag, London, United Kingdom, ISBN: 978-1-4471-2903-5. (accessed March 31, 2018).

Mulbry, W., Wilkie, A., 2001. Growth of benthic freshwater algae on dairy manures. J. Appl. Phycol. 13, 301–306.

Mulbry, W., Konrad, S., Pizarro, C., Kibede-Westhead, E., 2008. Treatment of dairy manure effluent using freshwater algae: algal productivity and recovery of manure nutrients using pilot-scale algal turf scrubber. Bioresour. Technol. 99, 8137–8142.

Muth, D.J., Bryden, K.M., Nelson, R.G., 2013. Sustainable agricultural residue removal for bioenergy: a spatially comprehensive US national assessment. Appl. Energy. 102, 403–417. Available from: https://doi.org/10.1016/j.apenergy.2012.07.028.

Nartker, S., Ammerman, M., Aurandt, J., Stogsdil, M., Hayden, O., Antle, C., 2014. Increasing biogas production from sewage sludge anaerobic co-digestion process by adding crude glycerol from biodiesel industry. Waste Manag. 34 (12), 2567–2571. Available at: https://www.sciencedirect.com/science/article/pii/S0956053X14003729 (accessed March 31, 2018).

Naumann, T., Çebi, Z., Podola, B., Melkonian, M., 2013. Growing microalgae as aquaculture feeds on twin-layers: a novel solid-state photobioreactor. J. Appl. Phycol. 25, 1413–1420. Available from: https://doi.org/10.1007/s10811-012-9962-6.

Negro, M.J., Manzanares, P., Ruiz, E., Castro, E., Ballesteros, M., 2017. The biorefinery concept for the industrial valorization of residues from olive oil industry. Olive Mill Waste. Elsevier, pp. 57–78. Available from: https://doi.org/10.1016/B978-0-12-805314-0.00003-0.

Nghiem, N.P., Montanti, J.M., Johnston, D., 2016. Sorghum as a renewable feedstock for production of fuels and industrial chemicals. AIMS Bioeng. 3 (1), 75–91. Available from: https://doi.org/10.3934/bioeng.2016.1.75.

Nichiporovich, A.A., Semenenko, V.E., Vladimirova, M.G., Spektorov, K.S., 1962. Some principles of intensification of photosynthetic productivity in unicellular algae. Istvesia Acad. Sci. USSR Ser. Biol 2, 163.

Nizami, A.S., Rehan, M., Waqas, M., Naqvi, M., Ouda, O.K.M., Shahzad, K., et al., 2017. Waste biorefineries: enabling circular economies in developing countries. Bioresour. Technol. 241, 1101–1117. Available at: https://doi.org/10.1016/j.biortech.2017.05.097.

Njakou Djomo, S., Ac, A., Zenone, T., De Groote, T., Bergante, S., Facciotto, G., et al., 2015. Energy performances of intensive and extensive short rotation cropping systems for woody biomass production in the EU. Renew. Sust. Energy Rev. 41, 845–854. Available from: https://doi.org/10.1016/j.rser.2014.08.058.

Obernberger, I., Biedermann, F., Widmann, W., Riedl, R., 1997. Concentrations of inorganic elements in biomass fuels and recovery in the different ash fractions. Biomass Bioenergy 12 (3), 211–224.

OECD, 2002. OECD Agricultural Outlook: 2002–2007—Annex II: Glossary of Terms. OECD Publications, Paris. Available at: http://www.oecd.org/about/publishing/35205971.pdf.

OECD, 2004. Biomass and Agriculture: Sustainability, Markets and Policies, OECD Workshop on Biomass and Agriculture. OECD. https://doi.org/10.1787/9789264105546-en.

OECD, 2015. OECD Climate Change Mitigation. Policies and Progress. OECD Publications, Paris9789264238787, 116 pp. https://doi.org/10.1787/9789264238787-en. Available at: http://www.oecd-ilibrary.org/environment/climate-change-mitigation_9789264238787-en.

OECD, 2018. Agricultural land (indicator) (accessed February 17, 2018).

OECD-FAO, 2017. OECD-FAO Agricultural Outlook 2017-2026, OECD Publishing, Paris. http://dx.doi.org/10.1787/agr_outlook-2017-en.

OECD-FAO, 2018. OECD-FAO Agricultural Outlook 2017–2026. OECD Agriculture Statistics. Available at: http://stats.oecd.org/viewhtml.aspx?datasetcode = HIGH_AGLINK_2017&lang = en# (accessed March 28, 2018).

OECD/IEA and FAO, 2017. How2Guide for Bioenergy Roadmap Development and Implementation. IEA Publications978-92-5-109586-7. Available at: http://www.oecd-ilibrary.org/agriculture-and-food/biomass-and-agriculture_9789264105546-en.

Oswald, W.J., Golueke, C.G., 1960. Biological transformation of solar energy. Adv. Appl. Microbiol. 2, 223–262.

Owens, J.M., Chynoweth, D.P., 1993. Biochemical methane potential of municipal solid waste (MSW) components. Water Sci. Technol. 27 (2), 1–14.

Ozkan, A., Kinney, K., Katz, L., Berberoglu, H., 2012. Reduction of water and energy requirement of algae cultivation using an algae biofilm photobioreactor. Bioresour. Technol. 114, 542–548.

Patil, P., Deng, S., Rhodes, J.I., Lammers, J.P., 2010. Conversion of waste cooking oil to biodiesel using ferric sulfate and supercritical methanol processes. Fuel 89 (2), 360–364. Available at: https://www.sciencedirect.com/science/article/pii/S0016236109002488 (accessed March 30, 2018).

Paturau, J.M., 1986. Alternative uses of sugarcane and its byproduts in agroindustries. Sugarcane as Feed. FAO, Rome, Italy. Available from: http://www.fao.org/docrep/003/s8850e/S8850E00.htm#TOC.

Pippo, W.A., Luengo, C.A., 2013. Sugarcane energy use: accounting of feedstock energy considering current agro-industrial trends and their feasibility. Int. J. Energy Environ. Eng. 4 (1), 1–13.

Podola, B., Li, T., Melkonian, M., 2017. Porous substrate bioreactors: a paradigm shift in microalgal biotechnology. Trends Biotechnol. 35 (2), 121–132.

Preechajan, S., Prasertsri, P., 2017. Thailand Biofuels Annual 2017. Global Agricultural InformationNetwork (GAIN) Report, Bangkok. Available at: https://gain.fas.usda.gov/Recent GAIN Publications/Biofuels Annual_Bangkok_Thailand_6-23-2017.pdf.

Rajkumar, R., Yaakob, Z., Takriff, M.S., 2014. Potential of the micro and macro algae for biofuel production: a brief review. BioResources 9 (1), 1606–1633.

Regassa, T.H., Wortmann, C.S., 2014. Sweet sorghum as a bioenergy crop: literature review. Biomass Bioenergy. 64, 348–355. Available from: https://doi.org/10.1016/j.biombioe.2014.03.052. Elsevier Ltd.

Ren, Y., Yu, M., Wu, C., Wang, Q., Gao, M., Huang, Q., et al., 2018. A comprehensive review on food waste anaerobic digestion: research updates and tendencies. Bioresour. Technol. 247 (September 2017), 1069–1076. Available at: https://doi.org/10.1016/j.biortech.2017.09.109.

Richards, K.M., Pearce, S., Domac, J., White, W., Elbe, S., Hohle, E.E., 2015. Fuel Poverty and Bioenergy. Discussion paper. IEA Bioenergy Task 43. http://www.iea-bioenergy.com/wp-content/uploads/2018/01/IEA-Bioenergy-TR2015-03ii.pdf (accessed March 14, 2018).

Rosentrater, K.A., 2015. Production and use of evolving corn-based fuel ethanol coproducts in the U.S. In: Biernat, K. (Ed.), Biofuels—Status and Perspective. InTech, Rijeka, Croatia, pp. 81−98. Available from: http://dx.doi.org/10.5772/59951.

Roxburgh, S.H., Berry, S.L., Buckley, T.N., Barnes, B., Roderick, M.L., 2005. What is NPP? Inconsistent accounting of respiratory fluxes in the definition of net primary production. Funct. Ecol. 19 (3), 378−382.

RSPO, 2018. Round Table on Sustainable Palm oil. Available at: https://rspo.org/certification (accessed February 20, 2018).

Ruiz, E., Romero-García, J.M., Romero, I., Manzanares, P., Negro, M.J., Castro, E., 2017. Olive-derived biomass as a source of energy and chemicals. Biofuels Bioprod. Biorefining 11 (6), 1077−1094. Available at: http://doi.wiley.com/10.1002/bbb.1812 (accessed March 29, 2018).

Sahoo, D., Yarish, C., 2005. Mariculture of seaweeds. In: Andersen, R.A. (Ed.), Algal Culturing Techniques. Elsevier, 0-12-088426-7pp. 219−237. Available from: https://doi.org/10.1016/B978-012088426-1/50016-0.

Scarlat, N., Dallemand, J.F., Monforti-Ferrario, F., Nita, V., 2015. The role of biomass and bioenergy in a future bioeconomy: policies and facts. Environ. Dev. 15, 3−34. Available at: https://doi.org/10.1016/j.envdev.2015.03.006.

Sebestyén, P., Blanken, W., Bacsa, I., Tóth, G., Martin, A., Bhaiji, T., et al., 2016. Upscale of a laboratory rotating disk biofilm reactor and evaluation of its performance over a half-year operation period in outdoor conditions. Algal Res. 18, 266−272.

Seiple, T.E., Coleman, A.M., Skaggs, R.L., 2017. Municipal wastewater sludge as a sustainable bioresource in the United States. J. Environ. Manag. 197, 673−680. Available at: https://www.sciencedirect.com/science/article/pii/S0301479717303808 (accessed March 31, 2018).

Senelwa, K., Sims, R.E.H., 1999. Fuel characteristics of short rotation forest biomass. Biomass Bioenergy 17 (2), 127−140. Available from: https://doi.org/10.1016/S0961-9534(99)00035-5.

Sheehan J., Dunahay T., Benemann J., Roessler P., Weissman J., 1998. Look Back at the U.S. Department of Energy's Aquatic Species Program: Biodiesel from Algae; Close Out Report. NREL Report No. TP"580"24190. 325 pp. Available at: http: www. nrel.gov"docs"legosti"fy98" 24190.pdf

Shi, J., Podola, B., Melkonian, M., 2005. Removal of nitrogen and phosphorous from wastewater using microalgae immobilized on twin layers: an experimental study. J. Appl. Phycol. 19, 417−423.

Shortall, O.K., 2013. 'Marginal land' for energy crops: exploring definitions and embedded assumptions. Energy Policy 62, 19−27. Available from: https://doi.org/10.1016/j.enpol.2013.07.048.

Show, K.-Y., Yan, Y., Ling, M., Ye, G., 2018. Hydrogen production from algal biomass—advances, challenges and prospects. Bioresour. Technol. 257, 290−300. Available from: https://www.sciencedirect.com/journal/bioresource-technology.

Sulaiman, C., Abdul-Rahim, A.S., Mohd-Shahwahid, H.O., Chin, L., 2017. Wood fuel consumption, institutional quality, and forest degradation in sub-Saharan Africa: evidence from a dynamic panel framework. Ecol. Indic. 74, 414−419. Available from: https://doi.org/10.1016/j.ecolind.2016.11.045.

Statista, 2018. Production of Biogas Worldwide From 2000 to 2014. Available at: https://www.statista.com/statistics/481791/biogas-production-worldwide/ (accessed March 11, 2018).

Tan, S.T., Ho, W.S., Hashim, H., Lee, C.T., Taib, M.R., Ho, C.S., 2015. Energy, economic and environmental (3E) analysis of waste-to-energy (WTE) strategies for municipal solid waste (MSW) management in Malaysia. Energ. Convers. Manage.

102, 111–120. Available at: https://www.sciencedirect.com/science/article/pii/S0196890415001156?via%3Dihub (accessed March 11, 2018).

Torén, J., Wirsenius, S., Anttila, P., Dees, M., Ermert, J., Paappanen, T., et al., 2011. Biomass Energy Europe—Executive Summary, Evaluation and Recommendations. doi:10.13140/2.1.2968.1443.

The United Nations Statistics Division (UNSD), 2018. UNdata—A Data Access System to UN Databases. Available at: http://data.un.org/ (accessed March 26, 2018).

Thy, P., Yu, C., Jenkins, B.M., Lesher, C.E., 2013. Inorganic composition and environmental impact of biomass feedstock. Energy Fuels 27 (7), 3969–3987. Available from: https://doi.org/10.1021/ef400660u. Available at: pubs.acs.

Tyagi, V.K., Lo, S.L., 2016. Energy and resource recovery from sludge: full-scale experiences. Environmental Materials and Waste: Resource Recovery and Pollution Prevention. Elsevier, Amsterdam, the Netherlands, pp. 221–244.

USDA, 2018. US bioenergy statistics. Available at: https://www.ers.usda.gov/data-products/us-bioenergy-statistics/ (accessed February 17, 2018).

UFOP, 2017. Supply Report 2016/2017. European and World Demand for Biomass for the Purpose of Biofuel Production in Relation to Supply in the Food and Feedstuff Markets. Available at: https://www.ufop.de/files/7814/9977/4144/UFOP_supply_report_20162017.pdf (accessed February 17, 2018).

United Nations, 2018. Energy Statistics Yearbook 2015. Available at: https://unstats.un.org/unsd/energy/yearbook/.

US Energy Information Administration (EIA), 2018. Electric Power Annual. Available at: https://www.eia.gov/electricity/annual/pdf/epa.pdf.

US Environmental Protection Agency, 2017. Livestock Anaerobic Digester Database. Available at: https://www.epa.gov/agstar/livestock-anaerobic-digester-database (accessed March 30, 2018).

Vanthoor-Koopmans, M., Wijffels, R.H., Barbosa, M.J., Eppink, M.H.M., 2013. Biorefinery of microalgae for food and fuel. Bioresour. Technol. 135, 142–149.

Velázquez-Martí, B., Fernández-González, E., López-Cortés, I., Salazar-Hernández, D. M., 2011. Quantification of the residual biomass obtained from pruning of trees in Mediterranean olive groves. Biomass Bioenergy 35 (8), 3208–3217. Available from: https://doi.org/10.1016/j.biombioe.2011.04.042.

Venendaal, R., Jørgensen, U., Foster, C.A., 1997. European energy crops: a synthesis. Biomass Bioenergy 13 (3), 147–185. Available from: https://doi.org/10.1016/S0961-9534(97)00029-9.

Volk, T.A., Abrahamson, L.P., Nowak, C.A., Smart, L.B., Tharakan, P.J., White, E.H., 2006. The development of short-rotation willow in the northeastern United States for bioenergy and bioproducts, agroforestry and phytoremediation. Biomass Bioenergy 30 (8–9), 715–727. Available from: https://doi.org/10.1016/j.biombioe.2006.03.001.

Vree, J.H., Bosma, R., Jansen, M., Barbosa, M.J., Wijffels, R.H., 2015. Comparison of four outdoor pilot-scale photobioreactors. Biotechnol. Biofuels 8, 215.

Wang, B., Lan, C., 2010. Microalgae for biofuel production and CO2 sequestration. Nova Sci Pub Inc, New York978-1-61668-151-7, 168 pp.

Warburg, O., 1919. Über die Geschwindigkeit der photochemischen Kohlensäurezersetzung in lebenden Zellen. Biochem Zeit 100:230–270 Cited by Nickelsen K. (2007). Otto Warburg's first approach to photosynthesis. Photosynth. Res. 92, 109–120.

Whittaker, R.H., Likens, G.E., 1973. Primary productivity: the biosphere and man. Human Ecol. 1 (4), 357–369.

Wielgosz-Collin, G., Kendel, M., Couzinet-Mossion, A., 2016. Lipids, fatty acids, glycolipids, and phospholipids. In: Fleurence, Joël, Levine, Ira (Eds.), Seaweed in Health and Disease Prevention.) Elsevier, 978-0-12-802772-1pp. 185–221.

World Energy Council, 2016a. World Energy Resources: Waste to Energy, p. 74. Available at: https://www.worldenergy.org/wp-content/uploads/2013/10/WER_2013_7b_Waste_to_Energy.pdf.

World Energy Council, 2016b. World Energy Resources Bioenergy. Available at: https://books.google.com/books?id = WLC7CdLOZosC&pgis = 1.

Wright, L., 2007. Historical Perspective on How and Why Switchgrass Was Selected as a "Model" High-Potential Energy Crop, ORNL/TM-2007/109. Bioenergy Resources and Engineering Systems, Oak Ridge, TN. Available from: https://doi.org/ORNL/TM-2007/109.

Wright, T., Rahmanulloh, A., 2017. Indonesia Biofuels Annual Report 2017. Global Agricultural Information Network (GAIN) Report, Jakarta. Available at: https://gain.fas.usda.gov/Recent.GAIN.Publications/Biofuels Annual_Jakarta_Indonesia_6-20-2017.pdf.

Yang, L., He, Q.S., Corscadden, K., Udenigwe, C.C., 2015. The prospects of Jerusalem artichoke in functional food ingredients and bioenergy production. Biotechnol. Rep. 5, 77−88. Available from: https://doi.org/10.1016/j.btre.2014.12.004.

Zabed, H., Sahu, J.N., Suely, A., Boyce, A.N., Faruq, G., 2017. Bioethanol production from renewable sources: current perspectives and technological progress. Renew. Sust. Energy Rev. 71, 475−501. Available from: https://doi.org/10.1016/j.rser.2016.12.076.

Zacharof, M.P., 2017. Grape winery waste as feedstock for bioconversions: applying the biorefinery concept. Waste Biomass Valorization 8 (4), 1011−1025.

Zegada-Lizarazu, W., Monti, A., 2012. Are we ready to cultivate sweet sorghum as a bioenergy feedstock? A review on field management practices. Biomass Bioenergy 40, 1−12. Available from: https://doi.org/10.1016/j.biombioe.2012.01.048.

Zhang, R., El-Mashad, H.M., Hartman, K., Wang, F., Liu, G., Choate, C., et al., 2007. Characterization of food waste as feedstock for anaerobic digestion. Bioresour. Technol. 98 (4), 929−935. Available at: https://www.sciencedirect.com/science/article/pii/S0960852406000940?via%3Dihub (accessed March 23, 2018).

Zhu, L., 2015. Biorefinery as a promising approach to promote microalgae industry: an innovative framework. Renew. Sust. Energy Rev. 41, 1376−1384.

Zittelli, G.C., Biondi, N., Rodolfi, L., Tredici, R., 2013. Photobiorectors for mass production of microalgae. In: Richmond, A., Hu, Q. (Eds.), Handbook of Microalgal Culture. Wiley Blackwell, Chichester, pp. 225−266.

Zörb, C., Lewandowski, I., 2018. Biobased resources. In: Lewandowski, I. (Ed.), Bioeconomy. Shaping the Transition to a Sustainable, Biobased Economy. Springer International Publishing, Cham, Switzerland, pp. 77−86. Available from: https://doi.org/10.1007/978-3-319-68152-8_5.

CHAPTER THREE

Liquid Biofuels

Mercedes Ballesteros and Paloma Manzanares
Biofuels Unit, Energy Department, CIEMAT, Madrid, Spain

Contents

3.1 Introduction 113
3.2 First-Generation Technologies for Liquid Biofuels Production 114
  3.2.1 Feedstocks: conventional sugar, starch, and oil-containing crops 114
  3.2.2 Conversion technologies 121
3.3 Second-Generation Technologies for Liquid Biofuels Production 130
  3.3.1 Advanced feedstocks 130
  3.3.2 Advanced conversion technologies 132
3.4 Prospective Technologies 138
3.5 Challenges, Opportunities, and Barriers 139
3.6 Concluding Remarks 141
References 141

3.1 INTRODUCTION

According to the International Energy Agency (International Energy Agency Statistics, 2016b), 29% of all carbon dioxide (CO_2) emissions in OECD countries come from transport and 23% in a world basis (International Energy Agency Statistic, 2016a). However, the global demand for transport seems to continue growing (40% by 2035), according to the World Energy Outlook (Energy Information Administration, 2016). Thus, policy makers need to put into action measures aiming at limiting emissions from transport in accordance with mitigation of global warming strategies agreed at the world scale. In this context, renewable liquid biofuels have been promoted as a feasible and promising way of reducing CO_2 emissions from the transport sector.

Liquid biofuels, such as fossil fuels, are presented in various forms to be used in different internal combustion engines. In the initial phase of biofuels development, they were divided based on the key properties

The Role of Bioenergy in the Emerging Bioeconomy.
DOI: https://doi.org/10.1016/B978-0-12-813056-8.00003-0

of the products: biodiesel and, more recently, hydrogenated vegetable oils (HVO) to replace diesel and bioethanol to substitute petrol. However, concerns about the impact of biofuels on food security have led to a new classification based on generation categories: first-, second-, third-, and fourth-generation biofuels. In general terms, first-generation (1G) biofuels are produced from feedstocks that can also be consumed as human food or animal feed such as sugar, starch, or vegetable oil (VO). Since 1G biofuels are easily obtained using conventional technology, they are also known as "conventional biofuels." Second-generation (2G) fuels are produced from sustainable feedstock, mainly of lignocellulosic type, not normally used for human consumption. Since these lignocellulosic feedstocks are more difficult to transform into useful biofuels and more complicated conversion technologies are needed, 2G biofuels are also known as "advanced biofuels." The term third-generation (3G) refers to biofuel derived from algae. Previously, biofuel production using algae as feedstock was included in 2G biofuels. However, some key advantages over other feedstock (extraordinary diversity and higher yields with lower resource inputs) have justified its own separate category. Fourth generation (4G) is the latest subdivision of biofuels. It makes reference to the revolutionary "solar to fuel" concept. 4G biofuels are produced by metabolic engineering of photosynthetic microorganisms using novel synthetic biology tools. This new generation of biofuels is just emerging at the basic research level.

3.2 FIRST-GENERATION TECHNOLOGIES FOR LIQUID BIOFUELS PRODUCTION

3.2.1 Feedstocks: conventional sugar, starch, and oil-containing crops

1G liquid biofuels are produced by conventional mature technologies at a commercial scale. They include sugar and starch-based ethanol and oil crop–based biodiesel and VO. This section gives a general view about the major crops containing sugar, starch, and oil used for biofuels production, including data of agronomic characteristics, main production regions, productivity, and other relevant information in relation to its role as biomass feedstocks in the energy market.

3.2.1.1 Sugar-containing crops

Sugarcane (*Saccharum officinarum* L.) is one of the most important commercial crops for the manufacture of sugar, bioethanol, and bioenergy in the world due to its strong agricultural production system and processing infrastructure (Long et al., 2015). Moreover, based on its high productivity and conversion yields, it shows a positive net energy balance and a lower CO_2 emission than fossil fuels when it is used to obtain ethanol fuel, making it a leading feedstock for this purpose. Sugarcane is a perennial C4 plant with high capability to convert solar radiation into biomass and many positive features such as high sugar content, good juice purity, and good ecological adaptability and disease resistance. The most convenient climate for the cultivation of sugarcane is that having two different seasons: one warm and humid to promote germination and vegetative growth, followed by other dry and cold to reach maturity and, thus, sucrose accumulation on the stalks. Although the crop requires large quantity of rainfall, well-distributed rainfall indexes over 800 mm are enough to reach good productivities (Horta Nogueira, 2008). These specific climate conditions hinder the cultivation of this crop in areas with temperate climate such as Europe.

According to recent FAO statistics (Food and Agriculture Organization of the United Nations (FAO), 2017), 1890 billion tons were produced in the world in 2016, corresponding to 26,774,304 ha, with Brazil, India, China, and Thailand accounting for 38.2%, 18.5%, 6.2%, and 4.8%, respectively, of the world crop area. Fig. 3.1 shows sugarcane production values in 2016 in the 10 top producer countries, with Brazil peaking at 768,678,382 tons (40.8% of world production).

The aerial part of the plant is mainly composed of stalks, in which sucrose is accumulated, and tips and leaves that constitute sugarcane straw or trash. Approximately one-third of the energy content in sugarcane is contained in the sucrose fraction present in stalk, while another third is present in the bagasse generated after juice extraction and the last third is the field straw (Long et al., 2015). Sugarcane composition varies widely depending on a series of parameters such as climatic and soil conditions, maturity stage, variety, and others. A general sugarcane mass balance is 8%–14% fiber and 86%–92% juice, the latter containing 10%–25% soluble solids (Long et al., 2015).

Sugar beet (*Beta vulgaris* L.) is cultivated mainly in industrialized countries to provide about 16% of the world sugar production. In 2016, world production was 277 million tons from close to 45 million ha

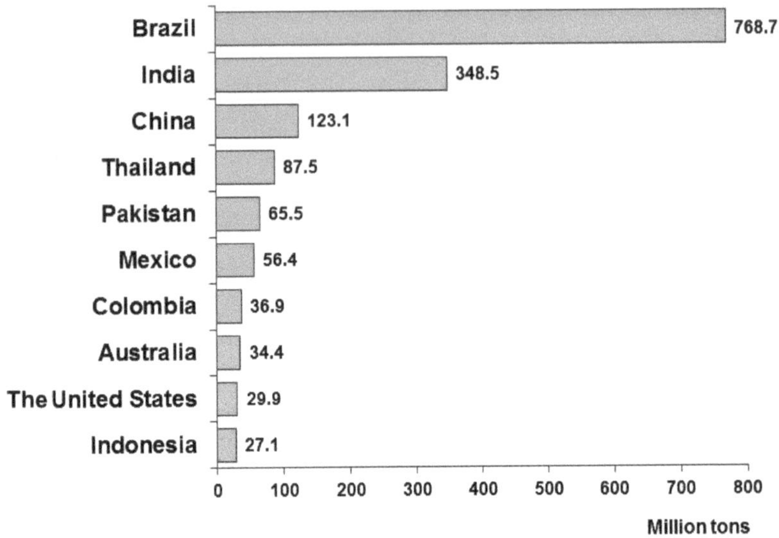

Figure 3.1 Top 10 sugarcane producing countries in the world (2016) (Food and Agriculture Organization of the United Nations (FAO), 2017).

(Food and Agriculture Organization of the United Nations (FAO), 2017). Russian Federation is the leader in sugar beet production accounting for 18.5% of the total, i.e., 51,366,830 tons in 2016, followed by France and the United States (12%), and Germany (9.2%).

Sugar beet is a biennial crop believed to originate from Asia than can grow in different climates (temperate and humid) and well-aerated deep soils. According to FAO data (Punda, 2009), sugar beet yield ranges between 55 and 65 tons/ha and the average sugar content is 14%–16%, which renders sugar yields from 6.6 to 7.8 tons/ha (80% sugar recovery after extraction) under ideal conditions.

Ethanol fuel from sugar beet is produced in European countries such as Germany and France (EurObserv'er Biofuels Barometer, 2017), while in other important producers such as the United States, no ethanol is currently produced from sugar beet. According to TEREOS (2017), the second largest European processor of sugar beet, 1 ton of beet produces 90–100 liters of alcohol on average and sugar beet pulp and molasses as by-products, widely used as feed supplements for livestock.

Sweet sorghum [*Sorghum bicolor* (L.) Moench] is a C4 annual plant native from Northeast Africa, very efficient in water use and tolerant to drought and flooding. Sweet sorghum has very good potential as feedstock for ethanol production in areas where natural precipitation does not fall below 350–450 mm. The optimum growing temperature is in the range from 27°C to 30°C. Some advantages of this crop is that it is a short-day plant with a growing period of about 4 months that makes it suitable to intercropping with early maturing crops (Cifuentes et al., 2014).

Sweet sorghum is widely cultivated in some regions of the world as the United States, Brazil, India, China, Mexico, Sudan, Argentina, and many other countries in Asia and Europe, mainly for use in sweeteners, primarily in the form of sorghum sirup. Regarding productivity, different values can be found in literature ranging from 49–56 tons/ha in the United States (Regassa and Wortmann, 2014) to 60–100 tons/ha in Mexico (Masera Cerutti et al., 2006). Sweet sorghum can accumulate juice up to 78% of the total biomass, whereas the sugar concentration has been estimated to range between approximately 16% and 18%, mainly comprising sucrose (70%), glucose (20%), and fructose (10%) (Cifuentes et al., 2014). The extraction process of sugar-rich juice from stalks is similar to that of sugarcane, and so a fiber-rich bagasse is also generated that could be used to produce power or cellulosic ethanol and other value-added coproducts (Nghiem et al., 2016).

3.2.1.2 Starch-containing crops

Maize and other cereals: At a global level, maize (*Zea mays* L.) is in the second position in the crop production sector, with a yield in 2016 of 1.060 million tons from close to 188 million ha (Food and Agriculture Organization of the United Nations (FAO), 2017). It currently constitutes one of the most important food sources for human consumption as well as the main crop for bioethanol production. The United States is the leader (36.3%), followed by China and Brazil, with 21.9% and 6.0%, respectively, of the total world production. Europe is the third maize-producing region in the world, accounting for 117 million tons in 2016, with France in the leading position (10.2%) (Food and Agriculture Organization of the United Nations (FAO), 2017).

The maize plant is a tall annual grass with a stout, erect stem up to 4 m tall and large narrow leaves. Different maize varieties can be grown in climates ranging from temperate to tropic during the period when mean daily temperatures are above 15°C and frost-free. Regarding water needs, according to the FAO (Food and Agriculture Organization of the United Nations (FAO), 2017), for maximum production, a medium maturity grain crop requires between 500 and 800 mm of water depending on the climate. The same source establishes that a good commercial grain yield under irrigation is 6–9 ton/ha (10%–13% moisture).

Currently, the United States is the world's largest producers of bioethanol based on maize grain. According to recent data from Renewable Fuels Association (Renewables Fuels Association, USA, 2017), in 2016 ethanol production picked at 15,250 million gallons, 95% based on maize starch and only 3% on other cereals such as sorghum, barley, or wheat starch.

Other grain cereals such as wheat (*Triticum aestivum* L.), barley (*Hordeum vulgare* L.), and oat (*Avena sativa* L.) can be used for ethanol production following the same process technology than for maize. Wheat production is the third largest cereal production in the world, after maize and rice, where cereals being the most important food source for human consumption. Wheat grain is mostly used for bioethanol in the European Union (EU), with significant capacity in Spain, Germany, France, and the United Kingdom (EurObserv'er Biofuels Barometer, 2017).

The use of food crops for energy purposes has created a strong debate at global level, and international organizations such as FAO have stated that the use of cereal feedstocks in the fuel ethanol sector have put additional pressure on prices, while creating incentives to increase production (FAO, IFAD, WFP, 2015). However, according to the European Renewable Ethanol Association (European Renewable Ethanol Association, 2016), in 2015, only 2%, or 10.3 million tons, of EU cereal supplies were used for ethanol production, representing 0.4% of global cereal production.

Cassava (*Manihot esculenta* Crantz) is a shrub with tuberous starchy roots that can be considered an interesting feedstock for ethanol production in tropical and subtropical countries. It is a perennial plant (1–5 m height), highly tolerant to acid soils and very efficient in the use of water and soil nutrients. The roots contain about 60% moisture and 20%–32% of starch at maturity, together with other minor components such as ash, protein, and fat (Howeler et al., 2013). As in the case of cereals, the use of cassava for nonfood purposes must be carefully evaluated, as it is among the most important food crop in Asia, Africa, and Latin America, after rice and maize.

3.2.1.3 Oil crops

Feedstocks used for biodiesel production are oils from VO crops cultivated worldwide for food or chemistry industry. The oil crop sector has been one of the most dynamic parts of world agriculture in recent decades, with an average global production in 2014–16 of 465.5 million tons (79% for food purposes and 13% for biofuels), according to OECD data (OECD-FAO, 2015). Major crops are soybean, rapeseed, sunflower, peanuts, cottonseeds, palm kernel, and coconut, being produced mainly in Indonesia, China, Malaysia, Europe, the United States, Argentina, and Brazil. Palm, soybean, rapeseed, and sunflower account for 80% of the VO produced worldwide while peanut, cotton, coconut, and olive constitute the next 11% share (Bart et al., 2010).

Currently, the production of biodiesel is basically based on a few VOs, mainly from rapeseed oil and, to a smaller extent, from soybean oil and sunflower seed oil in the EU; soybean oil in the Unites States, Argentina, and Brazil; and palm oil from Indonesia and Malaysia. The demand for VO from the biofuel sector has been considered one of the main drivers of consumption growth in the last few years (FAO. Food Outlook, 2015).

Soybean (*Glycine max* (L.) Merr.) is an annual shrub originating from China, which is nowadays one of the most important crops in the world for the production of oil and protein. In 2016, close to 335 million tons of soybean were produced globally from 121 million ha (Food and Agriculture Organization of the United Nations (FAO), 2017). The United States, Brazil, Argentina, and China are among the largest producers in the world.

Soybean plants are adaptable to a wide range of environments, including Europe, due to the existence of many different cultivars, each having specific characteristics and preferred conditions. Good yields can be obtained in warm areas of humid subtropic climate with temperatures 20°C–25°C and rainfall 500–700 mm. Under good conditions, soybean yields vary between 1.5 and 3.5 ton/ha seed (Food and Agriculture Organization of the United Nations (FAO), 2017). Seed oil concentration ranges from 83 to 279 g/kg with a mean of 195 g/kg, while seed protein concentration ranges from 341 to 568 g/kg of total seed weight, with a mean of 421 g/kg (Food and Agriculture Organization of the United Nations (FAO), 2017).

Rape (*Brassica napus* L.) is an annual plant that originated in the Mediterranean region that currently represents the major source for biodiesel production in Europe. At global production level, Canada, China, and India are the leading rapeseed-producing countries in the world, representing 58.8% of the total world production, estimated close to 69 million tons in 2016 (Food and Agriculture Organization of the United Nations (FAO), 2017). These seeds contain around 43% oil.

Sunflower (*Helianthus annuus* L.) is native from Central and North America. Lately, its importance as an oil crop has grown, and at present, it is the second most important oil crop together with soybean. It grows in warm climates on a variety of soils ranging from sandy to clay, with pH ranging from 6.5 to 8. Oil content of seeds ranges from 27% to 37% and can reach 45%. Due to the fact that sunflower oil is rich in polyunsaturated acids (oleic and linoleic), its use for human consumption has greatly increased. Total annual world production in 2016 reached 47 million tons of seed from 26 million ha (Food and Agriculture Organization of the United Nations (FAO), 2017), in which Ukraine, Russia, and Argentina are the leading countries in the world.

Palm (*Elaeis guineensis* Jacq.) is a prime feedstock for biodiesel in tropical countries and is produced for biofuels in Indonesia, Malaysia, and other countries of Southeast Asia. The plant is native from the tropical rain forest region of West Africa. During thousands of years, oil palm fruits have been processed for edible oil in Africa. According to GreenPalm (Green Palm: Supporting Sustainable Palm Oil, 2016), in 2015, Indonesia was the largest producer of palm oil (33.4 million tons), followed by Malaysia (19.9 million tons). Both countries account for 84% of the world palm production. There has recently been an increase in palm oil production in South America via Colombia, Ecuador, and Guatemala due to the global growing demand of edible VOs.

Palm oil was the highest consumed oil in 2012, reaching three billion people in 150 countries. Although its primary use remains for food, palm oil is increasingly being used as feedstock for biodiesel. However, this large development of the palm oil industry for biodiesel, particularly in Indonesia and Malaysia, has also generated considerable environmental and social costs and a great concern in nongovernmental organizations and public bodies. This has led to critical measurements against this expansion (i.e., Norway has recently banned palm oil—based biofuel in its public procurement) and the search of crop-specific sustainable certification standards to promote the growth and use of sustainable palm oil products such as the Roundtable on Sustainable Palm Oil (Round Table on Sustainable Palm Oil, 2018).

The use of nonfood crops for biodiesel production would provide the advantage of no competition in the food market and the possibility to enlarge cultivation areas and products with the associated social and economic benefits. In the last decades, intensive research has been carried out to identify and improve alternative oil-producing species that minimize competition with the food sector. One of the species that has raised greatest attention in this regard is *Jatropha curcas*. The plant, native from Central America and Africa, can reach up to 5 m height and yield 6–8 tons seed/ha (Galaz-Ávalos et al., 2012). It grows well in arid and semiarid soils, and in uncultivated land allows recovering nutrients and recovering soils affected by erosion. The jatropha industry covers a global area estimated at some 900,000 ha, more than 85% in Asia, chiefly Myanmar, India, China, and Indonesia. Africa accounts for approximately 120,000 ha, while Latin America has approximately 20,000 ha of jatropha, mostly in Brazil.

Jatropha as feedstock for biodiesel has raised great attention in the last years and countries such as India, the Philippines, and a number of African countries have started programs to grow this crop on a large scale. However, experts recognize the existence of drawbacks and many knowledge gaps concerning the best production practices for cultivation and the possible benefits and risks to the environment that make difficult to determine the true potential of this crop (Brittaine and Lutaladio, 2010).

Another example of alternative raw material for biodiesel production is the oil-containing seed plant *Camelina sativa*, native from the north of Europe. Camelina, also known as false flax, is an annual plant that grows optimally in temperate climates with reported yields from 450 to 1300 kg/ha in the United States and 2880 kg/ha in Eastern Europe (Obour et al., 2015). According to Ciubota-Rosie et al. (2013), the lipid content of individual seeds ranges between 35% and 45% with 27%–32% protein content. This high oil content and the fact that it is nonedible in terms of human consumption make camelina an attractive crop for biodiesel production via transesterification and as a renewable jet fuel through hydrodeoxygenation/hydrocracking (Ciubota-Rosie et al., 2013). Also, castor bean (*Ricinus communis* L.) used for many years as an industrial oilseed crop (~50% seed oil content) has been evaluated as an alternative feedstock for biodiesel production (Berman et al., 2011).

3.2.2 Conversion technologies

1G biofuels are currently the most developed and according to latest official statistics (Renewable Energy Policy Network for the 21st Century, 2017), annual global production of biofuels reached 98 billion liters of ethanol, 30 billion liters of biodiesel, and 4.9 billion liters of HVO. As it has been mentioned previously, typical feedstocks include food/feed crops such as corn, sugarcane, sugar beet, wheat, and VOs. Fig. 3.2 summarizes the technology routes for 1G biofuel production considering different feedstocks.

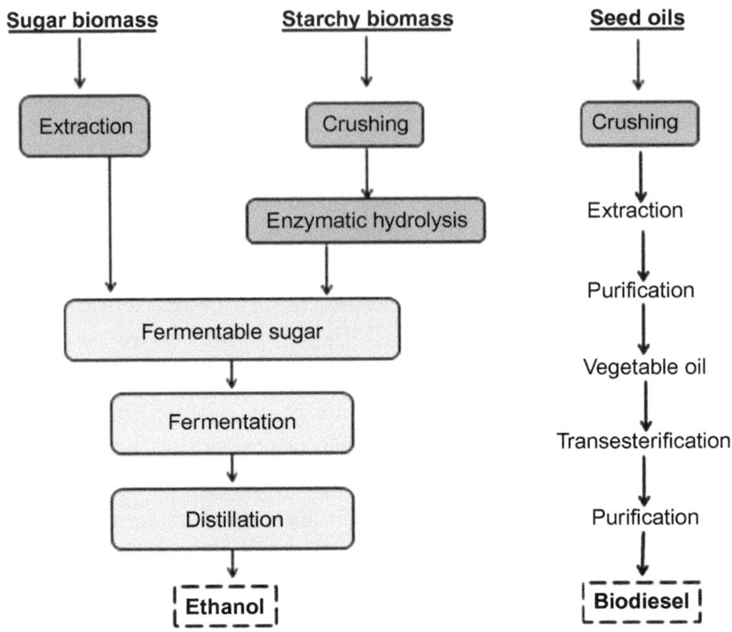

Figure 3.2 Technology routes for 1G biofuels production.

3.2.2.1 First-generation bioethanol

The production of 1G bioethanol is a well-established technology introduced at large scale, especially in the United States and Brazil. It can be produced by yeast fermentation from any organic raw material containing significant amounts of sugars or compounds that can be transformed into sugars, such as starch or inulin.

Ethanol from sugar: The simplest process to produce ethanol is based on the use of feedstocks containing sucrose such as sugarcane, beet, and sorghum. Sucrose (a nonreducing disaccharide composed of glucose and fructose) can be transformed directly into ethanol by conventional fermentative yeasts without any previous hydrolysis step. Usually, ethanol is commercially produced by the fermentation of sugarcane juice and molasses in a more or less integrated way with sugar production. The relative price of sugar and market demands determine the production ratio of each product. In Brazil, for instance, approximately 48% of the crop was diverted to sugar and 52% to ethanol in 2017 (Renewable Energy Policy Network for the 21st Century, 2017). In general, ethanol/sugar mills operate about 200 days/year and the average yield is about 90 L/ton sugarcane, which means around 7500 L/ha (Lopes et al., 2016). Ethanol installed industrial capacity depends on annual decisions made by individual plants to produce sugar and/or ethanol. Brazilian sugarcane industry responds to the theoretical ratio of 40:60 to switch between sugar to ethanol production or vice versa from harvest to harvest. As sugar prices increase, more sucrose can be derived to sugar production, and consequently fewer raw material remains to produce ethanol.

Various alternatives are used to obtain ethanol from sugarcane. Direct fermentation of primary juice, secondary or clarified sirups, molasses, and even mixtures of them are used as sugar-containing feedstocks. Process to produce ethanol is a well-known technology comprising basically the steps of extraction and treatment of juice, microbial fermentation of sugars, distillation, and, eventually, dehydration (PennState, 2017). The general production process of sugar and fuel bioethanol from sugarcane is shown in Fig. 3.3.

In summary, the cane is washed and shredded into smaller pieces of 20–25 cm. New advances in the cleaning technology focused on the use of "dry wash" are being implementing to reduce water consumption

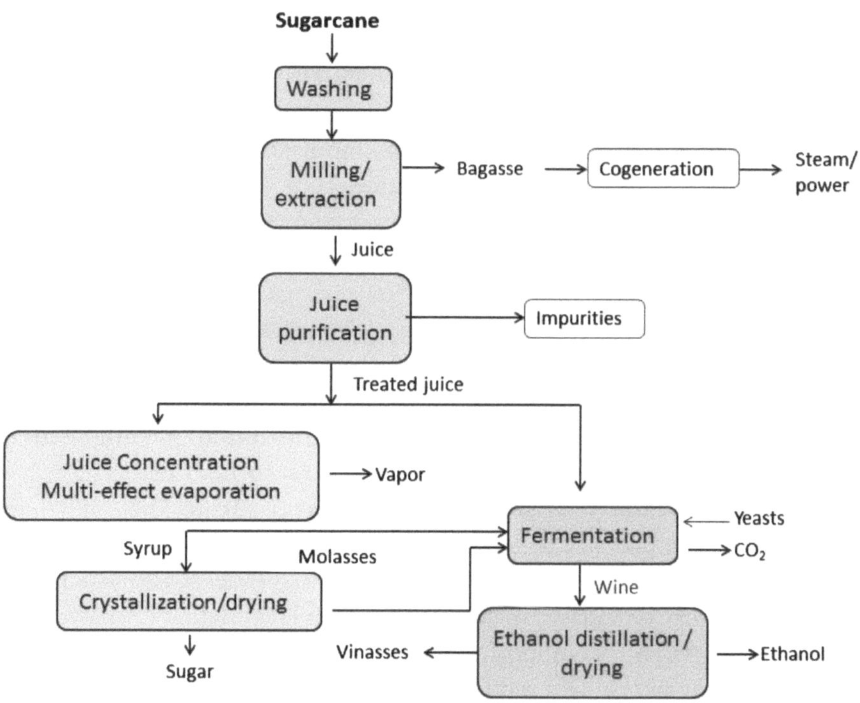

Figure 3.3 Sugar and fuel bioethanol production process from sugarcane.

(Goldemberg et al., 2018). The cane, prepared by the choppers, goes to the mills and, by pressure or diffusion, the juice is extracted, collected, and stored in tanks and the bagasse is removed. To improve the extraction of the sucrose contained in the stems, some water is added in the milling train. Subsequently, the cane juice is filtered and treated by chemicals. A secondary and increasingly important objective of the milling process is the production of bagasse with a low moisture content to be used as by-product as boiler fuel to produce steam and power.

Sugars are transformed into ethanol and CO_2 by the yeast *Saccharomyces cerevisiae* in the fermentation process. Fermentation time varies from 4 to 12 hours and process yield range from 80% to 90%, resulting in a mixture of diluted alcohol (7%–10% v/v) and yeast, called "wine." The "wine" is pumped towards the distillation columns where ethanol is concentrated to approximately 75% ethanol content and goes into the rectifying tower. Residual liquids and solids known as vinasse come out from the bottom of the column and are pumped into the treatment system. In the rectifying tower, the alcohol is concentrated to its azeotrope (96.5% v/v) and leaves the tower at the top. Optionally, the azeotropic ethanol enters in the dehydration section to remove the remaining water and increases ethanol purity up to 99.5%.

The use of coproducts (bagasse and vinasses) from ethanol industry is a key factor regarding the economy and sustainability of the production process. Bagasse is defined as the fibrous biomass remaining after sugarcane stalks are crushed to extract the juice. In a sugar–ethanol industry, bagasse is used to feed the boilers to supply enough electricity and steam for the process to be self-sufficient and export power to the grid. Heat and power generation from bagasse is one of the key factors that make sugarcane industry more competitive than sugar beet industry (Goldemberg et al., 2018).

Vinasse, the other final by-product from sugarcane ethanol, is the liquid fraction generated from the ethanol distillation. It is composed of 93% water and 7% solids, and presents from low to high dark color, bad smell, acid pH, high temperature, high levels of sulfate salts, and high COD values. Vinasse has been mostly used as a liquid fertilizer for crops (ferti-irrigation), but risks of soil salinization have been associated with the land disposal of vinasse in the long term. Treatment of vinasse (i.e., anaerobic digestion or composting) prior to recycling it as fertilizer and irrigation water potentially reduces the environmental impact of applying a nutrient-rich suspension to soils.

Brazil is the world's largest sugarcane ethanol producer and a pioneer in using ethanol as a motor fuel. In 1975, Brazil established the Proalcool program with the aim of substituting imported petroleum by domestic ethanol. To implement the program, the Government created a full institutional framework, providing incentives and guaranteeing competitive prices to bioethanol producers to build new distilleries, autonomous or annexed to existing sugar mills. By 1999, production and marketing of ethanol was no longer regulated but supported by public policies (obligation of blending between 18% and 27.5% of ethanol with gasoline) (Belincanta et al., 2016).

According to UNICA (Brazilian Sugarcane Industry Association UNICA, 2017), Brazil produced 27.25 billion liters in the 2016–17 campaign. Most of this production is absorbed by the domestic market as either pure ethanol fuel (hydrated ethanol) or blended (dehydrated ethanol) with gasoline.

Sugar beets are feedstocks for the production of bioethanol in some European countries such as France, Germany, Belgium, and the Czech Republic. Although the production process is similar to that developed for sugarcane, the process is less energy effective because sugar beets do not have a by-product such as bagasse that can be used for energy production (Lehnberger, 2015).

Ethanol from starch: In the conventional processes to obtain ethanol from cereal grains, the starch is enzymatically hydrolyzed and fermented into ethanol. There are two main routes to obtain ethanol from cereal grains that differ basically in how the grinding process is performed (Fig. 3.4). The wet mill process is more versatile than the dry mill process, and it produces a greater variety of products (germ oil, gluten, etc.), which allows the industry to react better to market conditions. However, due to lower capital costs of dry milling plants, most of the commercial starch–ethanol production plants use dry milling technology.

In the wet milling process, the grain is soaked in an aqueous sulfur dioxide solution (steeping) to facilitate the separation of the different compounds (germ oil, fiber, gluten, and starch). The remaining starch is mixed with α-amylases and heated up to 120°C–150°C to promote liquefaction and prevent microbial contamination. Then, the mixture is cooled and, by the action of the enzyme glucoamylase, the liquefied starch is transformed into glucose, a simple sugar that can be fermented by yeast. After fermentation, the medium, containing about 10% (v/v) of ethanol, is transferred to the distillation columns and ethanol (96% purity)

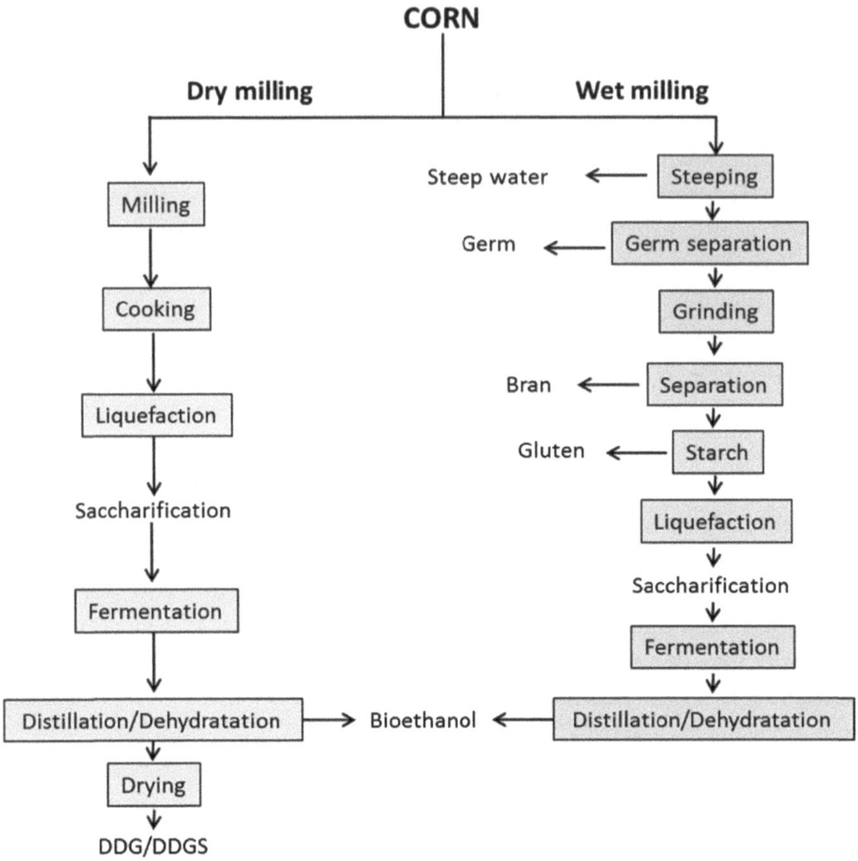

Figure 3.4 Process schemes for ethanol production from starch. *DDG*, dried distillers grains; *DDGS*, dried distillers grains with solubles.

is separated from the non-fermentable solids and water, and then, dehydrated in a molecular sieve unit. The solid is rich in protein and is used for animal feeding.

In the dry milling process, no separation of the different grain components is performed. Cereal grain is milled to obtain flour, mixed with water, and subjected to a cooking and liquefaction process to convert starch into dextrose. From here, the process is the same as those described above. Non-fermentable components (fibers, proteins, fats, vitamins, and minerals) contained in the cereal grain remain intact during fermentation process, and they are recovered in a highly concentrated form in the stillage from the distillation step. This mixture is centrifuged to separate granular solids from liquid (stillage). Stillage is evaporated to remove additional moisture resulting in condensed distiller's soluble (CDS), which contains approximately 30% dry matter. CDS can be sold locally as animal feed or combined with the solid fraction and dried to produce dried distiller's grains with solubles (DDGS). DDGS is a low-cost alternative animal feed ingredient due to its high energy, and high digestible protein and phosphorus content.

The United States is the world's largest corn bioethanol producer. Fuel bioethanol began to be used in the United States as a result of the approval in 1978 of the Energy Tax Act that introduced fiscal incentives for its production. However, the real boost to the bioethanol market was the environmental requirements demanded in the Clean Air Act of 1990 and the ban, in some states, to formulate gasoline with MTBE (methyl tertiary butyl ether). Currently, the use of ethanol is required by the federal Renewable Fuel Standard and the production capacity of ethanol is undergoing an unprecedented expansion, surpassing the Brazilian figures.

According to the Renewables Fuels Association, USA (2017), 200 ethanol plants were operating in 2016 (10% wet milling and 90% dry milling technology), producing more than 58 million of m^3, accounting for nearly 60% of global production and 42 million tons of high-protein animal feed.

The EU, although ranked third for world bioethanol consumption, is far behind the United States and Brazil. Wheat and corn are currently the main starch crops for bioethanol production in the EU. According to ePure (European Renewable Ethanol Association, 2016), 5.2 billion liters of ethanol, 78% of which was for fuel use, were produced in the EU in 2016.

The most common use of ethanol as a fuel is in mixtures of motor gasoline. If the ethanol content of motor gasoline does not exceed 10% by volume (E10), no modifications are required in the vehicle engine. On the other hand, flexible-fuel (flex-fuel) vehicles, with specially designed motors, can run on any mixture of ethanol and gasoline, from pure gasoline up to 100% ethanol. In North America and Europe, flex-fuel vehicles are optimized to use E85 (gasoline–ethanol blend containing 15% gasoline and 85% pure ethanol).

3.2.2.2 Biodiesel from VOs

VOs for biodiesel production are obtained by conventional methods from oleaginous seeds (sunflower, soybean, palm, and rapeseed), after a process including husking, cleaning, crushing, and conditioning. The cake obtained as residue of the pressing process has high protein content and, after conditioning and reprocessing, is commercialized for animal feed.

Although VOs can be used in adapted diesel engines, the high viscosity, poor thermal stability, and low cetane number make them unsuitable for use in conventional diesel engines. Biodiesel is a product derived from VOs with physicochemical and energy characteristics more similar to diesel and, therefore, with better engine performance and no modifications needed for diesel engines.

Biodiesel is made through a chemical reaction (transesterification) between VOs and fats and an alcohol, in presence of an acid, alkaline, or enzymatic catalyst. The transesterification process transforms large and branched triglycerol molecules into smaller and liner mono-alkyl esters of fatty acids with lower boiling point and viscosity (Fig. 3.5).

Alkaline-catalyzed transesterification (KOH, NaOH, and solutions of $NaOCH_3$ or $KOCH_3$) is the most widely used by the industry. However, the use of alkali implies the use of feedstock with low free fatty acid content (<0.5%) and anhydrous glycerides and alcohol (<0.06%) to avoid saponification reactions. Thus, alkaline-catalyzed transesterification is not suitable to be used on inexpensive feedstocks (i.e., fats and cooking oils) with high levels of free fatty acid. For free fatty acid raw materials, acid catalyst (H_2SO_4, HCl, or sulfonic acid) is most appropriate.

In both, base or acid catalysis, the utilization of homogenous catalysts, which dissolve in the liquid mixture in reaction, implies costly purification process (washing and drying) to ensure that the biodiesel meets

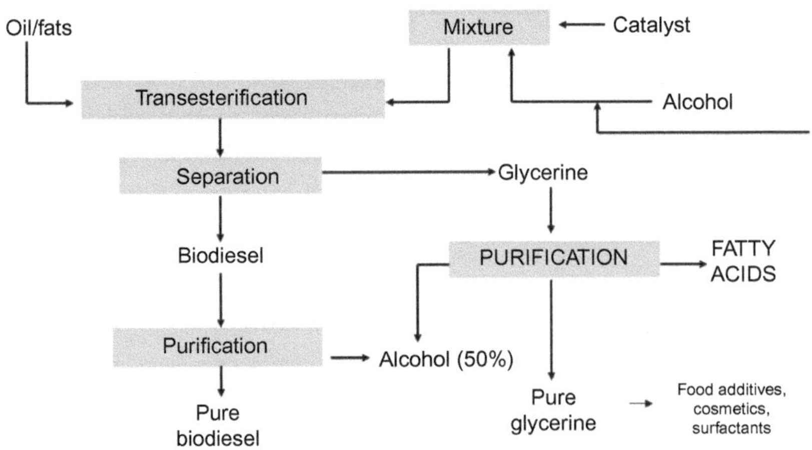

Figure 3.5 Process scheme for biodiesel production.

quality standards. On the other hand, glycerol has to be also purified to remove non-reactant alcohol and fatty acids, which can be reused as feedstock in the transesterification process or use the raw material to produce soap or other products.

Heterogeneous catalysis using alkaline earth alkoxides, mixed oxides, and immobilized lipases reduces the purification steps of biodiesel and glycerol and it is envisaged as a promising alternative to reduce capital and operation cost of biodiesel production (Semwala et al., 2011).

The EU is the world's largest biodiesel consumer. According to Eurobserver (EurObserv'er Biofuels Barometer, 2017), France, Germany, Italy, Sweden, and Spain are the top consumption countries. Other significant players are the United States, Brazil, Argentina, and Indonesia.

Apart from traditional biodiesel obtained from transesterification, a new biodiesel sector, HVO, came into the scene a few years ago derived also from VOs but using a different conversion process. HVOs, commonly referred to as renewable diesel and hydroprocessed esters and fatty acids (HEFA) are produced by a hydrogenation process using hydrogen as a catalyst instead of methanol (Fig. 3.6). Hydrogenation removes all the oxygen from the oils and HVO/HEFA are straight chain paraffinic hydrocarbons that are free of aromatics, oxygen, and sulfur and have a high cetane number. They can be mixed in any proportion with fossil fuel diesel.

Since gaseous biofuels and electrification are definitely no option for air transportation, liquid biofuels are the only renewable option to mitigate CO_2 emissions in the aviation sector. Hydroconversion of VOs is a potential processing route to obtain bio-jet fuel.

The 1G biodiesel and HVO/HEPA markets in Europe and the United States have reached a large production capacity in recent years, but its future growth remains limited by the availability of raw material. Therefore, feedstock diversification is the main driver for the biodiesel and HVO/HEPA industry.

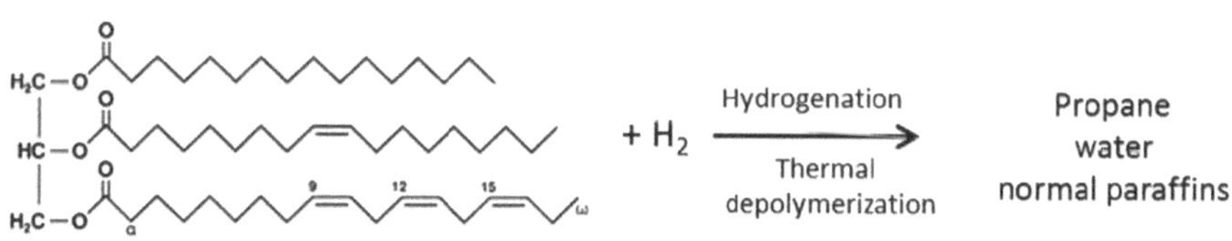

Figure 3.6 HVO production.

3.3 SECOND-GENERATION TECHNOLOGIES FOR LIQUID BIOFUELS PRODUCTION

3.3.1 Advanced feedstocks

Lignocellulosic biomass as woody and herbaceous crops specifically grown for bioenergy (energy crops), agricultural and woody residues, and some industrial wastes are the most promising advanced feedstocks for biofuel production. In fact, almost all biomass waste produced in agricultural, forest, and biomass-based industrial activities, including municipal solid wastes (MSWs), contain substantial lignocellulosic material that can be converted into biofuels.

Agriculture and forest residues are produced during production of food/feed crops and forest products (e.g., straw, corn stalks and leaves, or wood thinning from commercial forestry). Such biomass streams are typically found "in the field" and must be collected to be available for further use. In many instances, much of this residue needs to be left in the field to provide protection from erosion and it provides benefits such as micronutrient supplies, soil organic matter, and enhanced soil structure. However, in cases where land is relatively fertile and/or where conservation tillage methods are employed, a portion of the crop residues may be sustainably harvested (Searle and Malins, 2013).

Crop residues showing highest potential to be used for bioethanol production are corn stover, cereal (wheat/ barley and rice) straw and sugarcane straw, based on top crop production yield and cultivation area (Saini et al., 2015). Corn stover, predominantly generated in the United States from large corn cultivation, consists of cobs, husks, stalks, leaves, and tassel. According to Tan et al. (2012), corn stover may yield up to 7.6 t/ha, although it is necessary to consider the amount of residue required for either maintaining soil organic carbon level or preventing soil erosion.

Cereal straw, the aboveground part of the plant, comprises about half the total dry weight of the crop, and is currently used for feed and bedding of livestock or incorporated into the soil. However, surplus straw has been shown to have a high potential as biomass source in Europe based on its abundance and low cost (Kim and Dale, 2004).

Rice straw is the leftover of rice production, and it is one of the most abundant lignocellulosic waste materials in the world. Although it contains more silica than other agricultural residues as, i.e., wheat straw, it is considered as a potential feedstock for 2G bioethanol production. According to Saini et al. (2015), Asia alone produces more than 90% of annual global production of rice straw, which is estimated to be 731 million tons.

Sugarcane leaves and tops (straw, also called trash), which represents around one-third of the total primary energy of the sugarcane as a crop, has been normally burned after manual cane harvesting. However, manual harvest of sugarcane has been gradually substituted by mechanical operations with maintenance of the straw on the ground, in a system called green cane management, due to environmental, agronomic, and economic reasons (Regis et al., 2013). The possibility of using sugarcane and straw for 2G bioethanol is especially interesting for Latin American countries such as Brazil. While most Brazilian sugarcane mills use bagasse to produce bioelectricity by burning it in boilers making them energy self-sufficient, a growing number of mills also generates a surplus, which could be used for second ethanol production. Regis et al. (2013) have reported on the potential of sugarcane bagasse for 2G bioethanol production by different pretreatments and pathways for hydrolysis and fermentation, resulting in variable ethanol production yields from 180 to 190 L/ton of bagasse.

Forest residues from cleaning, thinning, or final felling of forest stands consist of small trees, branches, tops, and waste wood that are left in the forest after the harvesting operations. These residues, which are currently used worldwide as local fuel sources by combustion in wood burners or larger biomass boilers, could be considered potential feedstocks for biofuels production, especially in areas with high forest density. The main drawback is these forestry residues are bulky, difficult, and expensive to collect and transport and so nowadays have little commercial value. Searle and Malins (2013) estimated the potential of forestry residues in the EU in 80.7 million tons in 2011, with the Scandinavian countries being the only EU member states that harvest any significant quantities of them.

Most of the wood residues produced in the lumber industry are used to provide the energy needed for the production processes, although a part may be available for biofuel production. As it happens with sugarcane bagasse, these materials (bark, off-cuts, sawdust, shavings, etc.) are concentrated at processing plants, and thus no collection practices are needed. In the pulp and paper industry, a great part of the wood waste is used as boiler fuel. However, the black liquor residue that results from the pulping process requires expensive boilers for disposal and could be a potential source of biofuel feedstock. There has been an increasing interest in using black liquor residues for producing ethanol in the near term since there are substantial amounts of five-carbon (C5) and six-carbon (C6) sugars in this residue resulting from the hemicellulose portion of the woody feedstock used for pulping, being a good alternative to other higher value feedstocks (Wang et al., 2017).

The conversion of MSW into liquid biofuels has raised growing interest in the last decade due to its high production potential, low cost, and the possibility to offer a sustainable solution to its disposal. According to Malins et al. (2014), it is estimated that 44 million tons of MSW could be available in the EU in 2030. Several facilities at the industrial scale exist in industrialized countries to convert the organic fraction of MSW into ethanol based on both thermochemical (Enerkem, Fulcrum BioEnergy, and Solena Fuels) and biochemical (Fiberight, Perseo) technologies.

It is important to point out that, in general, biomass residues and wastes are linked with a complexity of markets. Many residues have useful applications as fodder, fertilizer, and soil conditioner, or are raw materials for different products, such as particleboard, medium-density fiberboard, and recycled paper. Thus, the availability of biomass residues and wastes has to be assessed considering those alternative uses.

Lignocellulosic biomass can be also obtained from dedicated energy crops such as perennial grasses and short-rotation coppice. Regarding herbaceous crops, tall grass species such as miscanthus (*Miscanthus sinensis*), switchgrass (*Panicum virgatum*), and reed canary grass (*Phalaris arundinaceae*) are examples of perennial crops that have been the focus of considerable interest in Europe and North America as feedstocks for 2G biofuels. Dedicated woody crops in Europe include species as poplar (*Populus* sp.), willow (*Salix* sp.), and eucalyptus (*Eucalyptus* sp.) in short rotation regime (also known as short rotation woody crops). In Latin America, forest species as pine and eucalyptus in short rotation cultivation are of special interest in countries with strong woody industry such as Chile, Brazil, and Uruguay. However, efforts to evaluate and develop energy crops are still in a relatively early stage of development, which reflects a situation where tremendous opportunities exist to use advanced plant science and agronomy to increase biomass yields.

3.3.2 Advanced conversion technologies

According to the European Industrial Biofuels Initiative (European Technology and Innovation Platform), advanced liquid biofuels are those

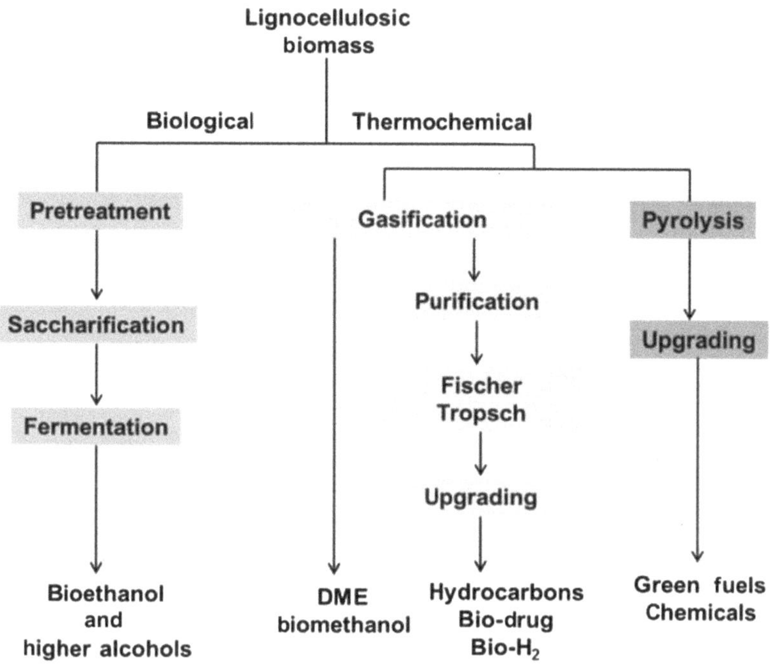

Figure 3.7 Processing routes for advanced biofuels production.

(1) produced from lignocellulosic feedstocks (i.e., agricultural and forestry residues, e.g., wheat straw/corn stover/bagasse, wood based biomass), nonfood crops (i.e., grasses, miscanthus, algae), or industrial waste and residue streams; (2) having low CO_2 emission or high greenhouse gas (GHG) reduction; and (3) reaching zero or low indirect land-use change (iLUC) impact.

The production of 2G biofuels from lignocellulosic feedstocks can be achieved through two very different processing routes (Fig. 3.7). In the biochemical route, enzymes and microorganisms are used to convert carbohydrates into sugars prior to their fermentation. The thermochemical route comprises two different processes: gasification to produce synthesis gas ($CO + H_2$) and pyrolysis to produce a bio-oil from which a wide range of long carbon chain biofuels that can be used as synthetic diesel or aviation fuel.

3.3.2.1 Biological processes

The most common biochemical conversion pathway used for biofuel production is yeast fermentation to ethanol. This technology involves the hydrolysis of biomass polysaccharides into fermentable sugars, and their subsequent fermentation to ethanol and other higher alcohols.

Cellulosic biomass has three primary components: cellulose, hemicellulose, and lignin. Cellulose is a polysaccharide consisting of chains of glucose monomers (1400–10,000 molecules), bundled into strong fibers that have an organized crystalline structure. Hemicellulose is a relatively amorphous component that is easier to break down than cellulose. It contains a mix of C6 and C5 sugars. The next most abundant polymer is lignin, which is composed predominantly of phenylpropane building blocks. Lignin performs an important role in strengthening cell walls by cross-linking polysaccharides, thus providing support to structural elements in the overall plant body. The crystallinity of cellulose and its association with hemicellulose and lignin are two key challenges that prevent the efficient breakdown of cellulose into glucose molecules that can be fermented into liquid biofuels.

Although cellulose hydrolysis can be performed by acids or enzymes, enzymatic hydrolysis has been demonstrated to be more suitable because no degradation components from sugars are formed and lower process temperature is required (Taherzadeh and Karimi, 2007). The basics of a whole enzymatic hydrolysis-based process comprises four main steps: (1) feedstock pretreatment, (2) enzymatic hydrolysis of cellulose to glucose, (3) fermentation of hexoses and pentoses to ethanol, and (4) distillation–dehydration of ethanol to standards required by fuel quality specifications.

Pretreatment step is vital to partially deconstruct the barrier constituted by the strong interactions among polymers of the cell wall, thus allowing the access of depolymerizing enzymes to cellulose. Development of an ideal pretreatment process is difficult, given that lignocellulosic biomass includes very different feedstocks (herbaceous, softwood, hardwood) with different compositions. Usually, pretreatment strategies have been categorized into biological (ligninolytic microorganisms or enzymes), physical (milling, extrusion), and chemical processes (acid or alkaline catalysts, ionic, liquids), or a combination of these approaches (steam explosion) (Alvira et al., 2010). Once the material has been pretreated, it is submitted to the enzymatic hydrolysis step by the action of specific enzymes called cellulases. The cellulases are divided into three sub-categories, representing three types of activity: *endoglucanases* (EGs), *exoglucanases* or cellobiohydrolases (CHBs), and *β-glucosidases* (βGs). For application to a biomass to ethanol process, cellulases are used as commercial enzyme preparations containing all the above-mentioned activities.

After enzymatic hydrolysis step, glucose released to the media is transformed by fermenting microorganisms. Historically, the most commonly used microorganism for ethanol fermentation has been *S. cerevisiae*. This yeast produces ethanol at a high yield and also has a high ethanol tolerance. In addition, *S. cerevisiae* has proven to be robust to other inhibitors, and hence it is suitable for fermentation of lignocellulosic materials. However, wild-type *S. cerevisiae* has limitations since it is able to ferment glucose coming from cellulose hydrolysis, but unable to metabolize xylose, which is the main hemicellulose-derived sugar from hardwood and agricultural residues. To overcome this drawback, during the last years, significant effort has been made to the development of engineered bacteria and yeast capable to ferment both type of sugars. Besides *S. cerevisiae*, microorganisms such as *Zymomonas mobilis* and *Escherichia coli* have been targeted through metabolic engineering for cellulosic ethanol production.

Cellulosic ethanol technology has reached an early commercial phase (BetaRenewables, DuPont, Poet-DSM, ST1, GranBio, Raizen, etc.), and the technology is ready to be replicated with competitive economics in the near future.

In the last decade, apart from cellulosic ethanol, there has been a growing interest in the production of higher alcohols (from 3-carbon propanol to 20-carbon phytol) from lignocellulosic biomass feedstocks, given their better fuel properties in comparison to lower alcohols (Rajesh Kumar and Saravanan, 2016).

3.3.2.2 Thermochemical processes

Biomass gasification—based pathway for liquid biofuels production is the most developed thermochemical process. Gasification is a process by which a carbonaceous, solid, or liquid material (biomass, coal, and oil) reacts with air, oxygen, and/or steam to produce a gas called synthesis gas (syngas). Syngas can be transformed into hydrocarbons via Fischer—Tropsch (FT) synthesis, or it can be further processed and used as feedstock for methane (Bio-SNG) or hydrogen (bio-H_2). Other conversion routes for syngas are the synthesis of bio-methanol or bio-di-methyl-ether (Bio-DME) (Fig. 3.8).

Before gasification, biomass is dehydrated, milled, and, in some cases, pretreated to ensure complete conversion. After pretreatment, the biomass is broken down into a syngas in the presence of heat, pressure, and a gasification agent (oxygen). Syngas is composed mainly of hydrogen (H_2),

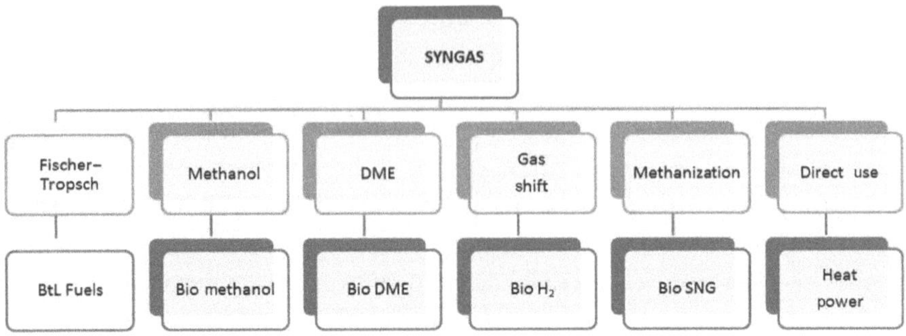

Figure 3.8 Conversion processes to produce biofuels from syngas.

carbon monoxide (CO), and carbon dioxide (CO_2). For biofuels production, CO and H_2 are the only valuable compounds and the presence of other hydrocarbons and inert components has to be kept as low as possible (Kavalov and Peteves, 2005).

Main gasification reactors (gasifiers) can be divided into three main groups: entrained flow gasifiers, fluidized bed gasifiers (bubbling/circulating), and fixed bed gasifiers. Fixed bed gasifiers are subdivided into counter-current (updraft), co-current (downdraft), and cross-current moving bed (Knoef, 2005). The main differences are related to the circulation of reactants and products in the reactor, and the resulting reaction conditions (atmospheric pressure or higher pressures).

Most syngas end-use pathways require the removal of contaminants to meet the requirements of the biofuels production process. The composition of the syngas depends on a number of factors such as both biomass and oxidizing agent characteristics, reactor design, and pressure and temperature during gasification. Depending on the further use of the syngas, certain H_2/CO ratios are needed, and this ratio can be increased via water−gas shift reactions, where CO and water are converted to hydrogen and carbon dioxide. CO_2 can be removed from the syngas by chemical and physical absorption with washing liquids or by adsorption with solid absorbents (Vessia, 2005). During the final step of the process, the cleaned syngas is transformed into biofuels. The best known synthesizing process is the FT synthesis, which consists of the formation of hydrocarbons of various molecular weights from CO and H_2, catalyzed by transition metals of Fe, Co, Ni, or Ru (Fe and Co at the industrial scale). The FT process is typically performed at 200°C−350°C and 25−60 bar (Opdal, 2006).

Methanol, di-methyl-ether (DME), methane, and $BioH_2$ can be also synthesized from syngas. Methanol production is currently done on an industrial scale exclusively by catalytic conversion of syngas at elevated temperatures and pressures. These processes use metallic catalysts (e.g., zinc, copper) to transform CO (and CO_2) with H_2 into methanol (Vessia, 2005).

DME is a novel high-cetane fuel, gaseous at ambient conditions originally synthesized from methanol in two steps via dehydration.

Biomethane is a by-product of gasification and, depending on process conditions, a gas with methane content of up to 30% can be obtained. Additionally, methane can be produced via catalytic (e.g., nickel) conversion of H_2 and CO at temperatures about 300°C (Vessia, 2005) or via anaerobic digestion (biogas).

Hydrogen (H_2) can be extracted from various compounds and via the syngas route. Currently, 77% of all hydrogen is produced mainly from natural gas and to a lesser extent from oil derivates. H_2 production from biomass is at an experimental stage due to the more sophisticated biomass conversion process, but also to the much lower relative hydrogen content of biomass compared to natural gas. Unlike all other automotive fuels, hydrogen is carbon-free and thus, its combustion and use in a fuel cell is very clean. Another advantage of hydrogen is the higher octane number, compared to petrol (Kavalov and Peteves, 2005).

Other promising technology for the conversion of large volumes of solid biomass into liquids is pyrolysis. In this process, biomass is thermally decomposed under inert atmosphere into solid (char), liquid (bio-oil), and gaseous products. Depending on the conditions of the pyrolysis process (residence time, heating rate, and temperature), the formation of one or the other product is favored. The use of high heating rates (fast pyrolysis), combined with moderate temperatures (around 500°C) and short residence times, leads to enhanced production of bio-oil. The pyrolysis oils to be used as fuels in diesel engines pose some problems still unsolved since it has low volatility and high viscosity and corrosiveness (Aubin and Roy, 1990). Consequently, biomass pyrolysis into liquid fuels suffers from important limitations that can be only overcome if the bio-oil is subjected to upgrading treatments to improve its properties. To improve their characteristics as engine fuel, bio-oils can be treated by hydro-deoxygenation (at temperatures between 300°C and 600°C with high pressure of H_2 in the presence of a heterogeneous catalyst), or by zeolites (crystalline microporous materials) to obtain advanced biofuels such as biogasoline and other hydrocarbon fractions (Sundaram et al., 2017).

3.3.2.3 Combination of thermochemical and biochemical technologies

Other interesting approach to obtain liquid biofuels from syngas is the combination of thermochemical and biological technologies. This combined process relies on an initial gasification step of wastes into synthesis gas or syngas and the subsequent transformation of the gaseous carbon into organic acids (acetate, formate, butyrate) and alcohols (ethanol, butanol) by carboxydotrophic acetogenic bacteria (i.e., *Acetobacterium woodii, Butyribacterium methylotrophicum, Clostridium carboxidivorans, Eubacterium limosu, Moorella,* and *Peptostreptococcus productus*) (De Tissera et al., 2017). The main advantage of the microbiological processes is the low operating temperature and pressure, the high tolerance of toxic gas compared to chemical catalysts, the high specificity to substrate, and the independent of H_2:CO ratio. The main disadvantage is the limited gas-to-liquid mass transfer rate requiring specific reactor designs.

Advanced liquid biofuels, due to their high specific energy, are especially interesting in the aviation sector. In fact, they are the only option for substituting kerosene and mitigate CO_2 emissions of jet fuels. In spite of the high-quality standards requirements for aviation fuels, ASTM-certified biofuels (HEFA, BtL, Iso-Paraffinic fuels, and Hydrogenated Pyrolysis Oils) represent no technical or safety problem as jet fuels.

3.4 PROSPECTIVE TECHNOLOGIES

3G and 4G biofuels from photosynthetic microorganisms are currently presented as a promising alternative to biofuels production by overcoming the major drawbacks associated with 1G and 2G biofuels. Microalgae are photosynthetic microorganisms with simple growing requirements (light, CO_2, N, P, and K and sugars) that can produce carbohydrates, lipids, and proteins over short periods of time.

Microalgae are cosmopolitan photosynthetic microorganisms and, compared with conventional crop plants, present superior photosynthetic performance and have a very short harvesting cycle ($\approx 1-10$ days depending on the process), allowing multiple or continuous harvests with significantly increased yields. Three main conversion processes, i.e., oil transesterification, ethanol fermentation, and anaerobic digestion and different methods for growing algae such as open ponds and closed photobioreactors, have been tested at laboratory and pilot scale. However, there are still many uncertainties in its implementation at a commercial scale, and it is unclear if current laboratory experiments and pilot programs can be scaled up (Ndubuisi Ezejiofor et al., 2014). Photobioreactors have proven to be most effective in producing high-quality algae, but they are expensive and the economic feasibility for commercial scale production has not been demonstrated. Open ponds are significantly cheaper but are susceptible to contamination by native algae species and evaporation. Many technical challenges, including photobioreactor design, biomass harvesting, and processing have to be addressed before the large-scale production of microalgae-derived biofuels can become a commercial reality. Beyond the technical challenges, issues such as land and water use will need to be addressed to commercialize liquid biofuels production from microalgae.

Other conversion routes are under development for the production of hydrocarbon fuels from biomass such as aqueous phase reforming (APR) and direct sugar to hydrocarbon (DSH) routes. APR was proposed for the first time by the group of Dumesic (Cortright et al., 2002) and involves the production of hydrogen and light alkanes from biomass-derived compounds such as glycerol, sugars, and sugar alcohols. The reactants of APR remain in liquid phase during the reaction and hydrogen production is feasible in a flow reactor at temperatures much lower ($\sim$225°C) than those required for pyrolysis or gasification, avoiding an energetically demanding vaporization step compared to processes such as steam reforming.

DSH processes transform sugars into paraffin molecule by fermentation. These processes are at early stage of development, being farnesene (branched hydrocarbon with multiple double bonds) the best developed and at commercial scale (MIT Technology Review, 2011).

3.5 CHALLENGES, OPPORTUNITIES, AND BARRIERS

The current global energy market, dominated by fossil fuels, had failed to offer long-term security of supply with minimal impact on the environment. While there are numerous renewable energy options for heat and electricity generation (wind, solar, hydro, etc.), transport sector is highly dependent on petroleum. Therefore, the development of alternative fuels for transport sector is currently a topic of growing interest and importance worldwide. Although renewable electricity is a promising option to decarbonize transport fuels, electric vehicles have a long way to go before making a significant contribution to GHG emission reduction targets (International Energy Agency, 2017). In this scenario, biofuels could technically substitute, in the short term, oil-derived fuels in all transport modes, since they are compatible with existing engine technologies and existing refueling infrastructures. Moreover, biofuels are the only means of displacing liquid fossil fuels in aviation, heavy duty vehicles, and maritime sectors where electro-mobility is not an option. However, the rapid expansion of 1G biofuels made from food/feed crops, and the need to produce raw feedstocks at a large scale, have become a source of concerns about negative environmental impacts such as deforestation, monocultures, loss of biodiversity, and direct and iLUCs. In addition, the debate over land competition for food and feed versus fuel, and the potential consequences of biofuel production on food security have questioned the sustainability of current biofuel systems in the long term. Nevertheless, liquid biofuels debate cannot ignore the current situation in which hundreds of millions of cars move around the world with petroleum derivatives and, whether we like it or not, that will continue to circulate for many years. It is essential, therefore, the introduction of renewable biofuels into transport fuels at the same time that minimize competition between food/feed and fuel crops and discourage expansion onto ecologically valuable lands to reduce agricultural and ecological risks.

The future evolution of liquid biofuels will be based on the use of nonfood feedstock that include agricultural, municipal, and forestry wastes as well as fast-growing, cellulose-rich energy crops. Beyond 2020, advanced biofuels will have a substantial share of new capacity development worldwide. However, the introduction of these advanced biofuels still requires some technological breakthroughs to reduce the production cost, produce large quantities, and/or improve biomass carbon utilization. In this context, the cascading use of biomass resources in a circular economy strategy can contribute to increase the productivity and efficient use of scarce and valuable raw material resources to increase the productivity and efficient use of biomass resources.

It should be highlighted that the promotion of sustainable biofuels production will create new economic opportunities in rural areas, and reduce local pollution and emissions of GHGs.

3.6 CONCLUDING REMARKS

In general, the production and use of biofuels in the transport sector presents a series of environmental, energy, and socioeconomic advantages over fossil fuels such as mitigation of GHG emissions, increase of security supply, and rural development. However, the current 1G biofuels situation have made imperative the development of advanced or 2G biofuels; or what is to say, those biofuels derived from dedicated plants or plant residues that do not enter into direct competition with land for growing food and feed crops.

Current technologies for the production of 1G biofuels will evolve towards more advanced 2G technologies based on the concept of biorefinery. The future biorefinery will integrate various processes of production of fuels, steam, electricity, and chemical products from biomass. Integrated biorefineries are the most promising way to create a new industry based on the use of biomass for obtaining food, safe and healthy feed, materials for the chemical industry, energy, and other products.

It is necessary to create the necessary instruments for a transition from a society based on fossil fuels to another one of a biological nature, with research and innovation as developing agents, in what has come to be called "bioeconomy"; that is, a more innovative and low-emission economy that merges the demands of sustainable management of agriculture and fisheries, food security, and the sustainable use of renewable biological resources for industrial purposes, while guaranteeing biodiversity and environmental protection.

REFERENCES

Alvira, P., Tomás-Pejó, E., Ballesteros, M., Negro, M.J., 2010. Pretreatment technologies for an efficient bioethanol production process based on enzymatic hydrolysis: a review. Biores. Technol. 101 (13), 4851–4861.

Aubin, H., Roy, C., 1990. Study on corrosiveness of wood pyrolysis oils. Fuel Sci. Technol. Intern. 8 (1), 77–86.

Bart, J.C.J., Palmeri, N., Cavallaro, S., 2010. Feedstocks for biodiesel production. Biodiesel Science and Technology: From Soil to Oil. Woodhead Publishing Series in Energy, Washington, DC, pp. 130–225.

Belincanta, J., Alchornel, J.A., Teixeira da Silva, J.M., 2016. The Brazilian experience with ethanol fuel: aspects of production, use, quality and distribution logistics. Brazilian J. Chem. Engin. 33 (4), 1091–1110.

Berman, P., Nizri, S., Weisman, Z., 2011. Castor oil biodiesel and its blends as alternative fuels. Biomass Bioenergy 35, 2861–2866.

Brazilian Sugarcane Industry Association (UNICA). Brazil, 2017. <http://www.unicadata.com.br/historico-de-producao-e-moagem.php> (accessed 15.07.17).

Brittaine, R., Lutaladio, N., 2010. Jatropha: A Smallholder Bioenergy Crop—The Potential for Pro-Poor Development, Integrated Crop Management, vol. 8. FAO, Rome, 96 pp.

Cifuentes, R., Bressani, R., Rolz, C., 2014. The potential of sweet sorghum as a source of ethanol and protein. Energy for Sustainable Development 21, 13–19.

Ciubota-Rosie, C., Ruiz, J.R., Ramos, M.J., Pérez, A., 2013. Biodiesel from *Camelina sativa*: a comprehensive characterisation. Fuel 105, 572–577.

Cortright, R.D., Davda, R.R., Dumesic, J.A., 2002. Hydrogen from catalytic reforming of biomass-derived hydrocarbons in liquid water. Nature 418, 964–967.

De Tissera, S., Köpke, M., Simpson, S.D., Humphreys, C., Minton, N.P., Dürre, P., 2017. Syngas biorefinery and syngas utilization. Advances in Biochemical Engineering/Biotechnology. Springer, Berlin, Heidelberg, pp. 1–34.

Energy Information Administration, 2016. International Energy Outlook <http://www.eia.gov/ieo> (accessed 12.11.17).

EurObserv'er Biofuels Barometer, 2017. <https://www.eurobserv-er.org/category/barometers-in-english/> (accessed 30.09.17).

European Renewable Ethanol Association, Belgium, 2016. European Renewable Ethanol—Key Figures. <http://epure.org/media/1610/2016-industry-statistics.pdf> (accessed 21.10.17).

European Technology and Innovation Platform, http://www.etipbioenergy.eu (accessed 15.08.17).

FAO, 2015. Food Outlook. Biannual Report on Global Food Markets. <http://www.fao.org/giews> (accessed 12.12.17).

FAO, IFAD, WFP, 2015. The State of Food Insecurity in the World 2015. Meeting the 2015 International Hunger Targets: Taking Stock of Uneven Progress. <http://www.fao.org/publications> (accessed 15.07.17).

Food and Agriculture Organization of the United Nations (FAO), 2017. <http://www.fao.org/faostat/en/#data/QC> (accessed 10.10.17).

Galaz-Ávalos, R.M., Avilez-Montalvo, R.N., Ucan-Uc, C.M., Chan-López, J.A., Loyola-Vargas, V.M., 2012. *Jatropha curcas* una alternativa para la obtención de biodiésel sin afectar al Sector Alimentario. BioTecnología 16 (2), 94–114.

Goldemberg, J., Coelho, S.T., Guardabassi, P., 2018. The sustainability of ethanol production from sugarcane. Energy Policy 36 (6), 2086–2097.

Green Palm: Supporting Sustainable Palm Oil, 2016. http://greenpalm.org/about-palm-oil/where-is-palm-oil-grown-2 (accessed 15.10.17).

Horta Nogueira, L.A., 2008. Sugarcane bioethanol. Sugarcane-Based Bioethanol: Energy for Sustainable Development. Banco Nacional de Deselvolvimiento Economico e Social and Centro de Gestao de Estudos Estrategicos, Rio de Janeiro, Brazil, pp. 65–68.

Howeler, R., Lutaladio, N.B., Thomas, G., 2013. Save and Grow: Cassava. A Guide to Sustainable Production Intensification. Food and Agriculture Organization of the United Nations (FAO), <http://www.fao.org/ag/save-and-grow/cassava/> (accessed 17.07.17).

International Energy Agency, 2017. Global EV Outlook 2017. <https://www.iea.org/publications/freepublications/publication/GlobalEVOutlook2017.pdf> (accessed 12.11.17).

International Energy Agency Statistic, 2016a. CO_2 Emissions From Fuel Combustion: Highlights. <http://www.iea.org/statistics> (accessed 12.11.17).

International Energy Agency Statistics, 2016b. Recent Trends in the OECD: Energy and CO_2 Emissions, <http://www.iea.org/statistics> (accessed 12.11.17).

Kavalov, B., Peteves, S.D., 2005. Status and Perspectives of Biomass-To-Liquid Fuels in the European Union EUR 21754. European Commission Directorate General Joint Research Centre, Petten, The Netherlands, 126 pp.

Kim, S., Dale, B., 2004. Global potential bioethanol production from wasted crops and crop residues. Biomass Bioenergy 26 (4), 361−375.

Knoef, H.A.M. (Ed.), 2005. Handbook Biomass Gasification. BTG: Biomass Technology Group, Enschede, The Netherlands, 400 pp.

Lehnberger, A., 2015. Challenges by transferring proven efficient design from beet to cane based on practical experience. Int. Sugar J. 878−882.

Long, S.P., Karp, A., Buckeridge, M.S., Davis, S.C., Jaiswal, D., Moore, P.H., et al., 2015. Feedstocks for biofuels and bioenergy. In: Souza, G.M., Victoria, R.L., Joly, C.A., Verdade, L.M. (Eds.), Bioenergy & Sustainability: Bridging the Gaps, 72. SCOPE, Paris, France, pp. 302−347.

Lopes, M.L., de Lima Paulillo, S.C., Godoy, A., Cherubin, R.A., Marcel Salmeron Lorenzi, M.S., Carvalho Giometti, F.H., et al., 2016. Ethanol production in Brazil: a bridge between science and industry. Brazilian J. Microbiol. 47, 64−76.

Malins, C., Searle, S., Baral, A., Turley, D., Hopwood, L., 2014. In: Harrison (coord), P. (Ed.), Wasted-Europe's Untapped Resource: An Assessment of Advanced Biofuels From Wastes and Residues. International Council of Clean Transportation (ICCT), <https://www.theicct.org/sites/default/files/publications/WASTED-final.pdf> (accessed 10.08.17).

Masera Cerutti O., Rodríguez Martínez N., Lazcano Martínez I., Horta Nogueira L.A., Macedo I.C., Trindade S.C., et al. In: Sener/BID/GTZ (Ed.), Potenciales y Viabilidad del Uso de Bioetanol y Biodiesel para el Transporte en México, México, DF, 2006. 260 pp.

MIT Technology Review, 2011. Sustainable Energy. Amyris Announces Commercial Production of Biochemicals. <https://www.technologyreview.com/s/423878/amyris-announces-commercial-production-of-biochemicals> (accessed 12.11.17).

Ndubuisi Ezejiofor, T.I., Enebaku, U.E., Ogueke, C., 2014. Waste to wealth—value recovery from agro-food processing wastes using biotechnology: a review. Br. Biotech. J. 4 (4), 418−481.

Nghiem, N.P., Montanti, J., Johnston, D.B., 2016. Sorghum as a renewable feedstock for production of fuels and industrial chemicals. AIMS Bioeng. 3 (1), 75−91.

Obour, K.A., Sintim, Y.H., Obeng, E., Jeliazkov, D.V., 2015. Oilseed Camelina (*Camelina sativa* L Crantz): production systems, prospects and challenges in the USA great plains. Adv. Plants Agric. Res. 2 (2), 1−10.

OECD-FAO, 2015. Agricultural Outlook 2006−2015, https://es.scribd.com/document/293993031/Food-and-Agriculture-Organization-OECD-OECD-FAO-BookFi-org (accessed 15.10.17).

Opdal, O.A., 2006. Production of Synthetic Biodiesel via Fischer-Tropsch Synthesis: Biomass to Liquids in Namdalen, Norway. Department of Energy & Process Engineering, Norwegian University of Science and Technology, Norway, 84 pp.

PennState, 2017. Colleague of Earth and Mineral Sciences. John A. Dutton e-Education Institute. EGEE 439: Alternative Fuels from Biomass Sources. Dr. Caroline B. Clifford, Instructor. <https://www.e-education.psu.edu/egee439/node/647> (accessed 10.08.17).

Punda, I., 2009. Agribusiness Handbook. Sugarbeet. White Sugar. Food and Agriculture Organization of the United Nations (FAO), Rome, 53 pp.

Rajesh Kumar, B., Saravanan, S., 2016. Use of higher alcohol biofuels in diesel engines: a review. Renew. Sustain. Energy Rev. 60, 84−115.

Regassa, T., Wortmann, C.S., 2014. Sweet sorghum as a bioenergy crop: literature review. Biomass Bioenergy. 64, 348−355.

Regis, M.L.V.L., Galdos, M.V., Scarpare, F.V., Seabra, J.E.A., Walter, A., Oliveira, C.O.F., 2013. Sugarcane straw availability, quality, recovery and energy use: a literature review. Biomass Bioenergy 53, 11–19.

Renewable Energy Policy Network for the 21st Century Renewables, 2017. Global Status Report. <http://www.ren21.net/status-of-renewables/global-status-report/REN21.net> (accessed 10. 08.17).

Renewables Fuels Association, USA, 2017. Annual Industry Outlook of Renewable Biofuels. <http://www.ethanolrfa.org/wp-content/uploads/2017/02/Ethanol-Industry-Outlook-2017.pdf> (accessed 10.11.17).

Round Table on Sustainable Palm Oil, 2018. <https://www.rspo.org/> (accessed 15.10.17).

Saini, J.K., Saini, R., Tewari, L., 2015. Lignocellulosic agriculture wastes as biomass feedstocks for second-generation bioethanol production: concepts and recent developments. Biotech 5 (4), 337–353.

Searle, S., Malins, C., 2013. Availability of cellulosic residues and wastes in the EU. White Paper. International Council of CleanTransportation (ICCT), <https://www.theicct.org/sites/default/files/publications/ICCT_EUcellulosic-waste-residues_20131022.pdf> (accessed 13.11.17).

Semwala, S., Arorab, A.K., Badonia, R.P., Tulib, D.K., 2011. Biodiesel production using heterogeneous catalysts. Biores. Tech. 102 (3), 2151–2161.

Sundaram, S., Kolb, G., Hessel, V., Wang, Q., 2017. Energy-efficient routes for the production of gasoline from biogas and pyrolysis oil—process design and life-cycle assessment. Ind. Engin. Chem. Res. 56 (12), 3373–3387.

Taherzadeh, M.J., Karimi, K., 2007. Enzyme based hydrolysis processes for ethanol from lignocellulosic materials: a review. Bioresources 2 (4), 707–738.

Tan, Z., Liub, S., Bliss, N., Tieszenb, L.L., 2012. Current and potential sustainable corn stover feedstock for biofuel production in the United States. Biomass Bioenergy 47, 372–386.

TEREOS, 2017. <http://tereos.com> (accessed 05.12.17).

Vessia, O., 2005. Biofuels From Lignocellulosic Material in the Norwegian Context 2010—Technology, Potential and Costs. Doctoral Thesis. NTNU, Norwegian University of Science and Technology, Trondheim, Norway, 101 pp.

Wang, W., Tan, X., Wang, Q., Yu, Q., Zhou, G., Qi, W., et al., 2017. Commentary on the reuse of black liquor for the enzymatic hydrolysis and ethanol fermentation of alkali-treated sugarcane bagasse. J. Microbiol. Biotechnol. 6 (2), 19–21.

CHAPTER FOUR

Solid Biomass to Heat and Power

Jorge Islas[1], Fabio Manzini[1], Omar Masera[2] and Viridiana Vargas[1]
[1]Instituto de Energías Renovables, Universidad Nacional Autónoma de México, Temixco, Morelos, Mexico
[2]Centro de Investigaciones en Ecosistemas, Universidad Nacional Autónoma de México, Morelia, Michoacán, Mexico

Contents

4.1 Introduction 145
4.2 Biomass for Solid Biofuels 150
  4.2.1 Wood biomass 150
  4.2.2 Herbaceous biomass 152
  4.2.3 Municipal waste 154
  4.2.4 Agro-industrial wastes 156
4.3 Solid Biomass Consumption 157
4.4 Solid Biomass Heat and Power Generation Technologies 158
  4.4.1 Combustion 160
  4.4.2 Co-combustion 162
  4.4.3 Gasification 163
  4.4.4 Combined heat and power 166
  4.4.5 Incineration 169
4.5 Thermal Energy and/or Electricity Generation Costs With Biomass 171
4.6 Conclusions 172
Acknowledgments 173
References 173

4.1 INTRODUCTION

Solid biomass is a biological material that stores energy through the process of photosynthesis in the presence of sunlight; it is derived basically from living organisms such as plants, crop residues, and animal excreta (Demirbaş, 2001; IRENA, 2015; McKendry, 2002a; Sansaniwal et al., 2017; Smith et al., 2014; Verna, 2013; WEC, 2016) and during its combustion can be used directly for heating or food cooking

The Role of Bioenergy in the Emerging Bioeconomy.
DOI: https://doi.org/10.1016/B978-0-12-813056-8.00004-2

(Demirbaş, 2001; WEC, 2016). Indirectly, solid biomass, or simply biomass from now on, is also used for electricity generation (WEC, 2016).

For the World Energy Council, bioenergy represents the transformation of organic matter, including biomass, into a source of energy, whether it is collected in a natural environment or cultivated specifically for this purpose (WEC, 2016).

Worldwide, biomass is the fourth most used energy resource, after oil, coal, and natural gas; it is estimated that it contributes between 10% and 14% to the world's energy supply (IPCC et al., 2012; McKendry, 2002a). It has been identified as a key fuel for the global transition to renewable energies. The inputs to produce bioenergy from solid biomass are very varied (agricultural and forest residues both primary and from the industries of these branches, byproducts of biological materials, wood, the organic fraction of municipal solid waste (MSW) and waste sludge from water treatment plants). Biofuels can be transformed, in turn, into heat and electricity through numerous routes or processes (Sansaniwal et al., 2017). On a global scale, the expansion of the use of biomass for energy purposes must take into account its lower energy density with respect to fossil fuels, the complexity of the supply chain (normally by-product of, but in case of dedicated crops in competition with the main uses of organic matter, such as food and materials), and other factors such as the conservation of biodiversity and land tenure (Bocci et al., 2014; IPCC et al., 2012).

Since ancient times, biomass, especially firewood, has been the main source of energy for cooking and heating. Currently, about 90% is used in the form of heat for cooking and heating in small industries, particularly in rural and peri-urban areas of developing countries (The use of distributed renewable energy in the cooking and heating sectors increased in 2015 due to advances in technology, increased awareness of deforestation, and increased government support. The most widely used technologies are the stoves (by the end of 2014, around the world, some 28 million households had adopted clean stoves, most of which were in Asia and Africa), and cooking with biogas (from of biomass waste that is produced in the biodigesters)), in a traditional way (firewood, charcoal, and agricultural waste) (WEC, 2016). It is estimated that in the near future biomass will continue to be the main source of primary energy for many of these countries, particularly those located in the sub-Saharan Africa (Sansaniwal et al., 2017). The traditional use of biomass for cooking includes 2600 million people (40% of the world population), a figure that represents approximately 9% of the global primary energy consumption,

55% of the world's wood harvest and an economic income of 74 billion dollars per year (Masera et al., 2015).

It should be noted that there is an additional classification of the sources of biomass that divide it into traditional and modern biomass, although the differences between both are not well defined. In several developed countries, firewood and other small-scale uses (traditional, in a certain sense) continue to be the main source of energy used for heating and cooking, a phenomenon that does not occur only in "isolated and rural areas" (Smith et al., 2014; WEC, 2016).

The so-called "modern" use of biomass differs from traditional use mainly because of two aspects: now it is sought that biomass comes from sustainable sources and that the technology used to obtain energy is efficient and clean to limit and mitigate emissions of combustion gases, as well as overcome the health and environmental problems associated with its inefficient combustion (IPCC et al., 2012; WEC, 2016).

It is expected that the efficient use of solid biomass will increase in the future, so there is an understandable concern about the sustainability of this energy worldwide. However, many initiatives are being implemented nationally and internationally to ensure that at least the biomass that is commercialized in the form of pellets, chips, charcoal, and other products is certified by an internationally recognized program (WEC, 2016).

The global production of solid biomass in 2014 was 27 million tons; North America and Europe accounted for 97% of all production volumes. In these regions, it is often also integrated with the pulp and paper industries. Modern biomass plants use fast-growing crops, trees, or other energy plantations, and forest residues collected in a sustainable manner. In some European countries, biomass is commonly harnessed with combined heat and power (CHP) technologies to increase energy efficiency. It should be noted that for the use of biomass in these cogeneration processes, a local consumer of the recovered heat is required, for example, an industrial plant or a heating district, as is currently done in the European districts heating (IPCC et al., 2012; WEC, 2016).

During 2013, the distribution of the biomass type for use as fuel, including waste, was 68% fuelwood, 10% charcoal, 7% black liquor, 4% bioethanol, 3% MSW, 2% biodiesel, 2% biogas, 2% forest residues, 1% pellets, 1% industrial residues, and 0.3% hydrogenated vegetable oil. The consumption pattern varies geographically, depending on the region and the country; however, the growing trade in pellets has transformed biomass into a consumer good (WEC, 2016). In 2014, of the total world energy supply, biomass provided 59.2 EJ; of this energy volume, 45.1 EJ were used in direct heat production, the rest was used in the production of electricity, equivalent to 493 TWh (Kummamuru, 2017). Currently, biomass is widely used as a residential heating source in most developed countries; in some of these countries it also represents one of the main energy sources; for example, in northern Europe (WEC, 2016).

There are numerous estimates of the maximum sustainable potential of biomass; with an average range of between 100 and 300 PJ/year (Smith et al., 2014). By 2050, the energy consumption of the biomass could reach 150 EJ according to some scenarios. For its part, the International Renewable Energy Agency (IRENA) estimated that 38%–45% of the total global supply of biomass will come from agricultural residues and waste, while the rest of the demand will be covered with energy plantations, forest products, and agroindustrial. The greatest potential for the use of agricultural residues will be observed in Asia and America due to the high production of rice and corn, respectively (WEC, 2016).

Regarding the technologies for the energy use of solid biofuels, the most commonly used pathway is that of thermochemical processes, which include direct combustion, gasification, and pyrolysis. Fluidized bed (FB) combustion, co-combustion, and the combined production of heat and electricity (CHP) are widely used technological options for the generation of heat and electricity. The co-combustion of biomass is particularly important as it displaces greenhouse gas (GHG) emissions from the use of coal by a marginal investment for boiler modernization (IRENA, 2015; IPCC et al., 2012).

Modern biomass power plants are more efficient than the first ones that were developed and their possible positive social impacts, such as the creation of jobs, particularly at the regional level, make biomass an interesting alternative for the future heat generation and for the cogeneration of heat and electricity (Carneiro, Ferreira, 2010). For its part, the replacement of traditional biomass conversion devices by more advanced technologies, such as high efficiency biomass stoves, brings with it several benefits, for example a lower demand for biomass, which implies the reduction of local emissions, especially of particulate material (García-Frapolli et al., 2010; WEC, 2016).

Among the benefits that have been found to the energy use of biomass, are the reduction of GHG emissions, regional development, the creation of new employment opportunities and the reduction of the external energy dependence of the countries (Carneiro and Ferreira, 2012). Data from the Renewable Energy Policy Network for the 21st Century, REN21 indicate that, during 2015, 822,000 jobs were generated in the solid biomass sector, globally, without counting several million generated in the traditional sector; in China there were 241,000; in the United States 152,000; in India 58,000; 49,000 in Germany and 48,000 in France (A power plant with biomass, generally, takes 2 years to be build; during construction and installation 14 jobs per year are generated per MW; in the stage of manufacturing the technology 2.9 jobs are generated per MW; in the operation and maintenance stage 1.5 jobs are created per MW; additionally, the supply of fuel for this technology generates 32.2 jobs per PJ. In the case of cogeneration plants, during the construction and installation stage 15.5 jobs are created per MW; during the manufacture of the technology 2.9 jobs are generated; in the operation and maintenance stage 1.5 jobs per MW and 32.3 jobs per PJ during the fuel supply phase are generated.) (REN21, 2017b).

It should be noted that the use of biomass may entail risks of overexploitation and other negative cases, so to ensure that the use of bioenergy is sustainable, several multicriteria certification frameworks have been developed that take into account environmental, social and environmental aspects of the use of biomass. These frameworks and mechanisms include the ENplus (The ENplus was first introduced in Germany in 2010, and since then, has had a rapid growth across Europe ("ENplus," http://www.enplus-pellets.eu/)), the Roundtable on Sustainable Biomaterials and its International Sustainability and Biomaterials Certification. (RSB, http://rsb.org), the WBA (World Bioenergy Association ("WBA—Worldbioenergy," http://worldbioenergy.org/)), the GBEP (Global Bioenergy Partnership ("GBEP—Website: www.globalbioenergy.org")), and the ISO (International Organization for Standardization ("ISO—International Organization for Standardization," https://www.iso.org/home.html)). These regulations should also be correlated with the UN Sustainable Development Goals (WEC, 2016). Criticisms of the energy use of biomass have focused on its possible use as liquid biofuels (ethanol and biodiesel), since they may require enormous extensions of dedicated crops; bioethanol and biodiesel can be produced from biomass waste, through second generation technologies. These critiques consider the following aspects (Sims et al., 2008): increase in food prices; relatively low reduction of GHGs; high costs of carbon reduction (dollar per ton carbon avoided); the need for government subsidies; and direct and indirect impacts on land-use change and related GHG emissions, but they are not directly applicable to solid biofuels used for heat and electricity production, since their supplies come in large part of managed forests and agricultural/forestry residues, both primary and secondary (e.g., sugarcane bagasse) that does not imply direct competition with food production, GHG mitigation is high, and the carbon costs competitive.

In this chapter it will be assumed that biomass sources have been obtained in a sustainable manner, excluding those cases and applications that generate competition for resources for food production: water consumption, soil degradation and deforestation, loss of biodiversity, and increase of GHG emissions, although this does not rule out that it can sometimes be interesting in the use of crops on marginal or contaminated lands or in depopulated areas, where biomass crops can have positive consequences for the environment and the rural population.

4.2 BIOMASS FOR SOLID BIOFUELS

Several studies, such as McKendry (2002a) characterize biomass types in different ways, but a simple method is used by the European standard EN 14961-1 (Bott, 2014), which classifies solid biofuels according to their origin and source in the following subcategories (These categories also include forestry, agricultural, and fishery products or municipal waste, as well as by-products and waste from agroindustry, the food industry, and food services (Demirbaş, 2001; FAO, 2009));

- Wood biomass, including forestry
- Herbaceous biomass
- Municipal waste
- Agro-industrial wastes

The description of each biomass type is presented in the following paragraphs:

4.2.1 Wood biomass

The production of wood biomass in agricultural land does not automatically displace the amount of food or fiber produced per unit area, since there are different alternatives for the production of both food and other products from the same soil. It is a common practice in many countries to establish multipurpose plantations that serve livestock, food crops, and pastures; at the same time, they capture carbon, reduce wind erosion and the loss of surface moisture. These well-planned plantations produce more biomass in a sustainable way, as well as wood for sawmills and other end uses, such as pulp and paper (WEC, 2016).

The woody biomass is used in the form of branches, chips, and pellets of sawdust and other residues; part of this wood comes from demolitions and constructions, from parks and gardens, and from industrial waste. Wood is the raw material in charcoal production, which is the main fuel in foundry furnaces; for example, worldwide, more than 52 million tons of charcoal used for the casting of iron and other metals are produced from wood (IPCC et al., 2012; WEC, 2016).

Worldwide, wood biomass is used in the production of electricity and heat for industry and the production of biofuels. It is estimated that the use of this biomass as energy worldwide represents 56 EJ, which means that it provides around 90% of the primary energy obtained annually from all forms of biomass; in particular, the wood biomass contributes with the 10% of the electric power supplied annually (WEC, 2016).

This type of biomass is used mainly as an input in combustion and co-combustion systems, where the technology is well developed, due to its high efficiency and low emissions; A modern wooden boiler is as good as a modern pellet boiler (WEC, 2016); it is also used, although to a lesser extent, in gasification systems in mixtures with other sources of biomass (Sansaniwal et al., 2017).

As an example of modern wood biomass, pellets are found. Actually, pellets are produced mainly from wood residues, although the volume produced from agricultural by-products such as straw, sunflower seed husks, stems, and corn husks is increasing. A key advantage of pellets, compared to unprocessed biomass, is the high density and the high energy content per unit volume, convenient for their transport over long distances. The largest pellet mills are in 21 countries, mostly in North America and Europe (WEC, 2016). The pellets can be used in the processes of direct combustion, co-combustion, or gasification, in mixture with other sources of biomass with higher moisture content.

There are three main sources of woody biomass for the production of heat and electricity: (1) direct (from forests, paper pulp, trees, chips, and forest residues); (2) indirect (from co-products, obtained after primary processing in sawmills, such as chips, bark, and sawdust); and (3) post-consumer material (wood recovered from construction, waste, storage, and packaging material). In general, woody biomass from forestry and

Table 4.1 Particle size of different types of woody biomass

Fuel	Typical particle size
Briquettes	>25 mm
Pellets	<25 mm
Powder fuel	<1 mm
Sawdust	1–5 mm
Wood chips	5–100 mm
Whole wood	>500 mm

Source: From Serup, H., Kofman, P., Falster, H., 2005. Wood for Energy Production, Irish edition. Dublin: COFORD, p. 72.

wood industry can be used as fuel for heat production, whether in wood stoves, boilers, heating plants, or cogeneration plants. Technologies that take advantage of this biomass have different requirements with respect to the physical properties of the biomass to be used, such as size, moisture content, ash content, and contaminants. Table 4.1 shows the typical particle size of woody biomass fuels (Serup et al., 2005).

In the production of wood, after the felling of the trunks, large quantities of forest residues remain in the area, mainly the canopies and branches; normally it is necessary to clean out these areas of residues to facilitate replenishment. The residues are collected and stored throughout the wood production area to facilitate the passage of vehicles; this waste takes between 5 and 10 years to decompose. It has been observed that these residues can be used profitably as fuel (Serup et al., 2005).

4.2.2 Herbaceous biomass

Herbaceous biomass is classified into cereal crops, pastures, oilseed crops, tubers and legumes, flowers, herbaceous biomass of gardens, parks, pruning, vineyards, orchards, and mixtures of all these. This biomass can be used raw (direct residues of the field) or processed (from the food industry) (Bott, 2014).

Residues from cereal crops include the use of the following parts of the plant: dried stems, pods, and husks, as well as their mixtures. Of the grasslands are considered stems, shells, and their mixtures. In oil crops, stems, leaves, pods, rinds, and their mixtures are considered. The residues of the tuber crops are integrated by stems, leaves, roots, and mixtures of these. The stems, leaves, pods, and their mixtures can be harvested from legume crops (Bott, 2014). It should be noted that the incorporation of legumes in an annual biomass cultivation system could reduce the nitrogen fertilizer requirements, which will reduce the inputs of bioenergy production (Mitchell et al., 2016); however, they also contribute to the anthropogenic emissions of nitrous oxide to the planet. In this case, nitrous oxide comes mainly from the decomposition of legume residues (Barton et al., 2011).

Finally, from the cultivation of flowers, the waste from the plant (when the quality of these is not adequate), the stems, leaves, as well as their mixture can be used (Bott, 2014).

For its part, processed residues can come from herbaceous residues without chemical treatment, as well as chemically treated herbaceous residues and its mixtures. This type of biomass is used mainly in combustion and co-combustion systems, where the technology is well developed, has high efficiency, and low emissions; nevertheless, herbaceous crops that have a very high K content may be detrimental for biomass combustion equipments. It is also used, albeit to a lesser extent, in gasification systems in mixtures with other sources of biomass.

The most abundant form of herbaceous biomass in the world, particularly in European countries, is straw, which is obtained during the annual production of cereals in a proportion of 0.6–0.8 tons of straw per ton of grain in the field. In countries where crop productivity is high (in the order of 8–10 tons of grain per hectare), the amount of straw per hectare can be between 6 and 8 tons and much of it has been removed for the next crop cycle. For example, Denmark's straw production is 6 million tons per year, more than 1 million is used for cogeneration, another million tons are used as feed for livestock and other agricultural uses, and about 2 million have commercial uses, including pellets (WEC, 2016).

Throughout the world, millions of tons of straw are available annually, but their use is very low. Less than 100 million tons are used as energy and the rest is usually burned freely or allowed to decompose, releasing GHGs as happens in Asia, Brazil, Argentina, or Egypt (WEC, 2016). However, through proper management of croplands, the reincorporation of nutrients from residues and straw can be allowed, particularly in unproductive or unused soils, protecting from erosion and loss of moisture (OECD/IEA, 2017).

Due to its characteristics, straw is used mainly in co-combustion systems and in mixtures in gasification systems.

The herbaceous biomass contains variable amounts of cellulose, hemicellulose, and lignin, and a small amount of other extractive, usually perennial, with looser fibers. The relative proportions of cellulose and lignin is one of the determining factors to identify the suitability of plant

Table 4.2 Syngas composition from different oxidants

Properties	**Herbaceous biomass type**	
	Wheat straw	**Barley straw**
Moisture (%)	16	30
Volatile material (%)	59	46
Fixed carbon (%)	21	18
Ashes (%)	4	6
Net calorific value (MJ/kg)	18.6	16.1
C (%)	48.5	45.7
H (%)	5.5	6.1
O (%)	3.9	38.3
N (%)	0.3	0.4
S (%)	0.1	0.1

Source: From McKendry, P., 2002a. Energy production from biomass (part 1): overview of biomass. Biores. Technol. 83 (1), 37–46. doi:10.1016/S0960-8524(01)00118-3.

species for further processing as energy crops. Table 4.2 shows some characteristics of types of herbaceous biomass (McKendry, 2002a).

The Switchgrass, a type of fine-stemmed herbaceous biomass, has been used in the United States as a herbaceous cultivation model, has a calorific value comparable to that of wood, but with a lower moisture content; it also has a low content of ash and alkalis, indicating that it could be an adequate fuel for combustion. The main difficulty of this type of biomass is the production of slag at the time of gasification, associated with the ash content. To avoid slag production, the temperature can be reduced, but this results in the loss of carbon. The gas formed in the gasifier contains impurities such as particles, tar, nitrogen compounds, sulfur compounds, and alkaline compounds (McKendry, 2002b).

4.2.3 Municipal waste

Waste is defined as any material that does not have a use and that does not represent any economic value for its owner. Depending on their physical state, the waste is divided into solid, liquid, or gaseous waste. In turn, solid waste is classified as municipal, hazardous, hospital waste, and radioactive waste. The generation of MSW, defined as any waste generated by human activity in homes, businesses, and institutions, and that are not dangerous, is a subject of global interest. The MSW are usually integrated by food waste, paper, cardboard, plastics, textiles, glass, metals, wood, parks, and pruning (Wang, 2014).

The amount and composition of MSW generated in any country depends on its economic development, population density, urbanization, and consumption patterns. Countries with greater purchasing power generate a greater number of waste per person compared to those with lower income; for example, it is estimated that in India the average generation per person is 370 grams per day; in contrast, in Denmark they produce 2200 grams per day. The World Bank estimates that by 2025, the average daily generation of MSW, only in Asia will be 1.8 million tons per day (Wang, 2014).

The generation of energy from MSW is an approach to the integral management of waste when it does not compete with other economic uses, because it entails benefits both from the point of view of waste management and its energy use. The implementation of this practice in the management of waste represents direct improvements in public health, as well as in the quality of the environment (Wang, 2014).

Given the large volume of waste generated in the world and its calorific value that can be used directly in energy generation, different technologies have been developed for obtaining energy; however, it is important to note that the selection of the appropriate technology will depend on the particular characteristics of the waste. In developing countries, such as India, MSW have a low calorific value (7.3 MJ/kg); In contrast, developed countries such as Japan and the United States have a calorific value exceeding 10 MJ/kg, although they also have a high content of inert materials that must be separated (Wang, 2014).

Considering these characteristics, both incineration and combustion have been used to generate energy from waste and reduce its volume significantly (up to 80%). On the other hand, MSWs can also be used in gasification and co-combustion technologies, mainly because of their capacity to mitigate environmental impacts due to the displacement of fossil fuel use, the need to reduce costs in industries, and to achieve an advantage of their thermal potential. The industries that tend to use this type of fuel most are steel industry, for displacing coal, and cement industry, for displacing oil (Bott, 2014). In the case of the food industry, for example, the heat generated by the burning of the waste can be used in steam turbines for the generation of electricity or as process heat (Wang, 2014).

It should be noted that the separation of waste in its organic and inorganic fractions is desirable because, on the one hand, the latter is more suitable for the gasification process by plasma, and on the other hand, if

Table 4.3 Calorific content of various materials present in the MSW (millions of BTU/ton)

Material	**Calorific content (millions of BTU/ton)**
Polyethylene terephthalate	20.5
High-density polyethylene	39
Polyvinyl chloride	16.5
Polypropylene	38
Polystyrene	35.6
Rubber	26.9
Leather	14.4
Textiles	13.8
Wood	10
Food waste	5.2

Source: From US Department of Energy, 2007. Methodology for Allocating Municipal Solid Waste to Biogenic and Non-Biogenic Energy. Washington, DC: Energy Information Administration: Office of Coal, Nuclear, Electric and Alternate Fuels, pp. 1–18. Available at: https://www.eia.gov/totalenergy/data/monthly/pdf/historical/msw.pdf.

the organic fraction of the waste is contaminated by metals and other chemical products, it generates a very polluting leachate, which is dangerous for the quality of the soil, the water, and the agricultural activities. However, mixed wastes (~28.7% wood and paper, ~17.8% plastics, rubber, textiles, ~3.6% metals, ~6.5 glasses, ~30.9% organic fraction, ~12.5% inert and other materials) are not suitable for incineration, because its valorization is recommended (reuse and recycling) prior to thermal processing (Ayres, 2007).

Table 4.3 shows the heat content of the materials present in the MSW.

Finally, within the classification of biomass are the agro-industrial wastes; its generalities are mentioned below.

4.2.4 Agro-industrial wastes

Agro-industrial waste is secondary or processed waste that is obtained during the production of food, such as bagasse and husks. These come from crops that produce valuable waste in their harvest, as well as from the portion of the waste available after processing; it is assumed that all quantities produced can be recovered and used for energy applications. For example, fruit biomass includes vegetable and garden waste, co-products and waste from the fruit canning industry, fruit mixtures, and mixtures of all previous residues (Bott, 2014).

Table 4.4 Composition of some types of agroindustrial biomass (kg/ton)

Agroindustrial biomass (crop)	N (kg/ton)	P (kg/ton)	K (kg/ton)	C (kg/ton)	Mg (kg/ton)	Lignin (kg/ton)	Polyphenol (kg/ton)
Coffee	17	1.3	29.0		1.8	39.6	13.8
Rice	6.3	1.4	3.8	0.8	0.4	166.6	0.1
Sugar cane	3.9	0.4	7.0	2.4	0.4	160.2	3.5
Water hyacinth	14.1	2.2	32.3	12.7	3.9	100	

Source: From Lichtfouse, E., 2012. Organic Fertilisation, Soil Quality and Human Health (vol. 9). Springer. doi:10.1007/978-94-007-4113-3.

The residues of vegetable gardens and vegetables include berries (pulp, seeds, mixtures, and whole fruits), residues from the canning industry (seeds, whole fruit, pulp, and mixtures thereof), all kinds of nuts (whole nuts, peels and pods, and mixtures thereof), and its mixtures.

Within the classification of co-products and waste of the fruit canning industry are fruit residues that have not been chemically processed and those that have already been processed. The first group includes berries, canning residues and seeds, nuts and pods, residual cakes from pressing for obtaining edible oils, as well as mixtures of all these. For its part, within the chemically processed fruit residues, berries, canning residues and seeds, nuts and pods, as well as the residual cakes exhausted from the pressing for the production of edible oils, as well as the mixtures of all these can be found (Bott, 2014).

Given the high moisture content of this type of biomass, it can be used directly mixed in gasification systems. Table 4.4 shows the composition of some types of agroindustrial biomass.

4.3 SOLID BIOMASS CONSUMPTION

In 2014, 445 TWh of electricity (about 2% of the world's electricity generation) were produced globally by biomass. In Europe and North America, the use of biomass in the production of electricity is remarkable, contributing with more than 70% of consumption, predominating the use of forest products and agroindustrial waste. On the other hand, in the developing countries of Asia and Africa, where an important part of the population does not have access to electricity, an increase in the use of biomass for the production of electricity has been observed in the last 5 years (IRENA, 2015; WEC, 2016).

In 2014, solid biomass covered approximately 7.6% of European energy demand (326.9 TWh); in particular, Germany is one of the European countries with the greatest development in the use of biomass; the German biomass facilities in the same year provided approximately 49.1 TWh of energy, which is 8% of the total electricity consumption of this country (Strzalka et al., 2017). Within solid biofuels, charcoal and wood are used in Europe and North America mainly in the production of heat and electricity in CHP processes (WEC, 2016).

As indicated in the previous paragraphs, solid biomass is an important input for the production of electricity that contributes to the displacement of fossil fuels and, therefore, represents an opportunity in the reduction of polluting emissions during the production of energy and the mitigation of climate change. In this sense, the technologies for the use of solid biomass are presented in the following sections.

4.4 SOLID BIOMASS HEAT AND POWER GENERATION TECHNOLOGIES

There are numerous technologies for the conversion of biomass into useful forms of energy. Bioenergy available as a solid fuel is mainly used to generate heat and power. According to the German bioenergy industry, processes for the production of energy from biomass can be basically divided into three main categories: *thermochemical, physical–chemical, and biological conversion routes*. Physical–chemical conversion processes can be applied for the production of biodiesel. The biological conversion routes of the biomass are used to convert the biomass into biogas or bioethanol (Jaap Koppejan and van, 2012; McKendry, 2002b; Sansaniwal et al., 2017; Strzalka et al., 2017). In this chapter, we will only focus on the thermochemical conversion of biomass, since from this route it is possible to obtain heat and electricity, the subject of this chapter analysis. The thermochemical biomass conversion processes include *combustion and co-combustion, incineration, gasification, and pyrolysis* (Jaap Koppejan and van, 2012; McKendry, 2002b; Sansaniwal et al., 2017; Smith et al., 2014; Strzalka et al., 2017). This last route of conversion will not be part of the study of this chapter since it is the conversion of biomass to liquid fractions (called bio-oils or bio-oils), gaseous, and, to a lesser extent, solid, heating the biomass in the absence of oxygen in temperatures at around 500°C. The process can be used to produce predominantly bio-oil if flash pyrolysis is used, which allows the conversion of biomass into bio-oil, with an efficiency of up to 80% (McKendry, 2002b; Sansaniwal et al., 2017). In this chapter, the pyrolysis process will not be considered since it is used primarily for the production of liquid and gaseous biofuels, which are preferably used in other sectors, while solid biofuels are generally used for the production of electricity.

Fig. 4.1 shows the main thermochemical conversion technologies of solid biomass in heat and power.

Due to its reliability, efficiency, and commercial availability, biomass technologies based on combustion and its variants (co-combustion) provide more than 90% of the world's bioenergy production for commercial use. One of the main advantages of these technologies is the use of non-selective biomass, which makes it possible to take advantage of a wide variety of raw materials as fuel. Additionally, other advantages are identified, including the widespread use of biomass kilns, their ease of operation, and high profitability. Biomass combustion systems can be found in

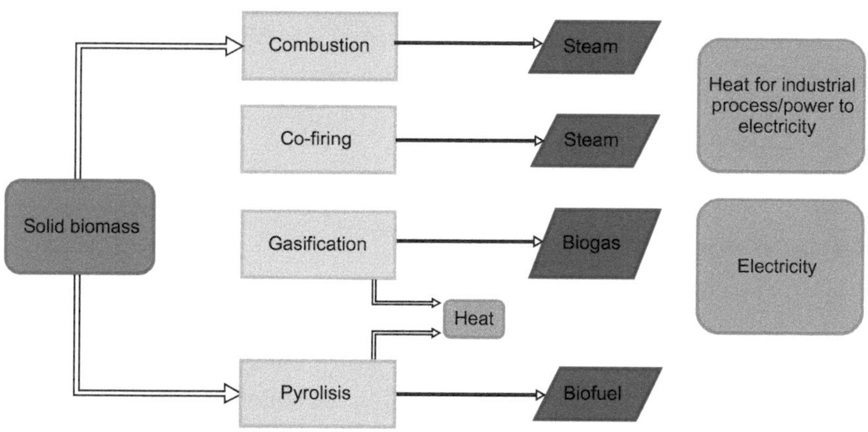

Figure 4.1 Biomass thermochemical conversion technologies. *Own design, with information from Jaap Koppejan, S. van L., 2012. The Handbook of Biomass Combustion and Co-firing. Routledge; McKendry, P., 2002b. Energy production from biomass (part 2): conversion technologies. Biores. Technol. 83 (1), 47–54. doi:10.1016/S0960-8524(01)00119-5; Sansaniwal, S.K., Pal, K., Rosen, M.A., Tyagi, S.K., 2017. Recent advances in the development of biomass gasification technology: a comprehensive review. Renew. Sust. Energy Rev., 72 (January), 363–384. doi:10.1016/j.rser.2017.01.038; Strzalka, R., Schneider, D., Eicker, U., 2017. Current status of bioenergy technologies in Germany. Renew. Sust. Energy Rev. 72 (January), 801–820. doi:10.1016/j.rser.2017.01.091.*

a wide range of sizes, from small systems—domestic—to large industrial facilities (Strzalka et al., 2017).

Modern technologies for the thermochemical conversion of biomass include the production of heat in boilers and cogeneration (joint production of heat and electricity) for industrial and residential purposes. CHP, in particular, together with the centralized urban heating and air conditioning systems (DC, located mainly in Europe), is an important part of the strategies to reduce GHG emissions in some countries, due to greater efficiency and less need of fossil fuels compared to independent heat, power, and cooling production systems, commonly used in developing countries (WEC, 2016).

It should be noted that pretreatment and biomass improvement technologies have been developed. The most common pretreatments include (1) drying to reduce the moisture content and transport costs of the biomass feedstock and improve the efficiency of combustion, (2) pelletizing and briquetting to mechanically compact voluminous biomass, such as sawdust or agricultural residues, and (3) torrefaction (for woody biomass) in which the biomass is heated in the absence of oxygen at 200°C–300°C and converted into coal, with a process similar to that of traditional coal production. After roasting, wood biomass is generally pelleted, reaching a volume and energy density 25%–30% higher than traditional pellets, which gives characteristics similar to those of coal (Adam Brown et al., 2017; ETSAP/IEA, 2015; IRENA, 2012b; OECD/IEA, 2017; REN21, 2017a; Ruiz et al., 2013).

4.4.1 Combustion

The combustion process is the burning of biomass in the presence of air, in which the chemical energy stored in the biomass is transformed into thermal, mechanical, or electrical, by means of equipment such as stoves, furnaces, boilers, or steam turbines (Jaap Koppejan and van, 2012; McKendry, 2002b; Sansaniwal et al., 2017). During the combustion of biomass, gases with temperatures between 800°C and 1000°C are produced; it should be noted that, in practice, combustion is only feasible for biomass with a moisture content of less than 50% (McKendry, 2002b).

Each biomass fuel or solid biofuel requires a certain amount of oxygen to carry out the combustion process (Demirbaş, 2001; IRENA, 2015; McKendry, 2002a; Sansaniwal et al., 2017). The stoichiometric ratio of air to fuel required for complete combustion of the biomass varies from

6.0:1 to 6.5:1, producing carbon dioxide (CO_2) and water vapor ($H2O_{(s)}$) as final products (Sansaniwal et al., 2017).

The net efficiencies of bioenergy conversion for solid biomass combustion plants range between 20% and 40%. The highest efficiencies are obtained in power generation systems of more than 100 MWe or when co-combustion of the biomass is co-fired in coal-fired power plants with FB burners. The gases generated during the combustion of biomass cannot be used as fuel in a gas engine by spark ignition or a gas turbine, unlike what happens with fossil fuels (McKendry, 2002b).

The main technologies of combustion of biomass on a small scale use burners, and these can be pellet boilers or chip burners. These systems are used commercially to produce hot water or steam. The main applications are domestic and urban heating, electricity generation, and CHP (EEA, 2008; Peterson, 2009).

Combustion technologies are commercially available throughout the world and play an important role in the production of energy from biomass; however, the technology must be optimized to meet the demands at lower costs, greater flexibility of fuels, lower emissions, and greater efficiency (IEA Bioenergy Task 32). The aim of combustion of biomass and co-combustion is to accelerate the introduction into the market of improved combustion systems through the exchange of technical and nontechnical information. This relates to both independent small- and medium-scale combustion plants and the co-combustion of biomass in existing coal plants (Jaap Koppejan and van, 2012).

After the combustion of the biomass, the nutrients that are obtained from the soil remain in the ashes (Table 4.5) and would be expected to return to the soil in which the biomass grows. Nitrogen is lost because it is emitted together with the combustion gases. When the ashes are deposited directly on the soil, the potassium is leached and available for the plants. For its part, phosphorus is insoluble and requires up to 20 years to be available for plants. These characteristics of the ash make it susceptible

Table 4.5 Ashes composition of certain biomass

Origin	Nutrients content
Clean wood pellets, bark	Ca and Mg
Agricultural residues, manure, forestry, wood trimmings, fast-growing biomass plantations	P and K

Source: From Bott, R., 2014. Solid biofuels for energy systems. In: Igarss 2014. doi:10.1007/s13398-014-0173-7.2.

to be used as raw material for the manufacture of fertilizers or for long-term purposes. The content of ash in the biomass varies from less than 1% to more than 20%; that of nitrogen varies from around 0.1% to more than 1%. For its part, the ash content in the woody biomass is usually low, but some wood species have ash content higher than 0.5% by weight (Bott, 2014).

Ash with high concentrations of potassium and phosphorus is most suitable for recycling and fertilizer use; biomass ash containing mainly calcium and magnesium can be used as soil improvers to balance the pH (Bott, 2014).

During the combustion of the biomass, nitrogen oxides (NOx) can be formed by the presence of nitrogen in the air and in the fuel. Both the nitrogen contained in the fuel and in the design of the combustion chamber of the system play an important role in the production of NOx; therefore, to reduce its formation, it is important to consider the following parameters: nitrogen content in the fuel, combustion by stages with low air content during the first stage, low temperature of the flame, and recirculation of the combustion gases (Serup et al., 2005).

4.4.2 Co-combustion

The co-combustion of biomass in traditional coal-fired power plants is basically the addition of one or more supplementary biofuels with the main fuel (coal) to be burned simultaneously (Bauen et al., 2009; Bott, 2014; IRENA, 2013).

The net electrical efficiency of a coal/biomass co-combustion power plant ranges between 36% and 44%, depending on the technology, size, quality, and proportion of plant biomass. Although with current technology it is feasible to integrate up to 20% (as energy content) of biomass in a co-combustion plant and technically it is possible to integrate more than 50%, currently the usual proportion of biomass is less than 5% and rarely exceeds 10% continuously. The co-combustion of biomass in coal power plants is an especially attractive option due to the high conversion efficiency of these plants (McKendry, 2002b).

Biomass raw materials include forest and agricultural residues, waste, and dedicated energy crops (IPCC et al., 2012; IRENA, 2013).

The co-combustion technologies used in the energy sector are classified, in general, into three types: direct co-combustion, indirect or parallel co-combustion, and co-combustion based on gasification.

1. Direct co-combustion: The previously pulverized biomass is fed to the boiler either separately or mixed with coal (Li et al., 2012). A single boiler with common or separate burners is used, which constitutes the simplest, cheapest, and most widespread approach (IRENA, 2013).
2. Indirect co-combustion: It involves the installation of a completely separate biomass boiler to produce low-quality steam which is fed to conventional power plants (based on fossil fuels) resulting in higher conversion efficiencies (IRENA, 2013; Li et al., 2012).
3. Co-combustion based on gasification: It involves the gasification of the solid biomass and the combustion of the fuel gas of the product in the coal boiler (Li et al., 2012).
4. Parallel co-combustion: Here a separate boiler is used for the biomass, and its steam generation is mixed with the steam from conventional boilers (IRENA, 2013).
5. Currently co-combustion is successfully applied in existing coal boilers for energy production or CHP. The scales of these plants range from 5 to 100 MWe (Bauen et al., 2009). Co-combustion can play an important role in increasing the use of biomass in power generation and the reduction of GHG emissions since the investment required to adapt existing coal plants is minimal with respect to building new plants. Co-combustion in CHP plants is the most competitive option to exploit the energy potential of biomass for the production of electricity and heat (IRENA, 2013).

The co-combustion technology has been used at different commercial scales and with diverse configurations. Among the challenges presented by this technology are the preparation, quality, storage and handling of fuel, the formation of deposits, and the increase in corrosion and erosion. In contrast, it has the following advantages: low cost of installation and maintenance of the plants, ease to take advantage of any type of biomass wood fuel, the use of biomass with high moisture content, suitable for fuels with high ash content, variable particle sizes, and low emissions (Maciejewska et al., 2006).

4.4.3 Gasification

Gasification is the thermochemical conversion of solid biomass to a liquid or gaseous biofuel, in the presence of oxygen (or any other oxidizing compound such as water vapor, CO_2, or its mixtures) in a lower proportion than in a complete combustion, a process that is carried out in a reactor called a gasifier (Bocci et al., 2014; Sansaniwal et al., 2017; Smith et al., 2014; WEC, 2016).

The gasification of biomass and the use of fuel gas for the production of energy and heat offer a high-generation efficiency and, therefore, it is one of the most promising technologies for the generation of distributed energy. Gasification plants can be applied in a wide range of power, from a few kilowatts to several megawatts (Ahrenfeldt et al., 2013; Puig-Arnavat et al., 2014). In addition, the gasification efficiency can be of the order of 70%–80% (Puig-Arnavat et al., 2010), depending on the gasifying agent used.

The main gasification product is synthesis gas, also called syngas, which consists of a gaseous mixture containing hydrogen (H_2), carbon monoxide (CO), methane (CH_4), and carbon dioxide (CO_2), which can be burnt in gas turbines for the production of electricity. Its main co-product is residual biomass, biochar, which can be used as a soil improver (Bocci et al., 2014; Sansaniwal et al., 2017). As can be seen in Table 4.6, the composition of the syngas varies depending on the gasifying or oxidizing agent: air, oxygen, or steam mixed with CO_2; of these, the most used gasification agent is air due to the high availability and its zero cost (Bocci et al., 2014).

Most gasifiers use wood and other lignocellulosic biomass; it can also be designed to convert other biomass materials and the dry organic fraction of MSW into energy (Demirbaş, 2001; Kirkels and Verbong, 2011; WEC, 2016).

The gasifiers are divided into two main groups (Sansaniwal et al., 2017): fixed bed (from which the mobile bed is derived) in which there is a bed of solid fuel particles (Beohar et al, 2012; Bocci et al., 2014), and FB, in which both the fuel and the inert bed material behave like a fluid (Arteaga-Pérez et al., 2014; Bocci et al., 2014).

Table 4.6 Syngas composition from different oxidants

	Composition (% vol)					
Oxidant	**H_2**	**CO**	**CO_2**	**CH_4**	**N_2**	**LHV (MJ/Nm^2)**
Air	9–10	12–15	14–17	2–4	56–59	3–6
Oxygen	30–34	30–37	25–29	4–6		10–15
Steam/CO_2	24–50	30–45	10–19	5–12		12–20

Source: From Bocci, E., Sisinni, M., Moneti, M., Vecchione, L., Di Carlo, A., Villarini, M., 2014. State of art of small scale biomass gasification power systems: a review of the different typologies. Energy Procedia 45, 247–256. doi:10.1016/j.egypro.2014.01.027.

Within the fixed-bed gasifiers, the biomass and the gasifying agent are fed to the reactor in different configurations (in parallel or countercurrent); the downstream gasifiers are generally used for the production of small-scale electricity (greater than 1 MW), On the other hand, a cross-flow gasifier has a shorter response time than any other type of fixed bed and requires a relatively simple gas cleaning system (Beohar et al., 2012; Bocci et al., 2014), which makes it an easy-to-operate reactor.

In FB gasifiers, the solid fuel, mixed with hot bed material (inert sand, catalyzer), is maintained in a semi-suspended condition (fluidized state). Thanks to the excellent gas–solid mixture and the high thermal inertia of the bed, the temperature and solid/gas concentration are uniform throughout the bed. For these reasons, the conversion of biomass in these systems is close to 100% and the yields (biomass flow rate per installed reactor area) are 2–10 times higher than in fixed beds (500–1000 $kg_{biomass}/hm^2$). FB gasifiers are generally more tolerant and suitable for large-scale installations (<10 MW), they operate at uniform and relatively low temperatures (700°C–900°C), they do not require mechanical components moving at high temperature (for example, moving grids as in the moving bed). Due to these advantages, most of the current development activities focus on large-scale FB gasifiers. However, they still have the following disadvantages: the operation, even if it is more flexible than the fixed bed, is more complicated. The concentration of particles in the gas is generally higher (from 10 to 100 g/Nm^3). The rapid movement of the bed material generates a high abrasive action (Bocci et al., 2014).

FB gasification has several advantages with respect to fixed/mobile bed or entrained flow (Beohar et al., 2012). The FB provides high mixing and reaction speeds, responds to the variation in fuel quality, and allows the process to be extended, which makes it ideal for processing biomass and waste (Gómez-Barea et al., 2013).

Among biomass-based power generation technologies, gasification is becoming one of the best technological routes for distributed cogeneration. In addition to high electrical efficiency, gasification offers many benefits for decentralized cogeneration: low investment cost, compact structure, and small system capacity (Asadullah, 2014).

One of the most successful biomass gasification plants is in Australia; the plant has a rotating FB gasifier that produces about 2.0 MW of electric power and 4.5 MW of thermal energy; once the gas is washed, it is used in a boiler with heat recovery for the production of steam and electricity. (Pereira et al., 2012).

The main advantage of the biomass gasification is the high energy to heat ratio that allows a considerable reduction of the polluting emissions of the power generation. Another advantage is the multiple possibilities for the use of produced gas that can be applied for the production of heat and power. The disadvantages of the biomass gasification are the complex operation of the plant, the gas cleaning, and the lack of experience (Strzalka et al., 2017). Table 4.7 shows the commercially available fixed-bed gasifiers available throughout the world (Sansaniwal et al., 2017).

In addition, research has been conducted on three types of catalysts for the use in biomass (Sutton et al., 2001):

1. Dolomite. Dolomite is a magnesium mineral, with the general formula $MgCO_3 \cdot CaCO_3$, it is used in the Pidgeon process for the production of magnesium by thermal reduction. The use of dolomite as a catalyst in the gasification of biomass has attracted much attention, since it is a cheap catalyst that can significantly reduce the tar content of gasification products. It can be used as a primary catalyst.
2. Alkali metals and other metals. The use of alkali metals for the removal of tar and improvement of the obtained products has also been investigated by several groups. The catalyst is generally added directly to the biomass in a dry mix or by wet impregnation. In order of activity, the following catalysts are found: potassium carbonate > sodium carbonate > trona ($Na_3H(CO_3)_2 \cdot 2H_2O$) > borax ($Na_2B_4O_7 \cdot 10H_2O$). Other catalysts that have been investigated are the K_2CO_3 and Na_2CO_3.
3. Nickel. Most of the research on the cleaning of gas from biomass gasification has been carried out on nickel. Many groups have investigated a gas cleaning system that includes the use of dolomite or alkali metals for the removal of tar and then nickel is used to adjust the composition of the gas (reformation of methane and remaining tar). Most research has been conducted on commercial catalytic conversion with nickel for hydrocarbon and methane reformation.

4.4.4 Combined heat and power

CHP generation, also known as cogeneration, is mainly applied to increase the overall efficiency of energy production. The CHP plants play an important role in the production of heat and electricity, and it is expected that they will maintain this position in the future (Gerssen-Gondelach et al., 2014).

Table 4.7 Commercial fixed bed gasifiers currently available

Country	Gasifier type	Used fuel	Capacity	Organization/Project
United States	Downward flow	Chips, logs	1 MW	CLEW
	Downward flow	Wood chips, corn cob	40 kW	Stwalley & Stwalley Engineering
Denmark	Upward flow	Residues of furriers and dangerous	2–15 MW	DTI
	Upward flow	Straw, chips, bark	1–15 MW	VOLUND R&D Center
	Downward flow	Wood residues	0.5 MW	Hollesen Engg.
New Zealand	Downward flow	Blocks of wood, shavings, scrub brush shavings	30 kW	Fluidyne
France	Downward flow	Wood, agricultural waste	100–600 MW	Martezo
United Kingdom	Downward flow	Wood chips, hazelnut peel, municipal solid waste (MSW)	30 kW	Newcastle University of Technology
	Downward flow	Agroindustrial residues	300 kW	Shawton Engineering
Switzerland	Stratified	Wood and agricultural biomass	50–2500 kW	DASAG
India	Downward flow	Wood chips, rice husks	100 kg/h	Associated Engg. Works
	Downward flow	Woody stems,corn cobs, rinds, rice husks		Ankur Scientific Energy Technologies
Belgium	Small size	Wood chips	160 kW	SRC Gazel
South Africa	Downward flow	Blocks of wood, shavings, briquettes	30–500 kW	SystBM Johansson gas producer
Finland	Upward flow BIONEER	Wood chips, straw, residues derived fuel (RDF) pellets, RDF peat	4–5 MW	Ahlstrom Corporation, VTT
Netherlands	Downward flow	Rice husk	150 kW	KARA Energy Systems

(Continued)

Table 4.7 (Continued)

Country	Gasifier type	Used fuel	Capacity	Organization/Project
China	Downward flow	Sawdust	200 kW	Huairou Wood Equipment
	Downward flow	Harvest waste	300 kW	Huantai Integrate Gas supply System
	Downward flow	Agricultural and forestry residues	200 kW	Tianyan Ltd
	FB	Agricultural and forestry residues	1,000 kW	Tianyan Ltd
	Circulating FB	Agricultural and forestry residues	200–1,200 kW	GIEC
	Circulating FB	Rice husk, straw, wood sawdust, peanut husk	5.5 MW	GIEC
India	Downward flow	Rice husk	25–100 kW	Husk Power Biomass Gasifier plant
	Downward flow	Dhaincha, corn cobs, wood, and other local plants	128 kW	Saran biomass gasifier power plant
	Downward flow	Ipomoea, corn residues, dhaincha	20–120 kW	Baharbari Village Power Plant, North Bihar
	Downward flow	Pine needles	9 kW	AVANI biomass gasifier power plant

Source: From Sansaniwal, S.K., Pal, K., Rosen, M.A., Tyagi, S.K., 2017. Recent advances in the development of biomass gasification technology: a comprehensive review. Renew. Sust. Energy Rev., 72 (January), 363–384. doi:10.1016/j.rser.2017.01.038.

In the case of biomass, energy cogeneration can be achieved when heat recovery systems are coupled to existing (or new) biomass combustion (or co-combustion) boilers. The total efficiency of cogeneration plants based on biomass for industrial or residential heating ranges between 70% and 90% (IRENA, 2012a). Usually, the recovered heat is used in Stirling engines for electricity production. Stirling engines are indirectly heated gas engines that operate by cyclical compression and expansion of gas, with air, helium, or hydrogen as the working medium. Stirling engines allow the generation of energy in a closed cycle with the advantage that several sources of heat can be used to power the system. The engine can be driven by almost any type of heat source with a flue gas temperature of approximately 1000°C with low particle and ash levels. The typical electrical capacities of the Stirling plants range between 1 and 100 kW, and the efficiency of the energy production is in the range of 12%–30% (Strzalka et al., 2017).

There is another application of CHP with biomass that has been used with great success and is called Organic Ranking Cycle (ORC). It is a more advanced decentralized cogeneration technology. One of the main advantages of this technology is that the energy generation mode is suitable for the typical energy demand structure in Europe (75% heat, 25% electricity) (Strzalka et al., 2017). As a result of its efficient operation at partial load and the reduction of operating costs, ORC systems have become a well-established technology within the bioenergy sector in Europe (Tańczuk and Ulbrich, 2013). There are more than 150 ORC plants in operation throughout Europe, which have mostly been installed in Germany (Strzalka et al., 2017) due to the favorable conditions of energy policies, especially the laws for the use of renewable sources. A large part of the ORC cogeneration plants based on solid biomass works driven by solid biofuels and feeds the heat in district heating networks (Erhart et al., 2011; Maraver et al., 2013).

4.4.5 Incineration

Incineration is a process of thermochemical conversion of waste used mainly in the treatment and/or thermal disposal of MSW (Bott, 2014; Fruergaard et al., 2010). In this process, controlled combustion of MSW is carried out in waste incinerators; most of these produce heat in combination with electricity, while some only produce heat to produce steam and electrical energy by steam turbines (Bott, 2014; Wang, 2014).

Modern incinerators can have electricity generation efficiencies (based on the net calorific value of waste) of between 20% and 30%, while heat production efficiencies range between 60% and 85% (Fruergaard et al., 2010).

It should be noted that the uncontrolled burning of waste and inadequate incineration contributes significantly to environmental pollution. For example, most food processing wastes are not suitable for burning due to the high moisture content, and their incineration is related to dioxin emissions into the atmosphere, which is why the use of incineration is restricted in the food industry (Wang, 2014).

The incineration of MSW has several advantages and disadvantages. In Table 4.8, some of these are presented.

The next section compares the costs of mature technologies for generating heat and electricity with biomass.

Table 4.8 Advantages and disadvantages of the incineration process of MSW

Advantages	Disadvantages
Waste management without prior treatment	Generates hazardous waste (air pollution control waste), which must be disposed of safely
Reduction of the demand for landfills of urban solid waste	Generates slags (ashes)
Waste volume reduction by 90%	Produces a large volume of combustion gases
Waste weight reduction by 70%	It has high investment and operating costs
Energy recovery possibility (electricity or heat)	It has high maintenance costs
If managed well, release of pollutants into the air is low	Requires qualified personnel
Potentially pathogenic and toxic organic pollutants are destroyed	Requires an adequate composition for self-combustion
It can be located near the MSW generation center	Still has a negative public perception
Waste transporting cost reduction	
Less space needed	
The chimney emissions are odorless	
Reduce organic matter mainly to CO_2 instead of CH_4 and other volatile organic compounds	

Source: From Quina, M.J., Bordado, J.C.M., Quinta-Ferreira, R.M., 2011. Air Pollution Control in Municipal Solid Waste Incinerators. In: Dr. Khallaf, M. (Ed.), The Impact of Air Pollution on Health, Economy, Environment and Agricultural Sources. InTech, ISBN: 978-953-307-528-0. Available from: http://www.intechopen.com/books/the-impact-of-air-pollution-on-health-economy-environment-and-agricultural-sources/air-pollution-control-in-municipal-solid-waste-incinerators.

4.5 THERMAL ENERGY AND/OR ELECTRICITY GENERATION COSTS WITH BIOMASS

Thermal energy and/or electricity generation technologies with biomass are mature and are a competitive generation option, provided that agroforestry waste or another type of low-cost biomass is available. In addition, new technologies are emerging that show significant potential for cost reduction. However, these reductions will not be significant because they are mainly related to the supply of fuel and not to the combustion process itself (REN21, 2017b).

The total investment cost (Table 4.9) of thermal energy and/or electricity generation systems with biomass varies significantly between technology and country; direct combustion, co-combustion with low percentages of biomass, and MSW incineration are biomass generation alternatives currently mature and commercially available; For their part, gasification is still under development, while gasification integrated with cogeneration is in the demonstration phase (IRENA, 2012a).

Table 4.9 shows the ranges of installation costs of different thermal energy and/or electricity generation technologies with biomass analyzed by several authors (IRENA, 2012a).

Table 4.9 Installation costs of thermal energy and/or electricity generation technologies with biomass, prepared with data from IRENA 2012a

Technology	Installation cost USD2010/kW
Direct combustion boilers	\$1390–\$3000
Boilers with cogeneration	\$2500–\$5080
FB boiler	\$1440–\$3000
FB boiler with cogeneration	\$4260–\$15,500
Bubbling bed boiler	\$2540–\$3860
Co-firing (conversion of traditional combustion systems)	\$100–600
Fixed-bed gasifier (with internal combustion engine)	\$1730–\$5074
Fixed-bed gasifier (with gas turbine)	\$3000–\$3500
FB gasifier (with gas turbine)	\$2470–\$4610
Fixed and fluidized bed gasifiers	\$2140–\$5700
Boilers with cogeneration	\$3550–\$6820
Gasifier with cogeneration	\$5570–\$6545
Co-combustion in existing plants	\$140–\$850

Source: From IRENA, 2012a. Renewable energy technologies: cost analysis series. Biomass Power Generation 1(1/5), 60. doi:10.1016/B978-0-08-098330-1.00011-9.

Table 4.10 Operation and maintenance costs (fixed and variable) of generation technologies with biomass, prepared with data from IRENA 2012a

Technology	Fixed O&M costs (% of investment cost)	Variable O&M costs (USD2010/MWh)
Any configuration of boilers, including co-combustion	3%–6%	$3.8–$4.7
Gasifiers	3%–%6	$3.7

Source: From IRENA, 2012a. Renewable energy technologies: cost analysis series. Biomass Power Generation 1(1/5), 60. doi:10.1016/B978-0-08-098330-1.00011-9.

Considering these aspects, it is observed, for example, that the investment cost in combustion boilers varies between $1880 and $4260 USD_{2010}/kW, while the costs for FB boilers are between $2170 and $4500 USD_{2010}/kW. For its part, the modernization of a combustion boiler to co-combustion requires minor investments, in the order of $400–$600 USD_{2010}/kW.

With regard to operation and maintenance (O&M) costs, the fixed and variable costs associated with the operation of biomass power plants are considered (Table 4.10). The fixed costs of O&M include labor, scheduled maintenance, routine replacement of components, and insurance, among others. They can be expressed as a percentage of the cost of the initial investment, with a range between 1% and 6% of the investments per year. The larger the size of the plant, the lower are the fixed costs of O&M (IRENA, 2012a). Table 4.10 shows O&M costs, both fixed and variable (IRENA, 2012a).

Variable O&M costs depend on the final configuration or output of the system and are usually expressed in currency per unit of output (USD/kWh), include fuel costs (non-biomass), disposal of the ashes, maintenance not planned, equipment replacement, and cost of services.

4.6 CONCLUSIONS

- The use of solid biomass as fuel to generate electricity and process heat has been increasing in recent years.
- The problems related to climate change have been strong drivers for a greater use of solid biomass as a source of energy generation; this is one of the reasons why various industries have decided to take advantage of their biomass waste.

- Thermochemical processes for the thermal utilization of biomass arise as a response to the problem of urban solid waste pollution.
- The main thermal conversion processes of solid biomass in thermal energy and electricity are combustion, co-combustion, gasification, cogeneration, incineration; however, this last process generates by-products harmful to the environment and public health.
- Biomass energy generation technologies are mature and are a competitive option, but they depend to a greater extent on the availability of the biomass resource.
- In the case of co-combustion, the investment costs are marginal, which allows to take advantage of biomass with competitive prices.
- An essential parameter of energy production from biomass is that the production of fuel is low cost.
- In general terms, cogeneration systems have a higher investment cost with respect to single-cycle generation technologies.
- The process that has better yields and lower emissions is gasification; however, it still presents important challenges to be economically viable.

ACKNOWLEDGMENTS

We appreciate the *SENER-CONACyT CEMIEBIO-24691 Cluster of Solid Biofuels (CBS) for Thermal and Electric Generation* project for the financing of this work. We also thank María de Jesús Pérez Orozco for the technical support to develop the document. In the same way we thank Emmanuel Gómez his support in the preparation of this chapter and Mariano Birlain for his support in the English edition of this chapter.

REFERENCES

Ahrenfeldt, J., Thomsen, T.P., Henriksen, U., Clausen, L.R., 2013. Biomass gasification cogeneration—a review of state of the art technology and near future perspectives. Appl. Therm. Eng. 50 (2), 1407–1417. Available from: https://doi.org/10.1016/j.applthermaleng.2011.12.040.

Arteaga-Pérez, L.E., Casas-Ledón, Yannay, Cabrera-Hernández, Jagni, Rodríguez, L., 2014. Gasificación de biomasa para la producción sostenible de energía. Revisión de las tecnologías y barreras para su aplicación. Afinidad Lxxii 5641, 138–145.

Asadullah, M., 2014. Barriers of commercial power generation using biomass gasification gas: a review prospects. Renew Sust. Energy Rev. 29, 201–215. Available from: https://doi.org/10.1016/j.rser.2013.08.074.

Ayres, R., 2007. Energy and economic growth. In: Ulgiati, S. (Ed.), Sustainable Energy Production and Consumption: Benefits, Strategies and Environmental Costing. Springer.

Barton, L., Butterbach-Bahl, K., Kiese, R., Murphy, D.V., 2011. Nitrous oxide fluxes from a grain-legume crop (narrow-leafed lupin) grown in a semiarid climate. Global Change Biol. 17 (2), 1153–1166. Available from: https://doi.org/10.1111/j.1365-2486.2010.02260.x.

Bauen, A., Berndes, G., Junginger, M., Londo, M., Vuille, F., Ball, R., et al., 2009. Bioenergy—a sustainable and reliable energy source: a review of status and prospects. IEA Bioenergy, Paris, France, p. 108.

Beohar, H., Gupta, B., Sethi, V.K., Pandey, P.M., 2012. Parametric study of fixed bed biomass gasifier: a review. Int. J. Therm. Technol. 2 (1), 134–140.

Bocci, E., Sisinni, M., Moneti, M., Vecchione, L., Di Carlo, A., Villarini, M., 2014. State of art of small scale biomass gasification power systems: a review of the different typologies. Energy Procedia 45, 247–256. Available from: https://doi.org/10.1016/j.egypro.2014.01.027.

Bott, R., 2014. Solid biofuels for energy systems. In: Igarss 2014. doi:10.1007/s13398-014-0173-7.2.

Brown, A., Le Feuvre, P., 2017. IEA Technology Roadmap: Delivering Sustainable Bioenergy. IEA, p. 94.

Carneiro, P., Ferreira, P., 2012. The economic, environmental and strategic value of biomass. Renewable Energy 44, 17–22. Available from: https://doi.org/10.1016/j.renene.2011.12.020.

Carneiro P, Ferreira, P., 2010. A contribution to economic evaluation of biomass energy. In: International Conference on Applied Business and Economics (ICABE), La Coruna, Spain.

Demirbaş, A., 2001. Biomass resource facilities and biomass conversion processing for fuels and chemicals. Energy Convers. Manag. 42 (11), 1357–1378. Available from: https://doi.org/10.1016/S0196-8904(00)00137-0.

ENplus®, History. Retrieved March 23, 2018, from https://enplus-pellets.eu/en-in/about-us-en-in/history-of-enplus.html.

Erhart, T., Eicker, U., Infield, D., 2011. Part-load characteristics of Organic-Rankine-Cycles. In: 2nd European Conference of Polygeneration. Tarragona, p. 11.

ETSAP/IEA, 2015. Biomass for heat and power. Technology Roadmaps (2), 1–41. Retrieved from http://www.etsap.org/E-techDS/PDF/E05-Biomass for HP-GS-AD-gct.pdf.

European Environment Agency (EEA), 2008. Maximising the environmental benefits of Europe's bioenergy potential. Technical Report 10/2008. Copenhagen, Denmark: European Environment Agency.

FAO, 2009. Biocombustibles y agricultura: panorama técnico. Biocombustibles: Perspectivas, Riesgos Y Oportunidades 11–25. Retrieved from ftp://ftp.fao.org/docrep/fao/011/i0100s/i0100s02.pdf.

Fruergaard, T., Christensen, T.H., Astrup, T., 2010. Energy recovery from waste incineration: assessing the importance of district heating networks. Waste Manage. 30, 1264–1272. Available from: https://doi.org/10.1016/j.wasman.2010.03.026.

García-Frapolli, E., Schilmann, A., Berrueta, V.M., Riojas-Rodríguez, H., Edwards, R. D., Johnson, M., et al., 2010. Beyond fuelwood savings: valuing the economic benefits of introducing improved biomass cookstoves in the Purépecha region of Mexico. Ecol. Econ. 69 (12), 2598–2605. Available from: https://doi.org/10.1016/j.ecolecon.2010.08.004.

GBEP (Global Bioenergy Partnertship), Working Together for Sustainable Development Since 2006. Retrieved March 24, 2018, from http://www.globalbioenergy.org.

Gerssen-Gondelach, S.J., Saygin, D., Wicke, B., Patel, M.K., Faaij, A.P.C., 2014. Competing uses of biomass: Assessment and comparison of the performance of bio-based heat, power, fuels and materials. Renew. Sust. Energy Rev. 40 (April), 964–998. Available from: https://doi.org/10.1016/j.rser.2014.07.197.

Gómez-Barea, A., Leckner, B., Villanueva Perales, A., Nilsson, S., Fuentes Cano, D., 2013. Improving the performance of fluidized bed biomass/waste gasifiers for distributed electricity: a new three-stage gasification system. Appl. Therm. Eng. 50 (2), 1453–1462. Available from: https://doi.org/10.1016/j.applthermaleng.2011.12.025.

International Renewable Energy Agency (IRENA), 2015. Biomass for Heat and Power—Technology Brief. IRENA.

IPCC, Chum, H., Faaij A., Moreira, J., Berndes, G., Dhamija, P., et al., 2012. Bioenergy. In: IPCC Special Report on Renewable Energy Sources and Climate Change Mitigation, 209−332. Retrieved from http://www.ipcc.ch/pdf/special-reports/srren/Chapter 2 Bioenergy.pdf.

IRENA (International Renewable Energy Agency), 2012a. Renewable Energy Technologies: Cost Analysis Series. Biomass for Power Generation 1 (1/5), 60. Available from: https://doi.org/10.1016/B978-0-08-098330-1.00011-9.

IRENA (International Renewable Energy Agency), 2012b. Technology Roadmap: Bioenergy for Heat and Power. Available at: www.iea.org/publications/freepublications/publication/bioenergy.pdf.

IRENA, 2013. Biomass co-firing: technology brief. In: Edenhofer, Ottmar, et al., (Eds.), IEA-ETSAP and IRENA© Technology Brief E21. IPCC Working Group III. Available from: https://doi.org/IEA-ETSAP.

ISO—International Organization for Standardization, n.d. https://www.iso.org/home.html (accessed 24.03.18).

Jaap Koppejan, S., van, L., 2012. The Handbook of Biomass Combustion and Co-firing. Routledge.

Kirkels, A.F., Verbong, G.P.J., 2011. Biomass gasification: still promising? A 30-year global overview Renew. Sustain. Energy Rev. 15 (1), 471−481.

Kummamuru B., 2017. WBA Global Bioenergy Statistics 2017, pp. 80. Available at: http://www.worldbioenergy.org/uploads/WBA GBS 2017_hq.pdf.

Li, J., Brzdekiewicz, A., Yang, W., Blasiak, W., 2012. Co-firing based on biomass torrefaction in a pulverized coal boiler with aim of 100% fuel switching. Appl. Energy 99, 344−354. Available from: https://doi.org/10.1016/j.apenergy.2012.05.046.

Lichtfouse, E., 2012. Organic Fertilisation, Soil Quality and Human Health, vol. 9. Springer. Available from: https://doi.org/10.1007/978-94-007-4113-3.

Maciejewska, A., Veringa, H., Sanders, J., Peteves, S.D., 2006. Co-firing of biomass with coal: constraints and role of biomass pre-treatment. Report of the Institute for Energy—JRC, European Commission.

Maraver, D., Sin, A., Royo, J., Sebastián, F., 2013. Assessment of CCHP systems based on biomass combustion for small-scale applications through a review of the technology and analysis of energy efficiency parameters. Appl. Energy 102, 1303−1313. Available from: https://doi.org/10.1016/j.apenergy.2012.07.012.

Masera, O.R., Bailis, R., Drigo, R., Ghilardi, A., Ruiz-Mercado, I., 2015. Environmental burden of traditional bioenergy use. Annu. Rev. Environ. Res. 40, 121−150. Available from: https://doi.org/10.1146/annurev-environ-102014-021318.

McKendry, P., 2002a. Energy production from biomass (part 1): overview of biomass. Biores. Technol. 83 (1), 37−46. Available from: https://doi.org/10.1016/S0960-8524(01)00118-3.

McKendry, P., 2002b. Energy production from biomass (part 2): conversion technologies. Biores. Technol. 83 (1), 47−54. Available from: https://doi.org/10.1016/S0960-8524(01)00119-5.

Mitchell, R.B., Schmer, M.R., Anderson, W.F., Jin, V., Balkcom, K.S., Kiniry, J., et al., 2016. Dedicated energy crops and crop residues for bioenergy feedstocks in the central and eastern USA. Bioenergy Res. 9 (2), 384−398. Available from: https://doi.org/10.1007/s12155-016-9734-2.

OECD/IEA, 2017. How2Guide for Bioenergy Roadmap Development and Implementation. IEA, ISBN 978-92-5-109586-7.

Pereira, E.G., Da Silva, J.N., De Oliveira, J.L., MacHado, C.S., 2012. Sustainable energy: a review of gasification technologies. Renew. Sust. Energy Rev. 16 (7), 4753−4762. Available from: https://doi.org/10.1016/j.rser.2012.04.023.

Peterson D., Haase S., 2009. Market assessment of biomass gasification and combustion technology for small- and medium-scale applications. Technical Report NREL/TP-7A2-46190.

Puig-Arnavat, M., Bruno, J.C., Coronas, A., 2010. Review and analysis of biomass gasification models. Renew. Sust. Energy Rev. 14, 2841–2851.

Puig-Arnavat, M., Bruno, J.C., Coronas, A., 2014. Modeling of trigeneration configurations based on biomass gasification and comparison of performance. Appl. Energy 114, 845–856. Available from: https://doi.org/10.1016/j.apenergy.2013.09.013.

Quina, M.J., Bordado, J.C.M., Quinta-Ferreira, R.M., 2011. Air pollution control in municipal solid waste incinerators. In: Khallaf, M. (Ed.), The Impact of Air Pollution on Health, Economy, Environment and Agricultural Sources. InTech 978-953-307-528-0. Available from: http://www.intechopen.com/books/the-impact-of-air-pollution-on-health-economy-environment-and-agricultural-sources/air-pollution-control-in-municipal-solid-waste-incinerators.

REN21, 2017a. Renewables 2017: global status report. Renew. Sust. Energy Rev. 72. Available from: https://doi.org/10.1016/j.rser.2016.09.082.

REN21, 2017b. Renewables Global Futures Report: Great Debates Towards 100% Renewable Energy. REN21, Paris, France.

RSB (Roundtable on Sustainable Biomaterials), International Sustainability and Biomaterials Certification. Retrieved March 23, 2018, from http://rsb.org/.

Ruiz, J.A., Juárez, M.C., Morales, M.P., Muñoz, P., Mendívil, M.A., 2013. Biomass gasification for electricity generation: review of current technology barriers. Renew. Sust. Energy Rev. 18, 174–183. Available from: https://doi.org/10.1016/j.rser.2012.10.021.

Sansaniwal, S.K., Pal, K., Rosen, M.A., Tyagi, S.K., 2017. Recent advances in the development of biomass gasification technology: a comprehensive review. Renew. Sust. Energy Rev. 72 (January), 363–384. Available from: https://doi.org/10.1016/j.rser.2017.01.038.

Serup, H., Kofman, P., Falster, H., 2005. Wood for Energy Production, Irish edition. COFORD, Dublin, p. 72.

Sims, R., Taylor, M., Jack, S., Mabee, W., 2008. From 1st to 2nd generation bio fuel technologies: an overview of current industry and RD&D activities. IEA Bioenergy 1–124. Retrieved from https://www.iea.org/publications/freepublications/publication/2nd_Biofuel_Gen.pdf.

Smith, P., Bustamante, M., Ahammad, H., Clark, H., Dong, H., Elsiddig, E.A., et al., 2014. Agriculture, Forestry and Other Land Use (AFOLU). In: Edenhofer, O., et al., (Eds.), Climate Change 2014: Mitigation of Climate Change. Contribution of Working Group III to the Fifth Assessment Report of the Intergovernmental Panel on Climate Change. Cambridge University Press, Cambridge, UK and New York, NY, USA.

Strzalka, R., Schneider, D., Eicker, U., 2017. Current status of bioenergy technologies in Germany. Renew. Sust. Energy Rev. 72 (January), 801–820. Available from: https://doi.org/10.1016/j.rser.2017.01.091.

Sutton, D., Kelleher, B., Ross, J.R.H., 2001. Review of literature on catalysts for biomass gasification. Fuel Process. Technol. 73 (3), 155–173. Available from: https://doi.org/10.1016/S0378-3820(01)00208-9.

Tańczuk, M., Ulbrich, R., 2013. Implementation of a biomass-fired co-generation plant supplied with an ORC (Organic Rankine Cycle) as a heat source for small scale heat distribution system—a comparative analysis under Polish and German conditions. Energy 62, 132–141. Available from: https://doi.org/10.1016/j.energy.2013.09.044.

US Department of Energy, 2007. Methodology for Allocating Municipal Solid Waste to Biogenic and Non-Biogenic Energy. Energy Information Administration: Office of Coal, Nuclear, Electric and Alternate Fuels, Washington, DC, pp. 1–18. Available at: https://www.eia.gov/totalenergy/data/monthly/pdf/historical/msw.pdf.

Verna, R.S.M., 2013. Symmetrical analysis of biomass gasification techniques. MR Int J Eng Technol 5 (1), 17–24.
Wang, L., 2014. Sustainable Bioenergy Production. CRC Press, Boca Raton, FL, p. 2014.
WBA (World Bioenergy Association), Bioenergy and Sustainability. Retrieved March 23, 2018, from http://worldbioenergy.org/news/329/56/Bioenergy-and-Sustainability.
World Energy Council, 2016. World Energy Resources Bioenergy. 60. doi:10.1016/0165-232X(80)90063-4.

CHAPTER FIVE

Biorefineries

Maria Hingsamer and Gerfried Jungmeier
Joanneum Research Forschungsgesellschaft mbH, LIFE—Centre for Climate, Energy and Society, Graz, Austria

Contents

5.1 Introduction 180
5.2 Description of Biorefineries 181
  5.2.1 Definition of biorefineries 181
  5.2.2 Motivations 183
  5.2.3 Classification of biorefinery systems 185
5.3 Concepts of Biorefineries 189
  5.3.1 Introduction 189
  5.3.2 Examples: energy-driven biorefineries 190
  5.3.3 Example: product-driven biorefinery 194
5.4 Integration in Existing Infrastructure 196
  5.4.1 Methodology 196
  5.4.2 Example: Austria 197
  5.4.3 Conclusions 201
5.5 Biorefinery Complexity Index 201
  5.5.1 Methodology 201
  5.5.2 Examples 203
  5.5.3 Conclusions 205
5.6 Economic and Environmental Aspects of Biorefineries 206
  5.6.1 Methodology 206
  5.6.2 Examples 208
  5.6.3 Biorefinery Fact Sheets 210
  5.6.4 Example: Biorefinery Fact sheet 211
5.7 Perspectives 217
  5.7.1 Outlook and future paths in biorefineries 217
  5.7.2 Discussion and conclusions 219
References 221

The Role of Bioenergy in the Emerging Bioeconomy.
DOI: https://doi.org/10.1016/B978-0-12-813056-8.00005-4

5.1 INTRODUCTION

A key driver to satisfy the energy and material demand of the society sustainably is the implementation of the bioeconomy, which is based on renewable bio-based resources for the production of materials and energy. Different motivations for policy and research to speed the transition towards the bioeconomy are, e.g., the policy goals of climate change mitigation, energy security, the circular economy concept, rural development, and reindustrialization, as well as stimulation of innovation and technology development. Biorefining is a main element in the framework of the emerging bioeconomy as the broad spectrum of biomass resources offers great opportunities for a wide-ranging product portfolio to satisfy the different needs of society.

Biomass (e.g., crops, algae, and residues) has to be sustainably produced and used as efficiently as possible. Therefore, biocascading and biorefining approaches have to be applied. Such approaches help to cover future demands for food, feed, chemicals, materials, transportation fuels, power, and heat.

Currently some biorefineries are operating on a commercial scale, e.g., pulp and paper industry, biofuel industry, and food industry. Furthermore, many different newly advanced biorefineries are under development. The main characteristics of a biorefinery are:

- The coupled generation of energy (e.g., gaseous or liquid biofuels) and materials (e.g., chemicals, food, and feed).
- A combination of several process steps (e.g., mechanical processes such as pressing, and thermochemical processes such as gasification).
- The use of different raw materials from both virgin and residual sources.

Biomass resources are fractionated into their composing intermediates (such as proteins, sugars, oils and fibers/lignins) that are further processed by biochemical and/or thermochemical (catalytically supported) pathways to marketable bio-based products and bioenergy. The biomass resources (land and aquatic crops; agro, forestry, and process residues; and postconsumer residues) have to be used in a resource-efficient, energy-efficient and sustainable way to produce bio-based products (food or feed ingredients, chemicals, and materials) and bioenergy (biofuels, power, and heat).

An aim of the biorefinery is furthermore to use all the synergies for a sustainable and efficient production, to maximize or optimize the economic, environmental, and social benefits. The future expected growth of the biofuel market and the development of new production processes for biofuels make it necessary to develop new integrated biorefineries. Biomass conversion plants will require similar plant concepts as today's chemical plants or today's refining of crude oil. The integration of new biorefinery concepts into existing industrial complexes has interesting prospects, e.g., to reduce the capital costs of the biofuel production facilities and thus to reduce the costs of the chemical and energy products produced. Necessary for the sustainable development are implementing strategies of biorefineries in the future emerging bioeconomy.

5.2 DESCRIPTION OF BIOREFINERIES

5.2.1 Definition of biorefineries

Currently many different biorefinery concepts have been developed and are already being implemented, producing a wide variety of products and playing a key role in establishing a bioeconomy. IEA Bioenergy Task 42 "Biorefining" developed a definition for biorefining:

> *Biorefining is the sustainable processing of biomass into a spectrum of marketable products (food, feed, materials, chemicals) and energy (fuels, power, heat)*
>
> ***Cherubini et al. (2009); Jungmeier et al. (2009).***

Energy and material recovery from biomass simplifies the conversion and expansion of the raw materials economy and thus also the development of an economic system oriented around sustainability criteria. To an increasing degree, companies are beginning to realize market opportunities in this field, mostly with a view to environmentally friendly production processes and developing new and renewable sources of raw materials. Biomass utilization can also help safeguard agricultural jobs and set the stage for a renaissance of rural space and a reinterpretation of its "supply role" (VDI, 2016).

An additional advantage of biomass utilization is that unused or only partially used material streams can be directed to industrial use, helping to develop innovative conversion and refining technologies to provide a competitive advantage. At the same time, social currents and opinion formation can be seen raising the topics of competition between the food and feed sectors and energy and material recovery from biomass, as well as the limited availability of land, speculation on raw materials, and consequences for biodiversity and conservation (VDI, 2016).

A major driver for the establishment of biorefineries is the sustainability. All biorefineries have to be assessed over the whole value chain on their environmental, economic, and social sustainability covering the whole life cycle (construction–operation–dismantling). Therefore, the methodology Life Cycle Sustainability Assessment (LCSA) can be used. LCSA is a combination of established methods for assessing environmental, economic, and social sustainability parameters. Three internationally approved methods are combined Life Cycle Assessment (LCA) for environmental impacts, Life Cycle Costing (LCC) for economic impacts, and Social Life Cycle Assessment (S-LCA) for social impacts. Relevant parameters for the assessment of sustainability of biorefineries include resource demand, greenhouse gas (GHG) emissions, primary energy demand, production costs, added value, and social impacts on the population/employees/stakeholders. This assessment should also take into account the possible consequences due to the competition for food and biomass resources, the impact on water use and quality, changes in land-use, soil carbon stock balance and fertility, and impact on biodiversity. Effects on international and regional dynamics, end-users, and investment feasibility are also important aspects to take into account. As the results of the LCSA are not absolute indicators, it is made in comparison to conventional systems providing the same amount of products and services.

A biorefinery is the integrated upstream, midstream, and downstream processing of biomass into a range of products. A biorefinery can use all kinds of biomass from forestry, agriculture, and aquaculture, and residues from industry and households including wood, agricultural crops, organic residues (plant and animal derived), forest residues, and aquatic biomass (algae and sea weeds). A biorefinery is not a totally new concept. A lot of traditional biomass-converting technologies such as those used in the sugar, starch, and pulp and paper industry can be (partly) considered as biorefineries. Several economic and environmental drivers such as global warming, energy conservation, security of supply, and agricultural policies have also directed those industries to further improve their operations in a biorefinery way (de Jong and Jungmeier, 2015).

A biorefinery should produce a spectrum of marketable products and energy. The former can be both intermediates and final products, and include food, feed, materials, and chemicals; whereas energy includes fuels, power, and heat. The main focus of biorefinery systems, which came into operation in the last years, was the production of transportation biofuels. This development is changing actually, as higher value products (e.g., biochemicals, biomaterials) derived from biomass are coming more and more into the focus of policy, industries, research, and further stakeholders.

Biorefineries are expected to improve the competitiveness and wealth of the countries by responding to the need for supplying a wide range of bio-based products and energy in an economically, socially, and environmentally sustainable manner. Biorefineries show promise in both industrialized and developing countries. New competences, job opportunities, and markets are also expected (de Jong and Jungmeier, 2015).

5.2.2 Motivations

In Fig. 5.1, the trade-off between market volume and price of transport fuels, biomaterials, and products is shown. Due to the current low price of fossil fuels, the market price of transportation biofuels is lower compared to the market price of chemicals and materials. However, the market volume is significantly higher for transportation fuels. Currently it is difficult to identify the most promising biorefinery products with the most interesting economic potential, since many framework conditions must be considered, e.g., raw material prices, energy prices, regulatory conditions like 10% renewable transportation fuels in Europe 2020 according to the Renewable Energy Directive (RED). Furthermore, the RED requires the EU to fulfill at least 20% of its total energy needs with renewables by 2020.

The European Commission is actually working on a new proposal for a RED during 2021–30 (REDII). The latest RED II proposal, dated

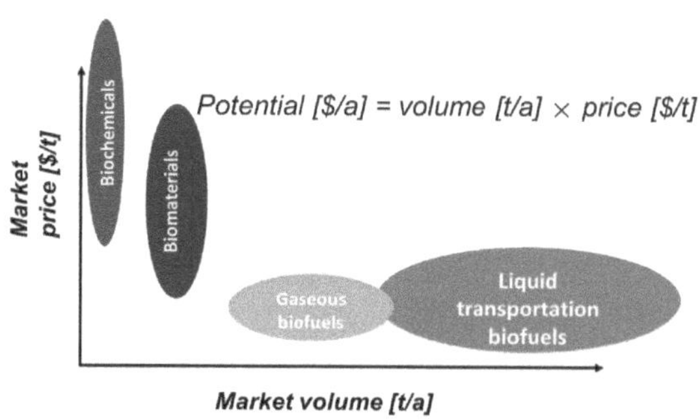

Figure 5.1 Trade-off: Market volume and price of different biorefinery products.

January 2018, includes the binding of EU-level targets of an 35% improvement in energy efficiency, a minimum 35% share of energy from renewable sources in gross final consumption of energy, and a 12% share of energy from renewable sources in transport, by 2030. The main elements of the proposal on transport are to cap the food and feed-based biofuels ("first-generation biofuels") to 2017 levels, with a maximum of 7% in road and rail transport. Furthermore, biodiesel made from palm oil will not count towards the renewable targets from 2021. The share of advanced biofuels (having a lower impact on land use), renewable transport fuels of nonbiological origin, waste-based fossil fuels, and renewable electricity will have to be at least 1.5% in 2021, rising to 10% in 2030 (European Parliament, 2018).

In principle two different motivations for biorefineries have been distinguished in IEA Bioenergy Task 42 (Jungmeier et al., 2013b, van Ree and van Zeeland, 2014):

- "product-driven" biorefineries (e.g., pulp and paper) and
- "energy-driven" biorefineries (e.g., road transportation biofuels).

In "energy-driven" biorefineries the main goal is to produce large volumes of relatively low-cost energy from biomass. Here, biomass is primarily used for the production of transportation biofuels, power, heating, cooling, or biogas. Process residues are sold or even further upgraded to value-added bio-based products or value-added feed to provide further economic and environmental benefits. Full value chain infrastructures exist; however, their profitability is still questionable and they require significant financial governmental support or a regulated market to guarantee large-scale market deployment. The expectation is that the coproduced biochemicals and biomaterials give additional economic and environmental benefits by creating additional value and distributing the environmental burdens. But it is evident that overall optimization strategies are necessary as there is a trade-off between maximizing the amount of transportation biofuels production and additional benefits from biochemicals and biomaterials.

In "product-driven" (i.e., chemicals, materials) biorefineries, the main goal is to produce smaller amounts of relatively higher value-added bio-based products. Biomass is typically fractionated into a portfolio of bio-based products with the focus to derive the highest economic value from the biomass. A cascade approach is often used where the primary (agro) and secondary (process) residues are used to produce feed or energy (power/heat) for internal or external use. Currently, only a few product-driven biorefineries are in operation, mainly because of the fact that some key technologies are still in the R&D, pilot, and demo-phase. However, their potential is huge, and it is generally believed that a refocusing will take place from mainly energy (fuel) applications to chemical/material applications, and—depending on the raw materials used—even to biorefineries that use biomass for both food and nonfood applications.

The current energy/fuel infrastructures, and the expertise to implement full biomass value chains, will be used as a starting point in transition process. However, also on the longer term, bioenergy, including biofuels, will be produced in significant amounts from primary (agro), secondary (process), and tertiary (postconsumer) residues, and therefore, will remain the cornerstone of a future bioeconomy.

5.2.3 Classification of biorefinery systems

IEA Bioenergy Task 42 (Jungmeier et al., 2009, 2013b, Cherubini et al., 2009) has developed a systematic classification system and naming scheme to describe different biorefineries. The developed system is based on a schematic representation of the full "biomass to end-products" chains. The classification of a biorefinery consists of the following four main features:

1. platforms,
2. products,
3. feedstocks, and
4. processes.

With the combination of these four features, different biorefinery configurations can be described in a consistent manner.

In Fig. 5.2 a generic classification scheme of a biorefinery is shown on the left and an example of the production of biodiesel, glycerin, and animal feed from oilseed crops in a biorefinery on the right.

The most important feature in the classification of biorefineries is the platform (or platforms). Platforms might be (Fig. 5.3):

- intermediate products in the biorefinery structure which may be further transformed into other intermediate or end products,
- linkages between different biorefinery concepts, or
- final products of a biorefinery.

The number of platforms involved is an indication of the system complexity. The platforms might represent mixtures of compounds (C6 and lignin, C5 and C6 sugars) or more isolated compounds. For the platform

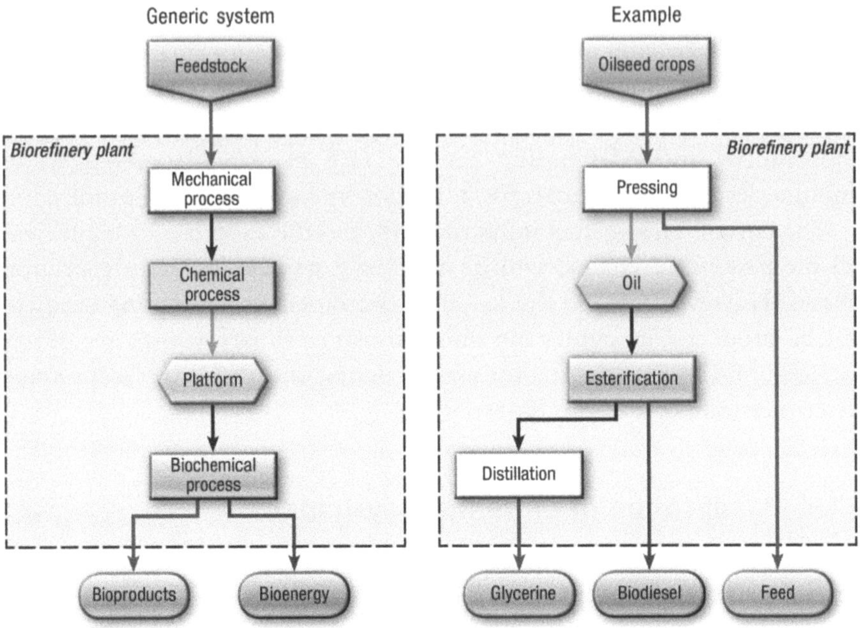

Figure 5.2 Classification scheme of a biorefinery: generic scheme (left), example (right) "1-platform (oil) biorefinery using oilseed crops for biodiesel, glycerin, and feed via pressing, esterification, and distillation."

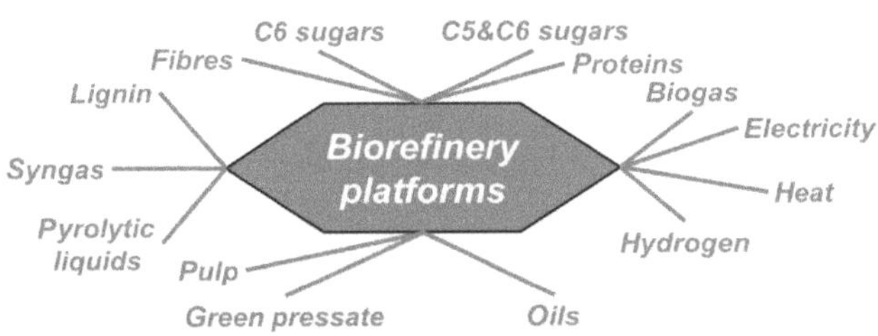

Figure 5.3 Examples for possible platforms in a biorefinery system.

"electricity and heat" in the different biorefinery concepts, it is important to define, if electricity and heat are produced from:

- process residues,
- directly from biomass feedstock,
- fossil fuels, or
- other forms of renewable energy (e.g., wind, solar).

Electricity and heat can be produced within the biorefinery plant, e.g., from process residues or can be covered by external supply.

The two product groups of biorefineries are energy (e.g., bioethanol, biodiesel, electricity, heat, and synthetic biofuels) and products (e.g., materials, chemicals, food, and feed). The biorefinery feedstocks can be energy crops from agriculture (e.g., starch crops, short rotation forestry) and aquaculture (algae, seaweeds), furthermore, biomass residues from agriculture, forestry, trade, and industry (e.g., straw, bark, wood chips from forest residues, used cooking oils, and municipal solid waste). A classification can be applied due to four main conversion processes: biochemical (e.g., fermentation and enzymatic conversion), thermochemical (e.g., gasification and pyrolysis), chemical (e.g., synthesis and esterification), and mechanical processes (e.g., fractionation and pressing).

When describing a biorefinery, both the platform(s) and the feedstocks(s) are taken into consideration and specified in the name; furthermore, the primary products may also be named. The naming of a biorefinery system consists of the following four elements:

1. number and name of the platform(s),
2. feedstock(s),
3. product(s), and
4. optionally processes.

The naming of the example in Fig. 5.2 (right) is the following:

"1-platform (oil) biorefinery using oilseed crops for biodiesel, glycerin, and feed via pressing, esterification, and distillation".

This classification system was also taken by the Verein Deutscher Ingenieure and forms the basis for the VDI Standard (VDI, 2016). The platforms are connected with the feedstock and products or other platforms through processes. Specific technologies are used, depending on the feedstock, platform, and product. It is possible to introduce additional product lines. It is also possible to add entire platforms, if, for example, a product should serve as the base material for further syntheses in the biorefinery (e.g., glycerin). This makes the classification model highly flexible, because arbitrary subgroups can be added at later point (VDI, 2016).

Biorefinery concepts can furthermore be combined at different levels to construct bigger "Biorefinery Complexes." Also, they can be colocated and integrated with oil industry infrastructure. The combination of the concepts can be done on the level of the same feedstock, same platforms, same products, or same processes. In Fig. 5.4 possible feedstocks, platforms, products, or processes of biorefineries for possible combinations

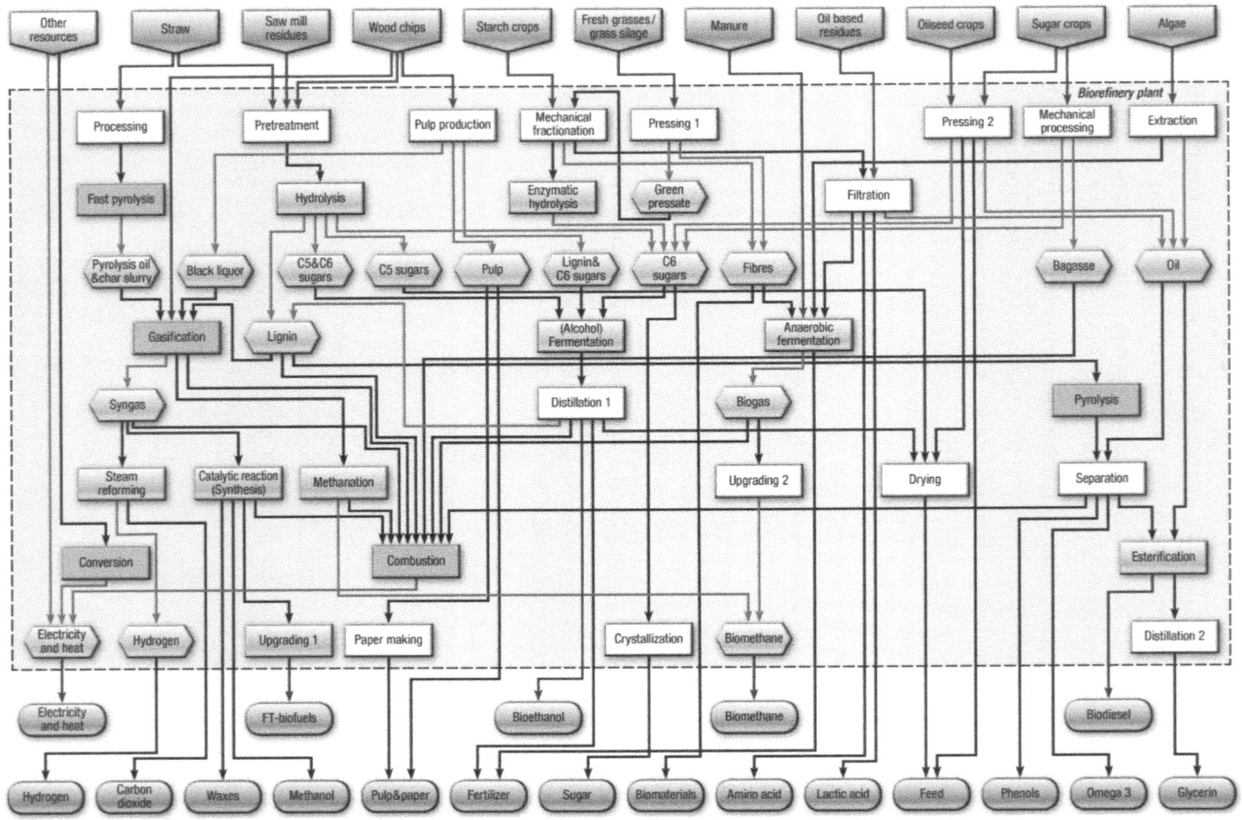

Figure 5.4 Possible feedstocks, platforms, products, and processes of energy-driven biorefineries.

are presented (Jungmeier et al., 2013b). Examples for feedstocks are oil crops, oil residues, sugar crops, starch crops, wood chips, straw, grass, saw mill residues (sulfite), liquor, and algae. For processes it is possible to distinguish between four main groups: these are physical (including mechanical) processes, thermochemical processes, chemical processes, and biotechnological processes. From these processes, transportation biofuels can be produced (e.g., biodiesel, bioethanol, Fischer–Tropsch (FT)-biofuels, biomethane from upgraded biogas, and synthetic natural gas). Next to the biofuels, biomaterials/bio-products are produced (e.g., animal feed, glycerin, phenol, pulp and paper, amino acids, lactic acids, fertilizer, waxes, hydrogen, and carbon dioxide).

5.3 CONCEPTS OF BIOREFINERIES

5.3.1 Introduction

Biorefineries are not a new development; they are already applied for example in the food industry for a long time. Conventional biorefineries exists for some considerable time. Industrial biorefining was initiated by the introduction of steam-driven paper machines in the 19th century. Developments in food production (e.g., crystalline sugar, potato starch, wheat and corn starch, soy oil, proteins, and vitamins) form the basis for most innovations in biorefining (de Jong and Jungmeier, 2015). Industrial processing techniques, which were developed mainly in Europe and North America, are applied worldwide and are examples for biorefining evolvement.

As mentioned before actually a significant number of biorefinery concepts are developed. A lot of development is actually ongoing, but research, development, and demonstration of biorefineries are very complex. Next to research activities in the field of integrated biorefinery concepts, a lot of research projects are focusing on one or more aspects or components of biorefineries.

The state of technology and the commercialization of biorefinery concepts are quite different for biorefineries. Three different states of technologies for energy-driven biorefineries can be distinguished (Jungmeier et al., 2013b):

- Commercial scale energy-driven biorefineries: These biorefineries are state of the art and are worldwide in commercial operation under current economic conditions (e.g., "a 1-platform (C6 sugars) biorefinery using sugar and starch crops for bioethanol and feed", "a 3-platform (C6 sugars, bagasse, and electricity and heat) biorefinery using sugar cane for bioethanol, electricity, heat, sugar, and fertilizer").

- Demonstration scale energy-driven biorefineries: These biorefineries or their main processes are demonstrated on a technical scale at one or more locations worldwide, but they need further technical optimization and cannot be operated under current commercial conditions. It is expected that these biorefineries can be commercially operated in 2025 (e.g., "a 3-platform (C5 and C6 sugars, electricity and heat, and lignin) biorefinery using wood chips for bioethanol, electricity, heat, and phenols", "a 5-platform (C6 sugars, C5 and C6 sugars, lignin, syngas, and electricity and heat) biorefinery using starch crops and straw for bioethanol, FT-biofuels, feed, electricity, and heat").
- Conceptual energy-driven biorefineries: These biorefineries are not demonstrated on technical scale so far, but it is expected that they will be further technically developed and demonstrated after further necessary R&D developments within the next years. It is expected that these biorefineries can be commercially operated in 2030 (e.g., "a 3-platform (pyrolysis oil, syngas, and electricity and heat) biorefinery using straw for FT-biofuels and methanol with oxygen gasification", "a 5-platform (C6 and C5 sugars, lignin and C6 sugars, electricity and heat) biorefinery using saw mill residues, wood chips, and sulfite liquor for bioethanol, pulp and paper, and electricity and heat").

5.3.2 Examples: energy-driven biorefineries

Advanced biorefinery concepts are still in the R&D, pilot, or small-scale demonstration phase with commercialization being further away. It is expected that a commercialization and implementation in the market of different concepts of advanced biorefineries will happen in the medium term (until 2030) in different countries (Bundesregierung der Bundesrepublik, Deutschland, 2012). Current economic conditions (e.g., relatively low oil prices) can cause delays in the market implementation of different biorefinery concepts. Furthermore, a multiproduct valorization is necessary to be fully market competitive and flexible, the biomass supply has to be sustainable, and the biomass sources have to be used in an efficient way.

Advanced biorefinery concepts are designed to use a wide range of biomass resources and for the production of a range of different products for different markets. VDI, 2016 summarized various biorefinery concepts in following groups:

- Sugar or starch biorefineries
- Vegetable oil biorefineries
- Lignocellulose biorefineries (cellulose/hemicellulose/lignin biorefineries)
- Green biorefineries
- Synthesis gas biorefineries

Three examples of concepts for energy-driven biorefineries are described below (Jungmeier, 2014b).

5.3.2.1 Three-platform (C5 and C6 sugar, electricity and heat, and lignin) biorefinery using wood chips for bioethanol, electricity, heat, and phenols

The wood chips (without bark) are transported to the biorefinery, where the wood chips are pretreated for the hydrolysis to separate the sugars and the lignin. The C5 and C6 sugars are fermented to bioethanol and the lignin is used to produce bio-oil via a pyrolysis step. The phenols from the bio-oil are separated, and the residues are combusted to produce electricity and heat.

This biorefinery system (Fig. 5.5) is partly demonstrated, the production of bioethanol is demonstrated in Sweden, and the pyrolysis of the lignin and removal of phenols were tested on a laboratory scale. This concept was tested in the European project "RENURESIN" (European Commission, 2018). R&D results show that the integration of a bioethanol production from wood in a pulp and paper production plant offers promising synergies like handling and logistic of wood, water and waste water treatment, electricity and steam infrastructure, and personell. The existing infrastructure in the pulp and paper production can be used for the bioethanol production as well. Realizing these synergies would enable a commercial bioethanol production from wood by 2025.

5.3.2.2 Four-platform (biogas, green juice, green fibers, and electricity and heat) biorefinery using grass silage and food residues for bioplastic, insulation material, fertilizer, and electricity

The grass is mechanically pressed and then separated in a liquid phase ("Green juice") and solid phase ("Fibers"). The fibers are used as insulation material or are further pelletized to be used as an ingredient for bioplastic. The green juice is used to produce biogas in an anaerobic

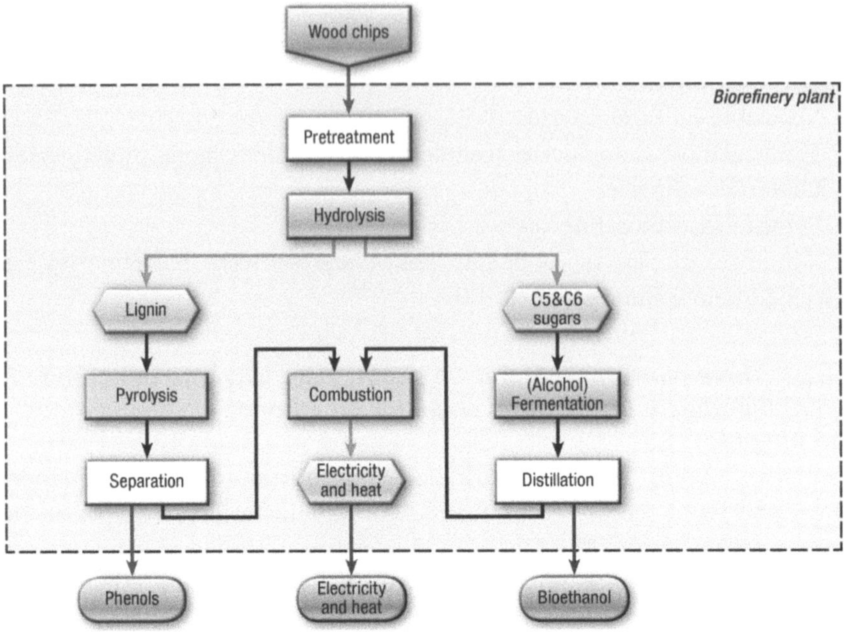

Figure 5.5 Three-platform (C5 and C6 sugars, electricity and heat, and lignin) biorefinery using wood chips for bioethanol, electricity, heat, and phenols.

fermentation. Food residues are used as an additional feedstock for the biogas fermentation. The digestate of the anaerobic fermentation can be used as a fertilizer. The biogas is used in a combined heat and power plant with an internal combustion engine to produce electricity and heat. The heat demand of the biorefinery is higher than the heat produced from biogas, so additionally natural gas is used to supply the heat. For electricity it is vice versa, so more electricity is produced than the electricity demand of the biorefinery is, so the excess electricity is sold to the grid. This type of biorefinery is already realized in several countries (Fig. 5.6).

5.3.2.3 Three-platform (pyrolysis oil, syngas, and electricity and heat) biorefinery using straw for FT-biofuels and methanol with oxygen gasification

In the fast pyrolysis the straw is used to produce pyrolysis oil and char in several decentralized locations close to the origin of the straw supply (Fig. 5.7). The oil and the char are mixed together and are transported as a slurry to one central gasification plant. In the gasification a syngas is

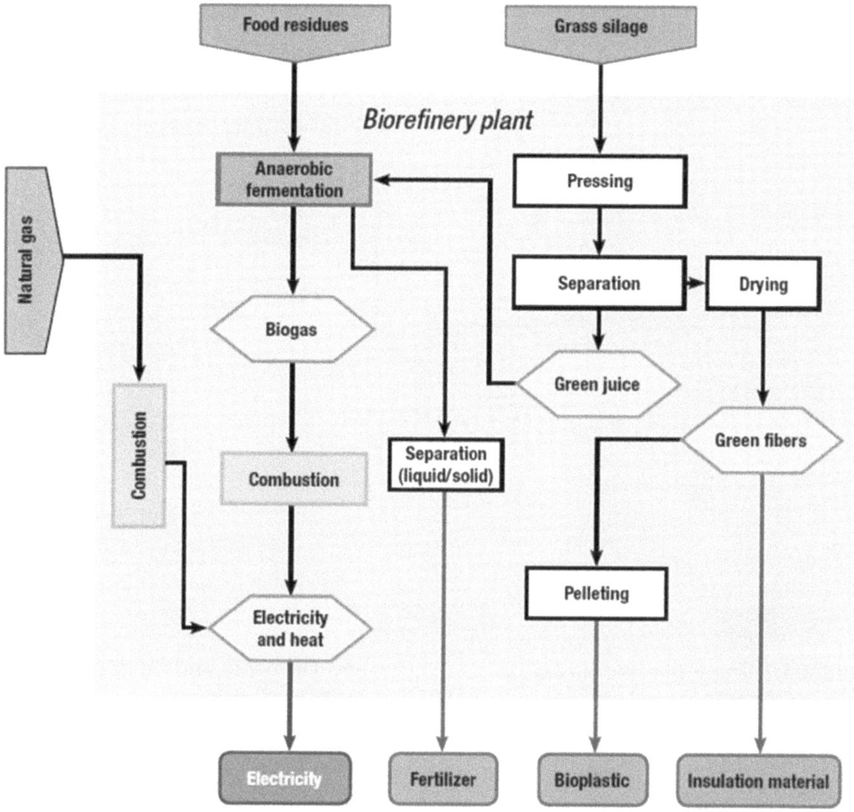

Figure 5.6 Four-platform (biogas, green juice, green fibers, and electricity and heat) biorefinery using grass silage and food residues for bioplastic, insulation material, fertilizer, and electricity.

produced by using oxygen as a gasification media. This syngas is then converted to FT-biofuels in the FT-synthesis and to methanol in the methanol synthesis. The main difference of the FT- and the methanol synthesis is on pressure, temperature, catalyst, and the ratio between CO and H_2 in the synthesis gas; FT-synthesis: 200°C–250°C, 20–30 bar with Fe and/or Co as a catalyst; methanol synthesis: 220°C–300°C, 50–100 bar with Zn/Cr and/or Cu/Zn as a catalyst. The methanol is mainly used as a chemical. Process residues (waste gas with heating value) are used to produce electricity and heat.

After the successful further development and demonstration of fast pyrolysis of straw in future further applications and uses for the pyrolysis, oil might become interesting, e.g., the direct integration of pyrolysis oil in an

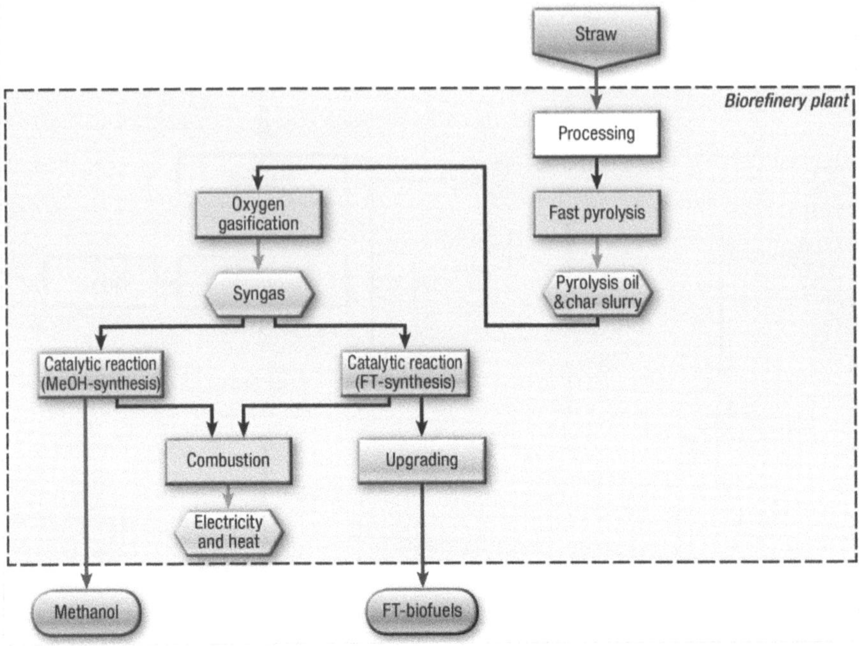

Figure 5.7 Three-platform (pyrolysis oil, syngas, and electricity and heat) biorefinery using straw for FT-biofuels and methanol with oxygen gasification.

existing oil refinery via upgrading to a renewable diesel fuel. In addition, the char from pyrolysis can be used to produce other products for chemical industry to substitute fossil-based products, e.g., activated char.

5.3.3 Example: product-driven biorefinery

Following is an example for a product-driven biorefinery. The focus within this biorefinery concept lies on the production of natural rubber and inulin from rubber dandelion (DRIVE4EU, 2017)[1].

Natural rubber is a renewable material that is used for more than 40,000 different products; among others, natural rubber is applied in transportation (tires, matting) industries, construction (sealants), and medicine (gloves). Today natural rubber is exclusively harvested from the rubber tree of which about 90% is grown in Southeast Asia. Because of the specific quality characteristics of natural rubber in many products, it cannot be replaced by synthetic rubber.

Within the project DRIVE4EU—"Dandelion Rubber and Inulin Valorization and Exploitation for Europe" (www.drive4eu.eu), a European value chain for the production and processing of natural rubber and inulin from cultivating *Taraxacum kok-saghyz* (TKS, Russian or Rubber dandelion) is set up to become less dependent on the import of natural rubber. The inulin, which is also part of the dandelion, can be used as the raw material for various green chemicals, such as furan-based polymers. The possible product portfolio is shown in Fig. 5.8. The inulin can be used as a raw material for the production of polyethylene furanoate (PEF) to substitute the fossil-based polyethylene terephthalate (PET) mainly for beverage packaging. In Fig. 5.9 the value chain of biorefining natural rubber and PEF from rubber dandelion is shown in comparison to the substituted conventional products (natural rubber from rubber tree and PET based on fossil resources) and their production processes. This new European value chain enables the EU to become less dependent on the import of natural rubber and oil and at the same time to respond to the threat of a global rubber shortage. Furthermore, the combination of natural rubber and inulin makes rubber dandelion very interesting as a production platform.

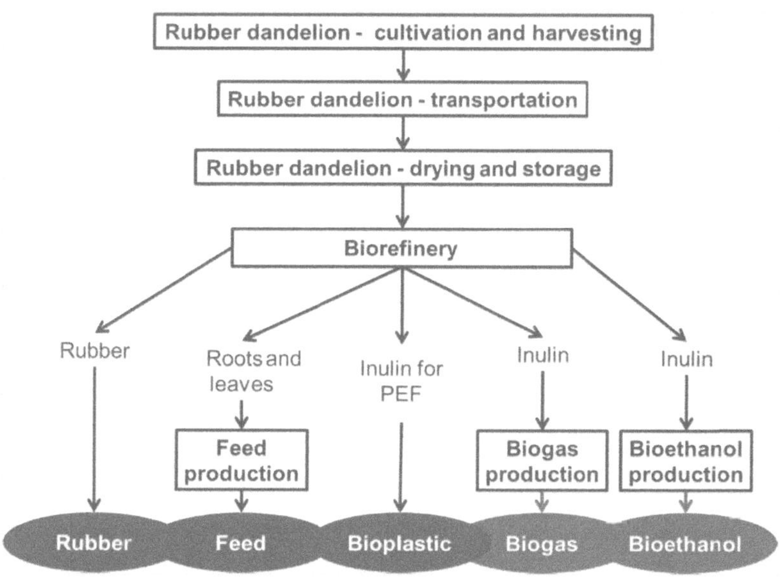

Figure 5.8 Value chain with the possible product portfolio from rubber dandelion.

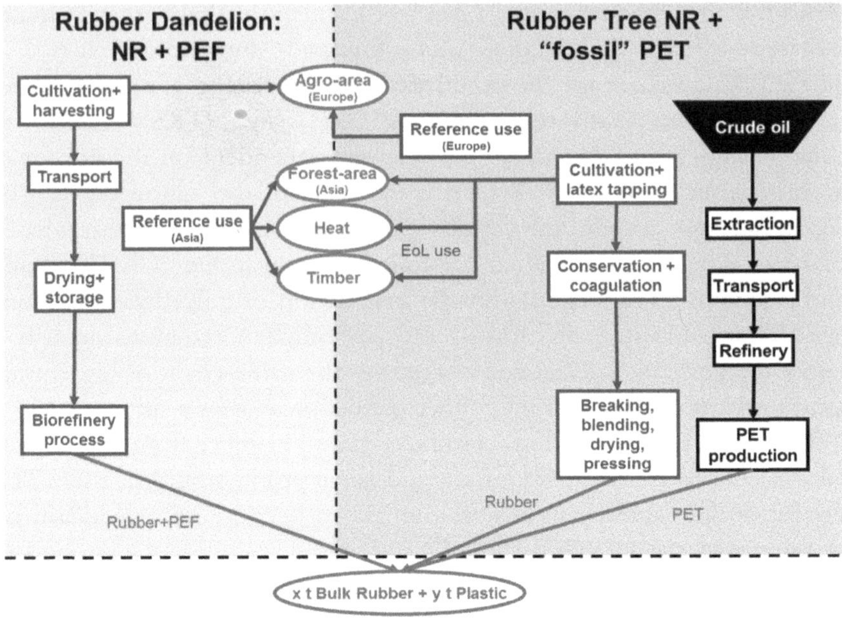

Figure 5.9 Value chain of biorefining natural rubber and PEF from rubber dandelion system in comparison to the conventional reference system of natural rubber from rubber trees and PET from crude oil. EoL, end of life.

5.4 INTEGRATION IN EXISTING INFRASTRUCTURE

5.4.1 Methodology

Existing industrial infrastructure (i.e., power plants, biofuel facilities, oil refineries, pulp/paper industry, and the food and feed industry) can be the starting point for further developing biorefineries. These industries and sectors therefore will play an important role in the development of future biorefineries and the transition towards a bioeconomy (Joergensen, 2015). It is necessary and relevant to analyze possibilities to integrate biorefineries in existing industrial infrastructure. Such an analysis provides support and information to industry, decision-makers, and investors for their strategies to invest and integrate resource-efficient biomass uses in the infrastructure of a future bioeconomy.

An approach to assess the integration and upgrading possibilities was developed under the framework of IEA Task 42 (Jungmeier and Buchsbaum, 2014) and based on Buchsbaum (2014). For this assessment, a selection of most interesting "energy-driven" biorefinery concepts by 2025 and their value chains, including the integration and deployment options in industrial infrastructures, was analyzed. The classification and description of existing infrastructure in combination with the available biomass resources and (future) market needs showed the matching points for a cost- and resource-efficient rollout of the bioeconomy by implementing various biorefineries. The identification of the upgrading opportunities for existing industrial infrastructure towards biorefineries in the bio-based industry was done the following way:

- The biorefineries were described based on their four features (platforms, products, feedstock, and processes).
- The existing industrial infrastructures were described based on their four features (platforms, products, feedstock, and processes).
- The matching features of a biorefinery and of the existing industrial infrastructure were identified.

5.4.2 Example: Austria

A verification of this approach by applying it to the Austrian situation was made as an example. The existing industrial infrastructure was classified into 10 different industry sectors with 31 subsectors as shown in Fig. 5.10 (based on Buchsbaum, 2014). Based on this systematic analysis, a database

1. Power and CHP plants
- 1.1 heating plant
- 1.2 CHP plant
  - 1.2.1 Biomass
  - 1.2.2 fossil (oil, gas, coal)
  - 1.2.3 waste

2. Biofuel facilities
- 2.1 biodiesel facilities
- 2.2 bioethanol facilities

3. Oil refineries

4. Pulp and paper industry
- 4.1 pulp
- 4.2 paper
- 4.3 pulp and paper

5. Wood industry
- 5.1 pellet facilities
- 5.2 wood based boards
- 5.3 saw mill

6. Biogas/Biomethane plants
- 6.1 biogas CHP
- 6.2 biomethane
- 6.3 green refinery

7. Chemical industry
- 7.1 anorganic and organic chemicals
- 7.2 industrial gases
- 7.3 fertiliser
- 7.4 candles and waxes
- 7.5 synthetic material production
- 7.6 food and animal feed additives

8. Waste treatment facilities
- 8.1 sewage treatment plants
- 8.2 landfill gas facilities
- 8.3 composting plants

9. Food industry
- 9.1 breweries
- 9.2 sugar mills
- 9.3 starch mills
- 9.4 oil presses
- 9.6 fruit + vegetable processing
- 9.7 creameries
- 9.8 bakeries

10. Others

Figure 5.10 Classification of existing industrial infrastructure in 10 sectors with 31 subsectors.

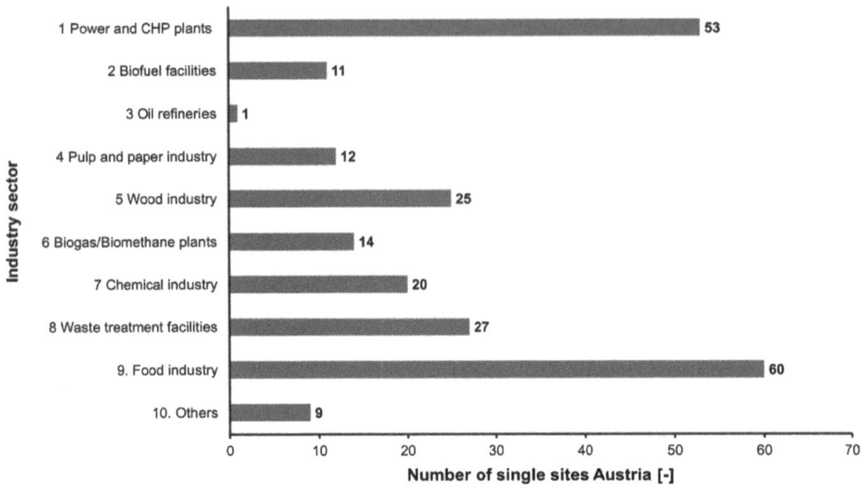

Figure 5.11 Database of the existing industrial infrastructure in Austria with 232 single sites in the 10 industry sectors.

for the existing industrial infrastructure was built. In Austria so far 232 existing industrial sites have been described with their main characteristics based on the four features (Fig. 5.11).

The results of this analysis showed which and how many features of the biorefineries matched the features of the existing industrial infrastructure. Two examples of a biorefinery for the possible integration in the existing infrastructure are given below. Fig. 5.12 shows how the biorefinery "1-platform (oil) biorefinery using oilseed crops for biodiesel, glycerin and feed" can be integrated in the 31 analyzed subsectors of existing infrastructure in Austria. The coverage for each of the four features is defined to be a maximum of 25% if all single features of the biorefinery match the features of the considered industry, e.g., all platforms are the same. As shown, a total coverage of 67% is reached in the biodiesel and vegetable oil industry and for some other industries the coverage is relatively low (8% to 25%). However, for most of the considered industries, the coverage is 0%, which means that there is no reasonable opportunity for integration. Fig. 5.13 displays how the biorefinery "5-platform (C5 and C6 sugars, lignin and C6 sugars, and electricity and heat) biorefinery using saw mill residues, wood chips and sulfite liquor for bioethanol, pulp and paper, electricity and heat" can be integrated in the 31 analyzed subsectors. With coverage between 55% and 65%, this biorefinery has very

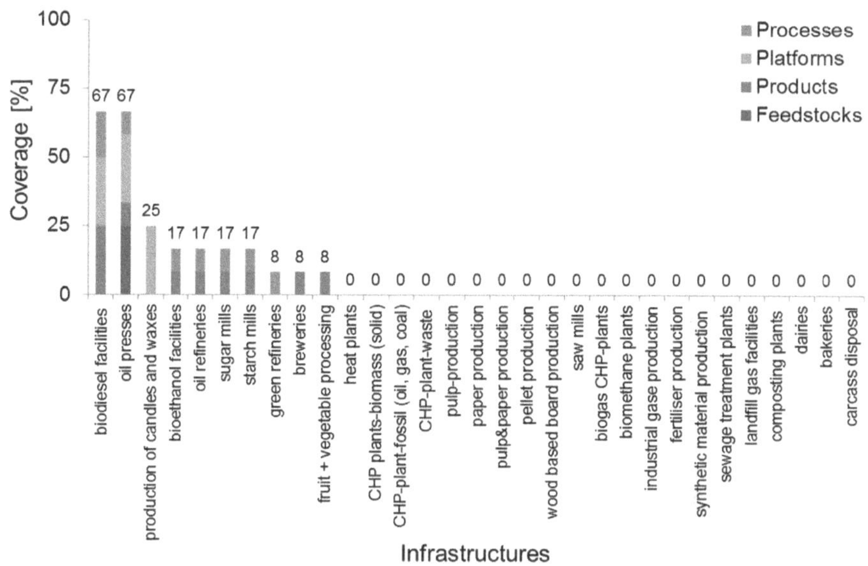

Figure 5.12 Example I: 1 Biorefinery ("1-platform (oil) biorefinery using oilseed crops for biodiesel, glycerin, and feed") and 31 industry subsectors (Buchsbaum, 2014).

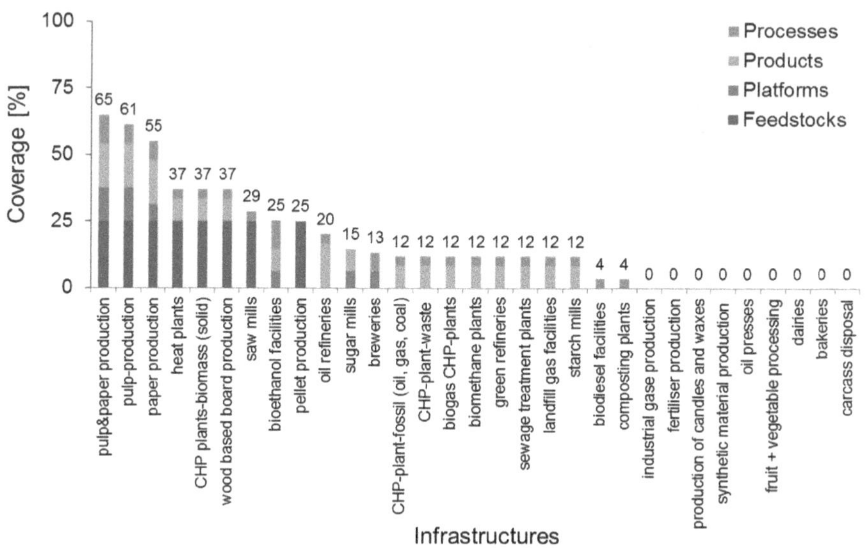

Figure 5.13 Example II: 1 Biorefinery ("5-platform (C5 and C6 sugars, lignin and C6 sugars, electricity and heat) biorefinery using saw mill residues, wood chips and sulfite liquor for bioethanol, pulp and paper, electricity and heat") and 31 industry subsectors (Buchsbaum, 2014).

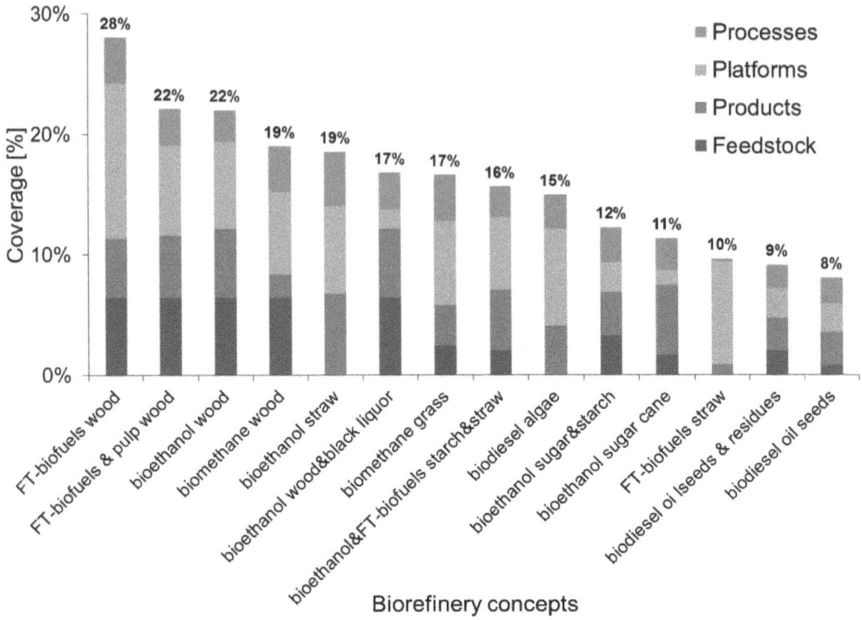

Figure 5.14 Coverage of the upgrading opportunities of 14 selected energy-driven biorefineries in the 31 industrial subsectors (Buchsbaum, 2014).

attractive opportunities to be integrated in the pulp, paper, and pulp and paper industries in Austria. Only 9 of the 31 industry sectors have no coverage with this biorefinery. The other 19 sectors have an integration opportunity with coverage between 4% and 37%.

The coverage of the overall integration opportunities in Austria of 14 selected energy-driven biorefinery concepts into the 31 industrial subsectors are shown and compared in Fig. 5.14. The index for the integration of the 14 selected energy-driven biorefinery concepts into the existing industrial subsector is applied per biorefinery concept and lies between 8% (biodiesel oil seeds) and 28% (FT-biofuels wood). Feedstocks and products offer the most attractive integration opportunities in different existing industrial infrastructures. "Hot spots" for favorable integration opportunities for the 14 selected energy-driven biorefineries in Austria were identified; these are pulp and paper and wood industry to additionally produce transportation biofuels and biochemical; food and feed industry to additionally produce bioenergy carriers; chemical industry to integrate biogenic resources to produce bio-based chemicals; and conventional oil refinery to integrate biofuel production.

5.4.3 Conclusions

Following are the main conclusions:

- A classification of different biorefineries and existing industrial infrastructure is possible via the four features: platforms, products, feedstocks, and processes.
- To analyze the possible integration of biorefineries in the existing industrial infrastructure, a systematic characterization of the infrastructure with 10 industry sectors and 31 subsectors was developed.
- The existing industrial infrastructure is described based on the four features—feedstock, platforms, products, and processes.
- The opportunity to integrate a biorefinery into existing industrial infrastructure is characterized by the type and amount of common features, where the feedstock and the platforms are more relevant for efficient integration than the platforms and processes.
- The most relevant features for integration are the feedstock and the products and are ranked the following:
  - Feedstocks and products $\geq$ platforms $\geq$ processes
  - Feedstocks: very strong
  - Products: very strong
  - Platforms: strong
  - Processes: possible

With the results of this analysis, the most promising industry sectors for upgrading towards biorefining can be identified. The further steps are to analyze and assess the individual opportunities of the relevant sectors via stakeholder involvement. The results assist various stakeholders in finding their position on biorefining in a future bio-based economy while minimizing unexpected technical, economic, and financial risks.

5.5 BIOREFINERY COMPLEXITY INDEX

5.5.1 Methodology

As indicated above, currently many different and various biorefinery concepts are being developed and are already implemented. Some of these biorefinery concepts are simple, using one feedstock (e.g., vegetable oil) and producing two or three products (e.g., biodiesel, animal feed, and glycerin) with current available commercial technologies. Other biorefinery concepts are sometimes very complex using many different feedstocks (e.g., algae, miscanthus, and wood chips from short rotation) to coproduce a broad spectrum of different products (e.g., bioethanol, phenol, omega 3 fatty acids, and biodiesel) using technologies that might become commercial in several years. Each of these different biorefinery concepts has a different degree of complexity, which makes it difficult for industry, decision-makers, and investors to decide which of these concepts are the most promising options on the short, medium, and long term, and to judge on the technological and economic risks.

Therefore, IEA Bioenergy Task 42 has published a document to present the current status of an approach to develop a "BCI" and to calculate the BCI for some selected biorefinery concepts (Jungmeier, 2014a, 2014b). The approach was developed since 2010 and started with the analogy to the "Nelson's complexity index" used for oil refineries.

The "Nelson's (complexity) index" was developed by Wilbur L. Nelson and published in the "Oil and Gas Journal" (1960–61) to quantify the costs of the refinery's components. The Nelson's Index is an indicator for the investment intensity, the cost index of the refinery, the value addition potential of a refinery, and refers to the refinery's ability to process feedstocks, such as high-sulfur crude, into value-added products. The higher the complexity of the refinery, the more flexible it is.

Based on the classification system of biorefineries as discussed above and the "Nelson's complexity index" for oil refineries, a BCI is developed.

The following basic assumptions on the complexity of a biorefinery are used:

1. The number of different features of a biorefinery influences the complexity. The complexity increases by the number of features in a biorefinery.
2. The state of technology of a single feature influences the complexity. The complexity decreases the closer a technology is to a commercial application, meaning a high "Technology Readiness Level (TRL)" of a feature has lower technical and economic risks, and so a lower complexity.
3. For the products and feedstock the "Market Readiness Level" is applied in analogy to the TRL of the processes and platforms. Therefore, only the TRL is used.
4. This leads to the basic assumption for the calculation procedure of the BCI that the complexity is directly linked to the number of features and the TRL of each single feature involved.

5. The complexity of a commercial application, which means that all features are commercially available, is then only determined by the number of features, whereas in noncommercial application, the TRL increases additionally the complexity of the biorefinery system.

For each of the four features (platforms, feedstocks, products, and processes) of a biorefinery, the TRL can be assessed using level description between 1 ("basic research") and 9 ("system proven and ready for full commercial deployment"). Based on the TRL the Feature Complexity (FC) for each single feature of a biorefinery is calculated ($FC = 10 - TRL$; this means lower TRL results in higher FC and vice versa). With the number of features and the FC of each single feature, the Feature Complexity Index (FCI) for each of the four features (platforms, feedstocks, products, and processes) is calculated. The BCI is the sum of the four FCIs. To simplify the presentation, the Biorefinery Complexity Profile (BCP) is introduced. The BCP is a compact format to present the complexity of a biorefinery by giving the BCI and the four FCIs of each feature. The BCP, which includes the BCI and the four FCIs, has the following format:

$$\text{BCP: BCI } (\text{FCI}_{\text{platforms}}/\text{FCI}_{\text{Feedstocks}}/\text{FCI}_{\text{Products}}/\text{FCI}_{\text{Processes}}).$$

5.5.2 Examples

An example for a 1-platform (oil) biorefinery using oilseed crops for biodiesel, glycerin, and feed having a BCP of 8(1/1/3/3) is shown in Fig. 5.15. In this biorefinery 1-platform, one feedstock, three products, and three processes are involved, each of them with an FC of 1, as they are commercially available and provide a TRL 9. So the BCI is 8.

In Fig. 5.16 a second-generation biorefinery such as 3-platform (C5 and C6 sugars, electricity and heat, and lignin) biorefinery using wood chips for bioethanol, electricity, heat, and phenols has a BCP of 33(8/1/6/18) is shown. In this biorefinery the following features are involved:

- three platforms of which one (lignin) has an FC of 4, one (C5 and C6 sugars) has an FC of 3, and one (electricity and heat) has an FC of 1;
- one feedstock with an FC of 1;
- three products of which the phenol has an FC of 4 and the other products of 1; and
- seven processes with an FC between 4 and 1.

The BCI is 33.

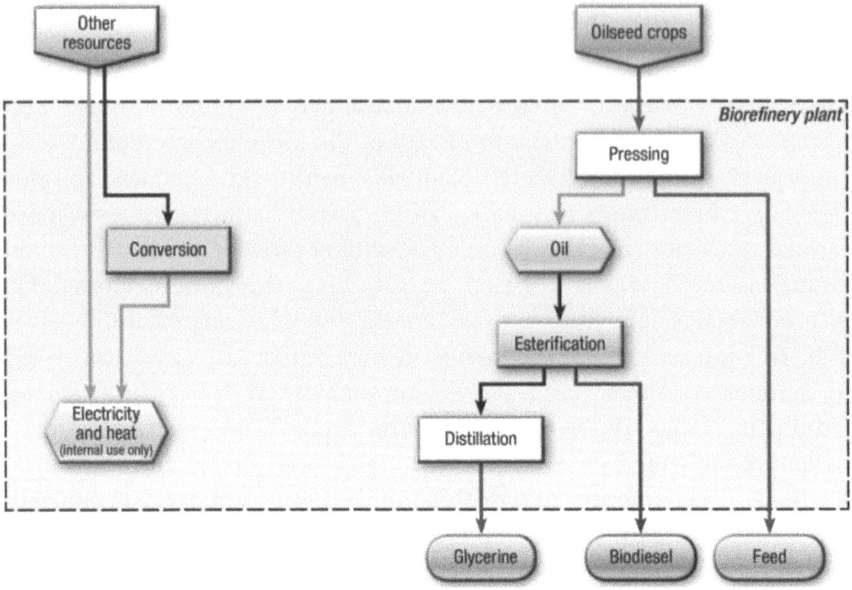

Figure 5.15 A 1-platform (oil) biorefinery using oilseed crops for biodiesel, glycerin, and feed with a BCP of 8(1/1/3/3).

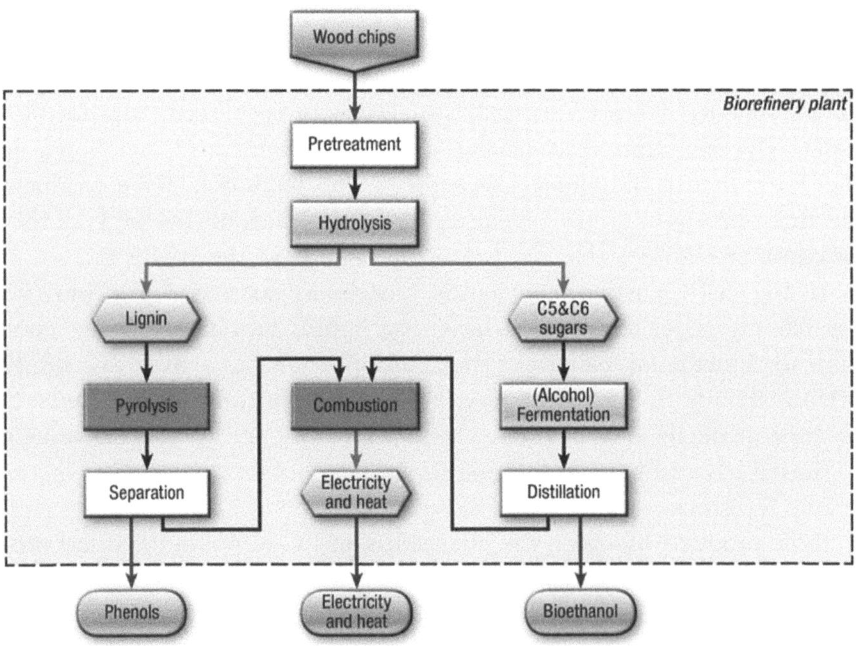

Figure 5.16 A 3-platform (C5 and C6 sugars, electricity and heat, and lignin) biorefinery using wood chips for bioethanol, electricity, heat, and phenols with a BCP of 33 (8/1/6/18).

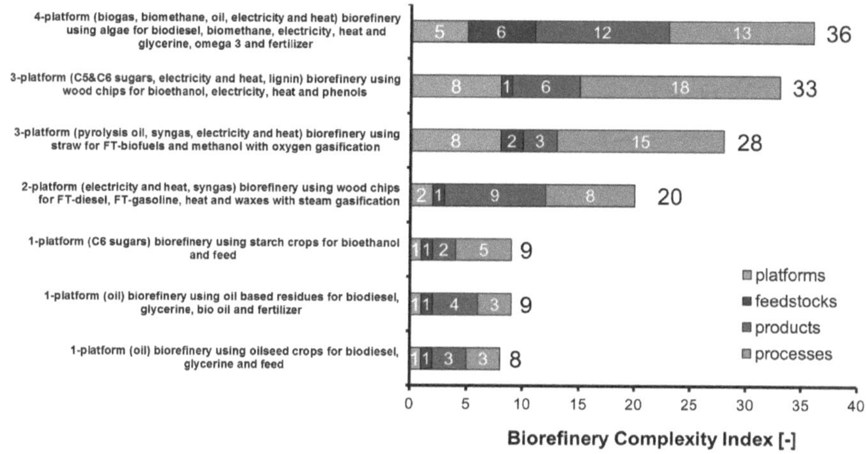

Figure 5.17 BCI of several biorefinery concepts.

The BCP and the BCI of several energy-driven biorefineries are shown and compared in Fig. 5.17. It can be seen that the BCI between 8 and 9 of the three commercial biorefineries producing biodiesel and bioethanol from conventional crops and residues are significantly lower compared to the biorefineries under development with a BCI between 20 and 36. Analyzing the BCP, it can be seen that the complexity of the processes and the platforms of the biorefineries under development significantly contributes to the BCI. In addition for the algae the complexity of the feedstock also has a strong contribution to the BCI.

5.5.3 Conclusions

The following conclusions on the BCI and BCP were drawn:

1. They give an indication for the relative comparison of different biorefinery concepts and their development potential.
2. They present a benchmark of the "complexity" of a biorefinery in terms of involved platforms, feedstocks, processes, and products, and their specific and overall "TRL."
3. The higher the BCI, the more beyond "state of the art" is the biorefinery.
4. The BCI of a biorefinery producing biodiesel from vegetable oil which is fully deployed, with 8(1/1/3/3) is a benchmark to compare the complexity of other current and future biorefinery systems.

5. The BCI will change in the future if the TRL has changed, e.g., if a pilot-plant, demonstration plant, and/or first-of-a-kind commercial plant will go into operation.
6. The BCP shows the most relevant features contributing to the complexity of a biorefinery.
7. The BCP of a biorefinery gives an indication on the technological and economic risks.
8. The first results and conclusions of a critical review by the country representatives in IEA Bioenergy Task 42 show that the "BCI" adds additional relevant information on the assessment and comparison of different biorefinery systems.

It was concluded that the results are potentially relevant for industry, decision makers, as well as investors, as additional information is generated to assist them in their strategies to implement the most promising biorefinery systems by minimizing technical and economic risks.

5.6 ECONOMIC AND ENVIRONMENTAL ASPECTS OF BIOREFINERIES

5.6.1 Methodology

The increased usage of biofuels in the transportation sector is the driving factor behind the development of advanced processes to produce biofuels in biorefineries efficiently. Economic and environmental benefits can be gained through system design, coproduct valorization, and the development of new conversion technologies.

The main question is what the advantages of a biorefinery as a multiproduct system have, as compared to conventional bioenergy systems (e.g., single-product systems). These advantages might be in terms of environmental, economic, technological, and social benefits. A consistent methodology is developed to establish a framework for such a comparison (Jungmeier et al., 2013b) using the LCSA methodology, which combines LCA for environmental impacts, LCC for economic impacts, and S-LCA for social impacts. It has already been applied to several biorefinery examples. For a comprehensive and uniform description of the key facts and characteristics of a biorefinery, IEA Bioenergy Task 42 developed a so-called "Biorefinery Fact Sheet" (Jungmeier et al., 2013a, Jungmeier, 2014b).

It is generally agreed that the whole value chain should be used as the basis of a comparative assessment of the environmental, economic, and social effects of a biorefinery relative to a conventional system. The value chain of a biorefinery includes all effects from the feedstock production on available land, to the supply of products, to the consumer until finally to the "end of life" of the products. The effects could be economic, socioeconomic, environmental, and social aspects. The following steps in the value chain must be considered in the assessment (Fig. 5.18):

- feedstock production (including land use)/residues collection,
- transport of feedstock/residues,
- processing of feedstock/residues into products in biorefinery plant,
- distribution of products,
- use of products, and
- "end of life" of products.

A conventional system includes the processing of biomass or fossil resources into one or more marketable products either energy carriers or (bio)products. These conventional systems are commercially available and used; e.g., the conventional processes for biomass are mainly used for heat and power supply via different processes. In Fig. 5.19 a biorefinery system is shown (left), from raw material to its products, in comparison to possible conventional systems, which could be purely fossil based (right) or a combination of using renewable and fossil resources (middle). In most of

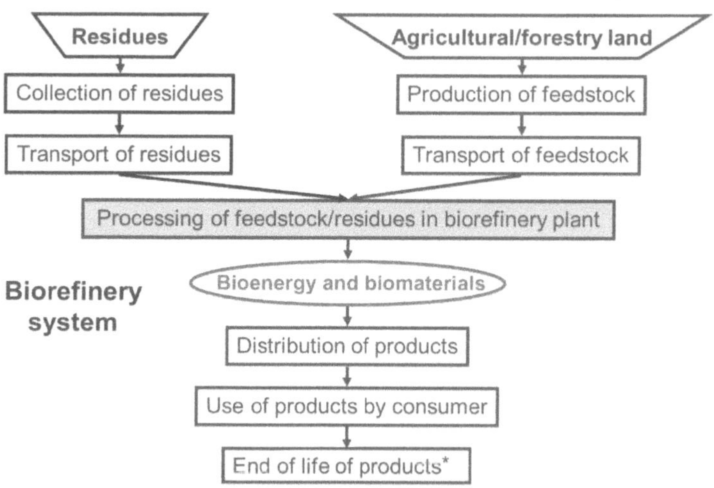

Figure 5.18 The value chain approach for biorefineries. *Collection and treatment: e.g., recycling, combustion.

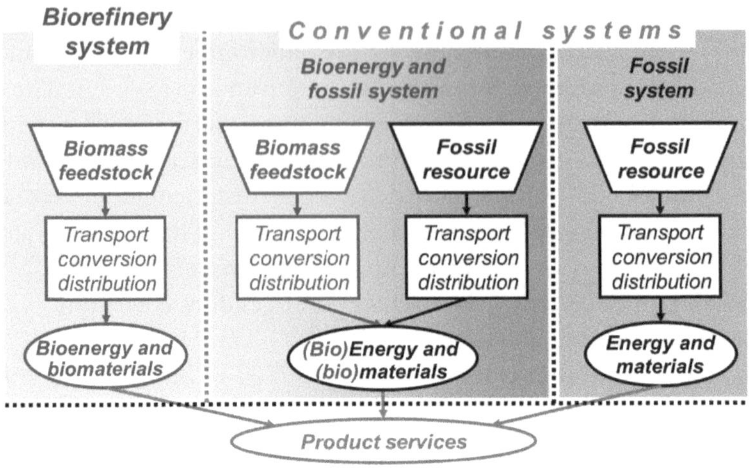

Figure 5.19 Framework for the comparative assessment of a biorefinery system to conventional systems.

the realistic applications the possible combination of fossil and renewable resources is most relevant.

For the comparison, the following framework is relevant:

1. The biorefinery system and the conventional systems should provide the same amount of products with the same services, e.g., same amount of road transportation service, same amount of materials or chemicals.
2. The same amount and type of biomass is considered in both systems and the following question should be addressed, "What happens to the biomass if it is not used as feedstock for a biorefinery?"
3. The same amount of agricultural and/or forestry area also should be considered in both systems. Here the following question is assessed, "What happens with the land if it is not used to produce feedstock for a biorefinery?"
4. The whole ("value") chain approach—from resource to products—is considered.

5.6.2 Examples

The following two examples for the determination of conventional systems in the framework of the sustainability assessment of energy-driven biorefineries are presented. The systems are typical of examples in Denmark and Austria:

1. Bioethanol from wood: "3-platform (C5 and C6 sugars, electricity and heat, lignin) biorefinery using wood chips for bioethanol, electricity, heat, and phenols".
2. Bioethanol from straw: "4-platform (C5 and C6 sugars, lignin, electricity and heat) biorefinery using straw for bioethanol, electricity, heat, and feed".

The type of the conventional system differs from case to case and can only be specified for individual cases.

5.6.2.1 Bioethanol from wood

The framework for the assessment of the energy-driven biorefinery "3-platform (C5 and C6 sugars, electricity and heat, lignin) biorefinery using wood chips for bioethanol, electricity, heat, and phenols", in comparison to a conventional system is shown in Fig. 5.20. The biorefinery system (right) uses 1 t of wood to produce 5 GJ of bioethanol as road transportation fuel, 1.2 GJ of electricity, 0.7 GJ of heat, and about 0.01 t of phenols. The conventional system (left) uses natural gas to produce the 1.2 GJ of electricity in a power plant and oil in the oil refinery to produce gasoline as road transport fuel, heat, and phenols. In the case of the conventional system the 1 t of wood was not used and remained in the forest, where natural oxidation took place.

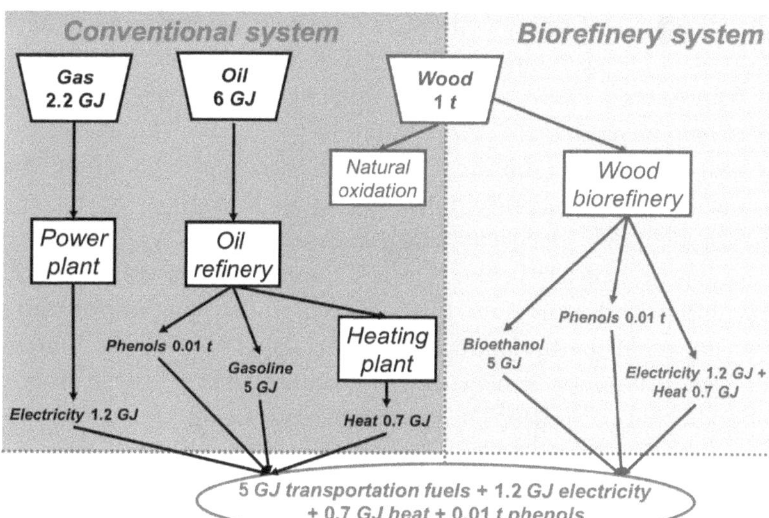

Figure 5.20 Framework for the comparison of a "3-platform (C5 and C6 sugars, electricity and heat, and lignin) biorefinery using wood chips for bioethanol, electricity, heat, and phenols" to a conventional system.

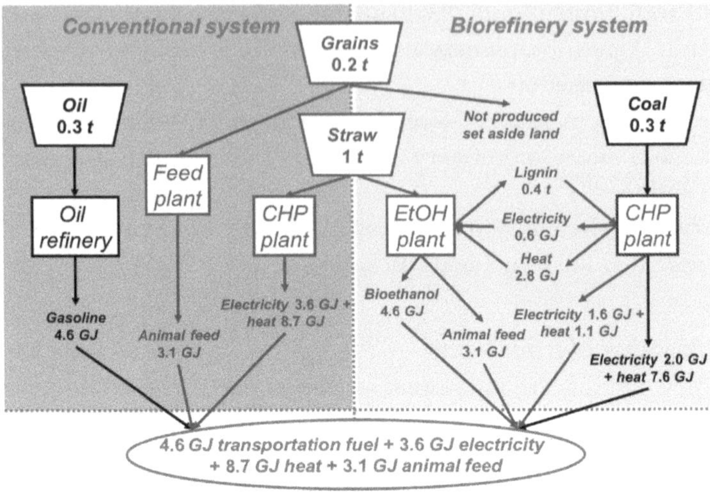

Figure 5.21 Framework for the comparison of a "4-platform (C6 sugars, C5 sugars, lignin, and electricity and heat) biorefinery using straw for bioethanol, electricity, heat, and feed" to a conventional system.

5.6.2.2 Bioethanol from straw

In Fig. 5.21 (right) the framework for the assessment of the energy-driven biorefinery "4-platform (C6 sugars, C5 sugars, lignin, electricity and heat) biorefinery using straw for bioethanol, electricity, heat and feed", in comparison to a conventional system is shown. The biorefinery system on the right hand uses 1 t of straw to produce 4.6 GJ of bioethanol as road transportation fuel, 1.6 GJ of electricity, 1.1 GJ of heat, and about 3.1 GJ animal feed (C5 molasses). The conventional system (left) uses raw oil to produce 4.6 GJ of gasoline and 1 t of straw is used to produce 3.6 GJ electricity and 8.7 GJ heat, referring to the situation in Denmark. As the amount of electricity and heat in the conventional system is more than the biorefinery system, the additional electricity and heat is produced in a coal CHP plant, which would be typically in the case of Denmark. In the conventional system the animal feed is provided by grain, whereas the amount of land used to grow the grain for the animal feed is set aside land in the biorefinery system. This assumption is necessary to guarantee that the same amount of agricultural land is considered in the compared systems.

5.6.3 Biorefinery Fact Sheets

In spite of the fact that biorefining is very promising for the sustainable valorization of biomass to both food and feed ingredients, bio-based products, and bioenergy, the whole biorefinery area is very broad; the concepts involved are often very complicated, and data on their (potential) technical, socioeconomic, and ecological performance are often very complex. To help the market deployment of biorefineries, one important critical success factor is clear knowledge dissemination to all stakeholders involved, so that they are finally able to speak the "same language," and to raise public support for implementation.

As the development status and the perspectives for implementation and development of the biorefineries are different, the IEA Bioenergy Task 42 has developed a "Biorefinery Fact Sheet" for the uniform description of the key facts and characteristics of a biorefinery (Jungmeier, 2014b, Jungmeier et al., 2013c, 2014). Based on a technical description and the classification scheme, the mass and energy balance is calculated for the most reasonable production capacity for each of the selected biorefineries. Then the three dimensions—economic, environmental, and social[2]—of sustainability are assessed for each biorefinery and documented in a compact form in the "Biorefinery Fact Sheet." Based on these sheets, an easy comparison of the different biorefinery systems is possible. The "Biorefinery Fact Sheet" assists various stakeholders in finding their position on biorefining in a future bio-based economy.

The "Biorefinery Fact Sheets" consist of two parts:

1. Part A: Biorefinery plant
2. Part B: Value chain assessment.

In Part A the key characteristics of the biorefinery plant are described by giving compact information on classification scheme, description of the biorefinery, mass and energy balance, share of costs and revenues. In Part B the sustainability assessment based on the whole value chain of the biorefinery plant is described by giving compact information on system boundaries, reference system, cumulated primary energy demand, GHG emissions, and costs and revenues.

5.6.4 Example: Biorefinery Fact sheet

The following is an example of a fact sheet for a "2-platform (electricity and heat, syngas) biorefinery using wood chips for FT-biofuels, electricity, heat and waxes with steam gasification" (Jungmeier, 2014b).

5.6.4.1 Part A: Biorefinery plant

The demonstration scale energy-driven "2-platform (electricity and heat, syngas) biorefinery using wood chips for FT-biofuels, electricity, heat and waxes with steam gasification" is shown in Fig. 5.22.

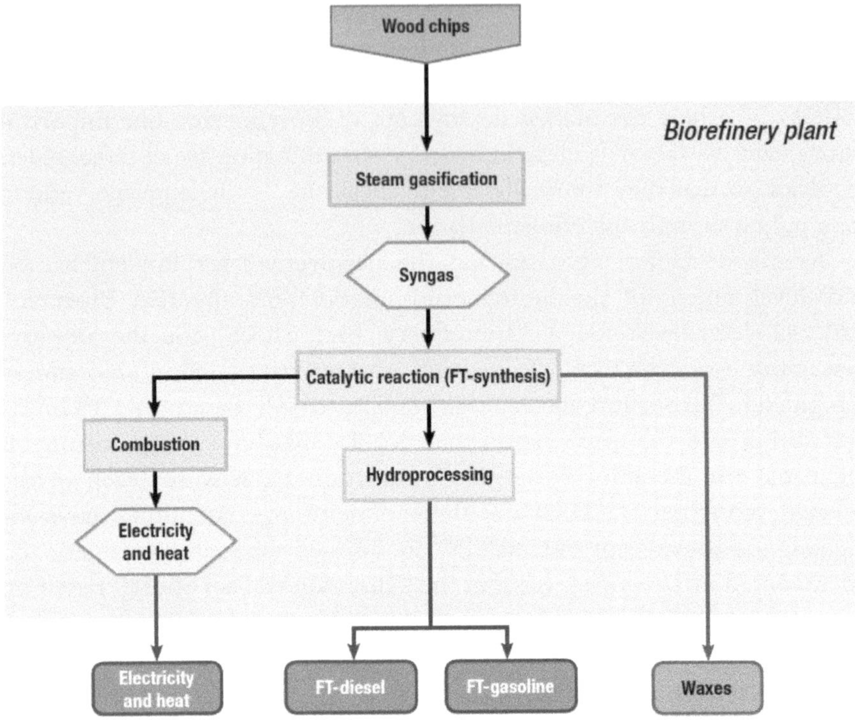

Figure 5.22 The 2-platform (electricity and heat, syngas) biorefinery using wood chips for FT-biofuels, electricity, heat, and waxes with steam gasification.

As the name suggests, in this biorefinery the wood chips are gasified with steam to produce a product gas, which is used to produce raw FT-biofuels via a catalytic reaction (FT-synthesis). The final quality of the transportation FT-biofuel is reached in the upgrading step (e.g., hydroprocessing). The process residues are combusted to produce electricity and heat. As a further step, product waxes are produced.

Depending on the further successful development, besides the steam gasification of wood, which is suitable for smaller to medium-sized gasifiers, the gasification with oxygen for large applications (e.g., entrained flow gasification) might also become interesting. The large amount of syngas will then be an optimal starting point to produce additional synthetic products depending on the market demand for biomass based chemicals (e.g., methanol).

Following information are shown below: description of the biorefinery, classification scheme (Fig. 5.22), key characteristics of the biorefinery plant (Table 5.1), mass (Fig. 5.23) and energy balance (Fig. 5.24), share of costs (Fig. 5.25) and revenues (Fig. 5.26).

Table 5.1 Key characteristics of biorefinery plant

"2-platform (electricity&heat, syngas) biorefinery using wood chips for FT-biofuels, electricity, heat and waxes with steam gasification"					
State of technology:	commercial 2013		Biorefinery Complexity Index		14.5
Country:	EU 27		(Products/Platform/Feedstock/Processes)		(4.5/3/2/5)
Main data sources:	BIOGRACE, JOANNEUM RESEARCH				
Products			**Auxiliaries (external)**		
FT-diesel	105 [kt/a]		electricity	0.00 [PJ/a]	
FT-gasoline	70 [kt/a]		heat	0.00 [PJ/a]	
waxes	10 [kt/a]				
heat	0.3 [PJ/a]		others: various	29.2	
Feedstock	[kt/a]	water [%]	**Costs**		
wood chips	1459	45.0%	investment costs	500 [Mio €]	
			feedstock costs	100 [€/t]	
			number of employees	35 [#]	
Efficiencies			mass	energy	
input to products			12%	30%	
input to transportation biofuel			7%	17%	

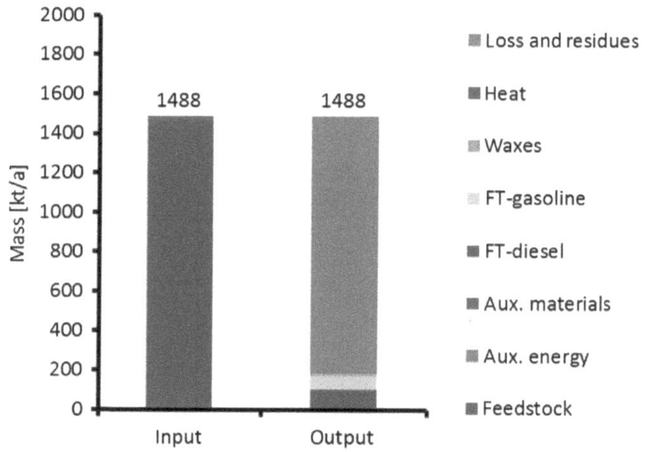

Figure 5.23 Mass balance of biorefinery plant.

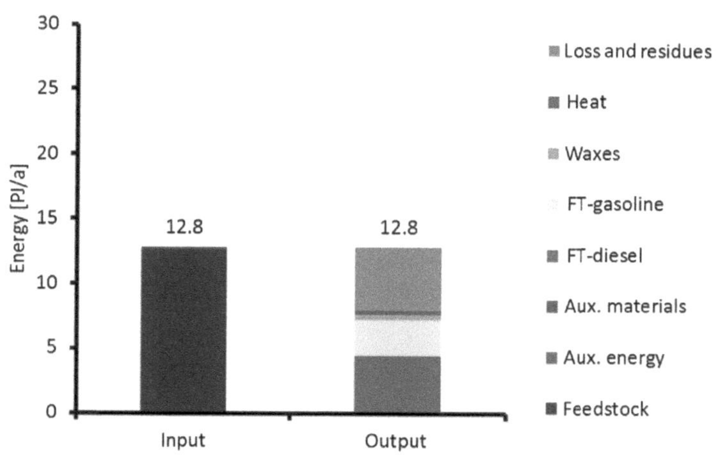

Figure 5.24 Energy balance of biorefinery plant.

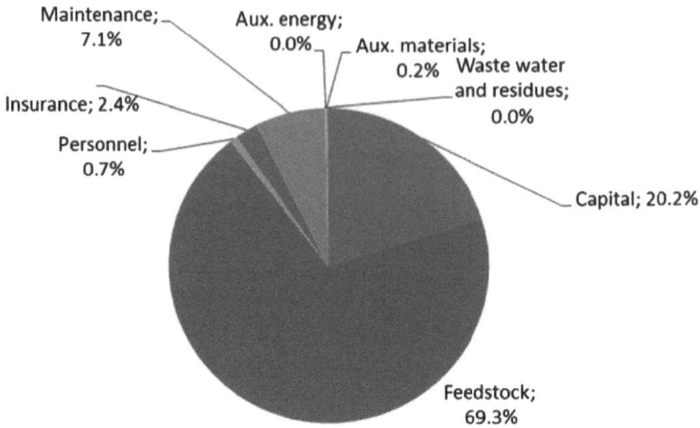

Figure 5.25 Share of costs.

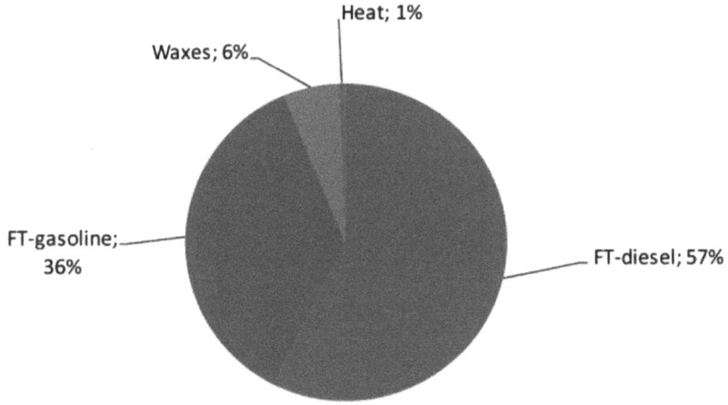

Figure 5.26 Share of revenues.

5.6.4.2 Part B: Value chain sustainability assessment

This part of the fact sheet provides information on the sustainability assessment based on the whole value chain of the biorefinery plant by giving compact information on system boundaries and reference system (Fig. 5.27), key characteristics of biorefinery value chain (Table 5.2), cumulated fossil primary energy demand (Fig. 5.28), cumulated primary energy demand (Fig. 5.29), GHG emissions (Fig. 5.30), and costs and revenues (Fig. 5.31).

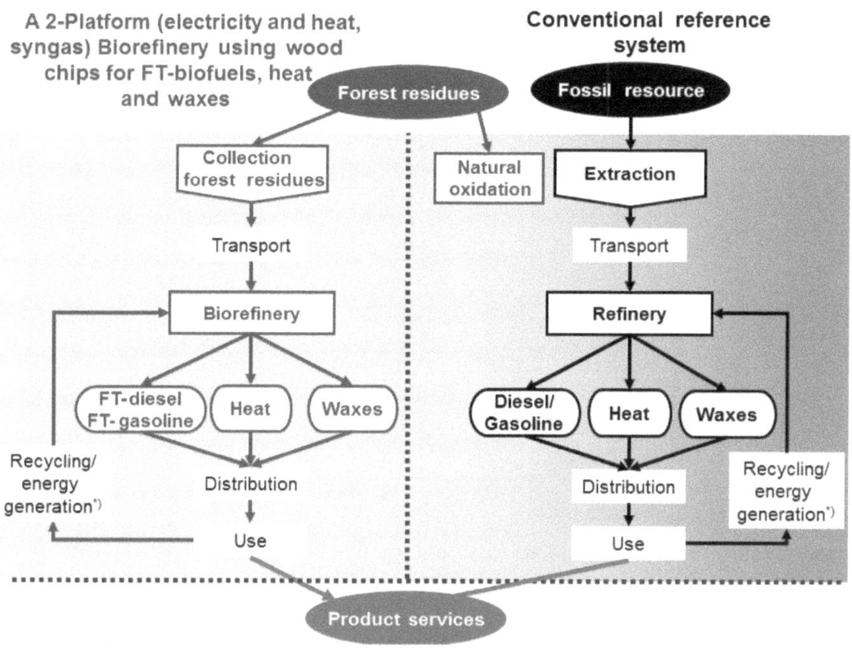

Figure 5.27 Comparison of the biorefinery with the conventional reference system on whole value chain (including "end of life management"). *Including transportation.

Table 5.2 Key characteristics of biorefinery value chain

Whole value chain		
GHG emissions	**Range**	
Biorefinery	59 (55 to 68)	[kt CO_2-eq/a]
Reference system	674 (630 to 770)	[kt CO_2-eq/a]
Saving	–91% (–85% to –100%)	[%]
Cumulated energy demand		
fossil		
Biorefinery	0.7 (0.63 to 0.78)	[PJ/a]
Reference system	8.7 (8.1 to 10)	[PJ/a]
Saving	–92% (–86% to –100%)	[%]
Total		
Biorefinery	13.5 (13 to 16)	[PJ/a]
Reference system	9.2 (8.5 to 10.5)	[PJ/a]
Change	47% (44% to 55%)	[%]
Agricultural area demand		
Feedstock	0 (0 to 0)	[ha/a]
Costs		
Annual costs	210 (200 to 240)	[Mio €/a]
Specific costs	1137 (1100 to 1300)	[€/t]
Revenues		
Annual revenues	218 (200 to 250)	[Mio €/a]
Specific revenues	1180 (1100 to 1400)	[€/t]

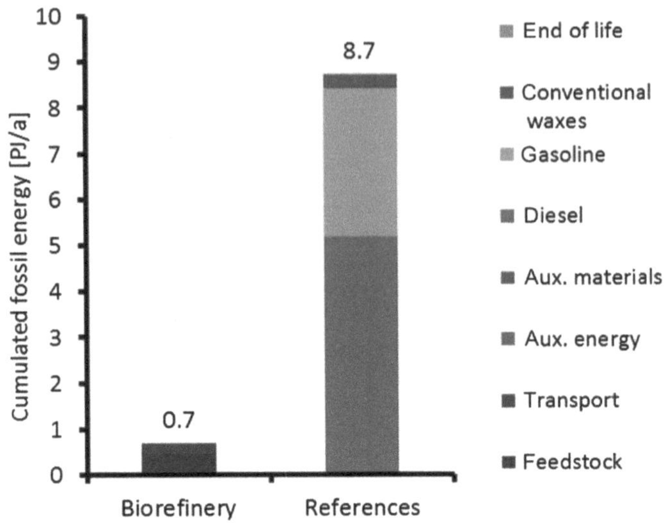

Figure 5.28 Estimated cumulated fossil energy demand of biorefinery and reference products.

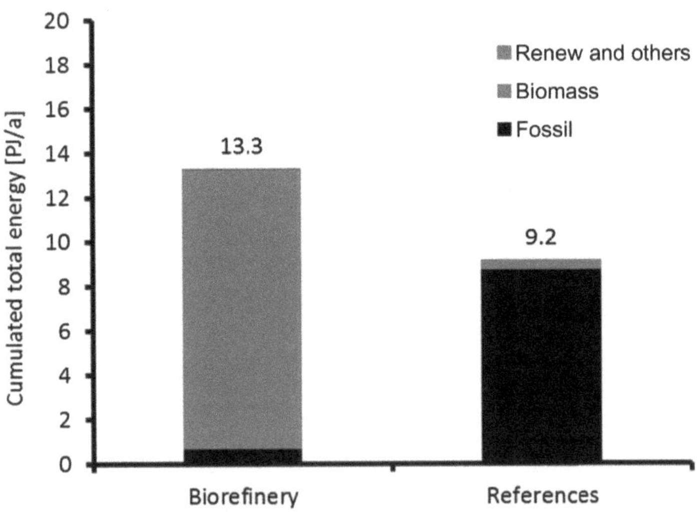

Figure 5.29 Estimated cumulated energy demand of biorefinery and reference products.

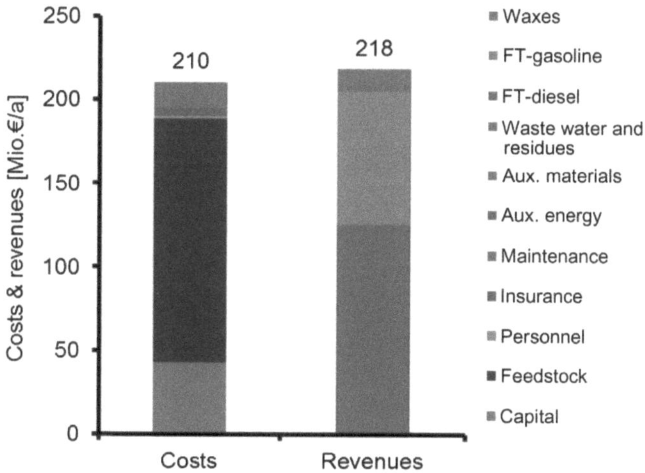

Figure 5.30 Estimated GHG emissions of biorefinery and reference products.

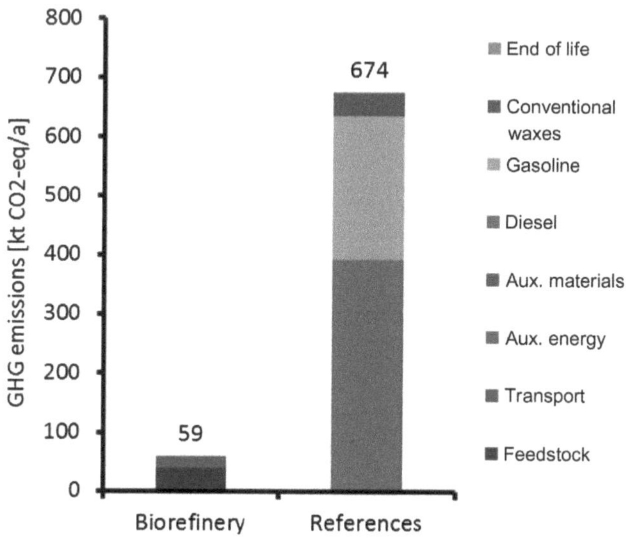

Figure 5.31 Estimated cost and revenues of biorefinery plant.

5.7 PERSPECTIVES

5.7.1 Outlook and future paths in biorefineries

Currently, biomass is mainly used for human food, animal feed, and the production of paper, fuels, power, and heat. In the future bioeconomy, biomass has to be used for the production of food, feed, bioenergy (power, heat, and biofuels), and bio-based products (chemicals, materials) in a sustainable and efficient way. The relatively limited availability of biomass under a sustainable development needs the development and implementation of very efficient conversion technologies to optimize valorization and to maximize the sustainable benefits of biomass supply chains.

The current expertise in biomass supply chains and facilities available could be a starting point for the development of more sustainable multi-product biorefineries. Residues from existing bio-based industries (mainly food industry) are an important biomass resource, which should be used in biorefineries to produce new types of bio-based products (food, chemicals, etc.). Current developments in biorefineries are building on the long-standing success of several industries, such as starch processing and pulp and paper, as well as biotechnology and conventional and advanced biofuel processes. The chemical industry and the biofuels sector could drive the development towards a bioeconomy.

The biofuels sector is already widely considered as biorefineries, especially in the case of those producing biofuels based on lignocellulosic material. In the short-term the overall economics in the energy sector could be improved by valorization of added-value bio-based products from available residues, whereas in the longer term, the energy sector will become an integral part of full biomass refining strategies.

Sustainability is a main attribute of biorefineries. Measuring sustainability is not an easy task for several reasons. In the case of biorefineries, sustainability assessment should reflect the important renewability attribute; furthermore information on the contribution to social, environmental, and economic well-being has to be provided. Assessments should be carried out on a lifecycle bases as far as possible, starting from biomass resources and extending to the end-of-life of the products. Biorefineries are very much diverse in their feedstocks, processes, and products. Furthermore, due to the state of the art of advanced biorefineries, data on upscaled processes are partly not fully available yet. Sustainable indicators, e.g., GHG emissions and primary energy demand, are included in most environmental sustainability assessments of biorefineries. Further interesting indicators, e.g., water demand, eutrophication, acidification, and impacts on biodiversity, should be included in the sustainability assessment as well, to gain a more holistic view on the environmental sustainability.

The food vs fuel debate has shifted the focus from agricultural biomass to nonfood biomass to be used as raw materials for biorefineries. However, this debate may be somehow misleading because the reasons of limited access to food is quite complex. The main issue is not the availability of food but imbalanced purchasing power of consumers in different countries. The restricted use of biomass alone is not enough to solve this problem. The main challenge in using biomass is that several goals have to be met simultaneously, e.g., ensuring sufficient food availability, preserving soil fertility, and also sufficient biomass availability if the transition to a bioeconomy is the ultimate goal.

Further biorefinery improvements are expected to allow the use of more feedstocks, using different technologies, and generating a broader portfolio of new products, inevitably offering all kinds of economic opportunities. Research and development will accelerate agricultural and rural development, increase industrial development, and open existing and new markets.

5.7.2 Discussion and conclusions

Biorefining is an optimal strategy for (large-scale) sustainable use of biomass in the bioeconomy resulting in cost-competitive coproduction of food and feed, bio-based products, and bioenergy with optimal socioeconomic and environmental impacts.

In biorefineries bioenergy and biomaterials may be efficiently produced according to the potential of the different biogenic resources. Furthermore, the different needs of the society may be sustainably covered, e.g., transport services, electricity, food and feed, chemicals, and other materials. The products of the biorefinery may substitute the products from fossil raw materials, e.g., diesel, plastic, and this leads to a reduction of GHG emissions (e.g., CO_2 from burning fossil fuels). As biomass availability is limited, it should be used efficiently, effectively producing materials and energy in multipurpose biorefineries. The use of renewable materials and energy in biorefineries ensures the increase of renewable energy.

The coupled production of bioenergy and biomaterials opens up characteristics and the product potentials of the raw materials and contributes to an increasing energy and material efficiency. Extending the value chain—from the biogenic resources to energy and material services—increases income and secures high-quality jobs in all sectors of this value chain (e.g., agriculture and forestry, industry, and trade). The development and application of innovative technologies in the field of biorefineries ensures and generates higher value-added production and stimulates the export of sustainable products and technologies to further expand the competitiveness.

Based on the future demand for biogenic fuels (e.g., EU target of 10% in 2020 according to the current EU Directive; proposed target on a 12%

share of energy from renewable sources in transport by 2030 with a cap to food and feed-based biofuels and a main focus on advanced biofuels suggested by RED II proposal in January 2018), biorefineries have a very large market potential in the medium term. The combined production of biofuels and high-quality biomaterials makes optimum use of the raw material potential, whereby the biomaterials and biochemicals can be used to expand the value-added chain to be able to offer competitive biofuels on the energy market.

The use of the "BCI" may add additional important information on the assessment and comparison of different biorefinery systems. The BCI is potentially relevant for industry, decision makers, as well as investors as supplementary information could assist them in their strategies to implement the most promising biorefinery systems by minimizing technical and economic risks. The conflict between energy and food production can be decreased by developing technologies based on lignocellulosic materials, but this currently results in a much higher BCI. Biorefining requires further innovation but offers opportunities to all economic sectors.

The scale of biorefineries is a key challenge, and it is currently necessary to build on existing infrastructure. The efficient integration of biorefineries in existing industrial infrastructure can be a key driver for a sustainable biorefinery and bioeconomy. Regional biorefineries linked to existing oil refineries, petrochemical clusters, chemical factories, pulp/saw mills, and district heating systems would enable a smooth transition. This integration could enable a very efficient conversion of biogenic resources to fuels, chemicals, and materials at a scale that would make a difference, using the advantage of investments and permits of existing infrastructures. The further optimization of biomass conversion and the combination for food and nonfood applications could even further raise the impact of sustainable biomass use.

ENDNOTES

1. DRIVE4EU has received funding from the European Union's Seventh Framework Programme (FP7/2007-2013) under grant agreement no 613697. The project is coordinated by Wageningen University and Research. Thanks to all project partners: Wageningen University and Research, KeyGene, Institute of Botany ASCR, Tereos Syral, GEA Westfalia Separator Group GmbH, Apollo Tyres Global R&D BV, QEW Engineered Rubber BV, InExCB-KZ, Mitas a.s, Agricultural Innovation and Knowledge Centre Rusthoeve, ILVO—Institute for Agricultural and Fisheries Research, NETZSCH Feinmahltechnik GmbH).
2. The integration of social aspects is under development.

REFERENCES

Buchsbaum, M., 2014. Possibilities for the future integration of biorefineries in the existing infrastructure in Austria. In: Master Thesis at Graz University of Technology and JOANNEUM RESEARCH, 2014.

Bundesregierung der Bundesrepublik, Deutschland (Ed.), 2012. Roadmap Bioraffinerien—im Rahmen der Aktionspläne der Bundesregierung zur stofflichen und energetischen Nutzung nachwachsender Rohstoffe. Fachagentur Nachwachsende Rohstoffe e.V. (FNR), Rostock, Germany. Available from: https://www.bmbf.de/pub/Roadmap_Bioraffinerien.pdf.

Cherubini, F., Jungmeier, G., Wellisch, M., Willke, T., Skiadas, I., van Ree, R., et al., 2009. Toward a common classification approach for biorefinery systems. Biofuels Bioprod. Bioref. 3, 534–546.

de Jong, E., Jungmeier, G., 2015. Biorefinery concepts in comparison to petrochemical refineries. In: Pandey, A., Hofer, R., Larroche, C., Taherzadeh, M., Nampoothiri, M. (Eds.), "Industrial Biorefineries and White Biotechnology". Elsevier B.V, Amsterdam, Netherlands, Hardcover 9780444634535.

DRIVE4EU, 2017. Dandelion rubber and inulin valorization and exploitation for Europe, FP7 project, www.drive4eu.eu.

European Commission, 2018. RENURESIN—Renewable resins for industrial wood based panel production, January 2018. https://cordis.europa.eu/project/rcn/67234_en.html.

European Parliament, 2018. MEPs set ambitious targets for cleaner, more efficient energy use, January 2018. http://www.europarl.europa.eu/news/en/press-room/20180112IPR91629/meps-set-ambitious-targets-for-cleaner-more-efficient-energy-use.

Joergensen, H., 2015. The role of industry in a transition towards the BioEconomy in relation to biorefinery, IEA Bioenergy Task, 42. Technical University of Denmark.

Jungmeier, G., 2014a. The Biorefinery Complexity Index, IEA Bioenergy, Working Paper, Task 42.

Jungmeier, G., 2014b. The Biorefinery Fact Sheet, Working Paper, IEA Bioenergy Task 42.

Jungmeier, G., Buchsbaum, M., 2014. Upgrading Strategies for Industrial Infrastructures—Integration of Biorefineries in Existing Industrial Infrastructure, Working Paper, IEA Bioenergy Task 42.

Jungmeier, G., Cherubini, F., Dohy, M., de Jong, E., Jørgensen, H., Mandl, M., et al., 2009. Definition and classification of biorefinery systems? The approach in IEA Bioenergy Task 42 Biorefineries. In: Presentation held at the Biorefinery Course Adding Value to the Sustainable Utilisation of Biomass. Ghent, Belgium, June 12, 2009.

Jungmeier, G., Stichnothe H., De Bari, I., Joergenson, H., van Ree, R., de Jong, E., et al., 2013a. A Biorefinery Fact Sheet for the sustainability assessment of energy driven biorefineries—efforts of IEA Bioenergy Task 42 Biorefining. In: Proceedings of the 21th European Biomass Conference, Copenhagen, 2013.

Jungmeier, G., Hingsamer, M., van Ree, R., 2013b. Biofuel-driven biorefineries. A selection of the most promising biorefinery concepts to produce large volumes of road transportation biofuels by 2025, IEA Bioenergy Task 42 Biorefining, Wageningen. http://www.iea-bioenergy.task42-biorefineries.com/upload_mm/b/6/4/8d4294a1-9021-4360-adb3-a76c8b309519_Biofuel-driven%20Biorefineries%20Report%20LR%20February%202013.pdf.

Jungmeier, G., Stichnothe, H., de Bari, I., Jørgensen, H., Van Ree, R., de Jong, E., et al., 2013c. A biorefinery fact sheet for the sustainability assessment of energy driven biorefineries. Efforts of IEA Bioenergy Task 42 "Biorefining." In: Conference Proceedings of the 21th European Biomass Conference, Copenhagen, Denmark, June 2013.

Jungmeier, G., Stichnothe H., De Bari, I., Joergenson, H., van Ree, R., de Jong, E., et al., 2014. Facts, figures and integration of biorefineries in a future bioeconomy—findings in IEA Bioenergy Task 42 "Biorefining." In: Proceedings of the 22th European Biomass Conference, Hamburg, 2014.

van Ree, R., van Zeeland, A., 2014. IEA Bioenergy Task 42 Biorefining, Sustainable and synergetic processing of biomass into marketable food & feed ingredients, chemicals, materials and energy (fuels, power, heat), IEA Bioenergy Task 42 Biorefining, Wageningen. https://www.iea-bioenergy.task42-biorefineries.com/upload_mm/4/2/9/591e5c11-6fac-4c7b-b917-f252536c164b_IEA%20Bioenergy%20Task42%20Biorefining%20Brochure%20SEP2014_HR.pdf.

Verein Deutscher Ingenieure (VDI), 2016. Klassifikation und Gütekriterien von Bioraffinerien (Classification and quality criteria of biorefineries, VDI 6310 Blatt1: 2016-01. Beuth Verlag, Berlin.

PART II

Sustainability

CHAPTER SIX

Sustainability of Bioenergy

Contents

6.1 State of the Art 226
  6.1.1 Sustainability: From Stockholm to Rio to New York—and Beyond 226
  6.1.2 Sustainable Biomass: Principles, Criteria, and Indicators 229
  6.1.3 Sustainability Challenges for Bioenergy: Open Issues 232
    6.1.3.1 Bioenergy and biodiversity 232
    6.1.3.2 Bioenergy and climate change mitigation 232
    6.1.3.3 Bioenergy, land, and the level of consumption 234
  6.1.4 Governance of Sustainable Bioenergy 235
  6.1.5 The Way Forward 235
6.2 Sustainability of Bioeconomy 241
  6.2.1 Social Considerations for the Bioeconomy Sustainability Assessment 241
    6.2.1.1 Bioeconomy transitions In societal perceptions 242
    6.2.1.2 Bioeconomy and research interests 243
  6.2.2 Social Methods in Bioeconomy 245
  6.2.3 Job Creation, Working Conditions, and Rural Development 247
  6.2.4 Food Production, GMOs, and Links to the Sustainable development goals 252
  6.2.5 Stakeholders, Gender, Skills, and Training 254
  6.2.6 Society Perception, Corporate Sustainability Reporting, and Monitoring 255
  6.2.7 Conclusions 256
6.3 Socioeconomic Pillar. Methodology and Case Study 259
  6.3.1 Overview 259
  6.3.2 Socioeconomic Implications of Bioenergy: A Review 259
  6.3.3 Methodologies to Estimate the Socioeconomic Implications of Bioenergy 265
  6.3.4 The IO Methodology 266
    6.3.4.1 A case study 270
  6.3.5 Discussion and Conclusions 276
6.4 Environmental Pillar: Methodology and Case Study 280
  6.4.1 Methodological Background 280
  6.4.2 Application to Bioenergy Chains 282
  6.4.3 Bourgogne Pellets: A Case Study Based on Perennial Biomass Crops 284
    6.4.3.1 Assessing the environmental impact of bioenergy for the full supply chain 284
    6.4.3.2 Optimizing the environmental impact of logistic chains 286

The Role of Bioenergy in the Emerging Bioeconomy.
DOI: https://doi.org/10.1016/B978-0-12-813056-8.00006-6

6.4.3.3 *Environmental impact of 1 GJ of miscanthus at BP facility gate and scale effect* 287
6.4.3.4 *Environmental performances and profitability of alternative logistic options* 291
6.4.4 Advanced Topics 292

6.1 State of the Art

Uwe R. Fritsche
International Institute for Sustainability Analysis and Strategy (IINAS), Darmstadt, Germany

6.1.1 SUSTAINABILITY: FROM STOCKHOLM TO RIO TO NEW YORK—AND BEYOND

It was in 1972 that the *UN Conference on the Human Environment* in Stockholm (UN, 1972) brought the linkages between environment degradation and human development into the international arena (Sachs, 2015).

In took until 1980 that the term *sustainable development* was coined (IUCN et al., 1980), and 7 more years before it was discussed broadly in the Brundtland Report *Our Common Future* (WCED, 1987).

On the level of the United Nations, the 1992 *UN Conference on Environment and Development* (UNCED) in Rio de Janeiro established the three "Rio" conventions (on biological diversity, climate change, and desertification), and the Agenda21 on sustainable development.

Twenty years later, the Rio + 20 Conference decided to develop a set of *sustainable development goals* (SDGs), and on September 25, 2015, the UN General Assembly formally adopted the universal, integrated and transformative 2030 Agenda for Sustainable Development, along with a set of 17 SDGs and 169 associated targets (UN, 2015).

In parallel to this political process, sustainability became an issue of science (de Vries, 2013), and concepts such as

- *limits to growth* (Meadows et al., 1972, 1992, 2004),
- *strong vs weak* sustainability (Ayres et al., 1998; Barbier and Burgess, 2017),
- *planetary boundaries* (O'Neill et al., 2018; Rockström et al., 2009; Steffen et al., 2015), and
- *degrowth* (Döring and Muraca, 2017) enriched the political discussion.

In the last years, science contributed to the development of the SDGs (UN-DESA, 2013) and, nowadays, supports the SDG implementation (Beisheim et al., 2015; EC, 2015; Griggs et al., 2017; Schmalzbauer and Visbeck, 2016; TFM, 2016).

The SDGs are increasingly considered an appropriate *normative framework to guide a sustainable bioeconomy* (Aguilar et al., 2018; El-Chichakli et al., 2016; FAO, 2017; GBS, 2015; Lamers et al., 2016).

Yet, the SDGs do not make any reference to bioenergy or biofuels, but the overall use of biomass (i.e., the "bioeconomy") is expected to increase globally.

This process could be driven by several SDGs but could also be subject to sustainability safeguards from other SDGs, as indicated in Fig. 6.1.1.

The central role of bioenergy with regard to SDG7 (sustainable energy) has been described by Nerini et al. (2018), and recent work of Global Bio-Energy Partnership (GBEP) identified the close relation of the GBEP sustainability indicators with the SDGs and their indicators (Fritsche et al., 2018).

Other research underlined that the SDGs have strong interlinkages (Griggs et al., 2017; Nilsson et al., 2016), implying that work on biomass sustainability could significantly influence other SDGs (Pradhan et al., 2017). Prominent links of biomass sustainability to the SDGs—beyond sustainable energy (SDG 7)—concern

- food (SDG 2), as biomass can both improve or decrease food security (Kline et al., 2017);
- water (SDG 6), as bioenergy can be provided from waste water treatment, and biomass cultivation can improve water supply (IEA Bio and GBEP, 2016), but may also interfere with access to clean water (Bonsch et al., 2016);
- growth and employment (SDG 8), sustainable consumption and production patterns (SDG 12), combat climate change (SDG 13), and life on land (SDG 15).

Using the SDGs as a normative framework for the sustainability of biomass implies not only strong interlinkages but also that the goals

SDG	Key wording	Driver	Safeguard
1	End poverty in all its forms everywhere	(✓)	(✓)
2	**End hunger, achieve food security and improved nutrition and promote sustainable agriculture**	✓	✓
3	Ensure healthy lives and promote well-being for all at all ages	(✓)	(✓)
4	Ensure inclusive and equitable quality education and promote lifelong learning opportunities for all		
5	Achieve gender equality and empower all women and girls	(✓)	(✓)
6	Ensure availability and sustainable management of water and sanitation for all	(✓)	(✓)
7	**Ensure access to affordable, reliable, sustainable and modern energy for all**	✓	(✓)
8	**Promote sustained, inclusive and sustainable economic growth, full and productive employment and decent work for all**	✓	(✓)
9	**Build resilient infrastructure, promote inclusive and sustainable industrialization and foster innovation**	(✓)	
10	Reduce inequality within and among countries	(✓)	(✓)
11	Make cities and human settlements inclusive, safe, resilient and sustainable	✓	(✓)
12	**Ensure sustainable consumption and production patterns**	✓	(✓)
13	**Take urgent action to combat climate change and its impacts**	✓	✓
14	Conserve and sustainably use the oceans, seas and marine resources for sustainable development	(✓)	(✓)
15	**Protect, restore and promote sustainable use of terrestrial ecosystems, sustainably manage forests, combat desertification, and halt and reverse land degradation and halt biodiversity loss**	✓	✓
16	Promote peaceful and inclusive societies for sustainable development, provide access to justice for all and build effective, accountable and inclusive institutions at all levels		(✓)
17	Strengthen the means of implementation and revitalise the global partnership for sustainable development	(✓)	(✓)

Figure 6.1.1 Role of the SDGs for sustainable biomass supply and use. Bold text indicates SDG related to energy; ✓, partially relevant. *From Fritsche, U., et al., 2018. Linkages between the sustainable development goals (SDGs) and the GBEP sustainability indictors for bioenergy (GSI). Technical Paper for the GBEP Task Force on Sustainability. IINAS & IFEU, Darmstadt & Heidelberg. http://www.globalbioenergy.org/fileadmin/user_-upload/gbep/docs/Indicators/IINAS_IFEULinkages_SDGs_and_GSIs_-_final.pdf.*

should be *achieved in an integrated* way to avoid tradeoffs. This will require to move beyond the "pillar" concept and to consider the SDG interlinkages in a positive way:

Biomass—with bioenergy as part of the overall bioeconomy—should be seen as a *means to implement* the SDGs. This would in turn imply that only those biomass systems would be sustainable which comply with (at least the majority of) the SDGs. To assess these opportunities, the issue of *measuring sustainability* becomes relevant.

6.1.2 SUSTAINABLE BIOMASS: PRINCIPLES, CRITERIA, AND INDICATORS

Since the mid-2000s, global interest in and use of bioenergy increased significantly, driven by concerns about climate change, energy security, rising oil prices, and shifting agricultural subsidies. Yet in parallel, concerns increased about the sustainability of bioenergy, and biofuels in particular—from negative impacts on biodiversity, food security, and greenhouse gas (GHG) emissions to land "grabbing" and water scarcity (Goetz et al., 2017; Walter et al., 2018).

As a reaction, the EU and the United States started introducing legislation to safeguard against negative impacts of biofuels developments, especially regarding GHG emissions. Other countries such as China and South Africa restricted biofuel feedstocks to nonedible crops.

In parallel, many UN organizations, the Global Environment Facility, the World Bank and the Inter-American Development Bank, as well as groups such as the WorldWatch Institute (Hunt et al., 2006) and WWF (Fritsche et al., 2006) developed sustainability standards for bioenergy and biofuels in particular (Box 6.1.1).

On the international level, the G7 Gleneagles Summit in 2005 initiated the GBEP, aiming at clarifying what sustainable bioenergy would be and how to measure it.

The GBEP held extensive consultations and discussions within its broad country membership and included many experts from international and national research institutions to compile a comprehensive set of sustainability indicators for bioenergy (GBEP, 2011), aiming at the national policy level (see Fig. 6.1.2).

BOX 6.1.1 Early Warnings From the Last Century

In the wake of the 1973 and 1978 "oil crises," bioenergy and biofuels were considered as potential alternatives to fossil oil. In 1980, the Office of Technology Assessment of the US Congress raised concerns that

> using agricultural land for energy crops can compete with feed and food production and thus lead to increased food prices *(OTA, 1980, p. x), and that*

> increased food prices caused by bioenergy production would fall disproportionately on the poor because the purchase of food takes a greater share of their disposable income. Increased food prices also would raise farmland prices, which could increase economic pressures on small farmers and further concentrate ownership of agricultural land *(OTA, 1980, p. 13).*

In a later study, OTA reflected quite critically on bioenergy crops (OTA, 1993), yet also considered potential benefits such as rural employment.

Environmental pillar	Social pillar	Economic pillar
1. Life-cycle GHG emissions	9. Allocation and tenure of land for new bioenergy production	17. Productivity
2. Soil quality	10. Price and supply of a national food basket	18. Net energy balance
3. Harvest levels of wood resources	11. Change in income	19. Gross value added
4. Emissions of non-GHG air pollutants, including air toxics	12. Jobs in the bioenergy sector	20. Change in consumption of fossil fuels and traditional use of biomass
5. Water use and efficiency	13. Change in unpaid time spent by women and children collecting biomass	21. Training and re-qualification of the workforce
6. Water quality	14. Bioenergy used to expand access to modern energy services	22. Energy diversity
7. Biological diversity in the landscape	15. Change in mortality and burden of disease attributable to indoor smoke	23. Infrastructure and logistics for distribution of bioenergy
8. Land use and land-use change related to bioenergy feedstock production	16. Incidence of occupational injury, illness and fatalities	24. Capacity and flexibility of use of bioenergy

Figure 6.1.2 The GBEP sustainability indicators for bioenergy. *From GBEP, 2011. The GBEP Sustainability Indicators for Bioenergy. Rome: Global Bio-Energy Partnership. http://www.globalbioenergy.org/fileadmin/user_upload/gbep/docs/Indicators/The_GBEP_Sustainability_Indicators_for_Bioenergy_FINAL.pdf.*

In parallel, private sector voluntary sustainability certification schemes for liquid biofuels emerged (Lewandowski and Faaij, 2006; van Dam et al., 2008, 2010), which were extended to solid bioenergy (Fritsche et al., 2014; Mai-Moulin et al., 2017) and biogas/biomethane (Liotta and Mainero, 2017).

This proliferation of different sustainability certification schemes for bioenergy on several governance levels and by various actors (Fig. 6.1.3)

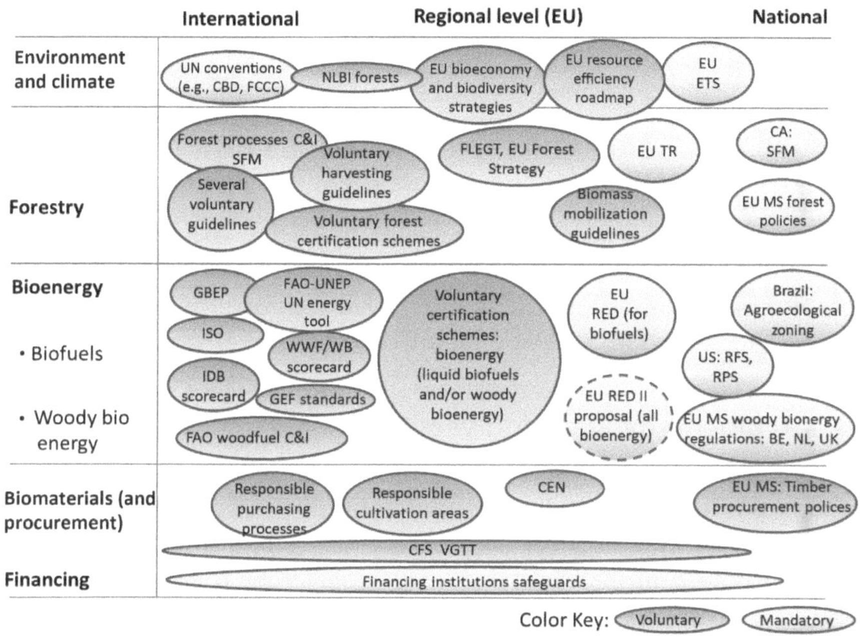

Figure 6.1.3 Sustainability initiatives and schemes for bioenergy. CBD, Convention on Biological Diversity; FCCC, United Nations Framework Convention on Climate Change; NLBI, Non-Legally Binding Instrument on All Types of Forests; ETS, EU Emissions Trading System; C&I, criteria and indicators; SFM, sustainable forest management; FLEGT, Forest Law Enforcement, Governance and Trade; EUTR, EU Timber Regulation; CA SFM, Canadian Sustainable Forest Management; MS, member states; FAO, Food and Agriculture Organization of the United Nations; UNEP, United Nations Environment Programme; WWF, World Wildlife Fund; WB, World Bank; RED, Renewable Energy Directive; RFS, US Renewable Fuel Standard; RPS, Renewable Portfolio Standards; IDB, Inter-American Development Bank; GEF, Global Environment Facility; BE, Belgium; NL, Netherlands; CEN, European Committee for Standardization; CFS, Committee on World Food Security; VGGT, Voluntary Guidelines on the Responsible Governance of Tenure of Land, Fisheries and Forests. *From: Iriarte, L., et al., 2015. Consistent cross-sectoral sustainability criteria and indicators. S2Biom Deliverable D5.4. Darmstadt, Madrid: IINAS. http://www.s2biom.eu/images/Publications/IINAS_2015_S2Biom_D5_4_Sustainability_C_I_proposal_Main_report_30_Mar.pdf.*

poses the risk of confusing market participants and can be seen as a barrier to sustainable bioenergy trade (Stupak et al., 2016).

Both in the EU and the United States, many research projects and initiatives continued to explore and deepen understanding of sustainability requirements for bioenergy, and compiled "umbrella" systems which are meant to provide comprehensive criteria and indicators for all biomass use, i.e., an *integration step* towards a sustainable bioeconomy (Iriarte et al., 2015, 2016; Diaz-Chavez et al., 2016).

6.1.3 SUSTAINABILITY CHALLENGES FOR BIOENERGY: OPEN ISSUES

Given the rich work on sustainable bioenergy and the SDGs, this seems to be a productive approach towards implementation. Yet, there are at least three key issues remaining unresolved and requiring further analysis, and integration (or explicit trade-off analysis).

6.1.3.1 Bioenergy and biodiversity

Biomass cultivation is typically land-intense (Fritsche et al., 2017) and requires continued human intervention in terms of suppressing competing plant growth, controlling weeds, and harvesting biomass. Agriculture is seen as a key contributor to reducing biodiversity (Betts et al., 2017; UNEP-WCMC and IUCN, 2016), and bioenergy crops cultivation and harvesting may contribute to further biodiversity loss unless specific practices (e.g., agroforestry, perennial crops) are encouraged (Dauber and Miyake, 2016; Haughton et al., 2016).

6.1.3.2 Bioenergy and climate change mitigation

6.1.3.2.1 Climate impacts from indirect land-use change

The global landscape has seen human-induced land-use change (LUC) over millennia, from formerly grass and forest land into cropland for food, feed and fiber, and settlements (Ellis et al., 2013), with significant implications for biodiversity, GHG emissions, and soil and water. LUC is defined as the change from one land use type to another, where land use is generally classified as forest, cropland, grazing land, and human settlement. Note that LUC, per se, is not necessarily negative—for example, integration of energy crops could improve the sustainability of large-scale monocultures of annual crops (Berndes and Fritsche, 2016).

Direct LUC refers to land-use conversion occurring where, e.g., a crop is cultivated. In the case of biofuels, this could be deforestation prior to establishing an energy crop or conversion of grassland to annual bioenergy cropping. Direct LUC is observable and measurable, e.g., through remote sensing to detect land clearings and subsequent land use (Gibbs et al., 2010).

Indirect LUC (ILUC) means LUC occurring elsewhere when biofuel crops displace the production of food or feed. As demand for these products remains, production may shift to somewhere else, which can result in deforestation or grassland conversion of other land (and respective GHG emissions). ILUC-related emissions can reduce or even negate GHG benefits of replacing fossil fuels. ILUC effects are primarily market-mediated; that is, driven by price reactions.

The quantification of ILUC is *possible only by modeling*, which compares a "counterfactual" (baseline) scenario with no bioenergy/biofuel to a scenario with bioenergy/biofuel, attributing the differences in LUC to bioenergy/biofuel (Fritsche et al., 2010). As such modeling is complex and subject to many uncertainties and data variation, the scale of ILUC remains highly contested, and the associated literature is extensive (Delzeit et al., 2016; De Rosa et al., 2016; Gerssen-Gondelach et al., 2017; Wicke et al., 2015, Woltjer et al., 2017).

Far less uncertainty is associated with strategies to minimize ILUC (Wicke et al., 2015) in ways such as using feedstock from residues and waste, from sustainable intensification and as feedstock from marginal or degraded land; these are valid options with low negative (adverse) ILUC risks, particularly if biodiversity and social safeguards are in place.

The implementation of policies to reduce deforestation and the inclusion of all LUC-related GHG emissions in a global accounting regime or a cross-sectoral certification system would reduce potential GHG emissions from ILUC to zero. This remains a long-term prospect at best, though.

6.1.3.2.2 Carbon-negative bioenergy?

Following-up on the 2015 Paris Agreement, the IPCC is working on a special report to assess opportunities and challenges of a 1.5°C climate target (IPCC, 2018). In respective preparatory work and already in IPPC's WG III contribution to the fifth Assessment report (IPCC, 2014), the so-called *negative emission technologies* (NETs) became increasingly prominent as a means to reduce the atmospheric concentration of GHGs, and among NETs, *Bioenergy with Carbon Capture and Storage* (BECCS) has received much attention.

The recent IEA (IEA, 2017) strongly argued BECCS is needed to stay below 2°C, and that to achieve the required increase in bioenergy supply and use, an appropriate approach to sustainability governance is needed. Yet, the potential scale of BECCS and the respective bioenergy supply have raised concerns (EASAC, 2018; Muri, 2018; Rosen, 2018). Key potential negative tradeoffs are biodiversity loss from large-scale monocultures, related land tenure problems, and water availability (Heck et al., 2018; Humpenöder et al., 2018; Stoy et al., 2018; Strapasson et al., 2017).

This indicates that some of the tradeoffs between, e.g., renewable energy (SDG 7), climate change mitigation (SDG 13), and biodiversity conservation (SDG 15) are yet unresolved, and possible negative implications for, e.g., food security and land tenure (SDG 2) would need consideration as well.

6.1.3.3 Bioenergy, land, and the level of consumption

From the viewpoint of strong sustainability (Barbier and Burgess, 2017), it is fundamental for sustainable bioenergy to consider a *maximum level of sustainable supply* and to restrict possible biomass demand and use accordingly.

Many argue that the biomass resource base is restricted by land (Searchinger et al., 2017). Yet, there are many land uses with significant opportunities to provide bio-energy as part of integrated approaches such as agroforestry (IRENA, 2017; Sharma et al., 2016), land rehabilitation (Mehmood et al., 2017), and sustainable rural development (Souza et al., 2015) while supporting food security goals (Kline et al., 2017). As part of implementing SDG 15.3 (land degradation neutrality), bioenergy may become quite instrumental (Cowie et al., 2018).

Still, bioenergy is limited, and sustainable bioenergy should prioritize integrated resource management and aim at transitioning to a circular and mainly waste-based bioeconomy (Zabaniotou, 2018).

6.1.4 GOVERNANCE OF SUSTAINABLE BIOENERGY

Last but not least it must be noted that sustainable bioenergy is not just an issue of appropriate metrics (criteria and indicators) and respective measurements but also depends on how those are defined, and implemented, and by whom (Müller et al., 2015; Stupak et al., 2016).

Recent work by de Man and German (2017) indicate that sustainability certification for biofuels is an inadequate substitute for public regulation, given that many "weak" schemes are operating which only consider GHG reduction targets and very few restrictions for biodiversity.

On the other hand, even ambitious sustainability schemes such as the Roundtable on Sustainable Biomaterials cannot overcome the problem of *leakage*: sectoral or commodity-related certification implies that there is biomass (and respective land use) *outside* of the scheme to which unsustainable practices can be shifted.

In a broader sense, sustainability governance of bioenergy must be part of a *sustainable bioeconomy governance* (El-Chichakli et al., 2016), which in turn must embrace practically all land use to avoid cherry picking and leakage (Filho et al., 2018).

Biomass sustainability governance should be seen as an integration opportunity (UNFCCC, 2018), but discussion is needed on appropriate mechanisms for different governance levels. As Müller et al. (2015) pointed out, this will have to include reviewing trade standards (e.g., WTO regulations), existing national legislations, and private governance schemes to identify where adjustments are needed and where sustainability aspects within and across existing regimes could be strengthened.

6.1.5 THE WAY FORWARD

Defining "boundaries" for biomass use, deciding on tradeoffs between SDGs, and designing governance and safeguards for sustainable bioenergy are not merely (transdisciplinary) scientific endeavors (Lewandowski, 2018). Those are political issues which in democratic societies need public discourse, engaging a wide range of stakeholders (Mustalahti, 2018), and decision-making in appropriate bodies (e.g., parliaments).

Science can—and should—help in this, especially in supporting information of and communication with stakeholders. Furthermore, science is needed to identify and strengthen existing sustainability governance mechanisms and agencies, rather than inventing new sets of institutions (Fritsche et al., 2017; Müller et al., 2015).

REFERENCES

Aguilar, A., Wohlgemuth, R., Twardowski, T., 2018. Perspectives on bioeconomy. New Biotechnol. 40, 181–184.

Ayres, R., van den Bergh, J., Gowdy, J., 1998. *Viewpoint: weak versus strong sustainability.* Tinbergen Institute Discussion Paper #98-103/3 (1998-09-28). http://papers.tinbergen.nl/98103.pdf.

Barbier, E., Burgess, J., 2017. Natural resource economics, planetary boundaries and strong sustainability. Sustainability 9, 1858.

Beisheim, M., et al., 2015. *Measuring sustainable development: how can science contribute to realizing the SDGs?* Background Paper with Results From the Conference, Berlin. http://www.dfg.de/download/pdf/dfg_im_profil/geschaeftsstelle/dfg_praesenz_ausland/nordamerika/2015/150421_dfg_unu_konferenz/backgroundpaper.pdf.

Berndes, G., Fritsche, U., 2016. May we have some more land use change, please? Biofuels Bioprod. Biorefineries 10, 195–197.

de Vries, B., 2013. Sustainability Science. Cambridge University Press, New York.

Betts, M., et al., 2017. Global forest loss disproportionately erodes biodiversity in intact landscapes. Nature 547, 441–444.

Bonsch, M., et al., 2016. Trade-offs between land and water requirements for large-scale bioenergy production. GCB Bioenergy 8, 11–24.

Cowie, A., et al., 2018. Land in balance: the scientific conceptual framework for land degradation neutrality. Environ. Sci. Policy 79, 25–35.

Dauber, J., Miyake, S., 2016. To integrate or to segregate food crop and energy crop cultivation at the landscape scale? Perspectives on biodiversity conservation in agriculture in Europe. Energy Sustain. Soc. 6, 25.

Delzeit, R., Klepper, G., Söder, M., 2016. An evaluation of approaches for quantifying emissions from indirect land use change. Kiel Working Paper No. 2035. http://hdl.handle.net/10419/130757.

Diaz-Chavez, R., Stichnothe, H., Johnson, K., 2016. Sustainability considerations for the future bioeconomy. In: Lamers, P., et al., (Eds.), Developing the Global Bioeconomy—Technical, Market, and Environmental Lessons from Bioenergy. Academic Press, pp. 69–90.

Döring, R., Muraca, B., 2017. From (strong) sustainability to degrowth: a philosophical and historical reconstruction. In: Caradonna, J. (Ed.), Routledge Handbook of the History of Sustainability. Routledge, London; New York, pp. 339–362.

EASAC, 2018. Negative emission technologies—what role in meeting Paris Agreement targets? European Academies' Science Advisory Council. EASAC policy report 35. Brussels, Belgium. https://easac.eu/fileadmin/PDF_s/reports_statements/Negative_Carbon/EASAC_Report_on_Negative_Emission_Technologies.pdf.

EC, 2015. The role of science, technology and innovation policies to foster the implementation of the sustainable development goals (SDGs). Report of the European Commission Expert Group "Follow-up to Rio + 20, notably the SDGs". Luxembourg. http://bookshop.europa.eu/en/the-role-of-science-technology-and-innovation-policies-to-foster-the-implementation-of-the-sustainable-development-goals-sdgs--pbKI0415809/downloads/KI-04-15-809-EN-N/KI0415809ENN_002.pdf.

El-Chichakli, B., et al., 2016. Policy: five cornerstones of a global bioeconomy. Nature 535 (7611), 221–223.
Ellis, E., et al., 2013. Used planet: a global history. Proc. Natl. Acad. Sci. USA 110 (20), 7978–7985.
FAO, 2017. Towards sustainable bioeconomy guidelines. International Sustainable Bioeconomy Working Group on Sustainable Bioeconomy. Food and Agriculture Organization of the United Nations, Rome. Available from: http://www.fao.org/3/a-bs923e.pdf.
Fritsche, U., et al., 2006. Sustainability Standards for Bioenergy. Prepared for WWF by Oeko-Institut, Darmstadt. http://www.iinas.org/tl_files/iinas/downloads/bio/oeko/2006_Sustainability_Standards_Bio-WWF.pdf.
Towards a sustainable bioeconomy: principles, challenges and perspectives. In: Filho, L., et al., (Eds.), World Sustainability Series. Springer International Publishing, Cham, Switzerland.
Fritsche, U., et al., 2018. Linkages between the sustainable development goals (SDGs) and the GBEP sustainability indictors for bioenergy (GSI). Technical Paper for the GBEP Task Force on Sustainability. IINAS & IFEU, Darmstadt & Heidelberg. http://www.globalbioenergy.org/fileadmin/user_upload/gbep/docs/Indicators/IINAS_IFEULinkages_SDGs_and_GSIs_-_final.pdf.
Fritsche, U.R., Sims, R., Monti, A., 2010. Direct and indirect land-use competition issues for energy crops and their sustainable production—an overview. Biofuels Bioprod. Biorefineries (4), 692–704.
Fritsche, U.R., et al., 2014. Extending the EU renewable energy directive sustainability criteria to solid bioenergy from forests. Nat. Resour. Forum 38, 129–140.
Fritsche, U.R., et al., 2017. Energy and land. Working Paper for the UNCCD Global Land Outlook. Darmstadt. https://global-land-outlook.squarespace.com/s/Fritsche-et-al-2017-Energy-and-Land-Use-GLO-paper-corr.pdf.
GBEP, 2011. The GBEP Sustainability Indicators for Bioenergy. Rome: Global Bio-Energy Partnership. http://www.globalbioenergy.org/fileadmin/user_upload/gbep/docs/Indicators/The_GBEP_Sustainability_Indicators_for_Bioenergy_FINAL.pdf.
GBS, 2015. Communiqué of the Global Bioeconomy Summit 2015—Making Bioeconomy Work for Sustainable Development. Berlin. http://gbs2015.com/fileadmin/gbs2015/Downloads/Communique_final.pdf.
Gerssen-Gondelach, S., Wicke, B., Faaij, A., 2017. GHG emissions and other environmental impacts of ILUC mitigation. GCB Bioenergy 9 (4), 725–742.
Gibbs, H., et al., 2010. Tropical forests were the primary sources of new agricultural land in the 1980s and 1990s. Proc. Natl. Acad. Sci. USA 107 (38), 16732–16737.
Goetz, A., et al., 2017. Do no harm? Risk perceptions in national bioenergy policies and actual mitigation performance. Energy Policy 108, 776–790.
Haughton, A., et al., 2016. Dedicated biomass crops can enhance biodiversity in the arable landscape. GCB Bioenergy 8, 1071–1081.
Griggs, D., et al., (Eds.), 2017. A Guide to SDG Interactions: from Science to Implementation. International Council for Science, Paris. Available from: http://www.icsu.org/cms/2017/05/SDGs-Guide-to-Interactions.pdf.
Heck, V., et al., 2018. Biomass-based negative emissions difficult to reconcile with planetary boundaries. Nat. Clim. Change 8 (2), 151–155.
Humpenöder, F., et al., 2018. Large-scale bioenergy production: how to resolve sustainability trade-offs? Environ. Res. Lett. 13, 02401.
Hunt, S., et al., 2006. Biofuels for Transportation—Global Potential and Implications for Sustainable Agriculture and Energy in the 21st Century. Worldwatch Institute, London. Available from: http://www.iinas.org/tl_files/iinas/downloads/bio/oeko/2006_Biofuels_for_Transportation-WWI.pdf.

IEA, 2017. Technology Roadmap: Delivering Sustainable Bioenergy. International Energy Agency, Paris. Available from: http://www.iea.org/publications/freepublications/publication/Technology_Roadmap_Delivering_Sustainable_Bioenergy.pdf.

IEA Bio and GBEP, 2016. Examples of Positive Bioenergy and Water Relationships. Rome: IEA Bioenergy & Global Bioenergy Partnership. http://www.globalbioenergy.org/fileadmin/user_upload/gbep/docs/2015_events/AG6_workshop_25-26_August_2015/AG6_Examples_of_Positive_Bioenergy_and_Water_Relationships_Final.pdf.

IPCC, 2014. Climate Change 2014: Mitigation of Climate Change. In: Edenhofer, O., et al. (eds.), Contribution of Working Group III to the Fifth Assessment Report of the IPCC. Cambridge, New York: IPCC. http://report.mitigation2014.org/report/ipcc_wg3_ar5_full.pdf.

IPCC, 2018. Global Warming of 1.5°C—an IPCC special report on the impacts of global warming of 1.5°C above pre-industrial levels and related global greenhouse gas emission pathways, in the context of strengthening the global response to the threat of climate change, sustainable development, and efforts to eradicate poverty. Intergovernmental Panel on Climate Change (forthcoming; http://www.ipcc.ch/report/sr15/).

IRENA, 2017. Bioenergy from degraded land in Africa. Abu Dhabi: International Renewable Energy Agency. http://www.irena.org/-/media/Files/IRENA/Agency/Publication/2017/Dec/IRENA_Bioenergy_Africa_degraded_land_2017.pdf.

IUCN, UNEP, WWF, 1980. World Conservation Strategy: Living Resource Conservation for Sustainable Development. International Union for Conservation of Nature and Natural Resources, UN Environment Programme & World Wildlife Fund. Gland https://portals.iucn.org/library/efiles/documents/wcs-004.pdf.

Iriarte, L., et al., 2015. Consistent cross-sectoral sustainability criteria and indicators. S2Biom Deliverable D5.4. Darmstadt, Madrid: IINAS. http://www.s2biom.eu/images/Publications/IINAS_2015_S2Biom_D5_4_Sustainability_C_I_proposal_Main_report_30_Mar.pdf.

Iriarte, L., et al., 2016. *BioTrade2020plus Approach to Sustainability.* BioTrade2020plus Deliverable D2.4. IINAS, Imperial College, Alterra & Utrecht University. Madrid, etc. http://www.biotrade2020plus.eu/images/BioTrade2020plus_D2.4_IINAS_Main_Report.pdf.

van Dam, J., et al., 2008. Overview of recent developments in sustainable biomass certification. Biomass Bioenergy 32 (8), 749—780.

van Dam, J., Junginger, M., Faaij, A., 2010. From the global efforts on certification of bioenergy towards an integrated approach based on sustainable land use planning. Renew. Sustain. Energy Rev. 14 (9), 2445—2472.

Kline, K., et al., 2017. Reconciling food security and bioenergy: priorities for action. GCB Bioenergy 9, 557—576.

Lamers, P., et al., (Eds.), 2016. Developing the Global Bioeconomy—Technical, Market, and Environmental Lessons From Bioenergy. Academic Press, San Diego, CA.

Lewandowski, I., Faaij, A., 2006. Steps towards the development of a certification system for sustainable bio-energy trade. Biomass Bioenergy 30, 83—104.

Lewandowski, I. (Ed.), 2018. Bioeconomy—Shaping the Transition to a Sustainable, Biobased Economy. Springer International, Cham, Switzerland. Available from: https://doi.org/10.1007/978-3-319-68152-8.

Liotta, F., Mainero, D., 2017. Recent developments of biomethane sustainability voluntary certifications: a global overview. Procedia Environ. Sci. Eng. Manage. 4 (2), 77—84.

Mai-Moulin, T., et al., 2017. Toward a harmonization of national sustainability requirements and criteria for solid biomass. Biofuels Bioprod. Biorefineries . Available from: https://doi.org/10.1002/bbb.1822 (published online 9 Nov 2017).

Meadows, D., et al., 1972. The Limits to Growth. Universe Books, New York, NY.
Meadows, D., Meadows, D., Randers, J., 1992. Beyond the Limits: Confronting Global Collapse, Envisioning a Sustainable Future. Chelsea Green Publishing Company, White River Junction, VT.
Meadows, D., Randers, J., Meadows, D., 2004. Limits to Growth: The 30-Year Update. Chelsea Green Publishing Company, White River Junction, VT.
Mehmood, M., et al., 2017. Biomass production for bioenergy using marginal lands. Sustain. Prod. Consumption 9, 3–21.
De Rosa, M., Knudsen, M., Hermansen, J., 2016. A comparison of land use change models: challenges and future developments. J. Clean. Prod. 113, 183–193.
Müller, A., et al., 2015. The Role of Biomass in the Sustainable Development Goals: A Reality Check and Governance Implications. Institute for Advanced Sustainability Studies, Potsdam. Available from: https://www.iass-potsdam.de/sites/default/files/files/the-role-of-biomass_iass_working_paper_2nd_edition.pdf.
Muri, H., 2018. The role of large-scale BECCS in the pursuit of the 1.5°C target: an Earth system model perspective. Environ. Res. Lett. 13, 044010.
Mustalahti, I., 2018. The responsive bioeconomy: the need for inclusion of citizens and environmental capability in the forest based bioeconomy. J. Clean. Prod. 172, 3781–3790.
Nerini, F., et al., 2018. Mapping synergies and trade-offs between energy and the sustainable development goals. Nat. Energy 3, 10–15.
Nilsson, M., Griggs, D., Visbeck, M., 2016. Map the interactions between sustainable development goals. Nature 534, 320–322.
OTA, 1980. Energy from biological processes. Office of Technology Assessment, United States Congress, Washington, DC. Available from: www.princeton.edu/~ota/disk3/1980/8008/8008.PDF.
OTA, 1993. Potential Environmental Impacts of Bioenergy Crop Production. Office of Technology Assessment, US Congress. Background Paper OTA-BP-E-118. Washington, DC. http://www.princeton.edu/~ota/disk1/1993/9337/9337.PDF.
O'Neill, D., et al., 2018. A good life for all within planetary boundaries. Nat. Sustain. 1, 88–95.
Pradhan, P., et al., 2017. A systematic study of sustainable development goal (SDG) interactions. Earth's Future 5, 1169–1179. Available from: https://doi.org/10.1002/2017EF000632.
de Man, R., German, L., 2017. Certifying the sustainability of biofuels: promise and reality. Energy Policy 109, 871–883.
Rockström, J., et al., 2009. A safe operating space for humanity. Nature 461, 472–475.
Schmalzbauer, B., Visbeck, M. (Eds.), 2016. The Contribution of Science in Implementing the Sustainable Development Goals. German Committee Future Earth, Stuttgart, Kiel. Available from: http://www.dkn-future-earth.org/data/mediapool/2016_report_contribution_science_v7_final_low.pdf.
Rosen, J., 2018. The carbon harvest. Science 359, 733–737.
Sachs, J., 2015. The Age of Sustainable Development. Columbia University Press, New York, NY.
Souza, G., et al., 2015. Bioenergy & Sustainability: Bridging the Gaps. Scientific Committee on Problems of the Environment (SCOPE), Sao Paulo, p. 72. Available from: http://bioenfapesp.org/scopebioenergy/images/chapters/bioenergy_sustainability_scope.pdf.
Searchinger, T., Beringer, T., Strong, A., 2017. Does the world have low-carbon bioenergy potential from the dedicated use of land?. Energy Policy 110, 434–446.
Sharma, N., et al., 2016. Bioenergy from agroforestry can lead to improved food security, climate change, soil quality, and rural development. Food Energy Security 5 (3), 165–183.

Steffen, W., et al., 2015. Planetary boundaries: guiding human development on a changing planet. Science 347, 1259855.

Stoy, P., et al., 2018. Opportunities and tradeoffs among BECCS and the food, water, energy, biodiversity, and social systems nexus at regional scales. BioScience 68 (2), 100–111.

Strapasson, A., et al., 2017. On the global limits of bioenergy and land use for climate change mitigation. GCB Bioenergy 9 (12), 1721–1735.

Stupak, I., et al., 2016. A global survey of stakeholder views and experiences for systems needed to effectively and efficiently govern sustainability of bioenergy. WIREs Energy Environ. 5 (1), 89–118.

TFM, 2016. Harnessing the Contribution of Science, Technology, and Innovation For Achieving the 2030 Agenda and the 17 Sustainable Development Goals. Prepared by the 10-Member Group to support Technology Facilitation Mechanism June 4, 2016. New York, NY. https://sustainabledevelopment.un.org/content/documents/21201STI%20for%20SDGs%2010%20member%20group%20STI%20Forum%20final%20clean.pdf.

UN, 1972. Report of the United Nations Conference on the Human Environment. Stockholm, June 5–16, 1972. United Nations A/CONF.48/14/Rev.1, New York. http://www.un-documents.net/aconf48-14r1.pdf.

UN, 2015. Sustainable Development Goals. United Nations, New York. Available from: https://sustainabledevelopment.un.org/sdgs.

UN-DESA, 2013. Expert Group Meeting on "Science and Sustainable Development Goals. United Nations, New York. Available from: http://sustainabledevelopment.un.org/content/documents/1678EGM%20Science%20and%20SDGs-%20Concept%20note.final.pdf.

UNEP-WCMC, IUCN, 2016. Protected Planet Report 2016. United Nations Environment Programme World Conservation Monitoring Centre & International Union for the Conservation of Nature. Cambridge, Gland. https://wdpa.s3.amazonaws.com/Protected_Planet_Reports/2445%20Global%20Protected%20Planet%202016_WEB.pdf.

UNFCCC, 2018. Opportunities and options for integrating climate change adaptation with the sustainable development goals and the Sendai framework for disaster risk reduction 2015–2030. United Nations Framework Convention of Climate Change. Technical Paper by the Secretariat. Bonn. http://unfccc.int/files/adaptation/groups_committees/adaptation_committee/application/pdf/techpaper_adaptation.pdf.

WCED, 1987. Report of the World Commission on Environment and Development: Our Common Future. Oxford, UK. http://www.un-documents.net/wced-ocf.htm.

Walter, A., et al., 2018. Sustainability of biomass. In: Vaz Jr., S. (Ed.), Biomass and Green Chemistry—Building a Renewable Pathway. Springer, Cham, Switzerland, pp. 191–219. Available from: https://doi.org/10.1007/978-3-319-66736-2.

Wicke, B., et al., 2015. ILUC Prevention Strategies for Sustainable Biofuels—Synthesis Report From the ILUC Prevention Project. Utrecht University. Available from: http://www.uu.nl/faculty/geosciences/EN/research/institutesandgroups/researchinstitutes/copernicusinstitute/research/energyandresources/research/Documents/Synthesis%20report_ILUC%20prevention.pdf.

Woltjer, G., et al. (2017) Analysis of the latest available scientific research and evidence on indirect land use change (ILUC) greenhouse gas emissions associated with production of biofuels and bioliquids. Wageningen Economic Research, PBL, Wageningen Environmental Research & CENER. Study for the EC, Brussels. https://ec.europa.eu/energy/sites/ener/files/documents/20170816_iluc_finalstudyreport.pdf

Zabaniotou, A., 2018. Redesigning a bioenergy sector in EU in the transition to circular waste-based bioeconomy—a multidisciplinary review. J. Clean. Prod. 177, 197–206.

FURTHER READING

Brouwer, F., et al., 2018. Energy modelling and the Nexus concept. Energy Strategy Rev. 19, 1–6.

FAO, 2016. How Sustainability Is Addressed in Official Bioeconomy Strategies at International, National and Regional Levels—An Overview. Food and Agriculture Organization of the United Nations, Rome. Available from: http://www.fao.org/3/a-i5998e.pdf.

Fritsche, U.R., Iriarte, L., 2014. Sustainability criteria and indicators for the bio-based economy in Europe: state of discussion and way forward. Energies 7 (11), 6825–6836.

Nilsson, M., 2017. Important interactions among the sustainable development goals under review at the high-level political forum 2017. Stockholm Environment Institute Working Paper 2017-06, Stockholm. https://www.sei-international.org/mediamanager/documents/Publications/SEI-WP-2017-06-Nilsson-SDG-interact-HLPF2017.pdf.

Philp, J., 2018. The bioeconomy, the challenge of the century for policy makers. New Biotechnol. 40, 11–19.

Valin, H., et al., 2015. The land use change impact of biofuels consumed in the EU—quantification of area and greenhouse gas impacts. Study of Ecofys, IIASA and E4tech for EC DG ENER, Utrecht. https://ec.europa.eu/energy/sites/ener/files/documents/Final%20Report_GLOBIOM_publication.pdf.sss

Weigelt, J., Müller, A., 2016. Governing land restoration: four hypotheses. In: Chabay, I., Frick, M., Helgeson, J. (Eds.), Land Restoration: Reclaiming Landscapes for a Sustainable Future. Academic Press, Amsterdam, xix–xxii.

6.2

Sustainability of Bioeconomy

Rocio Diaz-Chavez
Stockholm Environment Institute Africa Centre, Nairobi, Kenya

6.2.1 SOCIAL CONSIDERATIONS FOR THE BIOECONOMY SUSTAINABILITY ASSESSMENT

This section will focus on the social implications of the different activities along supply chains and in general in bioeconomy and methods and tools used to asses it. Most of the impacts though are related to the first stages of the supply chain sourcing feedstocks and biomass to include in different bioeconomy processes. Some refer to socioeconomic issues from the point of view of employment, working conditions, rural development, governance, land rights, and gender. This section presents examples mainly from Europe with research projects' results funded by the European Commission research frameworks.

6.2.1.1 Bioeconomy transitions in societal perceptions

The work on bioeconomy in Europe was developed through the support of different policies that allowed to create a strategy to foster this sector. It's history is presented in the paper by Patermann and Aguilar (2018) that explains how since 1982 different research frameworks of the European Commission have supported biotechnology and natural sciences. One of the earliest research framework programs addressing these issues was the fifth framework which included key factors and one of them was the "Cell factory". This key factor, besides advancing science issues integrating innovations into living cells (microbial, plant, and animal cells) and their products, also included environmental aspects and products and processes of interest to the society. These societal interests included health, the environment, food, agriculture, agro-industries, and high value-added products (Patermann and Aguilar 2018). The different research programs addressing these innovations were the basis for more ambitious policy initiatives which included the Strategy on Biotechnology in 2002, the concept of the Knowledge-Based Bio-Economy, and the Strategy on Bioeconomy in 2012 (Patermann and Aguilar 2018).

Soon the concepts of biotechnology used with a more sustainability focus emerged. The document from the OECD defined bioeconomy as "A biobased economy is defined as a concept that uses renewable bioresources, efficient bioprocesses and eco-industrial clusters to produce sustainable bioproducts, jobs and income" (OECD, 2004). Later the same organization defined it more thoroughly to include all the process as "the economic activities related to the invention, development, production and use of biological products and processes" (OECD, 2010). Therefore, it also covers the activities related to it (feedstock production or origin, processes and conversion and final products). Nevertheless, as argued by Patermann and Aguilar (2018), these definitions focused mainly on the economic benefits of using biotechnology but not on social acceptance (or benefits). Different projects were funded during the 7th framework on bioeconomy such as the BIOCORE project that also presented a policy analysis of all the different regulatory instruments that shaped the concept of bioeconomy (Diaz-Chavez, 2012a). The evolution to the Bioeconomy Strategy in 2012 allowed to focus on new bio-based value through biorefining concepts and including new models with public–private partnerships with the BBI or Bio-Based Industry. The BBI (2018) operates with a 3.7 billion Public–Private Partnership between the EU and the Bio-based Industries Consortium, Operating under Horizon 2020. It is based on the principle to reduce Europe's dependency on fossil-based products, meet the EU climate change targets, and lead to a greener and more environmentally friendly growth (BBI, 2018). It currently funds over 60 projects in Europe regarding the main objectives of the BBI on bioeconomy.

6.2.1.2 Bioeconomy and research interests

Nevertheless, much of the literature focus on biofuels and their different feedstocks or technology development (e.g., second-generation biofuels or advanced technologies) making it difficult to differentiate between the terms of bioenergy, biofuels, and bioeconomy. The general interest on these topics has also changed in the last 12 years or more.

A review of trends in the literature and on the internet shows the evolution of the terms biofuels, bioenergy, and bioeconomy. Analyzing the results from 2004 to 2018, the peak of the term biofuels is shown in 2008, when indeed in Europe the Renewable Energy Directive (2009) was about to be released, while the term bioeconomy shows much lower interest (Fig. 6.2.1) (Google trends: Numbers represent search interest relative to the highest point on the chart for the given region and time. A value of 100 is the peak popularity for the term. A value of 50 means that the term is half as popular. A score of 0 means there were not enough data for this term). Regarding the trends on peer-reviewed papers, the relative number of papers on biofuels is also bigger than the one on bioeconomy. For comparison, both terms were reviewed in Google Scholar and Web of Science (Figs. 6.2.2 and 6.2.3). The word bioeconomy was searched more in 2017 and the trend can be seen as increasing from 2014 after the Bioeconomy Strategy 2012 was published (Google Trends, 2017).

The status of bioeconomy strategies in the world is highly dominated in countries belonging to the Organisation for Economic Co-operation

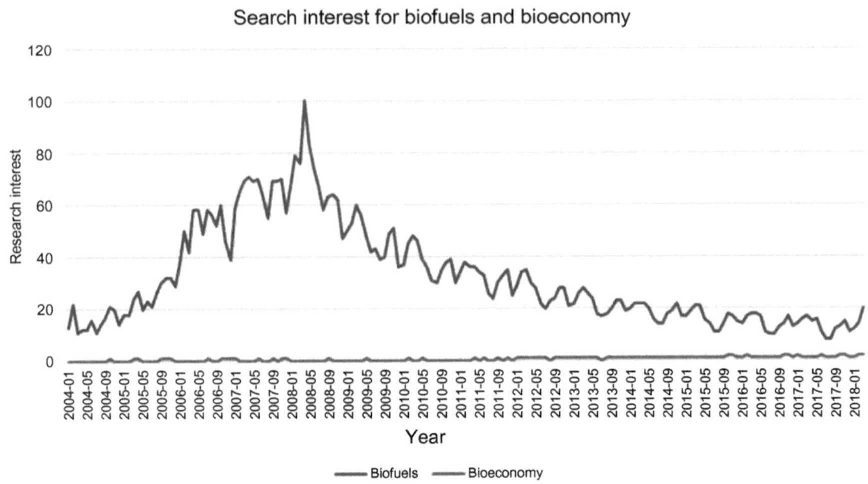

Figure 6.2.1 Search interest for biofuels and bioeconomy (Google trends, 2017).

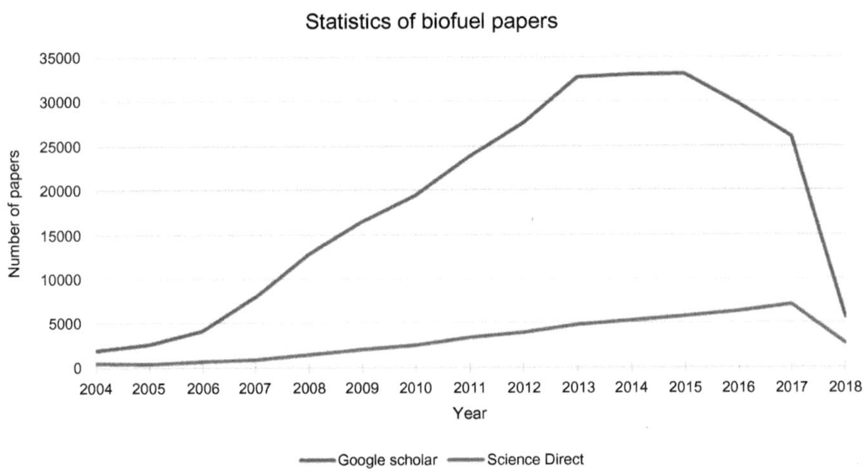

Figure 6.2.2 Biofuels research word in peer-reviewed papers (Google Scholar, 2018; Science Direct, 2018a).

and Development (OECD), particularly those belonging to the International Energy Agency (Hess et al., 2016). A review made on the original research of the IEA (Beerman et al., 2014 in Hess, 2016) explained that the approaches in most of the countries with a strategy tend to be top-down and policy-driven. The research topics mostly cover sustainable biomass supply and bioenergy production. Some of the countries in other regions covered by the study included Brazil, South Africa,

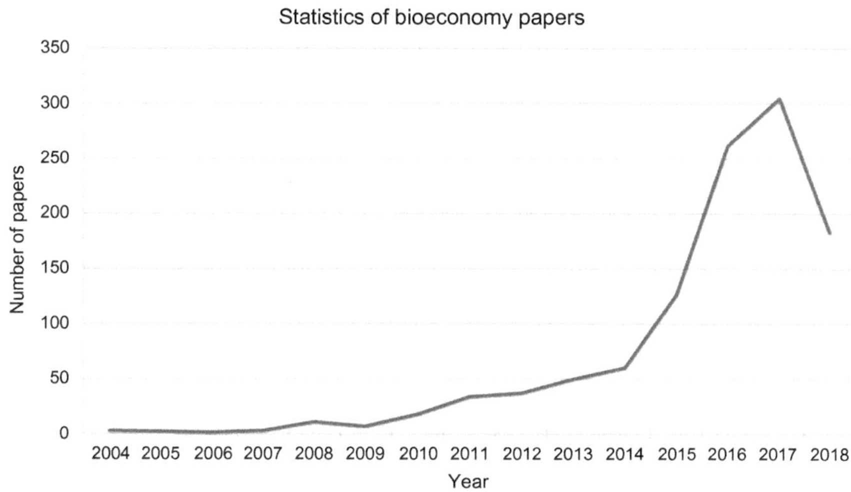

Figure 6.2.3 Bioeconomy research words in peer-reviewed papers (Science Direct, 2018b).

and Japan (Hess et al., 2016). A Bioresources Innovation Network exists in East Africa (BioInnovate Africa, 2018). This program in its first phase (2010–15) focused on building platforms on crop improvement technologies, value-added products from millet and sorghum, sustainable utilization of agro industrial waste, and bioscience innovation policies. The program extends to Burundi, Ethiopia, Kenya, Rwanda, Tanzania, and Uganda. This has a clearer social component as within the bioeconomy it considers an incentive for smallholder farming communities to produce more and for agro/bioprocessing enterprises to become more competitive and environmentally sustainable.

Nevertheless, many social issues are still to be considered also in OECD and EU countries to achieve a real sustainable bioeconomy.

6.2.2 SOCIAL METHODS IN BIOECONOMY

Looking at the evolution of the concept of bioeconomy, the links with biofuels, bioenergy, and biomass use can be seen in the literature. The social issues of topics related to biofuels, bioenergy, and bioeconomy have been investigated mainly in developing countries; much of them associated to the negative social impacts they produce (Diaz-Chavez,

2012b). These have been associated to food security (FAO 2008; Diaz-Chavez et al., 2010; Rosillo-Calle and Johnson, 2010; Kline et al., 2017), local livelihoods and reduction of poverty, land tenure, and large-scale production (Cotula and Leonard, 2010). Where opportunities existed, several issues were addressed for improving local conditions, through employment creation, business models that generate value for local (Vermeulen and Cotula, 2010) and social services and infrastructure (Diaz-Chavez, 2012a). Rural development for instance has been addressed in some voluntary standards associated to the production of biofuels and other value chains. For instance, the Roundtable on Sustainable Materials included a rural development guideline for the standard (Diaz-Chavez, 2012a). Some of the social considerations are presented below in Box 6.2.1.

Different methodologies have also been produced to assess biomass, bioenergy, and biorefineries, which are expected to play a key role in the large-scale implementation of bioeconomy worldwide (Lamers et al., 2016) as the main infrastructure with chemical paths and processes that can produce the bio-products desired for the market. This is under the idea of a low carbon industry that addresses the reduction of fossil fuels to mitigate climate change impacts. As such, different types of biorefineries

BOX 6.2.1 Social Issues in Bioeconomy

When analyzing the value chains in bioeconomy, the activities related to it include:

Feedstock production or origin	Processes and conversion	Products
Gender	Gender	Perceptions
Jobs	Jobs	
Working conditions	Working conditions	
Skills	Skills	
Rights (land, water)		
Food production		
Perceptions		

Feedstock production (origin) is the chain where most of the links to social considerations are included. Other aspects that have been investigated are related. Societal perceptions have been assessed but mainly on the topic of biofuels rather than on bioeconomy (Fallot et al., 2011).

have been developed, mainly the so-called "advanced biorefineries." Sustainability of biorefineries has been reviewed in the literature focusing some of the key challenges on food security, land-use change, genetically modified organisms (GMO), policy, biodiversity, and greenhouse gas balance (Diaz-Chavez et al., 2016). These issues are mainly based on the experience on biofuels and bioenergy. Diaz-Chavez (2015) reviewed some of the common methodologies to assess sustainability in bioenergy. Nevertheless, these methodologies can be applied to the bieoconomy sector. Table 6.2.1 indicates some of these methodologies applied to bioenergy and bioeconomy.

In the case of social issues, some projects have addressed them such as the BIOCORE. This project worked on new methods involving social impact assessment (SIA) and social life cycle assessment and implemented, thus assessing a large number of social issues and themes (Diaz-Chavez, 2012a, 2016).

Although one common method to assess sustainability in biorefineries is based on life cycle assessment (Diaz-Chavez et al., 2016), a weakness of it when applied to biomass sustainability is that it does not address economic or social factors. Currently there is no tool, method, or framework generally accepted for measuring biomass sustainability. According to Philp (2018), agreement on common social criteria is particularly difficult because of the qualitative nature of the assessment, for example, on land rights, issues of child labor. Nevertheless, several social methods have been attempted (Diaz-Chavez, 2012b, 2016).

6.2.3 JOB CREATION, WORKING CONDITIONS, AND RURAL DEVELOPMENT

The biogreen economy (biorefineries) also presents an opportunity for job creation and rural development by different parts of the supply chain beyond feedstock production, such as pressing, collecting, transporting, and storing (BIOCORE, 2012; Diaz-Chavez, 2012b). The BIOCORE (2012) project, for instance, reported that feedstock production for biorefinery may contribute to the creation of indirect jobs. In Europe, case studies have indicated the actual expected contribution of the sector in terms of employment. In the Beauce region in France, one company expected to create around 115 jobs in its biorefinery plant,

Table 6.2.1 Examples of assessment tools and methodologies

Social assessments	Brief description	Integrated assessments	Brief description
Social impact assessment	Environmental Management tool to "analyze, monitor, and manage the intended and unintended social consequences, both positive and negative, of planned interventions (policies, programs, plans, and projects), and any social change processes invoked by those interventions" (IAIA, 2018)	Sustainability assessment	It is an appraisal methodology formed that entails multi-disciplinarity (Sala et al., 2015). It is any process that directs decision-making towards sustainability (Bond et al., 2012)
Health impact assessment	"A combination of procedures, methods and tools by which a policy, program, or project may be judged as to its potential effects on the health of a population, and the distribution of those effects within the population." (Gothenburg Consensus Paper Page 4) (IAIA, 2018)	Integrated impact assessment	Is an assessment that includes environmental, social, and economic aspects in an analysis of environmental state and trends linked with policy analysis (UNEP, nd)
Environmental management systems (EMS)	An EMS is "a set of processes and practices that enable an organization to reduce its *environmental* impacts and increase its operating efficiency." (EPA, 2017)	Impact assessment	"It is the process of identifying the future consequences of a current or proposed action. The "impact" is the difference between what would happen with the action and what would happen without it" (IAIA, 2009)

Corporate social responsibility	Corporate social responsibility is a management concept whereby companies integrate social and environmental concerns in their business operations and interactions with their stakeholders (UNIDO, 2018)	Strategic impact assessment	"It is a set of analytical and participatory processes for incorporating environmental considerations, at early stages of decision-making, into policies, plans, and programs that affect natural resources. SEA evaluates, at the highest strategic level, a project's environmental impacts in the context of social and economic factors" (World Bank, 2018)
Social life cycle assessment (SLCA)	SLCA is a methodology to assess the social impacts of products and services (HSDB, 2013)		
Hot spot database (HSDB)	The HSDB is a simulation tool of data collection on prominent social issues that affect workers, businesses, communities, and societies worldwide. The Social Hotspots Database is a resource for Social Life Cycle Assessment (HSDB, 2013)		

Source: Based on Diaz-Chavez (2015).

whereas estimates for a similar plant in Hungary ranged from 250 direct jobs (some highly skilled) to up to 3000 indirect jobs (e.g., farmers and suppliers) (Diaz-Chavez, 2012b). From this report it was also concluded that the highest number of skilled jobs are produced the least is the number of these jobs. The literature on social issues in the bioenergy sector has produced estimates of job creation that vary according to region, and other factors in the supply chain. The IRENA report (IRENA, 2017) estimated that in 2016 the world gross employment was over 1.7 million in biofuel for transport, 723 thousand in solid biomass, and 333 in biogas.

In the EU, the "environmental goods and services sector" comprises production activities that generate environmental products such as in the bioeconomy. They can be for the purpose of environmental protection or resource management. This sector from 2002 increased employment by around 37%. In 2012, about 4.2 million people (in full-time equivalents) were employed in this sector. Between 2002 and 2011 employment (in full-time equivalents) in this area was almost double (Eurostat, 2013).

As per the type of job creation in bioeconomy, Philp (2018) indicated that the chemical and biotechnology sectors are those that will require more investment in the future. In terms of job creation and rural development, the advanced biorefinery concept proposed in the EU funded BIOCORE project (BIOCORE, 2012) proposed potential to have positive impacts, without generating significant risks, for example, for gender equity or health. Some of the issues were more related to the production of the biomass where for example less number of jobs is seen in women. While regarding health, some negative impacts were attributed to the physical labor in agricultural activities. Nevertheless, the results did underline risks due to competition for biomass. The project used three case studies in Europe (France, Hungary, and Germany) focusing on different types of biomass. In Germany, for example, stakeholders expressed concern about the reduction in the availability of wood for domestic use in heating and in Hungary concerns were expressed regarding competition with an existing bioethanol production facility. However, the extent to which these concerns and others might affect a biomass producer's willingness to sell biomass to an advanced biorefinery was not ascertained and thus this would need to be achieved in the case of an actual industrial project (Diaz-Chavez, 2012b). The use of the Hotspot database allows to analyze some of these indicators within the agricultural sector (feedstock production), the industrial sector (chemical), and the oil sector in Europe (Diaz-Chavez, 2012b). The case of the BIOCORE project is included in

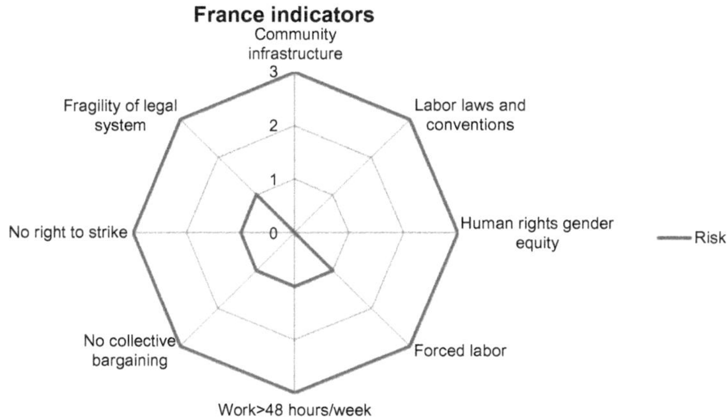

Figure 6.2.4 France's social indicators calculated with the HSDB (Diaz-Chavez, 2012b).

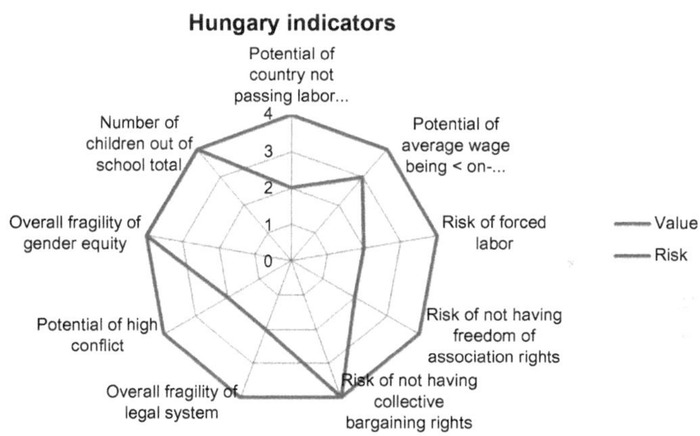

Figure 6.2.5 Hungary's social indicators calculated with the HSDB (Diaz-Chavez, 2012b).

Figs. 6.2.4 and 6.2.5. These figures show some of the indicators selected in the hotspots for the case of France and Hungary. It can be seen that the risks assessed the HSDB weights (the indicators to produce one single figure); the number of significant indicators are less in France than in the case of Hungary and allows to identify where risks might need to be considered. These results were used as a part of the combined methodology of SIA and social life cycle assessment (Diaz-Chavez, 2012b).

The EU funded project COSMOS (COSMOS, 2018) is currently assessing the social assessment for the value chains of crambe and camelina in Europe, while assessing the impacts of import reduction in three countries which are considered the main exporters of coconut oil (Philippines), palm kernel oil (Indonesia), and castor oil (India). The analysis in the countries in Europe (the Netherlands, Poland, Greece, and Italy) is being done considering different social impacts such as job creation, quality of jobs, rural development, land use, and working conditions, among others. The outcomes of this project will also entail the public acceptability of different value chains including the use of insects

6.2.4 FOOD PRODUCTION, GMOS, AND LINKS TO THE SUSTAINABLE DEVELOPMENT GOALS

A review of the sustainability issues of the global bioeconomy (Diaz-Chavez et al., 2016) indicated that it is still a major challenge in the issue of using food biomass based on the lessons learnt of biofuel production due to the perceived impacts on food security from direct and/or indirect land-use change. Although several authors including the work of Souza et al. (2015) counter these concerns, results are not yet widely distributed. The report indicated that while estimates of future sustainable bioenergy potential vary greatly based on dynamic considerations, numerous studies demonstrate the technical potential for increasing biomass production while simultaneously meeting other goals for food security, climate change mitigation, and ecosystem health (Souza et al., 2015).

Bioenergy from food crops can promote stable prices and thereby incentivize local production. Improved food security results from predictable and stable prices that create incentives for local investment in food production (IFPRI, 2015). Food (in)security strongly relates to household income, because welfare measurements are indicated by the fraction of marginal income spent on food (FAO, 2011). Other authors (Diaz-Chavez, 2010; Rosillo-Calle and Johnson, 2010) have indicated that there are other issues affecting food production, security, and availability. Concerns about the use of *Genetically Modified Organisms* (GMOs) as feedstocks or in biocatalytic conversion processes also present challenges as perceptions and definitions of GMOs vary widely in different parts of the world.

The continuous challenge on the use of crops for biomass has been slightly reduced in bioeconomy while considering the use of agricultural and forestry residues rather than the crop itself as it occurred with the biofuels development. On this, the links between food production, use of residues, and rural development could channel the activities of the bioeconomy sector in achieving several of the sustainable development goals (SDGs) (UN, 2018). Furthermore, the use of biomass residues for other activities including heating and power might have better perception and more social benefits. Table 6.2.2 shows the SDGs and their direct link to four stages of a simplified value chain in bioeconomy.

According to Schütte (2018), the success of the bioeconomy requires a societal discourse to reconcile economic growth and sustainability. Policy at all levels is deemed necessary to achieve this.

Table 6.2.2 SDGs and direct links to different stages of a value chain in bioeconomy

Number of goals	SDG	Feedstock production	Feedstock conversion	Logistics	Final product
Goal 1	No poverty	√	√	√	√
Goal 2	Zero hunger	√			
Goal 3	Good health and well-being for people				
Goal 4	Quality education	√	√	√	√
Goal 5	Gender equality	√	√	√	√
Goal 6	Clean water and sanitation				
Goal 7	Affordable and clean energy	√	√	√	√
Goal 8	Decent work and economic growth	√	√	√	√
Goal 9	Industry, innovation, and infrastructure	√	√	√	√
Goal 10	Reduced inequalities	√	√	√	√
Goal 11	Sustainable cities and communities				
Goal 12	Responsible consumption and production				√
Goal 13	Climate action	√	√	√	√
Goal 14	Life below water				
Goal 15	Life on land				
Goal 16	Peace, justice, and strong institutions				
Goal 17	Partnerships for the goals	√	√	√	√

Source: Author.

6.2.5 STAKEHOLDERS, GENDER, SKILLS, AND TRAINING

Different EU-Funded biorefinery projects have addressed issues on stakeholders, skills, and training. In Europe, little has been researched on the topic of gender and bioeconomy. The EU-funded BIOCORE project (BIOCORE, 2012), for instance, canvassed the views of stakeholders in three countries in Europe and India to assess skills and capacity in the green economy sector, particularly with reference to biorefineries. The opinion of different stakeholders varied as some view that required skills are already in place for feedstock production while others considered that the industrial sector (biorefineries) is not fully developed in terms of skills (BIOCORE, 2012). In the EU and other countries in Asia (such as China) and the Middle East, investment in capacity building is taking place in the areas of green chemistry and industrial biotechnology (Philp, 2018).

FAO's BEFSCI project compiled a set of environmental practices that recommended producers should adopt to minimize their negative environmental impacts (FAO, 2012). The document stresses the importance of training provision at all levels of the supply chain which incorporates an integrated approach to sustainable bioenergy. Likewise, the Global Bioenergy Partnership developed a set of 24 sustainability indicators for bioenergy, and Indicator 12 under the social pillar addressed the creation of skilled/unskilled jobs in the bioenergy sector (GBEP, 2011). The Bio-Tic biorefinery EU-funded project (Dupont-Inglis and Borg, 2018) proposed recommendations such as:

- improve public perception and awareness of industrial biotechnology and bio-based products Customers and end users are not sufficiently aware of what industrial biotechnology is or of the environmental and social benefits offered by its products;
- explore novel biomass usage avoiding competition with food production or land use; and
- focus on wastes and residues as alternative biomass sources.

A research in the south of Spain analyzed the gender issue in the olive sector as part of a proposed biorefinery for advanced chemicals. Nevertheless, this was more related to the feedstock production than the industrial part. Some of the findings indicate that at least with this feedstock there is a predominance of men throughout the sector: farming practices, cooperatives, olive oil mills ownership (Diaz-Chavez et al., forthcoming). In Europe the participation of women in agricultural activities varies considerably from one country to another. Nevertheless, inclusion of women in these activities may be possible to increase through better forms of education and participatory activities including subsidies that improve women participation.

In the Global South gender issues have been researched also with regards to the feedstock production because of the important role that women play in agriculture (Migot-Adholla et al., 1991); therefore, the main topic analyzed is on land rights as they exist in various forms in developing countries and have consistently discriminated against women (Carpano, 2011; Gomez and Tran, 2012; Whitehead and Tsikata, 2003; Diaz-Chavez et al., 2016).

Although in the EU and OECD member countries there are several countries that have bioeconomy strategies in place (Hess et al., 2016), there is little mention on the particular aspects of social benefits and almost nothing in particular on gender.

6.2.6 SOCIETY PERCEPTION, CORPORATE SUSTAINABILITY REPORTING, AND MONITORING

Lessons learnt from the biofuel sector showed how the media has focused on food insecurity, land grabbing, and deforestation (Delshad et al., 2010). Nevertheless, general public knowledge on bioenergy is still quite limited (Fallot et al., 2013). There is also little evidence on the general public or the media focusing on bioeconomy as a sector. Fig. 6.2.1 shows the difference between the search done for biofuels compared to bioeconomy. Linking the words biofuels and rural development or bioeconomy and rural development did not produce results.

The bioeconomy sector is still young to assess the effects of Corporate Sustainability Reporting (CSR) and the image of the sector. Again, this is more related to the feedstock production. For example, the Brazilian sugarcane sector has been at the forefront of CSR since 2005 making the sugar cane Union (UNICA) (Brazilian Sugar Cane Industry Association) to develop sustainability indicators for the whole sector.

6.2.7 CONCLUSIONS

Social aspects of the bioeconomy sector have been difficult to separate in the value chain from the feedstock production to the industrial part. Some socioeconomic assessments have included the number of jobs that exist in the production of the feedstock or in the industrial sector (food, chemical, and pharmaceutical industries). Nevertheless, several issues still need to be developed for this sector.

While research continues to advance on the chemical process, paths, products, co-products, and by-products, little research emerge on other environmental and social issues related to this sector. It is still difficult to separate them from the lessons learnt on the biofuels and bioenergy sector. And it will probably not be wise to separate them if real sustainability needs to be addressed. New methods to fully assess the whole value chain might need to emerge and not only on the technology side if a complete sustainability focus is to remain.

The governance system is not fully addressed. Despite there are many countries which already include a bioeconomy strategy, the individual stages of the value chain of a bioproduct are not fully addressed. The main focus remains on the feedstock production, even if it refers to the agricultural or forestry residues. Furthermore, there is not yet a harmonization system or framework for assessing the whole sustainability of a bio-based product.

The development of bioeconomy in developing countries is yet to start in some cases or to improve in others. Nevertheless, this could be a unique opportunity to link the bioeconomy sector with the global achievement of the SDGs which could provide greater benefits from the social point of view worldwide.

REFERENCES

BBI, 2018. Public-Private partnership. https://www.bbi-europe.eu/ (accessed February 2018).

BioInnovate Africa, 2018. Bioresources Innovations Network for Eastern Africa Development. http://bioinnovate-africa.org/

Bond, A., Morrison-Saunders, A., Pope, J., 2012. Sustainability assessment: the state of the art. Impact Assess. Proj. Apprais. 30 (1), 53–62. Available from: https://doi.org/10.1080/14615517.2012.661974.

Carpano. F., 2011. Opportunities and Challenges for Securing Women's Land Rights. Workshop Report. November 19, 2010. Maputo, Mozambique. Available at: http://www.ifad.org/english/land/women_land/WomenAndLand_Mz_Proceedings Report_Eng.pdf

COSMOS 2018. COSMOS, adding value to camelina and crambe oils. http://cosmos-h2020.eu/ Accessed February 2018.
Cotula, L., Leonard, R. (Eds.), 2010. Alternatives to Land Acquisitions: Agricultural Investment and Collaborative Business Models. IIED, United Kingdom, pp. 143.
Delshad, A.B., Raymond, L., Sawicki, V., Wegener, D.T., 2010. Public attitudes toward political and technological options for biofuels. Energy Policy 38 (7), 3414–3425.
Diaz-Chavez, R., 2012a. Rural and social development guidelines. The Roundtable on Sustainable Biofuels.
Diaz-Chavez, R., 2012b. Annex 7.4. Final assessment of the social/legal/political sustainability of the BIOCORE biorefining system. BIOCORE FP7 Project.
Diaz-Chavez, R., 2014. Indicators for socio-economic sustainability assessment. In: Janssen, R., Rutz, D. (Eds.), Socio-Economic Impacts of Bioenergy Production. Springer, Switzerland.
Diaz-Chavez, R., 2015. Assessing sustainability for biomass energy production and use. In: 2nd edition Rosillo-Calle, F., de Groot, P., Hemstock, S., Woods, J. (Eds.), The Biomass Assessment Handbook: Energy for a Sustainable Environment, 2015. Routledge, London: United Kingdom.
Diaz-Chavez, R., Stichnothe, H., Johnson, K., 2016. Sustainability considerations for the future bioeconomy. In: Lamers, P., Searcy, E., Hess, E.J., Stichnothe, H. (Eds.), Developing the Global Bioeconomy. Technical, Market, and Environmental Lessons From Bioenergy. IEA Bioenergy. Academic Press, San Diego, CA, pp. 69–90.
Dupont-Inglis, J., Borg, A., 2018. Destination bioeconomy—the path towards a smarter, more sustainable future. New Biotechnol. 40 (2018), 140–143.
EPA, 2017. Environmental Management Systems. US: Environmental Protection Agency https://www.epa.gov/ems (accessed March 2018).
Eurostat, 2013. Sustainable Development in the European Union. Monitoring Report of the EU Sustainable Development Strategy.
FAO, 2008. The Right to Food. Rome: FAO. http://publish.uwo.ca/~dgrafton/righttofood.pdf (accessed June 2011).
Fallot A., Chacón A., Rutz D., Janssen R., Haye S., Sugrue A., et al., 2011. Methodology and Guidelines for Assessing Public Perception on Biofuels and Bioproducts. Global-Bio-Pact Project. http://www.globalbiopact.eu/publications.html
GBEP, 2011. The Global Bioenergy Partnership Sustainability Indicators for Bioenergy, First Edition FAO, Rome.
Gomez, M., Tran, D.H., 2012. Women's Land and Property Rights and the Post-2015 Agenda. Official Background Paper—Global Thematic Consultation on Addressing Inequalities. Available at: http://www.worldwewant2015.org/node/283484.
Google Trends, 2017. Biofuels and Bioeconomy https://trends.google.com/trends/explore?date = all&q = biofuels,bioeconomy (accessed February 2018).
Google Scholar, 2018. Biofuels and bioeconomy https://csullender.com/scholar/ (Accessed February 2018).
Hess, J.R., Lamers, P., Stichnothe, H., Beermann, M., Jungmeier, G., 2016. Bioeconomy strategies. In: Lamers, P., Searcy, E., Hess, E.J., Stichnothe, H. (Eds.), Developing the Global Bioeconomy. Technical, Market, and Environmental Lessons From Bioenergy. IEA Bioenergy., 2016. Academic Press, San Diego, CA, pp. 1–9.
HSDB, 2013. Hotspot data base. http://socialhotspot.org/content/how-use-database (Accessed 2013).
IAIA, 2009. What Is Impact Assessment? International Association of Impact Assessment. http://www.iaia.org/uploads/pdf/What_is_IA_web.pdf (accessed March 2018).
IAIA, 2018. Social Impact Assessment Definition. International Association of Impact Assessment (IAIA). http://www.iaia.org/wiki-details.php?ID = 23 Accessed March 2018.

IRENAs, 2017. Renewable Energy and Jobs. Annual Review 2017. https://www.irena.org/DocumentDownloads/Publications/IRENA_RE_Jobs_Annual_Review_2017.pdf (Accessed February 2018).

Kline, K.L., Msangi, S., Dale, V.H., Woods, J., Souza, G., Osseweijer, P., et al., 2017. Reconciling food security and bioenergy: priorities for action. Bioenergy. 9 (3), 557–576. March 2017.

OECD, 2004. Biotechnology for Sustainable Growth and Development. Organisation for Economic Co-operation and Development, Paris.

OECD, 2010. Towards the Development of OECD Best Practices for Assessing the Sustainability of Bio-Based Products. OECD, Paris, p. 30.

Patermann, C., Aguilar, A., 2018. The origins of the bioeconomy in the European Union. New Biotechnol. 40 (2018), 20–24.

Philp, J., 2018. The bioeconomy, the challenge of the century for policy makers. New Biotechnol. 40 (2018), 11–19.

Rosillo-Calle, F., Johnson, F., 2010. Food Versus Fuel. An Informed Introduction to Biofuels. Zed Books, London, United Kingdom.

Science Direct, 2018a. Results bioeconomy. https://www.sciencedirect.com/search?qs = bioeconomy&show = 25&sortBy = relevance (accessed February 2018).

Science Direct, 2018b. Results biofuels https://www.sciencedirect.com/search?qs = biofuels&origin = home&zone = qSearch&lastSelectedFacet = years (accessed February 2018).

Schütte, G., 2018. What kind of innovation policy does the bioeconomy need? New Biotechnol. 40 (2018), 82–86.

UN, 2018. Sustainable Development Goals. United Nations Knowledge Platform. https://sustainabledevelopment.un.org/sdgs (accessed February 2018).

UNEP, nd. Guidelines for Conducting Integrated Environmental Assessments. United Nations Environment Programme. https://wedocs.unep.org/bitstream/handle/20.500.11822/16775/IEA_Guidelines_Living_Document_v2.pdf?sequence = 1&isAllowed = y (accessed March 2018).

UNIDO, 2018. What is Corporate Social Responsibility? https://www.unido.org/our-focus/advancing-economic-competitiveness/competitive-trade-capacities-and-corporate-responsibility/corporate-social-responsibility-market-integration/what-csr (Accessed March 2018).

Vermeulen, S., Cotula, L., 2010. Making the Most of Agricultural Investment: A Survey of Business Models That Provide Opportunities for Smallholders. IIED, London, United Kingdom, p. 116.

Whitehead, A., Tsikata, D., 2003. Policy discourses on women's land rights in sub–Saharan Africa: the implications of the re–turn to the customary. J. Agrar. Change 3 (1-2), 67–112.

World Bank, 2018. Strategic Environmental Assessment. Guidance Notes on Tools for Pollution Management. http://siteresources.worldbank.org/INTRANETENVIRONMENT/Resources/244351-1279901011064/GuidanceNoteonSEA.pdf (Accessed March 2018).

FURTHER READING

European Commission, 2012. Innovating for Sustainable Growth—A Bioeconomy for Europe. Publications Office of the European Union, LuxembourgISBN 978-92-79-25376-8 . Available from: https://doi.org/10.2777/6462.

6.3

Socioeconomic Pillar. Methodology and Case Study

Cristina de la Rúa[1] and Yolanda Lechón[2]
[1]Technical University of Munich, Munich, Germany
[2]CIEMAT, Madrid, Spain

6.3.1 OVERVIEW

This chapter will first provide a general overview of the main socioeconomic implications of bioenergy based on a literature review. Next we will introduce the reader the methods used to estimate these socioeconomic implications and specifically the input–output methodology and its extensions and hybridations. Then we will illustrate the application of this methodology to a case study taken from the EU project LogistEC (http://www.logistecproject.eu/the-project/) and related to the production of pellets of *Miscanthus* biomass in the Burgundy region (Eastern France). Finally we will discuss the effects on the economy as well the effects on the generation of employment of different bioenergy chains based on the results obtained and the literature reviewed.

6.3.2 SOCIOECONOMIC IMPLICATIONS OF BIOENERGY: A REVIEW

The use of biomass for energy is likely to have important effects, both positive and negative, on socioeconomic aspects at the local, regional, national, and global level. In the positive side, bioenergy may induce job and income creation, import reduction, increased energy security and diversification, and improved health conditions. On the negative side, higher food prices and the consequent increase in poverty and food insecurity, undesirable effects on land tenure, and local livelihoods have been identified (Beal and Rossi, 2011; Domac et al., 2005; Persson, 2013). Policies addressing these negative aspects are needed to attain a fully sustainable bioenergy production.

From the main socioeconomic aspects that may be impacted by bioenergy production identified above, this chapter will focus on the ones that are usually considered as the driving forces of bioenergy production: employment creation and income generation.

Regarding employment creation, modern bioenergy may create new employment opportunities along the supply chain at the local, regional, national, and global scale. Especially important are the effects stemming from the feedstock production phase that takes place in rural areas that are usually characterized by high unemployment rates. According to IRENA (2017), liquid biofuels, solid biomass, and biogas were the largest employers among renewable energies after solar photovoltaics with a total of 2.7 million jobs in 2016, with jobs concentrated in feedstock supply in Brazil, China, the United States, and India.

Regarding wages, an increase in labor demand may also affect positively wages. However, care must be taken to avoid a pressure to reduce costs on wages and labor conditions in some stages of the supply chain such as feedstock production, usually the most vulnerable stage in this respect (Beal and Rossi, 2011).

Modern bioenergy development may also create new business and income-generating opportunities along the entire supply chain. However, economies of scale and attempts to reduce costs may lead to concentration in bioenergy feedstock production and the undesirable exclusion of smallholders and small entrepreneurs from these positive effects (Beal and Rossi, 2011; Petrini et al., 2017). Petrini and co-workers illustrated this effect by analyzing the expansion of sugarcane for bioethanol production in Brazil. They found that the competition arising from the introduction of sugarcane for biofuel imposes a farm size, minimum yield and scale of production that lead to land concentration and exclusion of small farmers from the agricultural activity, who tend to rent their lands to the agroindustry instead of maintaining their status as agricultural producers. In this sense, Mirata and co-workers (Mirata et al., 2005) demonstrated the ability of some bioenergy chains to be examples and building blocks of what it has been defined as *distributed economies*(Johansson et al., 2005). This concept seeks to avoid the socioeconomically and environmentally unsustainable dynamics usually associated with large-scale, centralized production units favoring the development of small-scale, flexible units that are connected with each other and concentrate on local value addition.

Employment and income created by bioenergy may be divided into direct and indirect (Domac et al., 2005; Persson, 2013). Direct effects would be those directly resulting from the activities of the bioenergy chain (feedstock production and bioenergy facilities), while indirect effects would be those created in the rest of the sectors of the economy indirectly stimulated by the bioenergy chain activities. The ratio of total impacts (direct + indirect) over direct impacts is known as the multiplier. There are also induced effects on employment and income creation originated by the consumption of the direct and indirect employees and the consequent economy stimulation.

However, it is important to note that the bioenergy chain activities may displace other economic activities and the jobs associated with them, and it might compete with other types of agricultural production (Beal and Rossi, 2011; Petrini et al., 2017). These displacement effects (Domac et al., 2005; Persson, 2013) may determine the final magnitude and direction of the impacts. Therefore, it is important to acknowledge and estimate not only the gross effects of bioenergy production on the economy and employment but also the net effects.

An increase in employment and income is more likely when bioenergy makes use of formerly idle factors being those land, capital, or labor than when it displaces these factors from other activities since these displacement effects will not take place (Steininger and Wojan, 2011).

There are also socioeconomic effects associated to energy price changes. If bioenergy is more expensive than conventional fuels, these price increase has a dampening effect on economic growth as they reduce the available budget of households for consumption and increase the energy costs of the industry sector. These effects affect the whole economy through the multiplier effect and generally have a negative sign (Duscha et al., 2014). Neuwahl et al. (2008) estimated the employment consequences of the EU biofuels policy using an input–output (IO) framework, considering price effects and potential household budget reduction among other variables.

Socioeconomic consequences from different bioenergy chains can differ significantly. There are important differences between bioenergy chains using waste streams as feedstocks and those using energy crops and important factors affecting the impacts produced are the labor demand, the production costs, and the land demand (Trink et al., 2010). There are also differences between small decentralized systems and big centralized systems (Van Eijck et al., 2014).

Socioeconomic impacts of modern bioenergy have been investigated in the literature mainly through case studies of solid biomass utilization for bioenergy production and also the production of liquid biofuels.

The Joint Research Centre published in 2014 a report that evaluates the macroeconomic impacts of four bio-based applications—biofuels, biochemicals, bioelectricity, and biogas—using a general equilibrium model in the European Union (Smeets et al., 2014). The results showed that the production of the second-generation biofuel and biochemicals were the only competitive sectors. This was mainly due to direct and indirect wages effects due to the quite labor-intensive nature of biomass production. Another reason is the lower oil and fuel price as a result of the substitution of oil-based fuel production by bio-based fuel production, which indirectly benefits the entire economy.

Trink et al. (2010) examined the macroeconomic consequences of the substitution of fossil fuels by different bioenergy chains in Austria. Wood-based biomass (wood logs, wood pellets, and wood chips) showed the highest results (in job creation and welfare and GDP gains) both in the short and long term. However, some agro-based chains showed negative employment impacts in the short term and welfare and employment negative effects in the long term. The reason of this behavior is the pressure on land demand that makes necessary to import the displaced agricultural goods. Additionally differences in labor and cost intensity among bioenergy chains are also important in the size and sign of the impacts. They found the highest positive employment effect for straw pellets since straw is a residual product and no extra crop land is needed, and found the highest negative effect for corn due to the highest costs and the lowest energy yield compared to wood and Miscanthus. The substitution of fossil diesel by biodiesel also renders negative impacts especially in the long term.

Lehtonen and Okkonen (2013, 2016, 2017) investigated the socioeconomic local, regional, and national impacts of bioenergy investment in peripheral areas of Finland. Their studies showed net positive impacts on employment and economy in the local economy that also leak to the regional and national economies and that are key to support development in remote areas of the country. They conclude that bioeconomy investments are to be promoted as one of the solutions to rural development in these areas.

Thornley et al. (2008) estimated expected gross employment impacts of bioenergy plants in the United Kingdom and found an average figure of 1.27 man-years per GWh. They estimated direct, indirect, and induced effects but neglected the displacement impacts.

Kulišic et al. (2007) performed an IO analysis, where they estimated the impacts from biodiesel produced in Croatia, including this activity as a new industry in the national table. They concluded that biodiesel production could have significant positive net impact on the Croatian economy despite the high level of subsidies for rapeseed growing.

Gan and Smith (2007) studied the impacts of producing electricity using logging residues in Texas. They found high employment multipliers suggesting that bioenergy production could have strong regional positive effects.

Mukhopadhyay et al. (2017) estimated the macroeconomic effects of bioethanol in Canada using IO analysis and including two new industries and four new commodities in the Canadian Make and Use Table. Their results showed that agriculture sector is affected because of feedstock use in the biofuel sector and that the mining and manufacturing industries also show a considerable impact. In addition, they demonstrated that the impact on commodity prices cannot be ignored.

A growing body of literature exists on the socioeconomic effects of bioenergy in developing countries (Openshaw, 2010; Ewing and Msangi, 2009; Petrini et al., 2017; Van Dam et al., 2009, Kemausuor et al., 2016; Van Eijck et al., 2014; Silalertruksa et al., 2012; Souza et al., 2016).

In many developing countries biomass is the main source of energy. Openshaw (2010) quantified, using surveys, the people employed in the biomass supply chain in Malawi and found a figure of 133,000 full time people in 2008. This figure contrast with the 4600 people employed in the supply chain of other fuels. In those countries and especially for crop-based biofuel production, there is a growing concern about the implications on food security due to increases in global food prices. Ewing and Msangi (2009) showed that there is strong correlation between food insecurity, captured by a hunger index developed by the authors, and the rate of biomass use for meeting household energy needs. The poorest countries with a high hunger index are the one that rely more heavily on biomass for energy. The authors argue that in these countries, biofuel development may create opportunities to generate income and expand agricultural production technology that can increase welfare and reduce vulnerability to prices volatility. This is demonstrated by a case study in Ghana by Kemausuor and co-workers (2016) that analyzed a biogas system that uses crop residues and manure as feedstock to produce biogas for cooking. The system would be able to create employment in the community of Zambrama and reduce the consumption of firewood and save the time to collect it. However, as stated by the authors, the project would need some institutional economic support. In Tanzania, Van Eijck and co-workers (2014) analyzed two business models for jatropha biofuel production, one of them being a decentralized smallholder hedgerow model and the other one being a centralized monoculture plantation model owned by a big company. After the analysis of those systems against several indicators, they concluded that the smallholder system scores better on land rights and have lower risk of negative impacts while the plantation model, even though it creates more employment and higher benefits, can lead to decreased food security and loss of land rights and biodiversity affecting the livelihood of local population.

Van Dam and co-workers (2009) analyzed the Argentinian bioenergy production from soybeans and switchgrass against a set of sustainability indicators including some socioeconomic impacts. They analyzed several scenarios and found that the most socioeconomic benefits can be achieved when bioenergy uses the most advanced agricultural technologies available and is carried out in a region with the lowest and environmental performance and prevents leakages in land and biodiversity changes using abandoned or degraded land. Souza et al. (2016) quantitatively estimated the social effects of different sugarcane biorefinery supply chains. They found that the first-generation scenario leads to higher job creation levels over the supply chain mainly because of the workers employed in agricultural sugarcane manual operations, but the second-generation scenario is more reliant on sectors with a lower level of occupational accidents, higher average wages, higher education level, and more participation of women in the workforce.

Silaletruksa and co-workers (2012) studied the impacts of biofuels production in Thailand socioeconomic development. Although they found net positive socioeconomic impacts on employment and GDP creation, they warned about the need to protect small-scale farmers from unfair trade and poor families from food price volatility.

Recently, Song et al. (2015) developed a dynamic IO optimization model to evaluate energy, environmental and economic impacts of renewable energy (RE) policies scenarios introducing six energy conversion technologies, as well as of bioenergy scenarios in Japan. They found that regional economic development and greenhouse gas (GHG) mitigation can be potentially promoted simultaneously by bioenergy utilization and a proper GHG emission constraint.

A recent IEA report (Diaz-Chavez et al., 2017) focuses on socioeconomic impacts from the production of forestry that is a proxy to the production of pellets for bioenergy production in Brazil potential supplier of solid biomass (pellets) to Europe. They found that, overall the forest sector grew due to partially the close relationship between the sector and local communities but that the continuation of socioeconomic impacts of this sector will be conditioned by the business model adopted by firms. The growing trend towards automation may limit the scope for job creation but can create more and better employment opportunities for women which are recognized as a key factor in local development.

6.3.3 METHODOLOGIES TO ESTIMATE THE SOCIOECONOMIC IMPLICATIONS OF BIOENERGY

There are several methodologies available to estimate the socioeconomic implications of bioenergy, and the selection will depend on the type of question that wants to be answered.

Breitschopf and co-workers (2012) analyzed the available methodologies to estimate RE deployment impacts and distinguished between two types of questions and methodologies associated:

Question 1. What is the relevance of the RE industry (and its upstream industries) measured in sectoral employment and growth?

Question 2. What is the impact on economy-wide employment and growth (in all economic sectors)?

The first type of assessment is called a *gross* impact study and is a partial impact analysis restricted to the RE industry (and its upstream industries) and only considers positive effects. Two methodological approaches are recommended by the authors: the Employment Factor approach to quantify the number of jobs that are directly involved in manufacturing, construction, and installation of RE technologies or related services; and the gross IO model that allows the estimation of the economic and employment impacts in the RE industry and its upstream industries that are directly and indirectly stimulated.

The second type of assessment is called a *net* impact study and provides results about changes in the entire economy considering positive and negative, direct, indirect, and induced effects of RE technologies deployment.

It includes the effect of investments, operation and maintenance, and other related activities along the value-added chain, the avoided impacts due to avoided investments, and operation and maintenance of fossil technologies and also allows for the estimation of the effects of changes in electricity and fuel prices and income. The authors recommend two approaches: the net IO model in which the interactions between industries are governed by fixed coefficients and is recommended for simplified assessments of future impacts, and the full economic model which is a dynamic model with several feedback loops and multipliers and allows the estimation of all positive and negative effects.

These model-based approaches serve make ex ante estimates of the possible impacts. Additionally observed empirical studies that make use of surveys and questionnaires are able to make ex post analysis of the observed impacts and somehow validate the modeling approaches (Steininger and Wojan, 2011).

Most of the literature on the estimation of socioeconomic studies of bioenergy use IO modeling (gross or net) (Lehtonen and Okkonen, 2013, 2016; Okkonen and Lehtonen. 2017, Gan and Smith, 2007; Domac et al., 2005; Thornley et al., 2008; Neuwahl et al., 2008; Kulišic et al., 2007; Mukhopadhyaya et al., 2017) and few of them use social accounting matrixes (Gan and Smith, 2007) or general equilibrium models (Trink et al., 2010; Smeets et al., 2014). Some of the literature reviewed make use of surveys and questionnaires to obtain primary data on the effect produced (Petrini et al., 2017; Openshaw, 2010).

6.3.4 THE IO METHODOLOGY

The first work attempting to describe an economic system was conducted by François Quesnay, a French economist and physician (1694–1774). Quesnay represented flows of money and goods in an analytical way to explain the relationships between farms in a region. The flows were allocated in a table, *Tableau économique*, which showed how the incomes were distributed from farmers to other farmers and to landowners.

Two centuries later, in 1973, Wassile Leontief was awarded with the Nobel Prize in Economics for his contribution in the development of the IO analysis. His model represented the various inputs required to produce a unit of output in each economic sector based on surveyed census data of purchases and sales of industries. By describing the main economic sectors in a structural table, Leontief was able to trace the purchases required to produce outputs in each sector, including all the supply chain until raw materials extraction. Thanks to Leontief's development, governments could better understand the requirements of, and plan for, activities like war planning, procurement, or the economic needs for building infrastructure such as roads and railways.

The IO analysis has been used to estimate the economic impacts of industries within an economy, from a macroperspective. IO analysis describes, using symmetrical tables, the relationships and dependences between activity sectors within an economy (Leontief, 1941; Miller and Blair, 2009). The IO tables are built based on statistical data for a given economy that could be at regional or national level and describe the flows of goods and services in monetary terms between different activity sectors and demands or between producers and consumers. The IO tables can be divided into two main components: the interindustry flows or transaction matrix, which describes the flows from Sector i to Sector j, and the final demand. Intermediate goods and services are further processed by other sectors (Feng and Hubacek, 2015).

Table 6.3.1 shows a simplified IO table.

An IO table represents the production cost components in columns, accounting for the resources consumed from other sectors to obtain a certain production in each sector, while the rows describe how the production from one sector is distributed to other sectors. The value added row includes other nonprimary inputs, such as labor, taxes, and subsides.

Table 6.3.1 Example of IO table

		Processing sectors (intermediate demand)				**Final demand**	**Total output**
		1	**2**	...	***n***		
Processing sectors	1	z_{11}	z_{12}	...	z_{1n}	y	x
	2	z_{21}	z_{22}	...	z_{2n}		
	...	z_{31}	z_{32}	...	z_{3n}		
	n	z_{n1}	z_{n2}	...	z_{nn}		
Payment sectors	Value added	v_1	v_2	...	v_n		
	Import	m_1	m_2	...	m_n		
Total outlays		x					

The final column demand includes the sales of each sector to final consumers: household consumption and investment and government consumption.

Therefore, the total output from each sector is defined by:

$$x_\mathrm{i} = z_{i1} + z_{i2} + \ldots + z_{in} + y_i \tag{6.3.1}$$

This equation will be set for all sectors included in the IO table. We can use the matrix notation to describe the different elements of the equations system as:

$$x = \begin{bmatrix} x_1 \\ \ldots \\ x_n \end{bmatrix}; \quad Z = \begin{bmatrix} z_{11} & \cdots & z_{1n} \\ \vdots & \ddots & \vdots \\ z_{n1} & \cdots & z_{nn} \end{bmatrix}; \quad y = \begin{bmatrix} y_1 \\ \ldots \\ y_n \end{bmatrix} \tag{6.3.2}$$

where x is a vector that expresses the total output, Z is the IO matrix, and y is the final demand vector. Total output can also be measured as the sum of the intermediate inputs and the value added.

As mentioned before, Leontiefs model represented the inputs required to produce a unit of output in each economic sector. These ratios are the technical coefficients that represent the ratio of input to output. The technical coefficients are denoted as:

$$a_{ij} = z_{i1}/x_j \tag{6.3.3}$$

The technical coefficients can be expressed as a matrix, the A matrix. We substitute z_{ij} in Eq. 6.3.1 for the technical coefficients and the total output is defined by the following matrix equation:

$$x = Ax + y \tag{6.3.4}$$

Reorganizing Eq. 6.3.4, we get the following expression:

$$\mathrm{x} = (I - A)^{-1}\mathrm{y} \tag{6.3.5}$$

where $(I - A)^{-1}$ is the Leontief inverse matrix, or the multiplier matrix, that expresses the total production of each sector required to satisfy the final demand. That is the direct and indirect requirements per unit of final demand.

Through the IO analysis, we can estimate the total output in monetary terms that will be produced by the different sectors in the economy to satisfy the intermediate and final demand of goods and services. Eq. 6.3.5 can be also decomposed as follows:

$$x = (I + A + AA + AAA + \ldots + A^n)y \tag{6.3.6}$$

where Iy are the direct impacts and $Ay + AAy + \ldots + A^n y$ are the indirect impacts.

The IO analysis began as a method to analyze national or regional economies. However, production processes have become less domestic, and national economies are part of a global economy. Supply chains are increasingly fragmented across borders and this fundamentally modifies the nature of international trade with important consequences for the location of production as well as other related impacts. Multiregional IO (MRIO) modeling provides the opportunity to analyze the consequences of this fragmentation in a comprehensive way by including different regions and their trade relationships.

The MRIO tables follow a similar structure than the IO tables, but there are interregional and intraregional transactions, being possible to identify the purchases that one sector in Region 1 makes from other sector in Region 2.

Collecting the required data to build an IO table is a high demanding process both in time and resources. This effort is even higher when the table aims to describe a multiregional economy. In the last decade, many institutions and research groups have hardly worked on the development of harmonized MRIO databases. The Organisation for Economic Cooperation and Development (OECD) produced a database that comprises IO tables for 48 countries, covering 37 sectors. This database covers years from 2002 to 2006 (Yamano and Ahmad, 2006). The EORA database, developed by Lenzen and his team, is a MRIO database that works in native classification of each country. It includes data for 187 countries and 25 sectors (Lenzen et al., 2013). Other well-known database is the EXIOBASE, a global MRIO database, which covers 43 countries and 5 "rest of the world (RoW)" regions (Tukker et al., 2013). This database is available for any practitioner and the reference year is 2007. The developers already have a more updated EXIOBASE but it is only for internal use.

The choice of one database or other could influence the final results. Moran and Wood analyzed the reliability of different global MRIO databases, comparing the results of four of them, showing that the degree of disagreement varies by country and by model (Moran and Wood, 2014). This effect might be even more relevant if the analysis is extended to other aspects beside the economy due to the database structures and construction techniques, as Anne Owen explained in her book (Owen, 2017).

But IO analysis has not been only used to calculate impacts on the economy. In the early 1970s, already Leontief, among others, extended the analysis to cover environmental aspects within the IO analysis (Isard, 1969; Leontief, 1970). Although it did not have a high repercussion at the time, its use has increased in the last decades due to the higher concern about environment and climate change. The Environmentally Extended Input–Output (EEIO) analysis is widely used and it is constantly under improvement through the hybridization with other methods such as Life Cycle Assessment (LCA) (Hendrickson et al., 1998; Suh et al., 2004; Wiedmann et al., 2011). The combination of EEIO analysis and LCA aims to take advantages of the strengths of both methods, while minimizing their weaknesses.

To include any extension, environmental or social, we need to have an additional matrix or vector that provides the amount of pollutants, i.e., emitted by each activity sector per monetary unit of output. Including this vector into Eq. 6.3.7, we get the final expression that we have used in this study:

$$e_i = l_i(I - A)^{-1}y \tag{6.3.7}$$

where l_i is the vector describing the direct impacts coefficients (per unit of output) and e_i is the total impact, direct and indirect, associated with the total output that satisfies the final demand.

Although IO analysis is not a novel methodology, its use in the energy sector has increased in the last years, particularly in biofuels production.

6.3.4.1 A case study

This chapter presents the socioeconomic study of the production of pellets from Miscanthus. It is based on the case study of Bourgogne Pellets (BP) analyzed in the LogistEC project (http://www.logistecproject.eu). This cooperative comprises around 350 farmers in the municipality of Aiserey, southeast of Dijon, in France. During the last 40 years, the plant has produced granulates from sugar beet pulp, since sugar beet was an important crop in the area. In 2008, the European Union decided to lower the sugar production quotas in France, and the BP plant had to find new resources to continue with the operation. Wood (coniferous) and perennial grasses became the new resources for BP. Now, BP grows 320 ha of miscanthus and 60 ha of switchgrass. However, the relatively high miscanthus production costs and the pellets' low price in the market have forced BP to move to other market, and it currently sells its biomass for animal bedding.

The case was analyzed by the Logistics for Energy Crops biomass (LogistEC) project (http://www.logistecproject.eu), funded by the seventh Framework Programme (FP7). LogistEC aimed to develop new or improve technologies of biomass logistics chain by considering the environmental, economic, and social aspects of sustainability. To include all aspects, several analyses were conducted. Firstly, the most probable locations of future fields and distances for the cultivation of miscanthus were identified using an economic optimization model. Then, based on data collected from the real cultivation areas, the potential environmental and socioeconomic impacts were estimated, taking into account the different life cycle stages from "cradle to gate."

In 2014, the total pellet production volume in France was about 1.03 Mt (Thrän et al., 2017). The production and use of biomass from wood has increased in the last years as shown in Fig. 6.3.1.

Biomass is usually consumed in boilers and stoves, and its use is expected to increase in the future. On October 28, 2016, the French government published the "Plan de programmation pluriannuelle de l'Energie (PPE)," which defines the targets for RE capacity to be installed by 2023. The PPE proposes two scenarios: a high and low scenario for

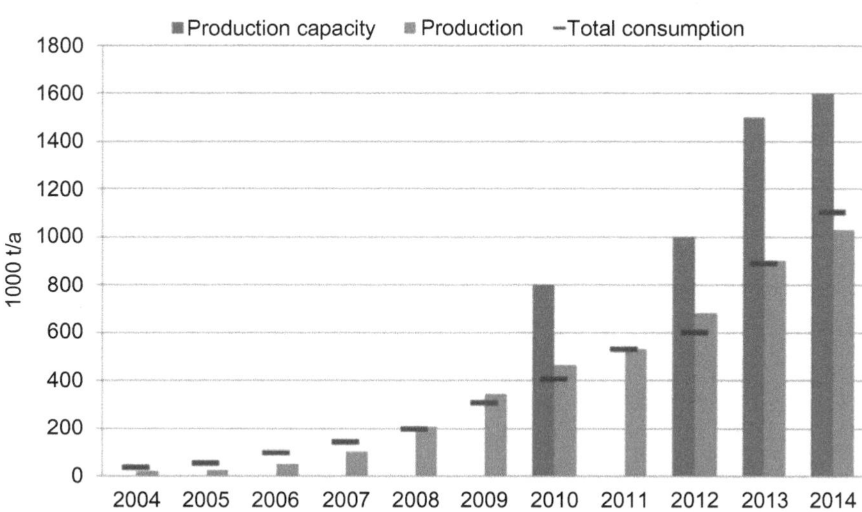

Figure 6.3.1 Wood pellet market development in France. *From (Thrän et al., 2017).*

electricity and heat. Biomass plays an important role for heat, in which the low objective aims to produce 13,000 ktep using biomass in 2023. In 2015, 23% of the bioenergy capacity in France came from wood.

For this chapter, it was assumed that 23% of the heating target from biomass will be supplied by wood and from this, 20% will be supplied by miscanthus in 2023. Considering a low heating value of 16 GJ per ton of miscanthus, an analysis will be performed for the potential socioeconomic impacts of the production of 1840 ktons of miscanthus pellets in France, so that 4.6% of the heating target is using this feedstock.

The methodology to estimate the socioeconomic effects of the production of bioenergy of miscanthus is the IO analysis, extended with sectorial employment vectors as well as sectorial GHG emissions vectors and considering the trade between the different economies of the world.

Four criteria were was considered to choose the database: (1) its availability for any user; (2) the year of reference, (3) the number of regions and countries, and (4) the number of activity sectors described.

Based on these criteria, we have chosen the World Input–Output Database (WIOD) to carry out this study. At the time when this study was designed, WIOD appeared as the best option among the others, covering the three main issues relevant for the study. The WIOD provides IO tables for 40 countries worldwide from 1995 to 2011 and a model for the RoW, describing 35 activity sectors (Timmer et al., 2015). It was developed in a FP7 project. This database contains detailed information on national production activities for 27 EU countries, 13 major non-EU countries, and 1 region referred to RoW. All data used in the WIOD have been obtained from official national statistics, mainly from National Accounts published by the National Statistical Institutes. Moreover, the WIOD also provides satellite accounts with socioeconomic and environmental data, necessary to extend the MRIO.

The WIOD provides data on labor and capital inputs as well as on pollution indicators at the industry level that can be simultaneously used to enlarge the scope of possible applications including the estimation of job creation and some environmental impacts. In this study, besides the economic impacts—total output activity and value added—we have included job creation.

Considering the limited number of sectors in the WIOD, analyzing the impacts associated to an increase in the final demand of goods produced by the Agriculture sector would lead to "average" socioeconomic impacts but not to the special case of miscanthus or any other crop. The problem of sector aggregation is one of the limitations of IO analysis also in its extensions, and it has been studied by several authors (Piñero et al., 2015; Bouwmeester and Oosterhaven, 2013). With the objective of minimizing sector aggregation problems, we have done a step back, defining all goods and services required to get our final product, pellets from miscanthus at BP plant. This assumption considers intermediate requirements as final goods and could underestimate the real impacts.

The intermediate requirements that describe the final demand vectors as costs in monetary units comprise the following activities: cultivation, harvesting, chip, and bale miscanthus production as well as its transport and storage and the final treatment to get pellets. Based on these phases, we have divided the analysis in several stages:

- Machinery and other investments, which includes the machinery costs, insurances and maintenance costs. The type of machinery required in each agricultural operation has been described along the project based on the results of the demonstration work packages.
- Agricultural operations that comprise the application of fertilizers and other agrochemicals, as well as their transport to the farmers, the rhizomes used in the first year, the consumption of fuel for the agricultural operations, and other costs such as loan interests. Most data have been directly provided by the farmers' cooperative Dijon Céréales. All details concerning the agriculture phase can be found in De la Rúa and Lechón (2016).
- Bales and chips transport and storage, considering the distances and costs per ton resulting from the economic optimization model.
- Pelletizing, considering the equipment and consumption of goods, such as energy consumed in the operation.

The salaries earned by the workers will be partially spent in other goods and services, generating a new final demand and, therefore, new impacts. This impact is defined as induced impact. In order to account for it, we have estimated the salaries and wages as well as the social contribution based on the working hours required in each phase. We have assumed that the workers have a tendency to save of 5% of the net salary, and they demand goods and services following the average expenditure of final demand described in the French IO table in 2011.

The final demand vector, to be used in the IO analysis, describes as well the economic performance of miscanthus pellets. The aggregated costs of the main stages are shown in the next table, as net present value, using a discount rate of 4% (Table 6.3.2).

Table 6.3.2 Miscanthus pellets production costs

Stages	**Costs €/t**
Machinery and other investments	8.71
Labor operations	26.92
Transport and storage	12.30
Pelletizing	54.80
Total	102.72

Table 6.3.3 Socioeconomic impacts from miscanthus pellets production

Stages	**Total output**	**Multiplier effect**	**Value added**	**Employment**
	Million €		**Million €**	**People**
Machinery and other investments	30.02	3.45	11.58	204
Labor operations	95.72	3.56	33.49	649
Induced effect	7.15		3.40	4810
Transport and storage	35.16	2.86	16.10	275
Pelletizing	218.86	3.99	69.00	1299
Total	386.94		130.17	2426

More than 50% of the total costs occur during the pelletizing stage, since it is a high energy demanding process. Agricultural operations take the second place, 26%, while transport and storage contributes with 12% to total costs, minimized by the use of the shortest paths.

Taking into account these costs and the sectors that would provide each good and service required along the supply chain of the pellets production from miscanthus, the socioeconomic impacts have been calculated using the IO analysis, and the scenario described previously for 2023 (Table 6.3.3).

To comply with the 2023 heating targets in France, different sources will be used and miscanthus could be one of them. Taking into account the current mix of energy carriers and the potential of miscanthus, we have assumed that 4.6% of the heating target could be supplied by this biomass. The production of miscanthus pellets and its whole upstream processes would lead to a total economic output around 290 million €, from which 45% would contribute to the value added. Value added is defined as the value of total output less the value of intermediate inputs. Each industry included in the economy contributes to the total value added and therefore to the gross domestic production of the country The indirect and direct activities around the miscanthus pellets production would create 2400 jobs. It is necessary to emphasize that these impacts, both the economic and the employment, are related only to the activities until the gate of the pellets factory. All downstream activities have not been included in the analysis and, therefore, these impacts could be even higher if the whole process, i.e., distribution and selling, would be considered. The highest contribution to both economic and employment impacts is due to the pelletizing stage, which could be expected since it is also the activity with the highest costs. Based on this argument, the activities linked to miscanthus growing, which are included under the labor operations phase, should also contribute significantly to both economic and employment impacts. This is the case, as the table shows. More than 95 million € will be produced to fulfill the requirements from the labor activities, being around one-third of them added value to the economy. Besides, 650 new jobs might be created in response to these activities. However, it is clear that the multiplier effect in this case is lower than in the case of the pelletizing activities. The table also includes the so-called induced effects derived from the salaries earned by the workers in the agriculture activities, which will be partially spent in other goods and services, generating a new final demand and, therefore, new impacts.

The current analysis has been conducted using a MRIO database, which allows to distinguish between domestic and nondomestic effects. Fig. 6.3.2 shows the distribution of the total output, value added, and employment among France, European Union (except France), and RoW.

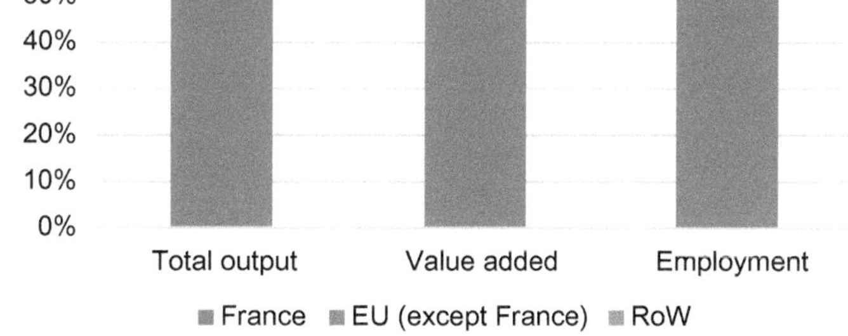

Figure 6.3.2 Distribution of socioeconomic impacts by regions.

Concerning total output and value added, more than 75% of the impacts would happen under the French economy and around 90% would be located within the European Union boundaries. However, in terms of employment, 30% of the new jobs would be created outside the European Union. This is consequence of the indirect relationships between sectors and countries in a global economy. Many basic goods and services are located nowadays at the beginning of the supply chains. Their production is usually associated to very low value added but high labor requirements. Under a global economy in the current situation, the production of these goods is located outside the European Union boundaries, mainly in Asia, due to the low cost of labor. This effect is reflected in the MRIO analysis.

6.3.5 DISCUSSION AND CONCLUSIONS

This chapter has provided an overview of the main socioeconomic implications of bioenergy deployment based on a literature review. Based on the evidence found we can conclude that modern bioenergy development may create opportunities for growth and employment both in developed and developing countries. There are important differences between different bioenergy chains, concentration and size of the investments, and country-specific context in their ability to generate socioeconomic benefits, and support policies need to properly recognize them to avoid undesirable outcomes. Additionally, it is important to have due consideration of the socioeconomic impacts of the displaced activities when introducing bioenergy projects in a region or country as well as other macroeconomic impacts that support policies may entail and that may threat the attainment of the intended benefits.

The chapter has also introduced the reader to the methods used to estimate the socioeconomic implications and specifically the IO methodology and its extensions and hybridations and have illustrated the application of this methodology to a case study in France. The example shown in this chapter represents an estimation of what reality could become if biomass and bioenergy are promoted and boosted. Although the socioeconomic impacts estimated in this research reflect a very specific case study, these results could easily be generalized to other bioenergy projects in France with similar agriculture cycle and therefore similar cost structure. A similar exercise should also be conducted to analyze the other two pillars of sustainability. A comprehensive picture of all impacts would help to successfully design measures to foster its production and use.

Main conclusion from this chapter would be that bioenergy deployment, if properly designed and promoted, could play an important role in the transition towards an RE system, while supporting the economic growth and the generation of new employments.

REFERENCES

Beal, E., Rossi, A., 2011. Good Socio-economic Practices in Modern Bioenergy Production. Minimizing Risks and Increasing Opportunities for Food Security. Food and Agriculture Organization of the United Nations (FAO). Available at: http://www.fao.org/docrep/015/i2507e/i2507e00.pdf (accessed January 2018).

Bouwmeester, M.C., Oosterhaven, J., 2013. Specification and aggregation errors in environmentally extended input–output models. Environ. Resour. Econ. 56 (3), 307–335.

Breitschopf, B., Nathani, C., Resch, G., 2012. Methodological guidelines for estimating the employment impacts of using renewable energies for electricity generation. Study commissioned by IEA's Implementing Agreement on Renewable Energy Technology Deployment (IEA-RETD). Available at: http://iea-retd.org/wp-content/uploads/2012/12/EMPLOY-Guidelines.pdf (accessed January 2018).

De la Rúa, C., Lechón, Y., 2016. An integrated multi-regional input-output (MRIO) analysis of miscanthus biomass production in France: socio-economic and climate change consequences. Biomass Bioenergy 94, 21–30. Available from: https://doi.org/10.1016/j.biombioe.2016.08.003. ISSN 0961-9534.

Diaz-Chavez, R., Walter, A., Gerber, P., 2017. Socio-Economic Assessment of Forestry Production for a Developing Pellet Sector. The Case of Santa Catarina in Brazil. IEA Bioenergy, ISBN 978-1-910154-32-8.

Duscha, V., Ragwitz, M., Breitschopf, B., Schade, W., Walz, R., Pfaff, M., et al., 2014. Employment and growth effects of sustainable energies in the European Union. Final report of the project "Support Activities for RES modelling post 2020". The project is funded by the European Commission, DG Energy, under contract ENER/C1/428-2012. Available at: https://ec.europa.eu/energy/sites/ener/files/documents/EmployRES-II%20final%20report_0.pdf. (accessed January 2018).

Domac, J., Richards, K., Risovic, S., 2005. Socio-economic drivers in implementing bioenergy projects. Biomass Bioenergy 28 (2), 97–106. Available from: https://doi.org/10.1016/j.biombioe.2004.08.002.

Ewing, M., Msangi, S., 2009. Biofuels production in developing countries: assessing tradeoffs in welfare and food security. Environ. Sci. Policy 12 (4), 520–528. Available from: https://doi.org/10.1016/j.envsci.2008.10.002. ISSN 1462-9011.

Feng, K., Hubacek, K., 2015. A multi-region input-output analysis of global virtual water flows. In: Ruth, M. (Ed.), Handbook of Research Methods and Applications in Environmental Studies. Edward Elgar Publishing Ltd, Cheltenham, United Kingdom.

Gan, J., Smith, C.T., 2007. Co-benefits of utilizing logging residues for bioenergy production: the case for East Texas, USA. Biomass Bioenergy 31 (9), 623–630. Available from: https://doi.org/10.1016/j.biombioe.2007.06.027. ISSN 0961-9534.

Hendrickson, C., Horvath, A., Joshi, S., Lave, L., 1998. Economic input-output models for environmental life-cycle assessment. Environ. Sci. Technol. 32 (7), 184a–191a.

IRENA, 2017. Renewable Energy and Jobs. Annual Review 2017, Abu Dhabi: International Renewable Energy Agency.
Isard, W., 1969. Some notes on the linkage of the ecological and economic systems. J. Region. Sci. Assoc. Int. 22 (1), 85–96.
Johansson, A., Kisch, P., Mirata, M., 2005. Distributed economies—a new engine for innovation. J. Clean. Prod. 13 (10–11), 971–979. Available from: https://doi.org/10.1016/j.jclepro.2004.12.015. ISSN 0959-6526.
Kemausuor, F., Bolwig, S., Miller, S., 2016. Modelling the socio-economic impacts of modern bioenergy in rural communities in Ghana. Sustain. Energy Technol. Assess. 14 (2016), 9–20. Available from: https://doi.org/10.1016/j.seta.2016.01.007. ISSN 2213-1388.
Kulišic, B., Loizou, E., Rozakis, S., Segon, V., 2007. Impacts of biodiesel production on Croatian economy. Energy Policy 35 (12), 6036–6045.
Lehtonen, O., Okkonen, L., 2013. Regional socio-economic impacts of decentralized bioeconomy: a case of Suutela wooden village. Finland. Environ. Dev. Sustain. 15, 245–256. Available from: https://doi.org/10.1007/s10668-012-9372-6.
Lehtonen, O., Okkonen, L., 2016. Socio-economic impacts of a local bioenergy-based development strategy—the case of Pielinen Karelia, Finland. Renew. Energy 85, 610–619. Available from: https://doi.org/10.1016/j.renene.2015.07.006. ISSN 0960-1481.
Leontief, W., 1941. The Structure of American Economy: 1919-1939. Oxford University Press, New York.
Leontief, W., 1970. Environmental Repercussions and the Economic Structure: An Input-Output Approach: The Review of Economics and Statistics, 52, 3. MIT Press, pp. 262–271. Available from: https://doi.org/10.2307/1926294.
Lenzen, M., Moran, D., Kanemoto, K., Geschke, A., 2013. Building EORA: a global multi-region input-output database at high country and sector resolution. Econ. Syst. Res. 25 (1).
Miller, R.E., Blair, P.D., 2009. Input-Output Analysis: Foundations and Extensions., Second Edition Cambridge University Press, Cambridge.
Mirata, M., Nilsson, H., Kuisma, J., 2005. Production systems aligned with distributed economies: examples from energy and biomass sectors. J. Clean. Prod. 13 (2005), 981–991. Available from: https://doi.org/10.1016/j.jclepro.2004.12.018.
Moran, D., Wood, R., 2014. Convergence between the EORA, WIOD, EXIOBASE, and OPENEU'S consumption-based carbon accounts. Econ. Syst. Res. 26, 245–261.
Mukhopadhyay, K., Chen, X., Thomassin, P., 2017. Economy wide impacts of ethanol and biodiesel policy in Canada: an input–output analysis. Int. J. Green Energy 14 (4).
Neuwahl, F., Löschel, A., Mongelli, I., Delgado, L., 2008. Employment impacts of EU biofuels policy: combining bottom-up technology information and sectoral market simulations in an input–output framework. Ecol. Econ. 68 (1–2), 447–460. Available from: https://doi.org/10.1016/j.ecolecon.2008.04.018. ISSN 0921-8009.
Okkonen, L., Lehtonen, O., 2017. Local, regional and national level of the socioeconomic impacts of a bio-oil production system—a case in Lieksa, Finland. Renew. Sustain. Energy Rev. 71, 103–111. Available from: https://doi.org/10.1016/j.rser.2017.01.003. ISSN 1364-0321.
Openshaw, K., 2010. Biomass energy: employment generation and its contribution to poverty alleviation. Biomass Bioenergy 34 (3), 365–378. Available from: https://doi.org/10.1016/j.biombioe.2009.11.008.
Owen, A., 2017. Techniques for Evaluating the Differences in Multiregional Input-Output Databases—2017. A Comparative Evaluation of CO2 Consumption-Based

Accounts Calculated Using Eora, GTAP and WIOD. Springer International Publishing AG, Cham, Switzerland. Available from: https://doi.org/10.1007/978-3-319-51556-4.

Persson, M., 2013. Socio-economic consequences of increased biomass demand. Systems Perspectives on Biorefineries. Chalmers University of Technology, Gothenburg, Sweden, ISBN: 978-91-980973-2-0.

Petrini, M.A., Vieira Rocha, J., Christopher Brown, J., 2017. Mismatches between mill-cultivated sugarcane and smallholding farming in Brazil: environmental and socioeconomic impacts. J. Rural Stud. 50 (2017), 218–227. Available from: https://doi.org/10.1016/j.jrurstud.2017.01.009. ISSN 0743-0167.

Piñero, P., Heikkinen, M., Mäenpää, I., Pongrácz, E., 2015. Sector aggregation bias in environmentally extended input output modeling of raw material flows in Finland. Ecol. Econ. 119, 217.

Silalertruksa, T., Gheewala, S.H., Hünecke, K., Fritsche, U.R., 2012. Biofuels and employment effects: Implications for socio-economic development in Thailand. Biomass Bioenergy 46, 409–418. Available from: https://doi.org/10.1016/j.biombioe.2012.07.019. ISSN 0961-9534.

Song, J.N., Yang, W., Higano, Y., Wang, X., 2015. Modeling the development and utilization of bioenergy and exploring the environmental economic benefits. Energy Conver. Manage. 103, 836–846.

Souza, A., Watanabe, M.D.B., Cavalett, O., et al., 2016. Sustainability assessment methodologies. Int. J. Life Cycle Assess. 1–12. Available from: https://doi.org/10.1007/s11367-016-1112-y.

Smeets E., Vinyes C., Tabeau A., Van Meijl H., Brink C., Prins A.G., 2014. Evaluating the macroeconomic impacts of bio-based applications in the EU. JRC Science and Policy Reports. JRC91385. EUR 26777 EN. ISBN 978-92-79-39536-9 (PDF). ISSN 1831-9424 (online). https://doi.org/10.2791/10930.

Steininger, K., Wojan, T., 2011. Economic impact of bioenergy development. Some evidence from Europe and the US. Eurochoices 10 (3), 31–37.

Suh, S., Lenzen, M., Treloar, G.J., Hondo, H., Horvath, A., Huppes, G., et al., 2004. System boundary selection in life-cycle inventories using hybrid approaches. Environ. Sci. Technol. 38 (3), 657–664.

Thornley, P., Rogers, J., Huang, Y., 2008. Quantification of employment from biomass power plants. Renew. Energy 33 (8), 1922–1927. Available from: https://doi.org/10.1016/j.renene.2007.11.011. ISSN 0960-1481.

Thrän, et al., 2017. Global Wood Pellet Industry and Trade Study 2017. IEA Bioenergy Task 40. June 2017.

Timmer, M.P., Dietzenbacher, E., Los, B., Stehrer, R., de Vries, G.J., 2015. An illustrated user guide to the world input–output database: the case of global automotive production. Rev. Int. Econ. 23, 575–605. Available from: https://doi.org/10.1111/roie.12178.

Trink, T., Schmid, C., Schinko, T., Steininger, K.W., Loibnegger, T., Kettner, C., et al., 2010. Regional economic impacts of biomass based energy service use: a comparison across crops and technologies for East Styria. Austria. Energy Policy 38 (10), 5912–5926. Available from: https://doi.org/10.1016/j.enpol.2010.05.045. ISSN 0301-4215.

Tukker, A., de Koning, A., Wood, R., Hawkins, T., Lutter, S., Acosta, J., et al., 2013. EXIOPOL—development and illustrative analyses of a detailed global MR EE SUT/IOT. Econ. Syst. Res., 25 (1), 50–70.

Van Dam, J., Faaij, A.P.C., Hilbert, J., Petruzzi, H., Turkenburg, W.C., 2009. Large-scale bioenergy production from soybeans and switchgrass in Argentina: part B.

Environmental and socio-economic impacts on a regional level. Renew. Sustain. Energy Rev. 13 (8), 1679–1709. Available from: https://doi.org/10.1016/j.rser.2009.03.012. ISSN 1364-0321.

Van Eijck, J., Romijn, H., Smeets, E., Bailis, R., Rooijakkers, M., Hooijkaas, N., et al., 2014. Comparative analysis of key socio-economic and environmental impacts of smallholder and plantation based jatropha biofuel production systems in Tanzania. Biomass Bioenergy 61 (2014), 25–45. Available from: https://doi.org/10.1016/j.biombioe.2013.10.005. ISSN 0961-9534.

Wiedmann, T., Suh, S., Feng, K., Lenzen, M., Acquaye, A., Scott, K., et al., 2011. Application of hybrid life cycle approaches to emerging energy technologies--the case of wind power in the UK. Environ. Sci. Technol. 45 (13), 5900–5907. Available from: https://doi.org/10.1021/es2007287.

Yamano, N., Ahmad N., 2006. The OECD Input-Output Database: 2006 Edition. OECD Science, Technology and Industry Working Papers. OECD Publishing. https://doi.org/10.1787/308077407044.

FURTHER READING

Décret n° 2016-1442 du 27 octobre 2016 relatif à la programmation pluriannuelle de l'énergie. Available at: https://www.legifrance.gouv.fr/eli/decret/2016/10/27/2016-1442/jo/texte

6.4

Environmental Pillar: Methodology and Case Study

Benoît Gabriel and Aurelie Perrin
EcoSys Joint Research Unit, AgroParisTech, INRA, Paris, France

6.4.1 METHODOLOGICAL BACKGROUND

Bioenergy production interacts with a host of environmental and ecological issues, ranging from human health to biodiversity, water, air, and soil quality through the eutrophication of ecosystems (see Section 6.1). Its impacts may be either negative or positive, depending on the systems and counterfactual situations considered. Bioenergy crops may increase or decrease biodiversity, as well as deplete or replenish soil organic matter stocks (Chum et al., 2011). To guide the implementation of bioenergy projects and optimize their environmental costs and benefits, a wealth of methods were developed over the last decades to provide relevant quantitative indicators of environmental impacts. Their main objective lies in the comparison with fossil-based alternatives, but they also seek to capture effects specific to bioenergy and feedstock production, in particular in relation to the use of natural resources such as land and water. These assessments of environmental sustainability have been high on the policy agenda for bioenergy, assigning for instance mandatory performance targets for the climate mitigation potential of biofuels (Gnansounou and Pandey, 2017).

While this section focuses on life cycle assessment (LCA), currently the most commonly used framework to assess bioenergy chains (Cherubini, 2010; Hellweg and Mila i Canals, 2014), other methods are worth mentioning here insofar as they complement this approach. From the point of view of the use of natural resources, "Emergy assessment" provides a "donor's" perspective (as opposed to user's with LCA) to the efficiency of a bioenergy chain. It computes its capacity to convert incoming solar radiation into utilizable energy, accounting for the same life cycle stages as LCA (Perrin et al., 2017). The ecological footprint pursues a similar objective via a single metric, the use of global hectares of biosphere that supply both feedstock and sinks for the pollutants emitted by the system (de Oliveira et al., 2005). Lastly, the ecosystem services framework appears particularly relevant to bioenergy since it can account for the implications of developing nonfood biomass on ecosystems in general (Koh and Ghazoul, 2010), and their capacity to regulate climate and biogeochemical cycles, to provide food, feed, fiber, and fuel, along with cultural services. However, this framework focuses on feedstock production and does not consider the whole life cycle of bioenergy systems. Growing bioenergy feedstock or extracting it from ecosystems is likely to impact biodiversity, either positively or negatively depending on the former use of land (Chum et al., 2011). However, a recent review of literature studies addressing this issue underlined an overall negative trend in relation to bioenergy development, but also a lack of consensus on the methodology used to estimate its effects, with metrics ranging from species richness to proxies such as habitat quality change (Gaba, 2018). On-going work under the auspices of the International Programme on Biodiversity and Ecosystem Services is likely to shed light on this particular matter.

Among the diversity of environmental assessment methods, LCA is predominanty used for policy purposes, eco-design of products, or the information of consumers. It aims to estimate the impacts resulting from the use of a particular product through its entire life cycle "from cradle to grave" (i.e., from the extraction of raw materials to the recycling or disposal of the product considered). The results are expressed relative to a measure reflecting the usefulness of the product system, called "functional unit." For bioenergy systems, functional units usually involve final usages (e.g., 1 km traveled in a passenger car) or the supply of energy carriers (e.g., 1 kWh of electricity delivered to the grid). These choices are directly related to the goals of the assessment, which may be diverse: benchmarking a given product against an alternative or reference product (e.g., bio-based versus fossil-derived, or first-versus second-generation biofuels), identifying "hotspots" in the life cycle of a product to improve its environmental performance or reporting the environmental footprint of a product or a portfolio of products to consumers or regulation authorities.

LCA is an ISO-standardized (ISO 14044, 2006) method consisting of four stages:

- goals and scope of the study, including the definition of system boundaries and the functional unit(s);
- the inventory of resource consumption (e.g., fossil energy) and environmental emissions (e.g., greenhouse gases (GHGs)) occurring throughout the product's life cycle;
- the characterization of the potential impacts associated with the emissions of pollutants, which are grouped into broad impact categories such as global warming or human toxicity; and
- interpretation of impact results and system optimization.

The implementation of LCA will be illustrated in Section 6.4.3 in the context of miscanthus production in Eastern France, while the following section introduces key issues in its application to bioenergy systems.

6.4.2 APPLICATION TO BIOENERGY CHAINS

Application of LCA methodology to bioenergy started in the 1990's and pinpointed a range of specific issues compared to purely industrial systems or products (Cherubini, 2010; Muench and Guenther, 2013), both in terms of data collection and methodological assumptions.

The first set of questions involves feedstock production and its representation or modeling. There is a large diversity of biomass production systems, depending on the local context, the potential of land to grow crops or trees, and the current configuration of agricultural systems. Capturing this diversity is a challenge since there are usually no statistics available to describe it at the level of details required by LCA—building a typology based on field surveys is a promising although time-consuming avenue to work around this difficulty and construct representative systems (Perrin et al., 2015). When the biomass plants are not currently grown, data from field trials may be extrapolated in prospective scenarios, but with caution, since commercial deployment is likely to incur changes, in particular regarding biomass yields (Lesur-Dumoulin et al., 2016).

Framing realistic feedstock production systems is especially important since differences in management, from the selection of a biomass plant species to fertilizer management, may result in "swings" in the GHG balance of a bioenergy chain, from sink to net source of GHGs (Davis et al., 2013). Land-use change effects are paramount in this outcome for new plantations, whether direct (regarding the use of land prior to the establishment of the plantation) or indirect. Indirect effects correspond to market-mediated compensations for the diversion of land from food, feed, or fiber production to energy purposes. They are further addressed in Section 6.4.4.

LCA results are also highly sensitive to crop yields, as the following example will underline (see also Cherubini, 2010; Davis et al., 2013), which vary over time and space according to climate conditions and soil properties. This variability may be captured by robust statistics (if available), statistical or biophysical models (see Section 6.4.3; also Dufossé et al., 2013), or expert guess. The emissions of substances derived from agricultural inputs (often referred to as field emissions) follow a similar pattern, of which nitrous oxide (N_2O) is iconic. This potent GHG is released into the atmosphere as a result of fertilizer N application and may contribute as much as 40% of the life cycle emissions of bioenergy chains (Dufossé et al., 2013). Its emissions may be considered proportional to fertilizer inputs, but with a large confidence interval, resulting in a significant uncertainty for the GHG balance of bioenergy systems (Smeets et al., 2009).

Secondly, biomass processing may raise methodological issues since it generally involves more than one end-product. In that case the environmental burden of biomass production, collection, and transformation upstream of the branching into several products requires an adaptation of the analysis. The easiest way to work around this issue is to partition this burden between the various co-products based on a physical parameter: for instance, their respective masses, energy contents, or protein contents after processing of one unit amount of feedstock. The selection of a criterion for this so-called allocation method depends on the final uses of the co-products (e.g., whether for food, feed, or energy purposes)—with the problem that they may differ across co-products. In certain cases (widely different usages or large discrepancies in the market valuation of the products) the allocation may be based on monetary terms, for instance, the market prices of these co-products. The caveat with these allocation methods is that they are somewhat arbitrary, and the selection of one or the other is far from neutral in the outcome of the LCA (Muench and Guenther, 2013). If the goal is to compare two alternative energy chains, it is therefore recommended to expand the system analyzed "to include the additional functions related to the co-products" (Cherubini, 2010). For instance, if a biofuel chain produces animal feed as a co-product, the alternative fossil-based chain may be complemented with a subsystem producing an equivalent amount of feed.

Thirdly, biomass supply chains may involve a range of logistical options, from harvesting systems to transport and on-site storage. These will be developed in the following example.

6.4.3 BOURGOGNE PELLETS: A CASE STUDY BASED ON PERENNIAL BIOMASS CROPS

6.4.3.1 Assessing the environmental impact of bioenergy for the full supply chain

Bourgogne Pellets (BP) is a farmer's cooperative comprising about 350 members based in the municipality of Aiserey in the Burgundy region of Eastern France. Between 2012 and 2016, BP has been involved in a European project entitled Logist'EC focused on new or improved technologies of the biomass logistics chains. More details on this case study can be found in Perrin et al. (2017).

In 2015, the supply area of BP covered 400 ha of miscanthus, scattered across arable land in an approximately 70 km radius around Aiserey. The supply chain operated by BP includes six stages, namely agricultural production, harvest, handling, transport, storage, and processing (Fig. 6.4.1),

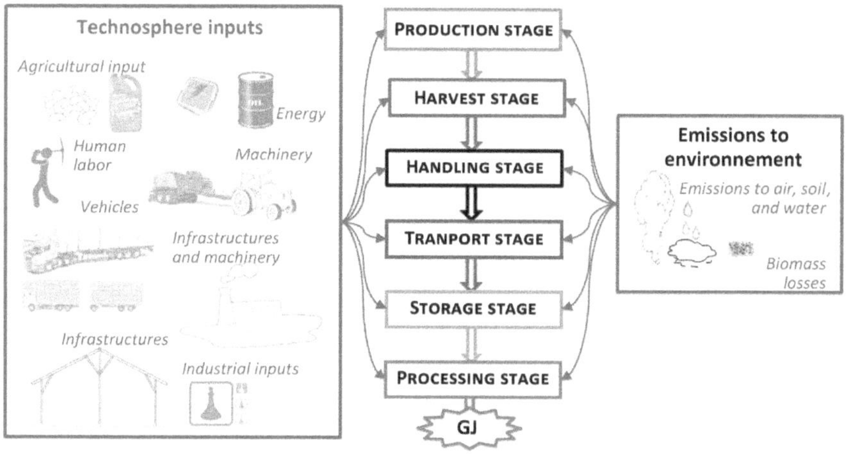

Figure 6.4.1 Miscanthus supply chain at BP.

and produces biomass feedstock products in a range of forms—chips, bales, and pellets. Miscanthus is a perennial crop with a life span of about 15 years thanks to rhizomes, which store starch, proteins, and other nutrients during winter, allowing for a regrowth in spring. From Year 2–15, the above-ground biomass is harvested once a year. Harvested products are either directly transported to the pelleting facility at BP or stored prior to transport to the BP facility. The share of miscanthus processed as pellets vary depending on the market demand and prices.

Quantities of inputs for each operation within the full supply chain were collected from field trials, direct interviews with farmer, and private companies involved in the supply chain. Processes inventoried in the Ecoinvent database v2.2 were used as background data for energy production (Dones et al., 2007), material production (Hischier, 2007), fertilizer production (Nemecek et al., 2007), and pesticide production (Sutter, 2010). LCA considers the emissions of pollutants to the environment. Reactive nitrogen emissions during miscanthus farming were estimated using the CERES-EGC model (Dufossé et al., 2013). The model takes into account soil, climate, and management conditions that influence N emissions. Values refer to the average for miscanthus in Burgundy. Reactive nitrogen emissions from the final removal of from the field were based on measurements performed at Grignon, France (Dufossé et al., 2014). Pesticides and phosphorus emissions were calculated following recommendations from Nemecek and Kägi (2007). We used the methodology recommended in International Life Cycle Data handbook (European Commission, 2010) to calculate impacts.

6.4.3.2 Optimizing the environmental impact of logistic chains

Six supply chain scenarios were defined to study the influence of keys variables (demand variations and alternative logistic solutions to supply the BP facility) on sustainability indicators (Table 6.4.1). The baseline (Scenario A) corresponds to the existing supply chain, with its current annual demand, production area, practices, and infrastructures. However, yields were based on estimated for mature miscanthus crops (Laurent et al., 2015) to work around the variability due to the various ages of fields within the supply area (from 2 to 7 years old in Burgundy). A scenario for alternative demand (Scenario B) represents expansion of supply from 6000 to 30,000 ton dry matter per year ($t\ DM \cdot year^{-1}$) based on a model of the potential miscanthus area expansion (Rizzo et al., 2014). Four additional scenarios represent alternative logistic solutions. They all

Table 6.4.1 Overview and main characteristics of studied scenarios

Cases	Supply chain scenarios	Key parameters	Unit	Value
Baseline	Scenario A	Fertilization	$kg \cdot ha^{-1}$	0
		Harvest time		Spring
		Harvest techniques		Chips, Bales S, Bales L
		Yield	$t \cdot DM \cdot ha^{-1}$	14.83
		Demand	$t \cdot DM \cdot year^{-1}$	6000
		Processing		Pelletization
		Capacity of pelletization	$t \cdot FM \cdot h^{-1}$	1.9
Alternative demand	Scenario B	Demand	$t \cdot DM \cdot year^{-1}$	30,000
Alternative yield	Scenario C Maximum	Yield	$t \cdot DM \cdot ha^{-1}$	16.68
Alternative harvest	Scenario D Autumn	Fertilization	$kg \cdot ha^{-1}$	67
		Harvest time		+ Autumn
		Harvest techniques		+ Shredder
		Yield Autumn	$t \cdot DM \cdot ha^{-1}$	18.69
Alternative processing	Scenario E Briquette	Processing		+ Briquetting
		Capacity	$t \cdot FM \cdot h^{-1}$	1.5

assume the same supply area as baseline (i.e., Scenario A). The scenario for alternative yield (Scenario C) is based on maximum values estimated for the Burgundy region with a statistical model derived from a meta-analysis of miscanthus yields (Laurent et al., 2015). The model was calibrated with local on-farm yields and integrates the effect of crop aging on yields. A scenario for alternative harvest timing (Scenario D) involves the possibility of harvesting part of the miscanthus crop in autumn. Data for harvesting green miscanthus in autumn were taken from field trials conducted in Italy using a shredder to cut the field instead of a silage harvester (Roncucci et al., 2015). A fertilizer input rate was adjusted to compensate for the higher nitrogen exportation compared to the spring harvest of miscanthus. In the baseline scenario we only considered the possibility of processing the biomass as pellets. However, miscanthus may also be processed as briquettes (Scenario E), which are similar to pellets but with a larger diameter (Ø75 mm instead of 8 mm for pellet). A mobile briquetting unit may reduce the need to transport bales as briquettes can be sold directly from intermediate storages. This option involves the transport of the briquetting press to the intermediate storage point where briquetting takes place.

6.4.3.3 Environmental impact of 1 GJ of miscanthus at BP facility gate and scale effect

Fig. 6.4.2 compares the impact of producing the equivalent of 1 GJ of miscanthus to nonrenewable reference system (1 GJ equivalent of light oil on the European market from Ecoinvent V3). For 9 out of the 14 impact categories, producing miscanthus was less impacting than light oil. One iconic category of impacts is climate change, for which miscanthus was twofold less impacting than the nonrenewable reference system. However, for five categories, miscanthus production resulted in higher impacts than the nonrenewable reference system. Some specificities of this agricultural product may explain this pattern.

Fig. 6.4.3 presents the contribution analysis for Scenario A standing for the existing supply chain compared to Scenario B standing for an higher biomass demand (30,000 tons instead of 6000 tons per year). Nitrogen emitted during the production phase was the main contributor to marine eutrophication. Similarly phosphorus emissions were the main contributor to freshwater eutrophication. Not surprisingly, the main contributors to land use were the production and the storage stages. These three categories often resulted in high impact for agricultural products due to biophysical

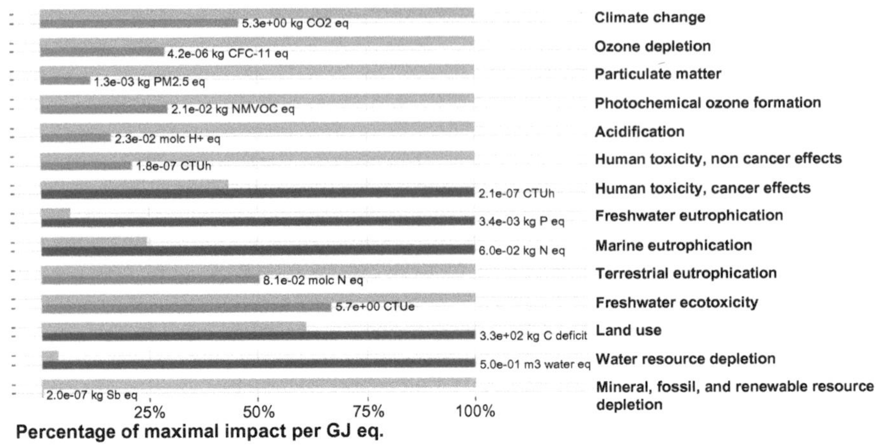

Figure 6.4.2 Benchmarking miscanthus biomass at facility gate (Scenario A) against a nonrenewable reference system (light oil, gray bars). Red bars indicate a higher impact for miscanthus than the reference; green bars indicate a lower impact for miscanthus than the reference. Please refer to online version of the text for the color interpretation of the figure.

processes involved in field emissions and the strong link with land use. The production phase was also the main contributor to freshwater ecotoxicity, but thanks to a low usage of pesticides and subsequent low pesticide emissions, this impact remained relatively low compared to the nonrenewable reference system. For human toxicity (cancer effect), the high impacts of miscanthus resulted from the large diesel consumptions of the harvesting equipment and the low efficiency of transport for such low-density feedstock. For water resource depletion, the difference resulted mainly from a particularly low impact of the nonrenewable reference system. For these two categories, the higher impact of miscanthus biomass compared to the nonrenewable reference system resulted from its low density, an important drawback of such renewable energy system.

Feedstock production contributed 40% of climate change impact of miscanthus biomass supply for both scenarios (Fig. 6.4.3). The main contributor was reactive nitrogen emissions (N_2O mainly) during the crop cycle. Biomass processing contributed a similar share overall. The remaining impacts were shared between harvest and transport stages in relation to fuel consumption. Acidification, terrestrial eutrophication, and photochemical ozone formation presented similar contribution patterns. Ozone depletion, particulate matter formation, and resource depletion presented a similar contribution pattern except for a particularly low contribution of the production phase. For those categories, low fertilizer inputs and

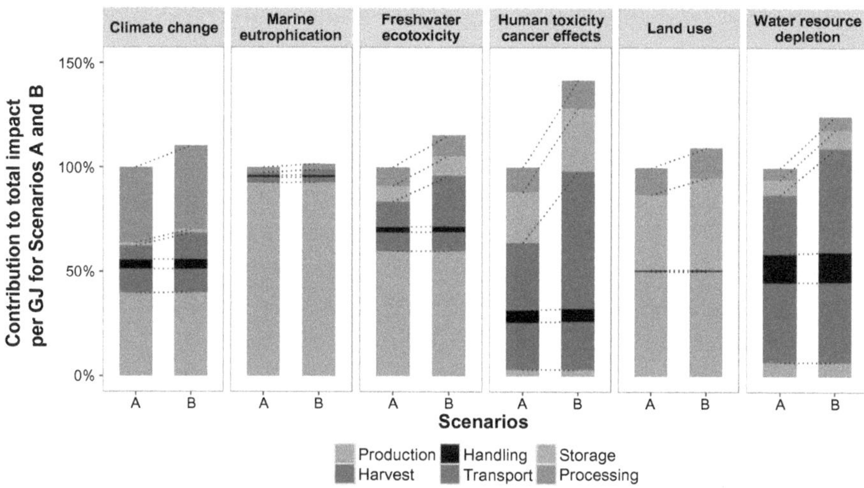

Figure 6.4.3 Identifying hotspot of a bioenergy chain for two scenarios: (A) actual production vs (B) plant maximal production.

the use of electricity in the processing phase resulted in relatively low impacts compared to the nonrenewable reference system. Benchmarking miscanthus biomass against a nonrenewable reference system (light oil here) demonstrated the need for multicriteria assessment methods. Indeed, on the only basis of climate change the biomass systems allowed a reduction of more than 50% of the impact, but the transfer of impact to other categories should not be neglected. The choice of the nonrenewable reference system is also of great importance. Here we selected heating oil because most uses of miscanthus biomass substituted this fuel in the studied region. However, future usages of miscanthus uses may lead to a different reference system (based on, e.g., natural gas or electricity from the grid).

For categories influenced by the change of scale, most impact variations across scenarios resulted from the variation of the contribution of storage and transport. Increasing the annual biomass supply from 6000 t · DM to 30,000 t · DM yielded higher impact per gigajoule of miscanthus produced. This was due to the trade-off between the increased impacts of the transportation stage and the economies of scale in the processing facility. The contribution of biomass transportation to the environmental impacts became so large that it could not be offset by the savings incurred in the biomass processing phase. Such discrepancy between the optimum size of biomass supply chain from the economic and environmental standpoints has already been observed (Steubing et al., 2014). For BP this phenomenon is mainly due to the specific spatial distribution of the miscanthus fields. Fields are scattered for two major reasons: (1) with a small number of operations over several years, miscanthus represents an opportunity of cropping mostly relevant for marginal lands and (2) converting only a small share of the farm area limits the financial risk of growing miscanthus. This study concludes that it would have been beneficial from economic (Fig. 6.4.4) and environmental stand points that miscanthus fields should be less scattered overall. It would be interesting to revisit this conclusion from the point of view of the food versus fuel

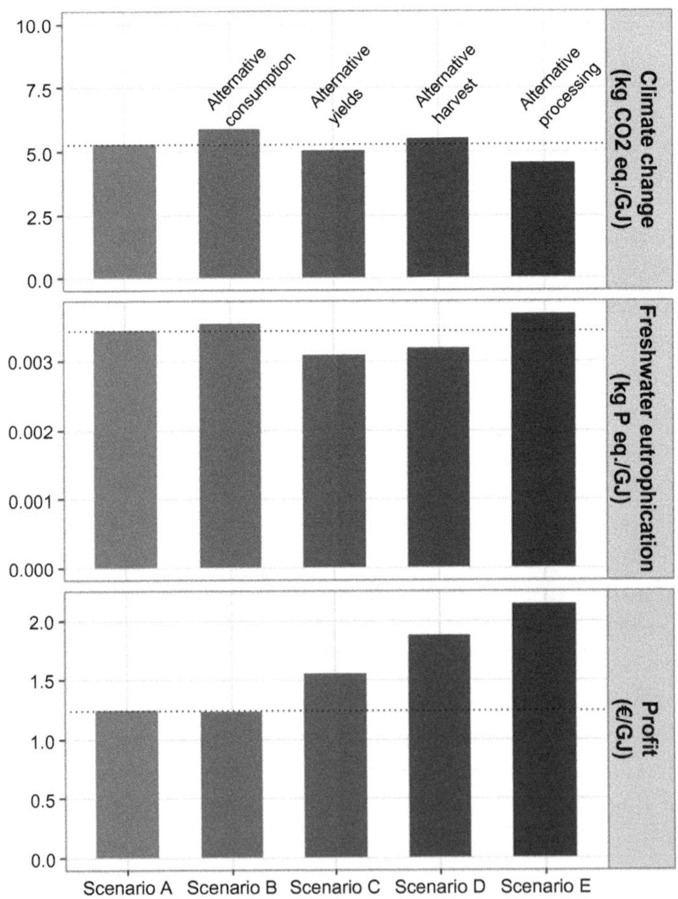

Figure 6.4.4 Climate change potential, freshwater eutrophication, and profit per gigajoule produced in each scenario. Dotted line corresponding to baseline (Scenario A).

competition, which may also warrant a scattered spatial distribution to prevent miscanthus from infringing on food crops.

6.4.3.4 Environmental performances and profitability of alternative logistic options

Maximizing yields (in Scenario C) led to an increase in profit and to a reduction in the impacts provided since it did not involve additional agricultural inputs (Fig. 6.4.4). In general, increasing yields per unit area through more input-intensive agricultural practices come at the cost of higher environmental burdens on ecosystems but reduces the land footprint of biomass. Therefore, there is a trade-off between land-use and environmental impacts per unit area.

Introducing the possibility of additional autumn harvesting (Scenario D) led to increased production when collecting biomass from the same fields as in the baseline scenario, due to the higher yields per hectare obtained in autumn. This came with additional costs due to the need for extra fertilizer inputs, but these costs did not undermine the increase in profit per gigajoule of biomass harvested. Moreover, autumn harvesting entailed a large reduction in storage costs and even a slight reduction in transportation costs due to lower needs for handling at intermediate storages. Spreading the harvest over two periods would almost eliminate the need for costly storage at the municipality level. Autumn harvesting can be considered as an economically viable alternative, especially if storage is a limiting factor and/or comes at considerable cost. Similarly, this scenario presents a lower freshwater eutrophication thanks to higher yields. However, including autumn harvest led to an increase in the climate change potential of biomass, due to the need for extra fertilizer inputs. Note that this type of harvest may also increase soil erosion risks as it reduces the rate of crop residue return to the soil, which imposes a constraint on this option.

Finally, briquetting (Scenarios E) reduced the total costs and drastically increased the profit of the supply chain. Looking closer at the various components, it appeared that briquetting reduced the transport but increased the storage. As mobile briquetting occurs in intermediary storage, often located far from the BP facility, more biomass had to be stored in farmsteads. It also reduced processing impact due to more efficient machinery. Determining whether briquetting should entirely substitute pellet production or remain more limited would require more detailed analysis, taking into consideration the investment capacity of BP and the amortization of existing infrastructure through a capital budgeting analysis. The option of using mobile briquetting should be considered as an alternative to a central pellet production facility, at least for areas without an existing pelletization facility. Mobile briquetting appears a viable economic alternative, especially in situations with a considerable scattering of the crop production and expensive transportation logistics as in this case study. Including briquetting at intermediate storage points appeared an environmentally friendly option to meet an expanding demand of biomass.

The strong relationship observed between environmental performance and the biomass collection area may be considered as a general tendency. However, it is unclear to which extent it was influenced by the modeling approach and notably by the economic optimization and the assumptions about market demand for each product. In a recent review, Lautala et al. (2015) identified the most prevalent challenges in the design and analysis of biomass supply chains: the availability and quality of data to analyze the various components of the supply chain, the lack of a common framework for sustainability indicators, and a deficiency in integrated analysis when developing the supply chains and individual components. Overall, under our assumptions, the potential economies of scale generated by increasing the biomass supply volume and hence its collection area were outweighed by the extra transportation and storage incurred. Such tendency results from local constrains which have been taken on board in this study such as small and scattered miscanthus plots on marginal lands and relatively low miscanthus yields resulting from extensive practices and low-quality soils where miscanthus was established. Such assumptions explaining the design of the local supply chain are not often taken on board on national or even larger scale studies. The framework proposed in this study is intended as a set of instruments aiding to navigate the potential complexities of supply chain design and providing quantitative examples of the consequences of agronomic or technological choices.

6.4.4 ADVANCED TOPICS

Despite its wide use in the scientific community as well as in the economic and policy arenas, LCA still faces considerable challenges in terms of its objectives of conveying a comprehensive and objective picture of the impacts generated by a particular product system. While the readers are referred to more detailed and specialized references on these matters (see, e.g., Gnansounou and Pandey, 2017, for biorefineries), this section introduces two of the points that emerged recently in sustainability sciences: the so-called "consequential" approach that capture ripple effects, and the need for a more refined description of the temporal and spatial dimensions of the systems assessed.

As its name indicates, consequential LCA seeks to track the environmental implications of the deployment of product systems, while the conventional approach (termed attributional) aims to ascribe impacts to the same particular products but ignores the potential effects of its development on other products, markets, or economic sectors (Gnansounou and Pandey, 2017). As far as bioenergy is concerned, consequential LCA has typically been used to factor in indirect land-use change effects related to the development of biofuels and to address food/feed/fuel competition (Muench and Guenther, 2013). Several approaches are available to estimate these consequences: expert judgments may be used to infer which food/feed crops are likely displaced by biofuel feedstocks as well as the origin of the products which will be imported to offset the decline in a given food/feed commodity; causal—descriptive approaches rely on simple representations of market mechanisms, based on retrospective trade data; partial or general economic equilibrium models simulate these mechanisms in more comprehensive and theoretically more-grounded manner (Ben Aoun et al., 2013). Regardless of the methods employed, land-use change effects are not only difficult to track and carry a large degree of uncertainty but also affect significantly the GHG balances (Muench and Guenther, 2013). Progress on these aspects of LCA may be expected in a near future as they are increasingly investigated. Note that strategies that valorize residual biomass are much less affected by land use—related issues.

As already underlined in Section 6.4.2, biomass feedstock production systems vary starkly over time and space, which results in another source of uncertainty in the life cycle inventory data and hampers comparisons between alternative systems. This also applies to the cause—effect chains modeled in the impact characterization stage of the LCA. The impact generated by a unit consumption of resources or emission of substance also hinges on the local context: using water to irrigate a crop in California will put a larger stress on already scant resources than it would in areas with ample water supply (for instance, Pfister et al., 2009). This also applies to the impacts on biodiversity and on landscape amenities, which occur on a local to regional scales, and are becoming increasingly important in decision-making at this level. Soil and air quality issues should also be considered based on local contexts and addressed by suitable methodologies at this scale. Regionalized approaches that attempt at mapping and assessing value chains within a geo-referenced information system are therefore needed (Hellweg and Mila i Canals, 2014). This calls for regionalized approaches that attempt at mapping and assessing value chains within a geo-referenced information system (Hellweg and Mila i Canals, 2014). While it may be relatively easy to delineate the production area for biomass feedstock supply (see example of Section 6.4.3), tracking the spatial details of inputs with long value chains is more challenging. However, trade data may be used to complement regional data to help delineate the actual spatial footprint of a product system. Impact characterization models are also increasingly regionalized.

In conclusion, the environmental sustainability of bioenergy systems is most commonly evaluated by means of LCA, which has become a dominant framework to provide guidance to project developers, policy-makers, and scientists. As a result of intense efforts in the past 25 years to tailor LCA to the specifics of bioenergy, it can be routinely applied to new systems and help in their design. A few issues remain pending, nonetheless, including the large variability of feedstock production systems and supply chains, the effect of the regional contexts, or the full inclusion of land-use changes. These warrant further research and development to fine tune the tools that are already operational and also calls for frameworks complementary to LCA, such as the inclusion of ecosystem services, or emergy assessment (see Section 6.4.1).

REFERENCES

Ben Aoun, W., Gabrielle, B., Gagnepain, B., 2013. The importance of land use change in the environmental balance of biofuels. Oilseeds Fats Crops Lipids 20, D505.

Cherubini, F., 2010. GHG balances of bioenergy systems—overview of key steps in the production chain and methodological concerns. Renew. Energy 35, 1565–1573.

Chum, H., Faaij, A., Moreira, J., Berndes, G., Dhamija, P., Dong, H., et al., 2011. Bioenergy. In: Edenhofer, O., Pichs-Madruga, R., Sokona, Y., Seyboth, K., Matschoss, P., Kadner, S., et al.,IPCC Special Report on Renewable Energy Sources and Climate Change Mitigation. Cambridge University Press, Cambridge, United Kingdom and New York, NY.

Davis, S.C., Boddey, R.M., Alves, B.J.R., Cowie, A.L., George, B.H., Ogle, S.M., et al., 2013. Management swing potential for bioenergy crops. GCB Bioenergy 5, 623–638.

de Oliveira, M.E.A.D., Vaughan, B.E., Rykiel Jr., E.J., 2005. Ethanol as fuel: energy, carbon dioxide balances, and ecological footprint. BioScience 55, 593–602.

Dones, R., Bauer, C., Bolliger, R., Burger, B., Faist Emmenegger, M., Frischknecht, R., et al., 2007. Life cycle inventories of energy systems: results for current systems in Switzerland and other UCTE countries. Final report Ecoinvent data v2. 0.
Dufossé, K., Drewer, J., Gabrielle, B., Drouet, J.L., 2014. Effects of a 20-year old miscanthus × giganteus stand and its removal on soil characteristics and greenhouse gas emissions. Biomass Bioenergy 69, 198–210.
Dufossé, K., Gabrielle, B., Drouet, J.-L., Bessou, C., 2013. Using agroecosystem modeling to improve the estimates of N2O emissions in the life-cycle assessment of biofuels. Waste Biomass Valorization 4, 593–606.
European Commission, 2010. International Reference Life Cycle Data System (ILCD) Handbook—General Guide for Life Cycle Assessment—Detailed Guidance. Institute for Environment and Sustainability, Ispra, Italy.
Gaba, S., 2018. Assessing the impact of bioenergy crops cultivation on biodiversity: review of studies including the causal chain "drivers of feedstock production—land use change—impact assessment". Sustain. Agric. Rev.
Gnansounou, E., Pandey, A., 2017. Life-Cycle Assessment of Biorefineries. Elsevier, Amsterdam.
Hellweg, S., Mila i Canals, L., 2014. Emerging approaches, challenges and opportunities in life cycle assessment. Science 344, 1109–1113.
Hischier, R., 2007. Life cycle inventories of packagings and graphical papers. Ecoinvent report 11, Swiss Centre for Life Cycle Inventories, Dubendorf.
ISO 14044, 2006. Environmental Management-Life Cycle Assessment—Requirements and Guidelines. International Standards Organization, 14044:2006 (E).
Koh, L.P., Ghazoul, J., 2010. Spatially explicit scenario analysis for reconciling agricultural expansion, forest protection, and carbon conservation in Indonesia. PNAS 107, 11140–11144.
Laurent, A., Pelzer, E., Loyce, C., Makowski, D., 2015. Ranking yields of energy crops: a meta-analysis using direct and indirect comparisons. Renew. Sustain. Energy Rev. 46, 41–50.
Lautala, P.T., Hilliard, M.R., Webb, E., Busch, I., Richard Hess, J., Roni, M.S., et al., 2015. Opportunities and challenges in the design and analysis of biomass supply chains. Environ. Manage. 56, 1397–1415.
Lesur-Dumoulin, C., Lorin, M., Bazot, M., Jeuffroy, M.-H., Loyce, C., 2016. Analysis of young miscanthus × giganteus yield variability: a survey of farmers' fields in east central France. GCB Bioenergy 8, 122–135.
Muench, S., Guenther, E., 2013. A systematic review of bioenergy life cycle assessments. Appl. Energy 112, 257–273.
Nemecek, T., Heil, A., Huguenin, O., Meier, S., Erzinger, S., Blaser, S., et al., 2007. Life cycle inventories of agricultural production systems. Final report ecoinvent v2.0.
Nemecek, T., Kägi, T., 2007. Life cycle inventories of agricultural production systems. Final report ecoinvent v2.0 no 15. Swiss Centre for Life Cycle Inventories Ecoinvent—ART, Zürich and Dübendorf, Switzerland.
Perrin, A., Basset-Mens, C., Huat, J., Yehouessi, W., 2015. High environmental risk and low yield of urban tomato gardens in Benin. Agron. Sustain. Dev. 35, 305–315.
Perrin, A., Wohlfahrt, J., Morandi, F., Østergård, H., Flatberg, T., De La Rua, C., et al., 2017. Integrated design and sustainable assessment of innovative biomass supply chains: a case-study on miscanthus in France. Appl. Energy 204, 66–77.
Pfister, S., Koehler, A., Hellweg, S., 2009. Assessing the environmental impacts of freshwater consumption in LCA. Environ. Sci. Technol. 43, 4098–4104.
Rizzo, D., Martin, L., Wohlfahrt, J., 2014. Miscanthus spatial location as seen by farmers: a machine learning approach to model real criteria. Biomass Bioenergy 66, 348–363.

Roncucci, N., Nassi O Di Nasso, N., Tozzini, C., Bonari, E., Ragaglini, G., 2015. Miscanthus × giganteus nutrient concentrations and uptakes in autumn and winter harvests as influenced by soil texture, irrigation and nitrogen fertilization in the Mediterranean. GCB Bioenergy 7, 1009–1018.

Smeets, E.M.W., Bouwman, L.F., Stehfest, E., Van Vuuren, D.P., Posthuma, A., 2009. Contribution of N2O to the greenhouse gas balance of first-generation biofuels. Global Change Biol. 15, 1–23.

Steubing, B., Ballmer, I., Gassner, M., Gerber, L., Pampuri, L., Bischof, S., et al., 2014. Identifying environmentally and economically optimal bioenergy plant sizes and locations: a spatial model of wood-based SNG value chains. Renew. Energy 61, 57–68.

Sutter, J., 2010. Life cycle inventories of pesticides. Final report Ecoinvent v2.2. Swiss Centre for Life Cycle Inventories, St. Gallen, CH.

FURTHER READING

Morandi, F., Perrin, A., Østergård, H., 2016. Miscanthus as energy crop: environmental assessment of a miscanthus biomass production case study in France. J. Clean. Prod. 137, 313–321.

CHAPTER SEVEN

Key Challenges and Opportunities

Contents

7.1 Contribution of Bioenergy to Rural Development 298
  7.1.1 The Status of World's Rural Population 298
  7.1.2 Energy Availability and Supply in Rural Areas 300
    7.1.2.1 Limited access to energy, particularly in developing countries *300*
    7.1.2.2 Energy demand and uses in rural areas *301*
    7.1.2.3 Energy carriers required for different uses *302*
  7.1.3 Access to Modern Energy Services by Means of Locally Produced Bioenergy 303
  7.1.4 Multifunctionality of Bioenergy in an Integrated Production System for Food, Feed, Energy, and Other Products 304
  7.1.5 Multiplier Effect of Bioenergy in Local Economic Development 305
    7.1.5.1 Jobs *305*
    7.1.5.2 Income *306*
    7.1.5.3 Economic activity growth and business diversification *307*
    7.1.5.4 Driver for innovation *308*
    7.1.5.5 Reduction of rural exodus *309*
  7.1.6 Environmental Benefits 309
  7.1.7 The Experience of "Bioenergy Villages" 311
  7.1.8 Specific Considerations for Developing Countries 312
    7.1.8.1 The move from traditional energy forms to modern bioenergy *312*
    7.1.8.2 Poverty alleviation *313*
    7.1.8.3 Improvements in health, education, and communication *314*
    7.1.8.4 Development opportunities for women *315*
  7.1.9 Potential Risks and Necessary Requirements for a Sustainable and Equitable Development 316
  7.1.10 Conclusions 317
7.2 Factors for Bioenergy Market Development 320
  7.2.1 Introduction 320
  7.2.2 Review of Bioenergy Consumer Studies 321
    7.2.2.1 Observations from wood pellet consumer studies *323*
    7.2.2.2 Observations from bioenergy and biofuel consumer studies *324*
  7.2.3 A Framework for Developing the Bioeconomy 325
  7.2.4 Bioenergy: Policy Context in Europe 326
  7.2.5 National Bioeconomy Strategies 328
  7.2.6 European Commission Market Pull Recommendations 330

The Role of Bioenergy in the Emerging Bioeconomy.
DOI: https://doi.org/10.1016/B978-0-12-813056-8.00007-8

7.2.7 B2B Marketing of Bio-Based Products 333
7.2.8 Public Procurement as a Market Pull Measure for Bioenergy 333
7.2.9 Conclusions 336
7.3 Bioenergy Governance 341
7.3.1 Climate Change and Bioenergy Governance 341
7.3.2 Developing a Climate Change Agenda 342
7.3.3 Bioenergy Opportunities: Setting the Agenda 344
7.3.4 Why Creating an International Agenda on Bioenergy? 347
7.3.5 Bioenergy Challenges: Governance Coordination and NDCs Implementation 348
7.3.6 Case Study: Colombia's Biodiversity Strategy as a Multilevel Approach to Governance 351
7.4 Bioenergy and Food Security: Synergies and Trade-offs 355
7.4.1 Introduction 355
7.4.2 The "Food Versus Fuel" Debate: A False Trade-off? 357
7.4.3 Rethinking Food Security and Bioenergy Linkages 361
7.4.4 Where Bioenergy Can Help (or Hurt) Food Security 366
7.4.4.1. Viable value chains and feedstock choice *367*
7.4.4.2. Prioritizing bioenergy development pathways and policies *368*
7.4.5 Conclusions 373

7.1

Contribution of Bioenergy to Rural Development

Francisco José Domínguez Pérez
Institute for Energy Diversification and Saving (IDAE), Madrid, Spain

7.1.1 THE STATUS OF WORLD'S RURAL POPULATION

There is not a single or agreed definition on rural areas. Due to the traditional differences between rural and urban ways of life, it has been usual to distinguish urban and rural areas according to the standard of living, as stated by UNStats (United Nations Statistics Division (UNStats)). Criteria like the availability of electricity and piped water in households or the existence of paved streets and sewerage systems can be useful in developing countries while in developed countries other elements are taken into account, the easy access to schools, hospitals, and recreation facilities among them.

As for economic activities, rural areas both in developing and developed countries are characterized by a relatively high percentage of population employed in agriculture.

Since in developed countries differences in the standard of living become less and less relevant, the classification of urban and rural areas is made based on population density and the existence of large cities. Even in this case, national peculiarities have to be regarded and areas which could be defined as rural in one country may be classified as urban in another one. Eurostat and the OECD, for instance, use several categories corresponding to predominantly rural, intermediate, and predominantly urban regions that are defined by means of settlement density considerations. Throughout this chapter, rural areas have to be understood as regions where criteria like those mentioned above for developed and developing countries are respectively met.

According to the last version of the UNDESA report *World urbanization prospects* (United Nations, Department of Economic and Social Affairs UNDESA, Population Division, 2015), globally, more people live in urban areas than in rural areas, with 54% of the world's population residing in urban areas in 2014. Nevertheless, urbanization levels vary considerably because Northern America, Latin America, and Europe are the most urbanized regions (more than 70%), while Africa and Asia remain mostly rural with 60% and 52% of their respective populations living in rural areas. The global rural population in 2014 amounted to 3.4 billion people.

Countries with higher per capita income are more urbanized. It is difficult to determine whether increases in GDP result in higher urbanization levels or, conversely, urbanization favors economic growth. Since industries and services are generally located within urban areas, skilled workers and financial capital are attracted to cities. Thus, urbanization could lead to economic progress and allow for the development of better infrastructures, educational and health services. But as a consequence of the economic growth, more people could also be interested in moving from rural to urban areas. Anyhow, there is a positive relationship between income and urbanization which in turn means that, overall, rural dwellers have lower earnings and less access to infrastructures, schools, hospitals, and other advantages of city life.

Finally, another distinguishing aspect of rural areas is the fact that a majority of the world's poor live in rural areas. The IFAD *Rural poverty report 2011* (International Fund for Agricultural Development (IFAD)) shows that 70% of people living in extreme poverty live in rural areas. Regarding the group known as the ultra-poor (well below the poverty threshold), around 80% live in sub-Saharan Africa and South Asia, and most of them are rural.

7.1.2 ENERGY AVAILABILITY AND SUPPLY IN RURAL AREAS

7.1.2.1 Limited access to energy, particularly in developing countries

As explained in the previous section, a key issue in rural communities is the lack of logistic infrastructure. This includes roads, railways, telecommunications, or piped water but also energy-related installations: production plants, electricity grid, gas pipelines, fuel storage facilities, and filling stations just to mention the major ones.

The relatively high investment required to provide energy access in sparsely settled regions (which means a high cost per inhabitant) is often cited as a critical barrier to the deployment of energy infrastructure in rural areas. Moreover, since in these territories the level of industrialization is low, both the government and the private sector may find that it is more effective to invest in electrification or grid extension in areas where industries are located (energy for productive uses) instead of investing in rural areas. In less developed countries private companies may as well be reluctant to invest in regions where incomes are low and people may not be able to pay the access to electricity or the cost of electric appliances. In fact, the inability to pay for energy services is another barrier that hampers the access to energy in rural areas where it is indeed available (supply exists) but people cannot afford the connection to the grid or the devices that make the energy useful (machinery, motors, pumps, etc.).

Again, in this sense there are huge differences between developed and developing countries. The special report of IEA *Energy access outlook 2017*

(International Energy Agency IEA, 2017) shows that in 2016 1.1 billion people, i.e., 14% of the world's population do not have access to electricity, 84% of which live in rural areas, mainly in Africa and Asia. In particular, sub-Saharan Africa is the region that concentrates the largest share of population without electricity access (590 million people which corresponds to 57% of the population, 80% of which live in rural areas). The report also highlights that nearly one-third of the world's population rely for cooking on the traditional use of solid biomass which in rural regions of developing countries remains the predominant source of energy for cooking and heating.

7.1.2.2 Energy demand and uses in rural areas

7.1.2.2.1 Energy demand

Regarding energy use in rural areas, there are also regional differences as depicted in the IEA report *Cities, towns, and renewable energy* (International Energy Agency (IEA), 2009).

In OECD countries rural dwellers usually consume more direct energy than urban residents. This is due to several factors related to the specific circumstances of the way of living in rural areas.

A relevant element is the higher demand for transport of people who live in remote areas where distances are large and relatively frequent displacements are needed for some basic services (e.g., school and health care). Private vehicles are often used because in rural communities there are not efficient public transport systems which are available in cities (for instance, subway, trams, buses, or rail). Furthermore, the kinds of vehicles that are used by rural residents are also different from those of people living in urban centers. Larger vehicles such as pick-up trucks or medium-duty trucks are driven in rural areas. In addition to the difference in consumption with respect to smaller cars that are used in cities, the lack of rural infrastructure makes it difficult a transition to alternative technologies (for instance, gas vehicles) which use fuels that could be cheaper than gasoline.

Another factor is the size of households which, on average, is quite bigger in rural communities than in towns. Moreover, they are usually detached houses that are more exposed to weather conditions than apartments in buildings and therefore they have higher energy losses. Just like in the case of transport, in the absence of pipelines for gas distribution or other energy infrastructure, alternative fuels are not available and more expensive heating and cooling systems have to be used.

In non-OECD countries energy consumption of urban dwellers is higher than that of rural inhabitants. This is due to the fact that modern energy services in developing countries are in practice available mainly in cities where, in addition, people can afford improved energy appliances as a result of their average higher income level.

7.2.2.2 Basic needs

Energy is required to cover basic needs at a household level and at a community level.

Typical basic domestic uses include lighting, cooking, space heating and cooling, and pumping potable water (if not supplied by public services). Electricity is also needed for refrigeration, television, radio, battery charging for telephone, and computers. And fuels for transport are also consumed by rural residents, as explained in previous sections.

Just to give a few relevant examples of community uses of energy, besides street lighting, health care and education can be named. Clinics and hospitals need lighting, safe and clean water supply, energy for medical instruments, sanitation and sterilization, as well as refrigeration for heat-sensitive products. Electricity in schools permits evening classes and the use of improved educational equipment such as audio–visual appliances and computers.

7.1.2.2.2 Economic activities

It has already been underlined that rural areas are characterized by the fact that agriculture is the predominant source of employment and income. Virtually all farming duties require energy. It is needed for tilling, irrigation, grinding, husking, drying, smoking, processing, and preservation of products but also (indirectly) for producing chemicals and fertilizers. Finally, transport is also needed to get the products to the market.

In addition to agriculture, other economic activities can be developed in rural communities. Industrial production usually requires energy for lighting, space heating, equipment, and process heat. Likewise, commercial businesses (retail, restaurants, etc.) demand energy for their activities.

7.1.2.3 Energy carriers required for different uses

Electricity is a very flexible energy carrier which allows to meet a wide range of rural population needs. It can be used for lighting, water pumping, and all kinds of appliances both at domestic and at community level. In the field of agriculture, machinery and irrigation systems may run on electricity too.

Heat is required by rural energy users to meet some basic needs such as cooking and household heating. It is also essential within the framework of economic activities since heat is extensively used in industrial processes and in several agricultural tasks, like drying.

Liquid or gaseous fuels may be used as intermediate carriers (for instance, in a backup diesel generator for electricity) or directly to run machinery (tractors, drying equipment, harvesters) as well as for transport.

7.1.3 ACCESS TO MODERN ENERGY SERVICES BY MEANS OF LOCALLY PRODUCED BIOENERGY

Bioenergy can provide all the energy carriers needed: electricity, heat, liquid, and gaseous fuels.

In rural areas, biomass is an available resource relatively affordable and easy to collect. As a matter of fact, biomass remains the main source of energy used by rural population in developing countries, albeit it is customarily used unprocessed (fuelwood, charcoal, and crop residues). Its availability along with the possibility of providing various energy carriers required in rural communities make the biomass an optimal option to improve the access to modern energy services that rural regions lack. Moreover, biomass fuels can be stored and used whenever they are needed which make them a reliable source of energy in remote areas where electricity shutdowns or disruptions in fuel supply are more likely than in cities.

Furthermore, locally grown biomass and residues may enable securing energy access in rural areas through different technologies, some of which are listed through examples hereafter. Small-scale digesters can be used to produce biogas (for electricity, heat, or transport) by means of anaerobic digestion from manure, crop residues, or other wastes. Pellets or briquettes for heat production are efficient alternatives to fuelwood due to their higher density and better combustion quality. Biodiesel obtained from vegetable or used cooking oils as well as from animal fats may be used in diesel engines for transport and for heating.

7.1.4 MULTIFUNCTIONALITY OF BIOENERGY IN AN INTEGRATED PRODUCTION SYSTEM FOR FOOD, FEED, ENERGY, AND OTHER PRODUCTS

Bioenergy production has a multifunctional nature. Although multifunctionality can be defined in a number of ways, in a broad sense it refers to a property of certain economic activities characterized by having multiple outputs. In the OECD report *Multifunctionality. Towards an analytical framework* (Organisation for Economic Co-operation and Development OECD, 2001), two key elements of functionality are pointed out: the production of diverse commodity and noncommodity outputs and the fact that some of these noncommodity outputs are externalities or public goods which may cause effects on welfare, for instance. For example, environmental benefits derived from the economic activity or impacts on employment that can lead to a reduction of migration.

The impact of multifunctionality of bioenergy is particularly relevant in rural areas since commodities related to this activity are produced and consumed locally and because effects of externalities are usually produced within the same region too.

The second element of multifunctionality, which deals with outcomes related to changes in economic, environmental, and social conditions as a consequence of bioenergy deployment will be addressed in following sections.

Regarding the first element, that is, the joint provision of different outputs, bioenergy projects, in particular in rural areas, can be implemented in a way that several commodities such as food, feed, and energy are produced. As described in the FAO report *Making integrated food-energy systems work for people and climate* (Food and Agriculture Organization of the United Nations (FAO), 2010), this kind of farming systems are designed to integrate and increase the simultaneous production of food and energy. Schemes like multiple cropping may be applied as well as techniques that allow utilization of by-products and waste, thus maximizing synergies between food crops, livestock, and energy products.

In developed countries there is growing interest in multiple cropping systems where the growth of energy crops and food crops are combined. By contrast, in developing countries, integrated systems usually consist of cultivation of food crops, livestock, and utilization of agricultural and animal waste to produce energy (typically biogas).

Since the implementation of integrated systems may lead to reduced costs due to the efficient use of resources and to savings in energy supply (for instance, locally produced biogas instead of imported fossil fuels), an increase in productivity can be achieved which, in turn, results in higher provision of different outputs obtained. Agricultural productivity may be increased because, as a consequence of cost savings, farmers can invest in better fertilizers, high-quality seeds, or mechanization (e.g., utilization of irrigation equipment which improves yields and facilitates multiple cropping).

7.1.5 MULTIPLIER EFFECT OF BIOENERGY IN LOCAL ECONOMIC DEVELOPMENT

Generally speaking, the introduction of new sectors of economic activity in a given area results in several impacts on economic development in the region concerned. These effects occur at three different levels: direct, indirect, and induced. Direct effects are produced as a consequence of the implementation of the specific project (for instance, on-site jobs created to build a conversion processing plant). The operation of this facility will require a regular supply of goods and services which means a higher demand in related sectors and, therefore, additional (indirect) jobs within those industries. Income generated by direct and indirect employment creation leads to an increased economic activity since wages earned by people (project workers and suppliers) can be spent in local retail businesses, for example, which in turn will result in new (induced) job growth needed in commercial activities. Thus, a multiplier effect in many economic indicators (employment among them, as illustrated) is derived from the initial investment for the new project.

7.1.5.1 Jobs

Bioenergy production has the potential to give rise to high employment multipliers because it is more labor-intensive than other sectors. In the IEA paper *Good practice guidelines—Bioenergy Project Development & Biomass Supply* (International Energy Agency (IEA), 2007), employment factors (jobs/GWh) for operation and maintenance of different technologies are estimated. For instance, liquid biofuels produced from energy crops would require more than 120 jobs/100 GWh, while photovoltaics would need around 55 jobs/100 GWh; wind projects would require less than 20 jobs/100 GWh and small hydro would require less than 10 jobs/100 GWh. According to the IRENA 2017 annual review report on *Renewable energy and jobs* (International Renewable Energy Agency IRENA, 2017), the bioenergy sector employed 2.7 million people (out of a total of 9.8 million jobs in renewables), directly and indirectly, in 2016 with jobs concentrated in feedstock supply.

Direct employment is of course generated to build the bioenergy production plant (and for decommissioning at the end of the life span). Nevertheless, these are short-term jobs.

By contrast, long-term direct jobs are created for all other activities involved in the bioenergy value chain. The first stage is the collection of biomass (cultivation, harvesting, etc.). Depending on the feedstock used in the conversion facility, employment can be created in agriculture (crops or agricultural residues), in silviculture (forest products or residues) or in waste collection and management (municipal or industrial waste). The second stage involves biomass transportation from plantations or collection points to the processing facilities, so additional job opportunities may arise in the transport sector. Finally, the third stage includes processing and conversion. Plant and equipment operators as well as maintenance workers will be required to carry out activities such as biomass pretreatment (chipping, grinding, milling, etc.) and conversion to energy carriers (biofuels, electricity, and heat). Many of these positions can be filled with low-skilled workers thus providing access to the labor market for people who otherwise could be at risk of social exclusion. But there is also a demand for skilled staff, typically in production plants or to operate sophisticated equipment.

In addition, indirect employment is generated in supporting industries that can be created in or attracted to the region where the bioenergy production takes place. And, as stated above, induced new jobs in all kinds of businesses may be needed due to the increased demand for products and services as a result of higher average income of local population.

7.1.5.2 Income

The impact on income at household level is twofold: on the one hand, increased earnings due to growing economic activity and, on the other hand, savings arising directly from the deployment of bioenergy.

Just as for employment, a multiplier effect occurs as a result of higher spending derived from additional income received by people involved in the bioenergy production chain. Part of the money earned will be used to purchase more goods and services thereby improving profits of businesses providing such products, which means added income growth in the corresponding sectors.

As previously mentioned, in rural communities energy access is often limited. Even in cases where supply is available rural residents may not be able to afford it because of high prices (as a consequence of higher costs of supplying energy in remote areas) or due to the lower average income level or rural population, as explained above. In this context, the income increase resulting from the implementation of bioenergy projects (including its multiplier effect) may enable rural inhabitants afford energy services or appliances which were out of their reach. This particular use of additional earnings leads to a growing demand for energy and, accordingly, to new investments in energy supply projects which, in turn, reinforce the income generation.

Furthermore, in less developed countries energy is not available at the household level in many instances. Locally produced bioenergy by means of small-scale systems can provide electricity or fuels which may be used to carry out daily chores or to avoid the need for some tasks that, in the absence of energy supply, are very time-consuming such as milling grain or collecting firewood. The gained time can thus be used in other income-generating activities.

Direct savings are achieved in those situations where bioenergy is cheaper than replaced alternative sources (for instance, biogas production from manure in farms instead of imported fossil fuels). This translates into extra available money for consumers, which can be regarded as equivalent to an income growth.

7.1.5.3 Economic activity growth and business diversification

Since agriculture is the main economic sector in rural areas, energy provision from biomass has a direct impact in the overall economic activity in such regions: firstly, because bioenergy production requires feedstock which can be provided by local farmers who, consequently, have access to a new market for their products; and in addition, due to the contribution of energy supply (by means of bioenergy) to the improvement of agricultural processes in a way that production becomes less vulnerable to adverse environmental conditions (for instance, in case of water scarcity or droughts).

Moreover, as outlined above, bioenergy deployment has the effect of boosting the local economy. In the first step, the new demand for components or equipment (electric switchgear, motors, farm machinery, etc.) and other inputs needed in the bioenergy production process (fertilizers, chemicals, etc.) can attract manufacturing companies. Besides, the increased income level of rural population favors expenditure on goods and services, thus fostering the creation and establishment of such enterprises and the expansion of retail businesses. Furthermore, the affordable supply of heat and electricity is key for the introduction of new sectors of economic activity, for example, rural tourism, which is a large potential source of income and jobs.

Another remarkable aspect resulting from the higher economic dynamism arising from bioenergy deployment is the strong increase in tax revenue. Companies have to pay taxes, some of which are collected by the national government while others are levied by local and regional authorities. In addition, the rising personal income concomitantly triggers the broadening of the tax base, and it also contributes to raise the fiscal revenue. The availability of a larger budget allows both the national and the local governments to invest in educational and health services, which are a key issue in remote regions, and in infrastructure that rural communities usually lack such as high-quality roads and better electricity networks. These improvements in services and infrastructure facilitate trade, communications, and the establishment of new businesses and are crucial for certain economic activities like the aforementioned development of rural tourism.

7.1.5.4 Driver for innovation

Bioenergy, as a core part of the bioeconomy, involves activities relying on a wide variety of sciences and technologies (agronomy, engineering, life sciences, biotechnology, and food science, among others). That is why, as stated in the EU Bioeconomy Strategy and Action Plan document entitled *Innovating for sustainable growth* (European Commission, 2012), bioenergy sectors and industries have strong innovation potential as they encompass the production of biomass as well as the management of waste streams and the conversion of these resources into value-added products through the application of the referred sciences and technologies.

In particular, the impact of innovation can be significant in rural areas because, to a large extent, it will be geared to the improvement of agricultural and forestry supply chains. Overall, new approaches and practices aimed at increasing productivity and resource efficiency while minimizing adverse effects will be developed following bioenergy deployment. A few examples have been reported in the EU *Review of the 2012 European Bioeconomy Strategy* (European Commission, 2017) like crops which are resource-efficient, land-based, and aquatic biomass for nonfood use, projects on biorefineries, and waste streams valorization.

Finally, innovation entails specialized and skilled activities. Companies providing innovative services contribute to enhance knowledge and to build capacity in the local community whose economy becomes more competitive thus leading to new business and employment opportunities.

7.1.5.5 Reduction of rural exodus

According to the International Labour Organization policy brief *Labour migration governance for rural development* (International Labour Office (ILO), 2011), over the last 50 years, 800 million people have migrated from rural to urban areas mainly as a result of lower incomes and human security in comparison with cities but also due to the degradation and decreased productivity of land, among other factors.

Indeed, economic and environmental issues are the major drivers for migration but, as reflected in the report *The role of sustainable energy access in the migration debate* published by the European Union Energy Initiative (European Union Energy Initiative Partnership Dialogue Facility (EUEI PDF), 2017), the lack of energy access exacerbates those problems. Therefore, they can be attenuated by improving energy supply, which, in rural areas where the biomass resource is available, may be achieved through bioenergy. In addition, as explained in previous sections, energy provision contributes to the economic development, the increase in income, and the deployment of infrastructure as well as educational and health services. All these aspects are key factors for a better quality of life and, therefore, their enhancement as a consequence of bioenergy deployment helps reduce rural exodus to urban areas.

7.1.6 ENVIRONMENTAL BENEFITS

This section is not intended to address overall effects on the environment due to bioenergy production and use, such as the potential reduction of greenhouse gas emissions which, of course, could have an impact in rural areas too. Instead, here the focus is on some specific local environmental problems that are often found in rural areas and which can be mitigated by means of bioenergy production systems.

For instance, the aforementioned cultivation of certain crops that can be used for bioenergy production may enable restoration of degraded or contaminated land which would not be suitable for growing food or feed crops. Restoration of marginal or degraded soils leads to a wide range of environmental benefits since it contributes to increase water infiltration and availability, reduce soil erosion, and prevent biodiversity loss, just to mention a few.

But without doubt, one of the major environmental problems specifically linked to activities carried out in rural areas is the management and disposal of agricultural and livestock waste.

As pointed out in the FAO report *Livestock's long shadow* (Food and Agriculture Organization of the United Nations FAO, 2006), organic wastes from animal production (slurries and manure) contain quite a few pollutants such as biological oxygen demand, nutrients (nitrogen, phosphorous, and potassium), methane, ammonia, heavy metals, pathogens, and drug residues. Pollutant effluents that are discharged into the environment threaten soil fertility and cause water contamination as well as air emissions. Even if they are stored in lagoons, they may leak into other water streams.

High concentrations of nutrients and BOD in water can lead to accelerated eutrophication, fuel algae growth, and damage the functioning of the ecosystem since some fishes and other aquatic plants may be killed. Most of the heavy metals ingested by livestock (which include copper, zinc, selenium, cobalt, arsenic, iron, and manganese) are excreted and returned to the environment. Excreta also contain zoonotic microorganisms and parasites that can be harmful to human health and a high amount of drug residues including antibiotics and hormones. In addition, gaseous emissions in the form of methane and ammonia are released from unmanaged livestock wastes.

Hence, bioenergy obtained from these residues helps protect the environment while providing other benefits. In view of the nature of the feedstock, biogas production is the most highly recommended technology option.

7.1.7 THE EXPERIENCE OF "BIOENERGY VILLAGES"

A bioenergy village is a community where a large share of the electricity and heat demand is met by means of locally produced energy from biomass, in some cases, even the whole needs are covered and the excess power produced is sold to energy providers. Quite a few bioenergy villages already exist in Europe, most of them in Germany but also in Austria, Denmark, and other countries.

Jühnde was first officially recognized as a bioenergy village in Germany. As described in the IEA Task37 document *The first bioenergy village in Jühnde/Germany. Energy self sufficiency with biogas* (International Energy Agency Bioenergy Task 37: Energy from biogas (IEA Bioenergy Task 37), the project started in 2006 with the aim of replacing all fossil fuels for electricity and heat production with biomass. The system which was implemented to supply this community (around 1000 residents) consisted of a biogas plant for combined heat and power production from liquid manure and whole plant silage of different crops, a boiler fueled with wood chips intended to provide additional heat in winter and a district heating network. As a consequence of the successful results achieved, the example has been followed in other regions and at present nearly 200 villages are applying the same concept (142 fully implemented and 45 on the way in January 2018, according to the German Agency for Renewable Resources, FNR (Fachagentur Nachwachsende Rohstoffe (Agency for Renewable Resources) (FNR)).

The characteristics of bioenergy villages (for instance, organization structure, financing schemes, or technologies used) may differ from one to another but some common criteria beyond the significant substitution of fossil fuels by local biomass must typically be met in these initiatives such as the involvement of local population in decision-making processes and the implementation of the project through a business model that allows farmers, forest owners, and consumers to share ownership of facilities.

Experience acquired over the years show that the bioenergy village concept is a feasible way for rural communities to reduce dependence on fossil fuels while ensuring a reliable and affordable energy supply.

Many benefits regarding environmental and economic aspects have been observed, including increased income, employment creation, diversification of economic activities, reduced migration to urban areas, and improved waste management.

Nevertheless, one of the most highly valued outcomes of bioenergy villages falls within the social field. Indeed, the empowerment of rural residents producing their own energy is viewed as a key aspect. Achieving self-sufficiency and becoming independent in terms of energy supply is cause for satisfaction among the population. All the more so as they are engaged in business management and involved in participatory processes related to community governance. Most bioenergy village projects are implemented by establishing cooperatives. These organizational structures are widely accepted because they encourage equity and transparency. Their members address decisions in a democratic manner, and they take into account common social and cultural needs besides economic considerations to maximize community welfare and local development. Thus, bioenergy village initiatives contribute to increase social cohesion and active citizenship.

7.1.8 SPECIFIC CONSIDERATIONS FOR DEVELOPING COUNTRIES

7.1.8.1 The move from traditional energy forms to modern bioenergy

In rural areas of less developed countries most people rely on traditional forms of energy. In fact, human and animal power are the major sources of energy used to carry out mechanical tasks such as farm operations (for instance, tillage, planting, or harvesting) and all types of activities that require transporting or lifting goods. In particular, human energy is primarily used in these areas for gathering and collecting biomass. And, of course, manual labor is the main way of performing other household duties like pumping water and washing. Energy needs for space heating and cooking are typically covered in rural regions of developing countries by means of biomass due to the lack of access to electricity or other fuels and because the biomass resource is available and affordable. Nevertheless, in most cases the technologies used to convert biomass into energy are inefficient and highly pollutant.

A shift from traditional to modern energy forms can result in a relevant improvement in the living standards of rural residents. As already mentioned, electricity supply by means of small-scale decentralized biogas production systems may be used both for agricultural tasks (for instance, irrigation) and household appliances (e.g., water pumping). Biofuels might also be used for farm machinery and transport. In these cases, human power is replaced by bioenergy thus avoiding drudgery and releasing time for income-generating activities. The issue of cooking is particularly interesting because, as stated above, in rural areas of developing countries people mostly rely on biomass for cooking but they use traditional stoves that are inefficient and poorly ventilated, which leads to high levels of indoor air pollution. So, the improvement in the quality of life as regards energy supply for cooking and other heating uses does not involve a shift to other energy source (bioenergy is already used since biomass is available and affordable) but to cleaner and efficient conversion devices (improved cookstoves) and modern fuel types (like briquettes).

7.1.8.2 Poverty alleviation

As explained in the first section of this chapter, the vast majority of world's poor population lives in rural areas. Most of them lack access to electricity and have a very limited availability of energy supply. Poverty cannot be eradicated without ensuring access to energy. And, as already mentioned, biomass (including residues and waste) is a resource that is usually available in rural areas which can be used to produce affordable energy in an efficient way.

In previous paragraphs it has been described how bioenergy can contribute to economic development and to an improvement in the quality of life. This is particularly true in the case of rural regions with crippling poverty where malnutrition and diseases afflict the lives of millions of people.

Since poor population in the least developed countries has no access to energy, they spend a considerable amount of time in activities such as collecting wood fuel and fetching water. Moreover, agricultural productivity is reduced because they cannot use machinery or irrigation systems. Therefore, household income is low as a consequence of low yields in agricultural production and because it cannot be increased by means of other economic activities given that daily chores are very time consuming. Due to low income available they cannot afford access to energy, thus turning into a vicious circle.

Energy supply by means of bioenergy may help to break the vicious circle of poverty, as explained in previous sections. It can result in economic development, employment creation, and income increase. This, in turn, means that farm equipment and household appliances can be purchased which leads to higher agricultural productivity and additional time available for other income-generating tasks. Furthermore, lighting provided by bioenergy systems enables that business activities can continue beyond daylight hours, making them more productive.

Besides economic development, energy access provided by bioenergy may contribute to alleviate other aspects of poverty. For instance, improved infrastructure to supply clean water can be implemented in poor rural communities, and this is a key issue in the fight against diseases. And regarding malnutrition, according to the FAO report *Global food losses and food waste. Extent, causes and prevention* (Food and Agriculture Organization of the United Nations (FAO), 2011), nearly one-third of the world food production is lost in the supply chain, including steps such as preservation, storage, and transportation, and a poor energy access is one of the factors behind these losses which exacerbate the problem of undernourishment in many poor regions of the world. Moreover, bioenergy deployment can help reduce hunger since, as outlined in previous sections, increases in agricultural productivity and higher yields in food production can be achieved as a result of improved energy access and income growth.

7.1.8.3 Improvements in health, education, and communication

Overall, better energy access (and, in particular, electricity supply) in rural regions brings a wide range of benefits. The most noteworthy, as illustrated in the World Bank report, *The welfare impact of rural electrification* (The World Bank, 2008), has to do with health, education, and communication.

The contribution of bioenergy to health improvement in rural areas has many facets.

Clearly, hospitals and clinics need a reliable energy supply. A highly relevant benefit of energy access in health facilities in developing regions is the preservation of the cold chain. Vaccines and other heat-sensitive products require that temperature is kept low from the point of manufacture to the point of use. Therefore, refrigerators are needed in hospitals for this purpose. Typically, they use electricity but some of them can also run on gas or other fuels. In addition, using bioenergy makes it possible to operate medical equipment such as X-ray machines to perform processes like sterilization of surgical instruments and to ensure a safe and clean water supply. Lighting, of course, is a basic need in health centers which may also be provided by means of bioenergy. In less developed countries, the availability of such services in hospitals has a huge impact in reducing child and maternal mortality as well as in fighting against all kinds of illnesses.

Another positive impact of modern bioenergy is the reduction of indoor air pollution which causes several respiratory and pulmonary diseases as well as pregnancy-related disorders, among others. As stated above, replacing traditional stoves by improved stoves for cooking and modern biomass fuels like briquettes results in decreased indoor pollution and, hence, in better health for rural population.

Finally, malnutrition, which is a major health problem in the least developed countries, can also be reduced through bioenergy deployment as described in the previous section.

Regarding education, bioenergy can provide electricity and space heating or cooling for schools. Access to electricity in rural communities allows schools to offer evening classes and households to have good study conditions when the sun sets. Taking into account that adults are working during the day, this can contribute to foster adult education too. Electricity in schools permits the use of better educational facilities, including audio–visual equipments, computers, and other appliances.

Information and communication in remote rural regions are also improved due to energy access provided by bioenergy. Modern communication services enable rural dwellers to receive information about business or market conditions, just to mention a few. And they are key for health-related actions such as calling to hospital requesting an ambulance or asking for medicines. Access to media (TV and radio) increases people knowledge and awareness (for instance, specific TV campaigns about health issues, vaccination, weather information, and educational programs, among others).

7.1.8.4 Development opportunities for women

In less developed countries, women are usually responsible for time-consuming tasks such as collecting firewood, fetching water, and milling cereals. The introduction of modern bioenergy services allows women to spend more time in educational and income-generating activities. Since bioenergy projects are very labor-intensive, previously unemployed rural women can get a job in the bioenergy production chain and earn a regular wage, which is usually difficult for them in developing countries. They may also develop small businesses due to available access to energy for lighting and other uses. Increased income helps empower women to contribute to the livelihood of their families and to become involved in community activities.

7.1.9 POTENTIAL RISKS AND NECESSARY REQUIREMENTS FOR A SUSTAINABLE AND EQUITABLE DEVELOPMENT

In previous sections many advantages of bioenergy production and use in rural areas have been presented. However, it should not be understood that the implementation of bioenergy initiatives always results in simultaneously achieving all those kinds of benefits. The effects of bioenergy deployment are context specific. It has the potential to produce positive impacts on economic, environmental, and social aspects but the specific local conditions will determine to a large extent whether main effects are related to employment and income or to health improvement, for instance. Typically, a combination of several benefits will occur. So, bioenergy should not be viewed as a panacea which will solve all the problems in rural areas but there is no doubt that it can significantly contribute to promote rural development.

Nevertheless, bioenergy can also pose potential risks if not developed properly. As regards negative impacts that could arise from bioenergy production in specific rural communities, mainly some adverse environmental and social effects might occur. Environmental risks include overexploitation of resources such as water (which could lead to water scarcity) or soil nutrients, biodiversity loss, and deforestation. As for social concerns, particularly in developing countries, issues related to land tenure and risk of displacement as well as changes in prices and availability of main staple crops are the more relevant challenges to be addressed. With respect to the latter, as described in previous sections, bioenergy production may indeed contribute to increase agricultural production, thus leading to a growth in local supply of staple crops. What this means is that bioenergy per se is neither good nor bad. What matters is the way it is produced. It should consequently be developed in a sustainable manner intended to maximize opportunities and to prevent potential risks.

In view of the peculiarities of rural areas, the successful implementation of bioenergy projects in such regions makes it advisable that certain requirements are met, in particular in developing countries. The FAO and UNEP document *A decision support tool for sustainable bioenergy* (Food and Agriculture Organization of the United Nations FAO and United Nations Environment Programme UNEP, 2010) points out recommended actions that should be taken to optimize the gains from bioenergy deployment while avoiding or minimizing the aforementioned potential risks.

According to that document, engagement of communities throughout the whole project implementation is paramount. Adopting a participatory approach so that local needs are effectively addressed helps in identifying, preventing, and mitigating social and environmental impacts that could arise. Involving stakeholders from the beginning results in more suited decisions where the scale of operations (large or small), ownership structure (e.g., large industrial plantations, large private farms, or small private farms) and the relationship between feedstock producers and processing companies (for instance, concession farming or contract farming) are taken into account and the potential risks and benefits of these schemes are assessed bearing in mind positive or negative impacts on local income generation, employment, labor conditions, land tenure rights, biodiversity loss, and availability of staple crops, among others.

Financing is a relevant aspect in bioenergy projects in general, but when it comes to rural areas of developing countries it becomes a critical issue since financial markets are not well developed and because, as stated in the ILO report *Developing the rural economy through financial inclusion: the role of access to finance* (International Labour Office (ILO), 2017), traditional banks usually do not offer services in rural areas due to high costs as a consequence of low population and lack of infrastructure. Therefore, alternative financing models such as microfinancing, community-based organizations, or supplier financing are required to unlock the potential of the rural economy.

7.1.10 CONCLUSIONS

Bioenergy can significantly contribute to rural development. And indeed bioenergy is commonly considered as a core element in most international, national, and regional strategies and plans for rural development.

A key characteristic of rural areas is the limited access to energy, in particular in developing countries. The lack of infrastructure in those remote regions hampers in many cases a reliable and affordable energy supply for rural residents who often face difficulties to cover basic needs (lighting, space heating) and to develop economic activities (agriculture, which requires energy in all steps, industry, or retail businesses).

Bioenergy has some comparative advantages over other options for energy provision in rural areas. Biomass resources are widely available. In addition, biomass fuels can be stored and used whenever needed and all energy carriers required (electricity, heat, gas, or liquid fuels for transport) can be produced from biomass.

Energy provision by means of bioenergy contributes to economic development in rural regions, where the income level is usually lower than in cities (most of the world's poor population live in rural areas). Since feedstocks for bioenergy production are obtained from agricultural and forestry activities, the implementation of bioenergy projects in rural areas has a direct economic effect at local level because farmers have access to a new market for their products. Improvements in productivity as a consequence of energy provision and the income increase dynamize local economy and foster job creation which leads to a better quality of life of rural dwellers thus contributing to reduce the exodus from rural to urban areas.

Integrated systems for food, feed, and energy production allow the management and use of farm waste, which implies relevant environmental benefits as well. The experience of "Bioenergy villages" shows that this kind of systems can be used to cover the energy needs in rural communities while social cohesion is reinforced due to the participatory approach adopted in most cases for the project implementation.

In less developed countries, bioenergy may have even more significant impacts since it can greatly contribute to poverty alleviation and to improve healthcare services, education, as well as to provide new development opportunities for women, who in developing countries usually find it difficult to get jobs and to earn their own income.

The implementation of bioenergy projects can lead to a wide range of benefits but it should be done in a sustainable manner to prevent potential environmental and social risks. In this sense, the engagement of communities throughout the whole process is a crucial requirement for a successful deployment of bioenergy in rural areas.

REFERENCES

European Commission, 2012. Communication From the Commission to the European Parliament, the Council, the European Economic and Social Committee and the Committee of the Regions. Innovating for Sustainable Growth: A Bioeconomy for Europe. European Commission, Brussels.

European Commission, 2017. Review of the 2012 European Bioeconomy Strategy. European Commission, Brussels.

European Union Energy Initiative Partnership Dialogue Facility (EUEI PDF), 2017. The Role of Sustainable Energy Access in the Migration Debate. EUEI PDF, Eschborn.

Fachagentur Nachwachsende Rohstoffe (Agency for Renewable Resources) (FNR). https://bioenergiedorf.fnr.de/.

Food and Agriculture Organization of the United Nations (FAO), 2006. Livestock's Long Shadow. Environmental Issues and Options. Rome: FAO

Food and Agriculture Organization of the United Nations (FAO), 2010. Making Integrated Food-Energy Systems Work for People and Climate. Rome: FAO.

Food and Agriculture Organization of the United Nations (FAO) and United Nations Environment Programme (UNEP), 2010. A Decision Support Tool for Sustainable Bioenergy. UN Energy.

Food and Agriculture Organization of the United Nations (FAO), 2011. Global Food Losses and Food Waste. Extent, Causes and Prevention. Rome: FAO.

International Energy Agency (IEA), 2007. Good Practice Guidelines—Bioenergy Project Development & Biomass Supply. Paris, France: IEA

International Energy Agency (IEA), 2009. Cities, Towns & Renewable Energy. Yes in My Front Yard. Paris, France: IEA.

International Energy Agency (IEA), 2017. Energy Access Outlook 2017. From Poverty to Prosperity. IEA.

International Energy Agency Bioenergy Task 37: Energy from biogas (IEA Bioenergy Task 37). The First Bioenergy Village in Jühnde/Germany. Energy Self Sufficiency With Biogas. IEA.

International Fund for Agricultural Development (IFAD). Rural Poverty Report 2011. New Realities, New Challenges: New Opportunities for Tomorrow's Generation. Rome, Italy: IFAD.

International Labour Office (ILO), 2011. Labour Migration Governance for Rural Development. Geneva, Switzerland: ILO.

International Labour Office (ILO), 2017. Developing the Rural Economy Through Financial Inclusion: The Role of Access to Finance. Geneva, Switzerland: ILO.

International Renewable Energy Agency (IRENA), 2017. Renewable energy and jobs—Annual review 2017. Abu Dhabi: International Renewable Energy Agency.

Organisation for Economic Co-operation and Development (OECD), 2001. Multifunctionality. Towards an Analytical Framework. Paris, France: OECD.

The World Bank, 2008. The Welfare Impact of Rural Electrification: A Reassessment of the Costs and Benefits. Washington, DC: World Bank.

United Nations Statistics Division (UNStats) https://unstats.un.org/unsd/demographic/sconcerns/densurb/densurbmethods.htm#B.

United Nations, Department of Economic and Social Affairs (UNDESA), Population Division, 2015. World Urbanization Prospects: The 2014 Revision (ST/ESA/SER.A/366). New York: UNDESA, Population Division.

7.2

Factors for Bioenergy Market Development

Paul O'Reilly
College of Business, Dublin Institute of Technology, Dublin, Ireland

7.2.1 INTRODUCTION

The global economy is on the verge of one of its most significant shifts in the last two centuries—the move from fossil fuel energy to renewable and sustainable energy. Imbalance in the production and consumption of fossil fuels and the emission of harmful greenhouse gases (GHGs) are at the core of this conversion, and as the world's largest source of such renewable energy, bioenergy is at the forefront of this revolution.

While the global economy is in the early stages of this major shift, in many countries bioenergy markets are still relatively small and immature, causing instability. This instability creates barriers for private enterprises to commit to large and long-term contracts. Bringing bioenergy products to market, and contributing to the development of a sustainable bioeconomy, involves following a route that commences with identifying and procuring unused or underexploited bioresources, then transforming these resources into replacements for fossil-based fuel and energy using new and often untested production technologies, and then bringing these replacement products to markets of which relatively little is known of their receptivity. This route sets out a value chain that is complex, high risk, and with relatively uncertain market projections.

Commercialization and adoption of new bioenergy products and their supporting technologies are observed as challenging both in a business to business context due to issues such as high switching costs or a lack of existing quality standards and towards final consumers (McCormick and Kautto, 2013; Golembiewski et al., 2015). As observed by Frewer et al. (1997) and confirmed in later studies, it has been shown that final consumers have been hesitant to embrace products generated from side or waste streams.

This chapter focuses on the market pull factors for the bioenergy market and in particular on those that relate to the consumer attitudes, behaviors, and perceptions. We observe that these consumer perceptions have significant implications for issues such as labels, public awareness, and tax incentives for bioenergy products, and incentive related to fossil carbon and GHG emissions. Consumer studies are reviewed and observations are drawn from EU and national bioeconomy strategies in particular. The chapter also addresses key policy interventions relating to public procurement.

7.2.2 REVIEW OF BIOENERGY CONSUMER STUDIES

In their review of a number of biofuel studies, overall Radics et al. (2016) observed that attitudes of the general public indicate a moderate to ambivalent support towards renewable energy including biomass-based energy. Some studies that show acceptance can relate to social benefits to community—including jobs—and ability to contribute to national security; however, threats to land availability for food production and increased food prices are seen as negative attributes to bioeconomy growth. Consumer studies also show that social welfare benefits and security of transportation—key motivations for government attempts to build the bioeconomy generally—have also been found to offer a positive perception towards biofuels (Binder et al., 2012).

In relation to energy solutions, a review by Radics et al. (2016) found numerous studies confirming that consumers are more willing to support and accept solar, wind, and hydro-energy sources relative to biomass-based energy. Their review explains that this is in large part due to the relatively recent introduction of bioenergy options in the market place compared to other options and the associated lack of knowledge of biomass-based energy impacts. Interestingly their review also found that the type of feedstock used for producing biomass-based energy also impacted on public support, with higher levels of support for energy produced from landfill wastes, wood waste, and grasses, with trees, genetically modified organisms, and corn-based bioethanol having lower levels of support.

Knowledge on the part of the consumer about the existence of a product is fundamental. Blackwell et al. (2002) explain that consumers must be aware of the new or innovative products before they become clients and that one is unlikely to acquire any new product if there is insufficient information about it or how to use it.

Studies by Hartmann and Apaolaza (2012), Selfa et al. (2011), and Qu et al. (2011) show that consumers are likely to support bioenergy options where there is a clear positive environmental impact compared to fossil fuels. These environmental factors have also been found to influence purchasing decisions for biofuels at the point of purchase (Delshad et al., 2010). As with other studies, low acceptance and support for bio-based energy were attributed to the lack of awareness and knowledge about this industry. The studies show that citizens are more likely to support solar, wind, and hydro energy sources relative to bio-mass energy. They conclude that this is in large part due to the relatively recent introduction of bioenergy alternatives. Negative perceptions identified in a study by Petrolia et al. (2010) were found to be associated with increased pollution (Selfa et al., 2011) and threats to food security (Savvanidou et al., 2010).

According to Wilson and Dowlatabadi (2007) social communication processes are essential with the mass media known to influence the decisions relating to the adoption of individual technology. Mass media communications create awareness and knowledge about new products such as bio-based products; however, Mahapatra and Gustavsson (2008) explain that personal channels are more effective in forming and modifying attitudes toward it and are likely to be more influential in decisions to take on new products.

A major five-country study by Sijtsema et al. (2015) explored perceptions regarding the concept of bio-based in general and in relation to specific bio-based products. Results of the focus groups indicated that the definition of bio-based is not clear with the focus group participants generally unfamiliar with concept of "bio-based" and "bio-based products." An inconsistent understanding among consumers of key bio-related terms was also observed.

The study by Sijtsema et al. (2015) found that consumer perceptions of bio-based products depend mainly on the answer to the question "what is in it for me?." The respondents were less convinced on perceived benefits of bio-based production methods. Environmental friendliness benefit relates to personal benefits such as feeling good, and "bio-based" was regarded as an additional value with other more important aspects requiring fulfillment before bio-based benefits came in for consideration. The focus groups demonstrated that when confronted by new products, consumers will buy what they see, need, and regard as beneficial. In this context, communication about bio-based needs to be as context-specific and concrete as possible.

The focus groups also confirmed that a lack of knowledge can create distrust. People were more positive towards bio-based products that use 100% bio-based materials, and distrust was noted for partially bio-based products. These findings underpin the requirement for bio-based products to be supported by coherent product concepts as bio-based claims bring high expectations and consumers do not like inconsistencies in a product's image.

7.2.2.1 Observations from wood pellet consumer studies

The survey by García-Maroto et al. (2015) found that Spanish consumers knew very little about pellets, and that the knowledge they did have about biofuels was directly related to their knowledge about pellets. Friends were the principal information source, followed by family members. While environmental concern had a limited weighting in the decision to adopt these heating systems, the existence of subsidies for their purchase was more important in consumers' ultimate decision.

The study by Mahapatra et al. (2008) of Swedish consumers found that the cost of pellet heating system investment is the most important factor for consumers, along with the price of the biofuel, the specific advantages of the product, and the need for information about the product and its relative advantages.

The study by Rouvinen and Matero (2013) of the stated preferences of Finnish private homeowners for residential heating systems confirmed that the cost of investment in home heating systems is very important, with environmental factors only having a very limited influence in choices.

In an attempt to better understand the psychological determinants involved in the adoption of wood pellet technology for home heating, Sopha and Klöckner (2011) surveyed 737 Norwegian households. Their findings suggest that for Norwegian adopters of wood pellet heating found early adopters more likely to be influenced by environmental concerns but this motivation moves to subsidies and operating costs with late adopters and non-adopters. However, the influence of these personal values and norms only has a weak impact on attitudes, and they conclude that strengthening environmental values and norms is not a promising driver for the diffusion of wood pellet heating.

7.2.2.2 Observations from bioenergy and biofuel consumer studies

In their own survey of transport fuel consumers in the United States, Radics et al. (2016) found that convenience and availability are key choice criteria. Additionally, there were concerns regarding compatibility with cars. The research concluded that the confusion was strongly related to a lack of knowledge among the transport fuel consumers and recommended that there is a need to have a consistent and simpler message delivered to consumers.

The technical makeup of biofuels is also found to influence potential consumers. Delshad et al. (2010), in their study of public attitudes toward political and technological options for biofuels, questioned consumers on their associations with the type of feedstock used for producing biofuels. Their study found high levels of support for energy produced from landfill and wood waste, with trees and corn-based bioethanol receiving lower levels of consumer acceptance.

Using household survey data, Ek and Soderholm (2008) showed the positive influence of social norms on participating in the Swedish green-electricity market. Their survey results suggested that the impact of choosing "green" on the household budget largely influences the choice between "green" and "brown" electricity. They also found that the degree of perceived personal responsibility for the issue and the felt ability to affect the outcome in a positive way impacted on choice. Lillemo and Halvorsen (2013) study of the impact of lifestyle and attitudes on residential heating choices in Norway found that urban lifestyles and concern for comfort has a strong negative impact on demand for firewood.

The lack of widely accepted quality standards for biodiesel is problematic. Van de Velde et al. (2010) investigated the perception of Belgian consumers with respect to the use of biodiesels and identified four consumer segments based on the perceived importance of different fuel characteristics. They found if companies want to convince the performance-oriented and environment-oriented consumers to use biofuels, they must supply information about biodiesel's quality and performance standards. Since biodiesel is produced from varying origins and qualities, it is necessary to install a standard of fuel quality to guarantee an engine performance without any difficulties and ensure better criteria of biodiesel for successful commercialization of biodiesel. Separately a Spanish study by Kallas and Gil (2015) found that car manufacturers were unwilling to recommend its use as it may lead to engine failure.

Interestingly, their survey also found that consumers were not willing to pay for biodiesel, especially when its production may negatively affect food prices. Ma et al. (2014) also report that quality and safety standard are pivotal factors in consumer choice of vehicle fuel. Arguments regularly put to consumers for biodiesel include (1) biodiesel prolongs engine life and reduces the need for maintenance and it is much safer; (2) biodiesel is better than conventional diesel in aspects of sulfur content, flash point, aromatic content, and biodegradability, and (3) biodiesel is nonflammable and cleaner.

Observing across German and British experiences, Bomb et al. (2007) note that equal price among competing fuel types is a necessary condition for the expansion of the biodiesel market.

Although biofuels are not new, usage remains low in certain western countries and public opinion of biofuels is still divided. According to Savvanidou et al. (2010), 80% of respondents who owned vehicles were willing to use biofuels, of which 45% were willing to pay more for them. These findings suggest that there is a need to educate people on the positive impacts of biofuels through campaigns (Savvanidou et al., 2010). Marra (2010) revealed that, in general, consumers were willing to pay for reductions in GHG emissions. Anderson (2012) and Petrolia et al. (2010) found that consumers were willing to pay a premium for ethanol and blended ethanol, respectively.

7.2.3 A FRAMEWORK FOR DEVELOPING THE BIOECONOMY

The European Commission has set a long-term goal to develop a competitive, resource-efficient and low-carbon economy by 2050, with various strategies, communications, directives, and declarations prioritizing bioenergy in the delivery of these ambitions. Scarlat et al. (2015) note that the transition toward a bioeconomy will rely on the advancement in technology of a range of processes, on the achievement of a breakthrough in terms of technical performances and cost effectiveness, and will depend on the availability of sustainable biomass. The emergence and confirmation of key bioeconomy products and markets, however, requires social acceptance of bio-based products and a willingness by consumers to pay for new and often technology-based products. In this context the

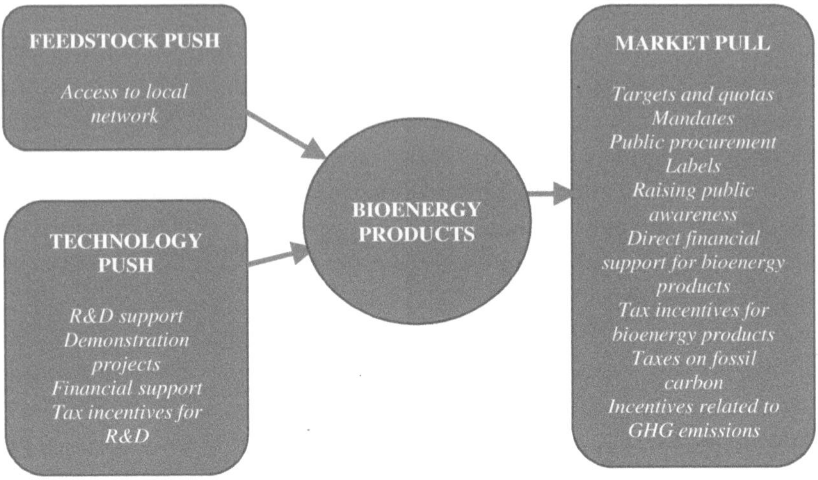

Figure 7.2.1 Framework for developing bioenergy markets.

transition to bioeconomy must be (1) technologically feasible, (2) economically viable, and (3) socially desirable. This transition is reflected across many national strategy documents for developing national bioeconomies and is often translated to three policy priorities: (1) feedstock push, (2) technology push, and (3) market pull (Fig. 7.2.1).

7.2.4 BIOENERGY: POLICY CONTEXT IN EUROPE

Much of the consideration of bioenergy markets in Europe is situated in the wider bioeconomy context. The knowledge-based bioeconomy, also referred to as a bio-based economy, offers the opportunity to make significant contributions to global challenges such as providing enough and healthy food for a growing global population, climate change, and the loss of soil fertility and biodiversity, as well as advancing the transition from an economy mainly based on fossil-based raw materials to an advanced economy based on renewable resources and efficiency in terms of raw materials. As the European Commission (2012a,b) explains "the bioeconomy's cross-cutting nature offers a unique opportunity to comprehensively address inter-connected societal challenges such as food, security, natural resource scarcity, fossil resource dependence and climate change, while achieving sustainable economic growth."

The *Energy 2020—A Strategy for Competitive, Sustainable and Secure Energy* identifies five energy priorities: (1) reduce energy consumption, (2) build internal market and develop infrastructure, (3) extend technological leadership, (4) ensure secure, safe, and affordable energy, and (5) reinforce external dimension. The 2009 *Renewable Energy Directive* targets the translation of renewable energy targets for 2020 into legally binding frameworks with renewable energy accounting for 20% and 10% in final energy consumption and transport sector consumption. The Renewable Energy Directive also includes a requirement for members' states of the European Union to establish and implement National Renewable Energy Actions Plans with roadmaps and targets. Under this directive bioenergy is expected to account for almost 60% of renewable energy by 2020 with longer term goals targeting creation of a competitive low-carbon economy. The 2009 *Fuel Quality Directive* sets a target of a 6% GHG reduction for the fuels used in transport in 2020. Both directives include criteria for sustainable biofuel production and verification procedures. Ambitions have been further expanded in the considerations for 2030 targets, with the European Council agreeing a new framework setting out a European Union target of at least 27% for the share of renewable energy consumed in the EU in 2030 (European Commission, 2014). Action is also sought through the Agenda for Sustainable Development which sees the bioeconomy having potential to contribute to as many as half of the sustainable development goals.

Bioeconomy development priorities—including bioenergy development priorities—are also addressed in the European Commission's research program Horizon 2020 and encapsulates its strategy for developing the bioeconomy in *Innovating for Sustainable Growth: A Bioeconomy in Europe* (2012) which seeks to give support to the development of an innovative, low-carbon, and more resource-efficient economy which is internationally competitive. Among a range of actions, this plan targets strengthening of markets and competitiveness with regards to innovative products.

In a review of policies and trade regimes, Lamers et al. (2011) identify common government policies to support the development of bioenergy sectors to include (1) promoting domestic consumption through consumption mandates such as biofuel content or minimum GHG savings through biofuels, (2) promotion of domestic production through production mandates, investment support, demonstration projects, infrastructure or R&D, feedstock support, or tax incentives, and (3) trade-related measures to shield local production such as import/export tariffs and standards. Alsaleh et al. (2017) observe that there are many determinants that can affect the development of international bioenergy markets. Junginger et al. (2011) identified restrictions to the development of bioenergy trade as including geographic location, import tariffs, sustainability certification systems, and logistics, while in Europe at least factors such as technical standards of development, high fossil-fuel process, and national policies to mitigate the effects of climate change had boosted the market.

In their review of international bioenergy trade, Lamers et al. (2011) note that many trade regimes design to support the development of bioenergy trade are unilaterally designed to support national interests and as a result cause market disruptions and trade inefficiencies at an international level. An imbalance in a significant domestic market can have a negative impact on the bioenergy international market's well-being. White et al. (2013) further observe that large unexpected changes in policy only serve to increase uncertainty and have potential to have significant negative effects on investments.

According to research by Wiesenthalm et al. (2009), partial or total tax exemptions for biofuels have been very effective in promoting the use of biofuels in the European Union, with fuel tax level compensated for the additional production costs of biofuels.

7.2.5 NATIONAL BIOECONOMY STRATEGIES

A 2015 study by the German Bioeconomy Council (German Bioeconomy Council, 2015) identified almost 50 national strategies pertaining to certain facets of the bioeconomy as well as a number of transnational and regional strategies. In 2011 over 60 countries had a national energy target or support policy with climate change mitigation, energy security, and protection of national industries in energy and agriculture providing supporting objectives (Kraxner et al., 2013).

A review of national strategies is complicated as they use their own definitions with some focused on narrow interpretations that equate the bioeconomy to biotechnology (e.g., OECD) and others taking account of the wider societal transformation offered by the bioeconomy (e.g., Germany). Across the different national strategies, a range of motivations for development of the bioeconomy are put forward. These include economic motivations including enhancing competitiveness of domestic industries, create new jobs, and contributing to the revitalizing of rural areas (Staffas et al., 2013; Vandermeulen et al., 2012).

De Besi and McCormick (2015) identify that national bioeconomy strategies can typically be separated into five development areas: (1) research and innovation, (2) biomass and land use, (3) economy and finance, (4) governance, and (5) social change. Social change represents the need identified by strategy authors for a "transformation in the mind-sets of society, industries and governments." In this context they observe a theme which explains "a transition to an economy predominately based on biomass will require cooperation of all sectors of society and a change toward sustainable consumption and production patterns, both on the demand side and on the supply side of the economy."

In the context of initiating the necessary social change to support the demand side of the bioeconomy, their study finds that a number of strategies highlight increasing dialogue with the public as a crucial requirement for addressing the "issue of overconsumption in society." The German strategy measures for addressing consumer behavior by providing information on sustainable consumption and food waste are an example of the required dialogue. Swedish and Finnish measures to communicate the benefits of the bioeconomy and bio-based products in society as a means of shifting consumption away from fossil-based products are also highlighted. The Swedish strategy extends its consideration of the societal challenge in a whole system perspective where a reduced societal consumption level requires a life cycle perspective that is needed for the production and consumption of bio-based products.

A review of national bioeconomy strategies by Priefer et al. (2017) found that relatively a few had any significant consideration of sustainable consumption issues (e.g., Germany and Sweden). Others have argued in response to the limited consideration of consumer behavior issues in national and transnational strategies that a societal transition towards bioeconomy will only provide solutions to the broader global challenges if it also includes changes in consumer behavior (Zwier et al., 2015; Birch et al., 2010).

From the review of German and Finnish national strategies (Davies et al., 2016), the BioSTEP project, funded by Horizon 2020 research and innovation program, identified a number of ways to encourage more active citizen participation in the bioeconomy: (1) a stronger focus on specific issues which directly affect citizens, (2) long-term communication campaigns (including the use of social media) to engage with citizens about the range and complexity of the bioeconomy, (3) funding for projects which encourage public awareness, (4) more discussion of sustainable development in the education system, (5) emphasis on the potential environmental benefits of the bioeconomy, and (6) genuine dialogue with citizens, aimed not only at informing but also at listening to and engaging with people's concerns.

While there have been several reviews of national strategies, including by this author, it is difficult to make an evaluation on the effectiveness of specific interventions, including public awareness interventions, because most of these interventions are still in the process of being rolled out or have only been relatively recently rolled out and there are few studies measuring their impact. For example, the *Austria Bioeconomy Background Paper* and the *German Policy Strategy on the Bioeconomy* were launched in 2013, the *UK Bioenergy Strategy* and the *Canada Blueprint beyond Moose and Mountains* were published in 2011, and the *USA National Bioeconomy Blueprint* was produced in 2012, as was *A Bioeconomy for Europe* by the European Commission was announced in 2011. At their time of publication, these national strategy documents represented the first national bioeconomy strategic plans for each country.

7.2.6 EUROPEAN COMMISSION MARKET PULL RECOMMENDATIONS

During the preparation of its bioeconomy strategy, the European Commission conducted a public consultation on the bioeconomy in Europe in 2011, which received over 200 submissions from organizations and individuals across most Members States of the EU. The consultation drew a number of findings. The majority of respondents offered an optimistic outlook on the bioeconomy with more than 60% thinking that potential benefits can be achieved by 2020 or 2030. The majority of respondents also believed that there are a number of risks associated with the bioeconomy, including potential overexploitation of natural resources and impacts on food security. A lack of public information and understanding of the bioeconomy were also considered important issues, especially regarding benefits and risks as well as ethical issues and sustainable patterns of consumption and production. In this perspective, more than 70% of the respondents called for actions related to communication and dissemination of information on the bioeconomy. Other key themes to emerge in the public consultation include fostering effective governance, promoting collaboration across disciplines and sectors, investing in interdisciplinary education and training, and ensuring robust linkages between research, innovation, and implementation.

In its report "Where next for the European bioeconomy?" (European Commission, 2014), the European Bioeconomy Panel and the Standing Committee on Agricultural Research Strategic Working Group set out a number of recommendations to support demand-side measures and increase awareness and understanding. These include:

- *Regulatory and public procurement measures*—Establishing an EU-wide public procurement program to boost awareness and uptake of bioeconomy products. It was recommended for consideration include the adoption of specific and binding targets for best performing, and most sustainable products could be considered in addition to enabling member states to grant tax incentives for certain sustainable EU bioeconomy products. The group proposed that environmental legislation could be used to promote the use of products from the bioeconomy and mandate the use of less environmentally impactful products. Provisions for the collection and recycling of biomass, residues, wastes, and bio-based products could also be incorporated into waste legislation, and the gradual substitution of more sustainable products and processes could be encouraged through the reform of existing legislation.
- *Labeling*—Clear European Standards on sustainability and bio-based content could be used to establish labels that might increase consumer and business ability to identify bio-based products. This includes business-to-business (B2B) and business-to-consumer communications facilitated through the development of sustainable product "ecolabels" and/or a "bio-based" label, linked to sustainability criteria. Certification and labeling could then be incorporated into the implementation of a coherent communications strategy to consumers.
- *Increasing awareness and understanding*—The group confirmed that there is a clear need to develop and implement a coherent communication strategy to raise consumer awareness around the bioeconomy and the opportunities for and barriers to its development. It recommends that this should be done in the context of the grand challenges facing future generations such as climate change, resource efficiency, energy, and food security. Among its recommendations are the dissemination of more case studies and establishing a European Bioeconomy Week.

The focus on public awareness and acceptance of bio-based products mirrors similar recommended interventions featuring in a number of national bioeconomy strategy considerations (e.g., the United Kingdom, Germany, Sweden, and Finland).

Identifying public perception as a major barrier to the development of the UK bioeconomy, Burns et al. (2016) make several recommendations to gain public acceptance of bio-based products: (1) clearer benefits relayed to customers, to put risk and benefit in perspective, (2) stronger regulation to improve public trust and ethical application of new technology, and (3) greater transparency and more genuine public engagement.

Communications relating to the bioeconomy and the bio-based industries has been identified as a key area for developing the bioeconomy by the European Commission Expert Group for Bio-based Products. Communication actions were included among the Expert Group's Lead Market Initiative priority recommendations. The Lead Market Initiative noted that communications for the bioeconomy refer to informing different stakeholders of bio-based products, their characteristics, their benefits, and their contribution to sustainable production and consumption. The Lead Market Initiative notes that communication should be based on scientific facts to help to increase transparency both within market and towards the wider public. They observe that the term "bio-based products" comprises a wide variety of innovative products; most of these products are not easily identified and recognized as such by users and consumers with the specific features of the bio-based products mostly invisible to the purchaser.

The Lead Market Initiative identifies this lack of awareness and knowledge as a major obstacle to increasing market uptake of bio-based products and sets out the following recommendations:

1. With ethical and environmental considerations playing an increasing role in purchasing decisions by consumers, there is a requirement to create awareness and knowledge of the characteristics of bio-based products among consumers. This requires addressing the lack of knowledge of bio-based products with meaningful labels and information campaigns.
2. It is important to communicate the benefits of bio-based products across the value-chain participants including producers, distributors, users and consumers, public authorities, and non-governmental organizations.
3. Develop trusted B2B guidelines and understandable labels for substantiation of chains in business-to-consumers communications.

7.2.7 B2B MARKETING OF BIO-BASED PRODUCTS

The Open-Bio Project (Peuckert and Quitzow, 2015a,b) carried out a Delphi survey involving 320 B2B buyers from 17 EU countries on the market acceptance of bio-based products. The most important market drivers observed among B2B buyers include (1) an environmentally friendly positive public image, (2) independence from fossil sources, (3) savings in CO_2 emissions, and (4) compliance with environmental regulation.

These market drivers strongly related to corporate social responsibility targets of businesses and a perception that consumers were shifting to supporting sustainable consumption. In this context, the B2B market uptake of bio-based products is driven by their positive public image as well as the potential environmental benefits. An example of such a corporate social responsibility initiative is the UK retailer Sainsbury's seeking to become a zero-waste company by processing its food and meat waste into bioenergy.

The Open-Bio Project (Peuckert and Quitzow, 2015a,b) found that the most important market barriers observed among B2B buyers include (1) higher cost of production, (2) uncertainty about future regulation, (3) uncertainty regarding volatility of feedstock prices, and (4) an unsupportive regulatory environment.

Knowledge on the part of the consumer about the existence of a product is fundamental. Blackwell et al. (2002) explain that consumers need to be made aware of new or innovative products and that one is unlikely to become a consumer of a product if there is insufficient information on how to use it. Thøgersen and Schrader (2012) observe that this is particularly the case for encouraging consumer to make pro-environmental actions in their consumer decision-making, with greater knowledge for initiation, diffusion, and enhancement of such actions.

7.2.8 PUBLIC PROCUREMENT AS A MARKET PULL MEASURE FOR BIOENERGY

With expenditure on the procurement of goods, works, and services by public authorities in the EU represents approximately 19% of EU GDP which equates to €2.1 trillion, it is not surprising that from its earliest policy submissions on developing a sustainable bioeconomy, the

European Commission has promoted the large potential of public procurement as an instrument for creating demand for innovative and environmentally friendly bio-based products (European Commission, 2008). The European Commission has also consistently encouraged Member States to give preference to bio-based products in tender specifications.

This potential has also been included within national bioeconomy strategies and in contributions to national strategy development. For example, Burns et al. (2016) recommend that the UK Government is underutilizing its public procurement power to drive market demand for bio-based products and recommend that public procurement targets for bio-based products would drive markets of bio-based products directly, and indirectly improve awareness of bio-based products.

The rationale for using public procurement to support the development of bioeconomy markets is similar to the reasons offered for using public procurement to support small to medium enterprise (SME) business innovation policies—that is to support the creation of a market with critical mass. This is not surprising given the common challenges on bio-based products and new technology-based products. These include (1) a lack of demand, (2) a lack of consumer awareness, (3) high switching costs, (4) the absence of standards and buyer's lack of knowledge, and (5) high initial costs (lower competitiveness).

The rationale for using public procurement to accelerate the development of bioeconomy markets is that it allows the public sector to act as launching customer, early adopter, or first buyer. The public sector also acts as market development facilitator by establishing a buyers' group for the market with critical mass that triggers industry to scale up its production chain to bring products on the market with desired quality/price ratio within a specific time.

In their recommendations, the Lead Market Initiative Advisory Group for Bio-based Products (European Commission, 2017) established by the European Commission's Directorate General for Industry proposed that a requirement or a recommendation to give preference can be laid down in a national action plan adopted by the government. Within this, buyer values can include criteria neglected by private consumers such as criteria related to climate or the environment. An additional advantage is that while public procurement is fragmented across thousands of buyers, public procurement processes tend to be codified from centralized principles. As well as creating markets through their own purchasing, public procurement also serves as a tool for raising awareness of bio-based products. This is demonstrated through use of biofuels in public transport and development of demonstration bio-initiatives.

The city of Gothenburg encourages the use of alternative fuels by (1) making it a requirement in public procurement, (2) allowing free parking for cars using biogas—the so-called "eco cars," and (3) by deploying a municipal car fleet made up of 90% "eco cars" (Hulthen, 2012). By 2020 the entire Gothenburg bus fleet will use this alternative fuel. Gothenburg's sewage water-treatment system produces 65 GWh of biogas which is almost exclusively sold for use in cars and buses in the Gothenburg area as a blend of 50% biogas/natural gas or as 100% biogas.

The European Commission report *Green Public Procurement—A Collection of Good Practices* on the 2009 Barcelona City Council's tender procedure for street cleaning and rubbish collection introduced a number of requirements to enhance the quality and sustainability of the service (European Commission, 2012a,b). Under the tender, it was an explicit requirement that service providers could only use vehicles which operate on noncontaminating combustible fuels and/or renewables, such as biofuels (bioethanol B-85, biodiesel, or biogas).

The Open-Bio Project (Peuckert and Quitzow, 2015a,b) survey of public procurement buyers were asked whether specifications on bio-based content could be utilized as a basis for public procurement in their own organization. One-third of the respondents said it was possible; however, the majority of responding public procurement buyers indicated that they were not sure or did not answer question. A second round of the survey sought an explanation for difficulties using public procurement, and the most common reasons included (1) the problem of verifying related claims, (2) the lack of information about biomass standards available, and (3) the lack of relevance as a product attribute.

The respondents also noted that other environmental benefits considered more important as criteria in green public procurement than aspects related to raw materials/bio-based content. A large majority of respondents consider a label for bio-based product as an important instrument for promoting the uptake of bio-based products. They also found that the incorporation of criteria for bio-based content in such labels can play an important role for their uptake in public procurement. A clear finding of the survey is that bioeconomy benefits require additional consideration in existing green public procurement processes.

The most important reasons identified by the public procurement buyers for the limited use of specifications on bio-based content in public procurement were that available bio-based products are frequently too expensive and that the specifications on bio-based content are difficult to verify. They also noted that information about available bio-based products as alternative for fossil-based products is not easy to find and that bio-based content is not considered a relevant product attribute.

On asking what were the most effective measures for promoting bio-based products in public procurement, the respondents identified political decisions to promote bio-based products via public procurement as most important. They also recommended the requirement for practical guidance for incorporating specifications on bio-based content in public procurement, a checklist to facilitate a systematic comparison of bio-based products and conventional products, and a database of bio-based products containing key product information.

When it comes to assessing the importance of environmental aspects for the practice of green public procurement, the priorities included (1) energy efficiency, (2) savings in CO_2, (3) recyclability, (4) reduction of environmental pollutants (other than CO_2), (5) use of recycled material/waste, and (6) bio-based content. These factors were considered likely to retain their importance in the future.

As previously mentioned, a lack of information was observed as an important inhibitor of public procurement of bio-based alternatives. Sought information in order of importance to facilitate purchasing bio-based products includes information pertaining to (1) toxicity, (2) environmental life cycle impacts, (3) percentage of bio-based content, (4) life cycle costs, and (5) CO_2 emissions.

7.2.9 CONCLUSIONS

There are numerous challenges for researchers investigating bioenergy market opportunities. In the first instance, official statistics tend to only report on traditional sectors with only limited distinction between synthetic and bio-based production. This results in many of the market-related indicators for bioenergy being estimated on the basis of a combination of multiple sources. In terms of estimating market sizes commercially available, bioeconomy market research reports yield a variety market estimates. The picture is further complicated by external influences—particularly political factors and movements in commodity prices—that have a strong impact on the status, feasibility, and growth momentum of bioeconomy markets. Further difficulties emerge from the fact that many bio-related markets have not yet been established and remain dependent on innovation outcomes to establish their feasibility. In this context, the development of markets will continue to depend on the development of new supporting technologies and regulatory reform.

The requirement for education of the general public and public awareness is supported. As observed in a number of the studies reviewed, public debate has led to the notion that biofuels might lead to engine problems and increased food shortage. These observations demonstrate that there is a need for public information regarding the benefits of biofuels, both for the climate and the economy of the EU.

At issue is that bio-based products are often based on new technologies and as an outcome suffer the same impediments in market entry as new generation technology products in most domains. Therefore, similar interventions are relevant. Given the value and variety of goods, works, and services procured by public authorities, the public procurement market provides significant opportunities for developing bioeconomy markets and be particularly beneficial for indigenous SMEs working in these markets. It is useful to observe that, as in the case of next generation technology products, the consumer market for bioenergy options can be segmented into early adopters and followers.

The research also shows that consumers are influenced by a variety of factors in their decision to purchase bio-based products. Social communication, from friends and family, are a key influencer with personal channels having a significant impact on consumer choices. Variations are also observed in the willingness of different consumers to commit to bio-based consumption, with a lack of knowledge leading to mistrust in bio-claims. Various studies of the attitudes of general public indicate a moderate and sometimes low to ambivalent support towards renewable energy including biomass energy. The "what is in it for me?" question is invariably more important than the potential "what is in it for the planet?" question.

REFERENCES

Alsaleh, A., Abdul-Rahim, A.S., Mohd-Shahwahid, H.O., 2017. An empirical and forecasting analysis of the bioenergy market in the EU28 region: evidence from a panel data simultaneous equation model. Renew. Sustain. Energy Rev. 80, 1123–1137.

Anderson, S., 2012. The demand for ethanol as a gasoline substitute. J. Environ. Econ. Manage. 63 (2), 151–168.

Binder, A.R., Cacciatore, M.A., Scheufele, D.A., et al., 2012. Measuring risk/benefit perceptions of emerging technologies and their potential impact on communication of public opinion toward science. Public Understanding Sci. 21, 830–847.

Birch, K., Levidow, L., Papaioannou, T., 2010. Sustainable capital? The neoliberalization of nature and knowledge in the European "knowledge-based bio-economy". Sustainability 2, 2898–2918.

Blackwell, P., Miniard, E., Engel, J., 2002. Comportamiento del Consumidor, 9th ed Thomson, Mexico.

Bomb, C., McCormick, K., Deurwaarder, E., Kåberger, T., 2007. Biofuels for transport in Europe: lessons from Germany and the UK. Energy Policy 35 (4), 2256–2267.

Burns, C., Higson, A., Hodgson, E., 2016. Five recommendations to kick-start bioeconomy innovation in the UK. Biofuels Bioprod. Biorefin. 10, 12–16.

Communication from the Commission to the European Parliament, the Council, the European Economic and Social Committee and the Committee of the Regions, 2014. A policy framework for climate and energy in the period from 2020 to 2030, COM/2014/015 final.

Davies, S., Griestop, L., Vironen, H., Bachtler, J., Dozhdeva, V., Michie, R., 2016. Case Studies of National Bioeconomy Strategies in Finland and Germany, BioSTEP.

de Besi, M., McCormick, K., 2015. Towards a bioeconomy in Europe: national, regional and industrial strategies. Sustainability 7 (8), 10461–10478.

Delshad, A.B., Raymond, L., Sawicki, V., Wegener, D.T., 2010. Public attitudes toward political and technological options for biofuels. Energy Policy 38, 3414–3425.

Ek, K., Soderholm, P., 2008. Norms and economic motivation in the Swedish green electricity market. Ecol. Econ. 68, 169–182.

European Bioeconomy Panel and the Standing Committee on Agricultural Research Strategic Working Group (SCAR), 2014. Where next for the European bioeconomy? European Commission.

European Commission, 2012a. Green Public Procurement: A Collection of Good Practices. Publication Office of the European Union, Luxembourg.

European Commission, 2012b. Innovating for sustainable growth: A bioeconomy for Europe. Communication from the Commission to the European Parliament, the Council, the European Economic and Social Committee and the Committee of the Regions, Brussels.

European Commission (2017) Final Report of the Commission Expert Group on Bio-based Products.

Frewer, L., Howard, C., Hedderley, D., Shepherd, R., 1997. Consumer attitudes towards different food-processing technologies used in cheese production: the influence of consumer benefit. Food Qual. Prefer. 8 (4), 271–280.

García-Maroto, I., García-Maraver, A., Muñoz-Leiva, F., Zamorano, M., 2015. Consumer knowledge, information sources used and predisposition towards the adoption of wood pellets in domestic heating systems. Renew. Sustain. Energy Rev. 43, 207–215.

German Bioeconomy Council, 2015. Bioeconomy Policy (Part II). Synopsis of National Strategies Around the World. Biookonomierat, Berlin, Germany.

Golembiewski, B., Sick, N., Bröring, S., 2015. The emerging research landscape on bioeconomy: what has been done so far and what is essential from a technology and innovation management perspective? Innov. Food Sci. Emerg. Technol. 29, 308–317.

Hartmann, P., Apaolaza, V., 2012. Consumer attitude and purchase intention toward green energy brands: the roles of psychological benefits and environmental concern. J. Bus. Res. 65 (9), 1254–1263.

Hulthen, A., 2012. From Waste to Wheels—Driving on Biogas in Gothenburg. Covenant of Mayors Office, Gothenburg, Sweden.

Junginger, M., Dam, J., Zarrilli, S., Mohamed, F., Marchal, D., Faaij, A., 2011. Opportunities and barriers for international bioenergy trade. Energy Policy 39, 2028–2042.

Kallas, Z., Gil, J.M., 2015. Do the Spanish want biodiesel? A case study in the Catalan transport sector. Renew. Energy 83, 398–406.

Kraxner, F., Nordström, E.-M., Obersteiner, M., Havlík, P., Gusti, M., Mosnier, A., et al., 2013. Global bioenergy scenarios—future forest development, land-use implications and trade-offs. Biomass Bioenergy 57, 86–96.

Lamers, P., Hamelinck, C., Junginer, M., Faaij, A., 2011. International bioenergy trade: a review of past developments in the liquid biofuel market. Renew. Sustain. Energy Rev. 15, 2655–2676.

Lillemo, S.C., Halvorsen, B., 2013. The impact of lifestyle and attitudes on residential firewood demand in Norway. Biomass Bioenergy 57, 13–21.

Ma, Z., Zhang, C., Chen, C., 2014. Analyzing the factors that influence Chinese consumers adoption of the biodiesel: the private vehicles owner s investigating in Beijing. Renew. Sustain. Energy Rev. 37, 199–206.

Mahapatra, K., Gustavsson, L., 2008. An adopter-centric approach to analyze the diffusion patterns of innovative residential heating systems in Sweden. Energy Policy 36, 577–590.

Marra, A., 2010. Classification of ethanol consumers and willingness to pay for reductions in greenhouse gas emissions through purchases of E85. Master's Thesis. University of Tennessee.

McCormick, K., Kautto, N., 2013. The bioeconomy in Europe: an overview. Sustainability 5, 2589–2608.

Petrolia, D.R., Bhattacharjee, S., Hudson, D., Herndon, C.W., 2010. Do Americans want ethanol? A comparative contingent-valuation study of willingness to pay for E-10 and E-85. Energy Econ 32 (1), 121–128.

Peuckert, J., Quitzow, R., 2015a. Acceptance of Bio-Based Products in the Business-to-Business Market. Open Bio.

Peuckert, J., Quitzow, R., 2015b. Acceptance of Bio-Based Products by Public Procurement Officials. Open Bio.

Priefer, P., Jörissen, J., Frör, O., 2017. Pathways to shape the bioeconomy. Resources 6, 10.

Qu, M., Ahponen, P., Tahvanainen, L., et al., 2011. Chinese university students' knowledge and attitudes regarding forest bio-energy. Renew. Sustain. Energy Rev. 15, 3649–3657.

Radics, R., Dasmohapatra, S., Kelley, S.S., 2016. Public perception of bioenergy in North Carolina and Tennessee. Energy. Sustain. Soc. 6, 17.

Rouvinen, S., Matero, J., 2013. Stated preferences of Finnish private homeowners for residential heating systems: a discrete choice experiment. Biomass Bioenergy 57, 22–32.

Savvanidou, E., Zervas, E., Tsagarakis, K., 2010. Public acceptance of biofuels. Energy Policy 38, 3482–3488.

Scarlat, N., Jean-François Dallemand, J.F., Monforti-Ferrario, F., 2015. The role of biomass and bioenergy in a future bioeconomy: policies and facts. Environmental Development 15, 3–34.

Selfa, T., Kulcsar, L., Bain, C., et al., 2011. Biofuels bonanza? Exploring community perceptions of the promises and perils of biofuels production. Biomass Bioenergy 35, 1379–1389.

Sijtsema, S., Onwezen, M., Reinders, M., Dagevos, H., Partanen, A., Meeusen, M., 2015. Consumer perception of bio-based products—an exploratory study in 5 European countries. NJAS Wagen. J. Life Sci. 77, 61–69.

Sopha, B., Klöckner, C., 2011. Psychological factors in the diffusion of sustainable technology: a study of Norwegian households' adoption of wood pellet heating. Renew. Sustain. Energy Rev. 15, 2756–2765.

Staffas, L., Gustavsson, M., McCormick, K., 2013. Strategies and policies for the bioeconomy and bio-based economy: an analysis of official national approaches. Sustainability 5, 2751–2769.

Thøgersen, J., Schrader, U., 2012. From knowledge to action—new paths towards sustainable consumption. J. Consum. Policy 35 (1), 1–5.

Van de Velde, L., Verbeke, W., Popp, M., Van Huylenbroeck, G., 2010. Trust and perception related to information about biofuels in Belgium. Public Understanding Sci. 20 (5), 595–608.

Vandermeulen, V., Van der Stee, M., Stevens, C.V., Van Huylenbroeck, G., 2012. Industry expectations regarding the transition toward a biobased economy. Biofuels Bioprod. Biorefin. 6, 453–464.

White, W., Lunnan, A., Nybakk, E., Kulisic, B., 2013. The role of governments in renewable energy: the importance of policy consistency. Biomass Bioenergy 57, 97–105.

Wiesenthalm, T., Leduca, G., Christidisa, P., Schadea, B., Pelkman, L., Govaert, L., et al., 2009. Biofuel support policies in Europe: lessons learnt for the long way ahead. Renew. Sustain. Energy Rev. 13 (4), 789–800.

Wilson, C., Dowlatabadi, H., 2007. Models of decision making and residential energy use. Annu. Rev. Environ. Res. 32, 169–203.

Zwier, J., Blok, V., Lemmens, P., Geerts, R., 2015. The ideal of a zero-waste humanity: philosophical reflections on the demand for a bio-based economy. J. Agric. Environ. Ethics 28, 353–374.

FURTHER READING

European Commission, 2010. Energy 2020—A Strategy for Competitive, Sustainable and Secure energy, SEC(2010)1346.

European Commission, 2011. Bio-based Economy for Europe: State of Play and Future Potential—Part 1; DG Research and Innovation. European Commission, Luxembourg, Belgium.

Radics, I.R., Dasmohapatra, S., Kelley, S.S., 2015. Systematic review of bioenergy perception studies. BioRes. 10, 8770–8794.

Sopha, B., Klöcknerc, C., Skjevrakb, G., Hertwicha, E., 2010. Norwegian households' perception of wood pellet stove compared to air-to-air heat pump and electric heating. Energy Policy 38 (7), 3744–3754.

Zhang, Y., Yu, Y., Li, T., Zou, B., 2011. Analyzing Chinese consumers' perception for biofuels implementation: the private vehicles owner's investigating in Nanjing. Renew. Sustain. Energy Rev. 15, 2299–2309.

7.3

Bioenergy Governance

Felipe García Cardona and Mateo Prada Quintero
Dirección General, Colombia Bio. Bogotá, Colombia

7.3.1 CLIMATE CHANGE AND BIOENERGY GOVERNANCE

In the last decades, the international community has developed a global climate change agenda. This ambitious project has opened a set of opportunities but also immense challenges for bioenergy governance, understanding by "governance" the myriad process through which actors set and enforce collective rules needed to achieve desired outcomes (Bazilian et al., 2014). On the one side, the adoption of the Paris Agreement and *The 2030 Agenda for Sustainable Development* has sent the strong political message to governments, business, and investors that it is imperative to transform the global energy sector—which is responsible for more than two-thirds of global greenhouse gas (GHG) emissions (The Royal Society of Chemistry, 2015)—to accomplish the goal of keeping global temperature below 2°C (International Energy Agency (IEA), 2016). As a consequence, the international community has started to close the gap between global climate change and energy governance, two closely link sectors, but historically separated by limited policy interaction (Heubaum and Biermann, 2015). This has opened a space in intergovernmental organizations (IGOs), such as the International Energy Agency (IEA), the International Renewable Energy Agency (IRENA), and the Organization for Economic Co-operation and Development (OECD), to discuss both the benefits and risk of bioenergy—generally seen by governments and business as a mean to move towards a low-carbon economy. On the other hand, as the global demand for bioenergy raises, and countries include, in their intended Nationally Determined Contributions (iNDCs) of the Paris Agreement, bioenergy as one of their main national strategies to mitigate climate change, it becomes imperative to enhance a multilevel governance framework to address the negative externalities of bioenergy and ensure the sustainability of bioenergy production. However, establishing an effective governance structure to enforce collective rules in the bioenergy sector represents a challenge to the international community, the nation states, and the enterprises due to the complexity of the energy market, the absence of a global government, and the lack of stakeholder coordination.

7.3.2 DEVELOPING A CLIMATE CHANGE AGENDA

The adoption of the United Nations Framework Convention on Climate Change (UNFCCC) in 1992 has set a precedent in the fight against climate change by establishing as a goal to "stabilize GHG concentrations at a level that would prevent dangerous anthropogenic interference with the climate system" (The United Nations, 1992). Later on, with the adoption of the Kyoto Protocol in 1998, an international agreement linked to the UNFCCC, the international community recognized the responsibility of developing countries on "the current high levels of GHG emissions in the atmosphere as a result of more than 150 years of industrial activity," and established internationally binding emission reduction targets (The United Nations, 1998).

Recently, the adoption of the Paris Agreement has put climate change again on the global agenda. This agreement, which builds upon the UNFCCC framework, brings together all nations for the first time to a common cause to undertake ambitious efforts to globally respond "to the threat of climate change by keeping global temperature rise this century well below 2 degrees Celsius above pre-industrial levels" (The United Nations, 2015a,b). As of 2017, 170 parties have ratified the Paris Agreement, and by doing so, they commit to put forward their best efforts through "NDCs," to reduce GHG emissions and put in place appropriate financial flow, new technological frameworks, and an enhanced capacity building framework to deal with climate change (The United Nations, 2015a,b). With the Paris Agreement, the international climate policy took a new approach, by creating a framework for voluntary pledges, which allows countries to establish their own levels of climate change mitigation, which will be compared and review internationally (Falkner, 2016). As a result, the debate around climate change has changed from a universalistic "bottom-down" approach to one centered on individual action subject to a certain degree of global coordination (Molina, 2016), commonly known as a "bottom-up" or a "building block approach" (Falkner et al., 2010).

Parallel to the UNFCCC and the agreements within its framework, world leaders have recently adopted *The 2030 Agenda for Sustainable*

Development. Over the next 15 years, countries are expected to mobilize efforts to "end all forms of poverty, fight inequalities and tackle climate change, while ensuring that no one is left behind" (The United Nations, 2016a,b). The 2030 Agenda sets 17 sustainable development goals (SDGs) and 169 targets in the "three dimensions of sustainable development: the economic, social, and environmental" (The United Nations, 2016a,b). Although the SDGs are not formally interrelated with the UNFCCC, they are inextricably linked (Leong, 2015). Understanding this relationship is key to have a comprehensive perspective of the United Nation's role in developing a climate change governance structure.

On the one hand, *Goal 13: Take urgent action to combat climate change and its impacts* acknowledges that "the United Nations Framework Convention on Climate Change is the primary international, intergovernmental forum for negotiating the global response to climate change" (The United Nations, 2016a,b), and that it is necessary to "implement the commitment undertaken by developed-country Parties to the UNFCCC to a goal of mobilizing jointly $100 billion annually by 2020 from all sources to address the needs of developing countries in the context of meaningful mitigation actions and transparency on implementation and fully operationalize the Green Climate Fund through its capitalization as soon as possible" (The United Nations, 2016a,b). On the other hand, as *Alvin Leong* notices, Article 4, paragraph 1 (c) of the UNFCCC requires all country parties to cooperate to reduce gas emissions in the energy, transport, industry, agriculture, and forestry sectors, while SGDs 7 (energy), 11 (cities), 9 (industrialization), 2 (agriculture), and 15 (forest) address those exact issues (Leong, 2015). Moreover, despite the fact of being two separate processes, both the UNFCCC, with the agreements within its framework, and the SDGs were the result of an international effort made in the framework of the United Nations.

This international organization has served as the main intergovernmental platform for countries to discuss and take action to tackle climate change. By providing a "forum for its members to express their views in the General Assembly, the Security Council, the Economic and Social Council, another bodies and committees" (The United Nations, 2015a, b), the United Nations has been able to become a mechanism for governments to discuss about climate change and find areas of agreement to solve problems together, as the UNFCCC and the agreements within its framework intend to do so. However, to tackle climate change it is necessary to align the outcomes sought by the energy sector, responsible for more than two-thirds of global GHG emissions and the ones set by the Global Climate Change Agenda. This process has been gradually carried out by the IEA and the IRENA in the last few years.

7.3.3 BIOENERGY OPPORTUNITIES: SETTING THE AGENDA

Despite the fact of being intrinsically linked, global energy consumption and climate change have had historically limited policy interactions (Heubaum and Biermann, 2015). Nevertheless, in the last years, while the United Nations has acknowledged the importance of the energy sector in the fight against climate change, the IEA has committed to support the transition towards a low-carbon economy. This, in addition to the recent creation of the IRENA, has had as a result that bioenergy, often seen as a mean to support this transition, gained visibility in these IGOs, leading to the progressive establishment of an international agenda on bioenergy and resulting in the strengthening of bioenergy governance.

As Heubaum et al. points out, the 1992 UNFCCC did not define the climate change problem as one closely related with energy use, and the 1998 Kyoto Protocol "mentions energy only 6 times (of which twice in the Annex), [and] there is no single mention of fossil fuels or coal, and oil and gas are only mentioned once in Annex A" (Heubaum and Biermann, 2015). Moreover, it is not until 2007 that the IEA, the world's leading IGO on energy issues actively started to work jointly with the UNFCCC to prepare the 13th meeting of the Conference of the Parties (COP13) at Bali and to produce periodic reports on climate change and energy consumption since then. Due to its original role, i.e., enable OECD countries to coordinate effective measures to deal with oil supply emergencies and price volatility, the IEA was, since it was established in 1974, essentially focused on tracking the development of oil markets (Florini and Sovacool, 2009).

However, nowadays the IEA actively supports the UNFCCC in promoting international collective action on climate change, attends the COP to provide technical support on energy policy, contributes to the UNFCCC negotiation process—including the Ad Hoc Working Group on the Paris Agreement, and provides technical collaboration to the UNFCCC Secretariat, including reviews of developed country annual GHG inventories (International Energy Agency (IEA), 2012). Even though the IEA has made an important progress in articulating the energy governance with the climate change governance during the last decade, the fact of being legally limited to OECD countries and excluding large economies such as China and India limits its capacity to align the energy sector objectives with the ones of climate change governance. This circumstance, added to the incomformity of some German, Danish, and Spanish interest groups and NGOs, that considered the IEA as a fossil fuels and nuclear lobby, let to the creation of the IRENA (Graaf, 2013), the "first international organization to focus exclusively on renewable energies, addressing the needs of both the industrialized and developing countries" (IRENA, 2009).

In a conference held in Bonn, Germany, in January 2009, 75 countries signed the agency's statue that created IRENA, and established as the main goal of the institution to "promote the widespread and increased adoption and the sustainable use of all forms of renewable energy," such as bioenergy, geothermal energy, hydropower, ocean energy, solar energy, and wind power (Graaf, 2013). By focusing on the transformation of the energy sector, IRENA acknowledges that fossil fuels have played an important role in climate change and intends to lead a radical departure from the current energy path. To accomplish this, IRENA has created new links between the energy sector and the climate change governance and has enhanced, in an indirect way, the commitment of the IEA with the energy transition towards renewables (Graaf, 2013).

Since its creation in 2009, IRENA has organized *Side Events* in the framework of the COP annual meeting of the UNFCCC. This *Side Events* have been devoted to the promotion of renewables energies (COP15—2009), the Nationally Appropriate Mitigation Actions (COP 18—2012), the role of renewable energy in addressing climate change (COP 20—2014), and the implementation of renewable energy targets in African Nationally Determined Contributions (COP 23—2017) (IRENA, 2010). Moreover, IRENA has been actively participating in the Technical Expert Meetings within the Ad Hoc Working Group on the Durban Platform for Enhanced Action (ADP), formed under the UNFCCC, drawing attention to the opportunity presented by renewable energies at Bonn on 2014, and sharing resources with delegates to help them with their national action plans, in the framework of the Paris Agreement negotiations on 2015. In addition to this new links between the renewable energy sector and the climate change governance, the creation of IRENA has indirectly enhanced the relationship of the IEA with the UNFCCC. As Thijs Van de Graaf suggests, the new role of IRENA in the energy governance structure may have produced a shock to the IEA, at the point of "upgrad[ing] its renewable energy "unit" to a "division", staffed by 9 full-time analysts," and boosting its relationship with the UNFCCC since then (Graaf, 2013).

The efforts made by the IEA to strengthen its relationship with the UNFCCC, along with the recent creation of the IRENA, and the recognition of the importance of the energy sector in the fight against climate change by the UN, have allowed the gradual setting of an international agenda on renewable energies. Within these, bioenergy has received particular attention on account of being the largest source of renewable energy, providing electricity, heat, transport, and fuels for millions of people (International Energy Agency, 2017a).

In that sense, and following what Shut et al. (2014) and Abbott and Snidal (2009) have defined as the multifactor governance phases, the efforts done by the IEA and IRENA have contributed to enhance the "development phase" by gradually setting an international agenda on bioenergy. This phase comprises "to capture public attention, frame issues in politically powerful ways, gather and disseminate information, and formulate appropriate ways to proceed" (Abbott and Snidal, 2009). In that regard, the UNFCCC, the IEA, and IRENA have taken advantage of the public debate around climate change, especially in the framework of the Paris Agreement, to gradually setting the agenda on bioenergy by capturing the attention of government and the general public around this topic. Through reports and data analysis (see, for example, IRENA, 2013, International Energy Agency (IEA), 2009), these intergovernmental institutions, together with the OECD (OECD, 2008), have remarked the challenges concerning the sustainable use of bioenergy. Moreover, they have framed the bioenergy issues in politically powerful ways by setting the agenda on the G20 and G8 meetings, hosting ministerial meetings, and making joint declarations. Furthermore, these IGOs have suggested ways to proceed towards a sustainable use of bioenergy (see, for example, International Energy Agency (IEA), 2017b, IRENA, 2016) providing policy makers and business a practical framework to improve their bioenergy sustainability standards.

This recent interest of IGOs to set the agenda on bioenergy above other renewable energies arises from well-founded concerns. Being the largest source of renewable energy—accounting for more than 70% of the world's renewable energy production (International Energy Agency (IEA), 2017b)—bioenergy entails a series of challenges and opportunities, concerning its sustainable production and climate change effects. This demands an international effort to strengthen bioenergy governance at all levels, to deliver sustainable bioenergy. The following section explores the reasons that led the international community to engage in the process of establishing bioenergy in the international agenda.

7.3.4 WHY CREATING AN INTERNATIONAL AGENDA ON BIOENERGY?

Creating an international agenda on bioenergy had become imperative due to both the potential benefits and risks associated with the production of energy derived from organic matter. On the one side, studies have shown that bioenergy has the potential of becoming a significant world-wide strategy to mitigate climate change (Purkus et al., 2012). If well regulated by governments and the international community, bioenergy could serve as a driver to accomplish the transition to a low-carbon energy system and to accomplish the Paris Agreement's targets, while ensuring access to renewable electricity and offering rural development opportunities (Johnson and Altman, 2014). However, on the other side, bioenergy produced unsustainably represents a risk to natural resources and could worsen the situation by increasing pressures on land use and contribute to CO_2 emissions (Popp et al., 2014).

Recent studies have showed that large-scale energy crop production commonly entails direct and indirect land-use change, displacing food and feedstock production, which could raise price levels of agricultural commodities, leading to put food security on risk, as the OECD and The UN Food and Agriculture Organization remarked in a joint report released in 2007 (OECD-FAO, 2007). Moreover, from a life cycle perspective, bioenergy production is not necessarily carbon neutral. Several factors, such as feedstock choice, growing methods, and technology used in the conversion process, can substantially affect the CO_2 bioenergy balance throughout the entire production cycle (Shubert and Blasch, 2010).

Furthermore, unsustainable production of bioenergy can be a cause of natural resources depletion (Food and Agriculture Organization of the United Nations, 2010). Large-scale crop breeding, especially in underdeveloped countries, could deepen the pressures on water resources due to an additional demand for irrigation. Additionally, a land-use change associated to bioenergy crops has been showing to affect water quality, although the magnitude of this impact depends on several aspects, like the crop type, prior land use, management practices, and topological characteristics of the land (Keerthi and Miller, 2017). Moreover, as demand for energy crop production increases and the opportunity cost of forest preservation raises, deforestation is more suitable to happen. Consequently, not only water production is affected but also CO_2 is released into the atmosphere since forests are natural carbon sinks (Popp et al., 2014). Biodiversity loss is also an expected consequence of deforestation owing to direct land-use change. This phenomenon could affect to a greater extent countries with high deforestation rates and biodiversity hotspots, like Brazil, Colombia, Peru, Indonesia, and the United States. Besides biodiversity loss due to deforestation, there are other factors that can intensify the pressures on biodiversity, such as the introduction of invasive species, habitat loss, and ecosystem fragmentation (Immerzeel et al., 2014).

The development of a Climate Change Agenda has gradually revealed to the general public and to policy makers the benefits and risk associated with bioenergy production. This process of awareness-raising has taken more than a decade, starting with the first active participation of the IEA in the COP13 of the UNFCCC in 2007, and the creation of IRENA in 2009, and has involved a large number of reports and data analysis from different stakeholders. As a result, the international agenda on bioenergy has been gradually developed, thus enhancing the "development phase" of bioenergy governance. However, with the adoption of the Paris Agreement, new challenges have been raised for bioenergy governance, particularly to the other two phases of governance: the "implementation" and "enforcement" phase.

7.3.5 BIOENERGY CHALLENGES: GOVERNANCE COORDINATION AND NDCS IMPLEMENTATION

The 2009 Copenhagen conference (COP15) was intended to "determine binding goals, targets and measures for the time beyond 2012, when the Kyoto Protocol's first commitment period ends" (Christoff, 2010). Instead, disagreements between the United States and China lead to set apart the idea of negotiating a "comprehensive, universal and legally binding treaty that prescribes in a top-down fashion, generally applicable policies based on previously agreed principles" (Falkner et al., 2010). After the COP15 failed to deliver a politically binding agreement on climate change ruled by a bottom-up logic, a new round of negotiations started in 2011, aiming to define a new agreement applicable to all developed and undeveloped countries, while assuring a more durable international cooperation (Falkner, 2016).

These negotiations lead to the signature of the Paris Agreement in 2015, an agreement with a new logic based on the principle of common but differentiated responsibilities (Maljean-Dubois, 2016). As a matter of fact, the Paris Agreement leaves behind the Kyoto's Protocol's idea of dividing countries into two categories (developed and developing countries) and legally binding only developed countries to reduce by at least 5% their overall GHG emissions below 1990 by the commitment period 2008 to 2012 (Leal-Arcas, 2012, The United Nations, 1998). Instead, the Paris Agreement acknowledges the primacy of domestic policy mitigation instruments, allowing countries, without differentiation, to set their own level of ambition and pathway for climate change mitigation (Falkner, 2016) by establishing their own NDCs. The NDCs are the core of the Paris Agreement, as they embody the domestic mitigation efforts post-2020 made by each country to reduce national emissions, aiming to achieve the Paris Agreement ultimate goal to keep global temperature rise this century well below 2°C above preindustrial levels (The United Nations, 2015a,b).

The Paris Agreement requests each country to submit to the COP their NDCs every 5 years, starting from 2020. However, the COP invited all countries to communicate to the secretariat their iNDCs before the Paris Agreement negotiations were finalized (World Resources Institute, 2016). By April 18, 2016, 190 parties had communicated their iNDC, in which they established their ambitions and mitigation strategies. In this context, bioenergy acquires a relevant role, as almost one-third of those 190 countries considers bioenergy production (biomass and biofuels in particular) as one of the main strategies to mitigate the emissions generated in the energy sector. For instance, pledges on bioenergy production range from creating a "national bioenergy strategy," promoting a "sustainable use of bioenergy" (Brazil, 2015), to "introducing biofuel blends" (Grenada, 2015), "promoting the use of biofuels" (Belize, 2015), and encouraging the "shift away from inefficient use of biomass in rural areas" (India, 2015).

This, together with the already on-going increase in the demand for bioenergy, creates immense challenges for bioenergy governance. First, countries willing to incorporate bioenergy as a national mitigation strategy in their NDCs will face a bioenergy governance structure that remains highly fragmented, despite the efforts of the last decade. Second, those who already have on track a national bioenergy or biofuel policy, such as Brazil and India, face the challenge of "implementing" and "enforcing" them in a sustainable way to accomplish their iNDC, thus requiring the active participation of other stakeholder asides from the IGOs, such as NGOs, enterprises, civil society, and government institutions.

On the one hand, bioenergy governance remains highly fragmented, despite the efforts of the IEA and the IRENA to construct a common project around the international climate change agenda. This creates an inefficient governance structure for bioenergy, in which multiple IGOs, namely the IEA, the IRENA, and the Global Bioenergy Partnership (GBEP), have different membership requirements and partners, often overlapping between each other. For instance, the IEA demands candidate countries to be a member country of the OECD, and meet other economic requirements, as having "crude oil and/or product reserves equivalent to 90 days of the previous year's net imports, to which government has immediate access" (International Energy Agency (IEA), 2015). This restrains IEA membership to an exclusive group of 30 countries with large oil reserves and excludes big economies, such as Russia and China, and many other countries with great bioenergy potential, as Brazil and India, among others. In contrast, the IRENA, since its creation in 2009, established less exigent membership requirements. Membership in the agency is open to all states that are members of the United Nations. This is why IRENA, as of 2018, counts with 156 active members and other 24 states in accession process (IRENA, 2010). By comparison with IRENA, the GBEP and its partners comprise 23 countries, which were defined in a consultation process held in 2006, before its formal creation later that year (GBEP, 2006). This scenario, in which the main bioenergy IGOs have different membership requirements, and often different agendas, generates administrative inefficiencies, and larger transaction costs to countries willing to have technical assistance to construct their NDCs. To overcome this fragmentation, it is urgent to strengthen the existing collaborative network among these IGOs, work towards a unified agenda on renewables beyond the Paris Agreement, and continue to share data and statistics.

On the other hand, coordination between the different phases and stakeholders remains a challenge to bioenergy governance. Despite the predominant role of IGOs and government institutions, bioenergy governance comprises not only the "development phase," in which the agenda is set but also the "implementation phase" and the "enforcement phase" of policies (Abbott and Snidal, 2009). In that regard, and despite the fact of playing an important part in setting an international agenda on bioenergy, international actors, such as the UN, the IEA, IRENA, the OECD, and the GBEP, have limited tools to implement and enforce policies. The "implementation phase" and the "enforcement phase" entail the active participation of other stakeholders, like enterprises, Academia, NGOs, and nation states (Shut et al., 2014). These intricate relationships constitute a broad governance spectrum with multiple nodes of authority and functions involving actors with different capacities and limitations (Bazilian et al., 2014). Articulating the three phases of governance, to ensure that the whole circle of policy-making is complete, is an immense challenge for bioenergy governance. This is only possible if there is a constant dialogue between the different stakeholders involved in the process of setting the agenda, developing international standards, and the ones in charge of implementing and enforcing them. Overcoming this challenge is key to ensure that those countries which already have a national bioenergy strategy can implement it and enforce it to accomplish their iNDC.

7.3.6 CASE STUDY: COLOMBIA'S BIODIVERSITY STRATEGY AS A MULTILEVEL APPROACH TO GOVERNANCE

Colombia is one of the countries with most biomass potential use for bioenergy production (Deng and Koper, 2015) and the second most biodiverse country in the world (Convention on Biological Diversity, 2005). This creates a scenario of immense opportunities and challenges for bioenergy governance, as unsustainable bioenergy production has been shown to be a driver for biodiversity lost (Kine et al., 2015).

Being aware of Colombia's immense biodiversity, the Presidency and the National Department of Science, Technology and Innovation created on 2015 Colombia Bio, a program that aims to better know, conserve, and use sustainably Colombian biodiversity through ST&I projects (Colombia Bio, 2015). Working together with public and private research centers, local governments, Academia, enterprises, and local communities,

Colombia Bio has created a multilevel governance structure to carry out scientific expeditions and open calls and to strengthen local "scientific tourism" initiatives, and the social appropriation of biodiversity in the country. Through these initiatives, Colombia Bio is changing the social imaginary of the general public, by showing that biological diversity can be an economic asset if it is well preserved, valued, and sustainably used.

In that sense, Colombia Bio is taking a different approach in regards to the relationship between biodiversity and bioenergy. Biodiversity is not seen as a passive asset, but an active one, that if well conserved and valued by local communities could foster bioenergy projects. To do this, Colombia Bio is constructing a bottom-up approach, in which local governments, communities, and regional research centers constitute the base of a multilevel governance structure supported by the national government and the international community. For instance, the scientific expeditions, apart from discovering more than 100 potential new species for science, have involved the local communities and former guerrilla members in the conservation of biodiversity, through scientific guidance courses and initiatives of social appropriation of local biodiversity. The next stage of this process, in which Colombia Bio is currently working, is to encourage local communities to sustainably use the biomass produced in Colombia's natural forests, which cover 52% of the countries' total surface (Ministerio de Ambiente, 2012), and to enhance, through regional open calls, R&D projects made by enterprises and academia to use native cassavas and tubers to produce biofuels (see, for example, some interesting research: Lizarazo et al., 2015, Cortes et al., 2010). As of 2018, Colombia Bio has invested, together with the local governments, and the international cooperation of the United Kingdom through the Newton Fund, more than 20 million pounds in 4 years to support R&D and innovation projects, scientific expeditions, and construct a multilevel approach to biodiversity governance.

REFERENCES

Abbott, K.W., Snidal, D., 2009. The governance triangle: regulatory standards institutions and the shadow of the state. The Politics of Global Regulation. Princeton University Press, Princeton, NJ, pp. 44–88.

Bazilian, M., Nakhooda, S., Van de Graaf, T., 2014. Energy governance and poverty. Energy Res. Soc. Sci. 1, 217–225.

Belize, 2015. Intended Nationally Determined Contributions (iNDCs).

Brazil, 2015. Intended Nationally Determined Contributions (iNDCs).

Christoff, P., 2010. Cold climate in Copenhagen: China and the United Stated at COP15. Environ. Politics 19, 637–656.

Colombia Bio, 2015. Colombia Bio, Colciencias.Available at: http://www.colciencias.gov.co/convocatorias/colombiabio (accessed November 2017).

Convention on Biological Diversity. 2005. Colombia—Overview CBD. Available at: https://www.cbd.int/countries/?country = co (accessed January 2018).

Cortes, S., López, C., Chavarriaga, P., 2010. Biocombustibles y biotecnología: la yuva (Manihot esculenta) como modelo de investigación. Acta Biol. Colomb. 15 (1), 3–24.

Deng, Y.Y., Koper, M., 2015. Country-level assessment of long-term global bioenergy potential. Biomass Bioenergy 74, 253–267.

Falkner, R., 2016. The Paris Agreement and the new logic of international climate politics. Int. Aff. 95 (5), 1107–1125.

Falkner, Robert, Stephan, Hannes, Vogler, John, 2010. International climate policy after Copenhagen: towards a "building blocks" approach. Global Policy 1 (3), 252–262.

Florini, A., Sovacool, B.K., 2009. Who governs energy? The challenges facing global energy governance. Energy Policy 37, 5239–5248.

Food and Agriculture Organization of the United Nations, 2010. Bioenergy Environmental Impact Analysis. Rome, Italy: FAO.

GBEP, 2006. GBEP Membership. Retrieved January 2018, from http://www.globalbioenergy.org/aboutgbep/partners-membership/en/.

Graaf, T.V., 2013. Fragmentation in global energy governance. Global Environ. Politics 13 (3), 14–33.

Grenada, 2015. Intended Nationally Determined Contributions (iNDCs).

Heubaum, H., Biermann, F., 2015. Integrating global energy and climate governance: the changing role of the International Energy Agency. Energy Policy 87, 229–239.

Immerzeel, D.J., Verweij, P.A., Hilst, F., Faaij, A.P., 2014. Biodiversity impacts of bioenergy crop production: a state-of-the-art review. GCB Bioenergy 6, 183–209.

India, 2015. Intended Nationally Determined Contributions (iNDCs).

International Energy Agency (IEA), 2009. Global Gaps in Clean Energy Research, Development and Demonstration. Paris, France: IEA.

International Energy Agency (IEA), 2012. IEA and the UNFCCC.Available at: https://www.iea.org/topics/climatechange/ieaandtheunfccc/ (accessed September 2017).

International Energy Agency (IEA), 2015. Member Countries. Retrieved January 2018, from https://www.iea.org/countries/membercountries/.

International Energy Agency (IEA), 2016. Energy, Climate Change & Environment. Paris, France: IEA.

International Energy Agency (IEA), 2017a. Bioenergy: Roadmap Development and Implementation. Paris, France: IEA.

International Energy Agency (IEA), 2017b. Technology Roadmap: Delivering Sustainable Bioenergy. Paris, France: IEA.

IRENA, 2009. International Renewable Energy Agency. Available at: http://www.irena.org/history/foundingconference (accessed September 2017).

IRENA, 2010. International Energy Agency. Available at: http://www.irena.org/Search?keywords = UNFCCC (accessed September 2017).

IRENA, 2013. Statistical Issues: Bioenergy and Distributed Renewable Energy. Abu Dhabi, United Arab Emirates: IRENA.

IRENA, 2016. Boosting Biofuels: Sustainable Paths to Greater Energy Security.

Johnson, T.G., Altman, I., 2014. Rural development opportunities in the bioeconomy. Biomass Bioenergy 63, 341–344.

Keerthi, S., Miller, S.A., 2017. Regional differences in impacts to water quality from the bioenergy mandate. Biomass Bioenergy 106, 115–126.

Kine, K., Martinelli, F., Mayer, A., 2015.). Bioenergy and biodiversity: key lessons from the pan American region. Environ. Manage. 1377–1396.

Leal-Arcas, R., 2012. Top-Down and Bottom-Up Approaches in Climate Change and International Trade. International Development Bank.
Leong, A., 2015. How the Paris Agreement and the SDGs Work Together. Available at: http://17goals.org/paris-agreement-sdgs/ (accessed September 2017).
Lizarazo, S., Hurtado, G., Rodríguez, L.F., 2015. Análisis técnico económico de la producción de bioethanol a partir de papa a nivel de laboratorio en Boyacá. Rev. Colomb. Cienc. Hortíc. 9 (1), 97–111.
Maljean-Dubois, 2016. The Paris Agreement: a new step in the gradual evolution of differential treatment in the climate regime? Rev. Eur. Compar. Int. Environ. Law 25 (2), 151–160.
Ministerio de Ambiente. (2012). *Importancia de los bosques, Colombia tercer país de la región en cobertura boscosa*. Available at: http://www.minambiente.gov.co/index.php/component/content/article?id = 1210:el-uso-sostenible-de-los-bosques-prioridad-de-minambiente-513 (accessed February 2018).
Molina, P., 2016. The changing climate of climate change economics. Ecol. Econ. 121, 12–19.
OECD, 2008. A Review of Policy Measures Supporting Production and Use of Bioenergy. Paris, France: OECD.
OECD-FAO, 2007. Agricultural Outlook 2007-2016. Paris, France: OECD-FAO.
Popp, J., Lakner, Z., Harangi-Rákos, M., Fári, M., 2014. The effect of bioenergy expansion: food, energy, and environment. Renew. Sustain. Energy Rev. 32, 559–578.
Purkus, A., Gawel, E., Thrän, D., 2012. Bioenergy governance between market and government failures: a new institutional economics perspective. UFZ-Diskussionspapiere 13, 1–29.
Shubert, R., Blasch, J., 2010. Sustainability standards for bioenergy—a means to reduce climate change risks? Energy Policy 2797–2805.
Shut, M., Soares, N.C., van de Ven, G., Slingerland, M., 2014. Multi-actor governance of sustainable biofuels in developing countries: the case of Mozambique. Energy Policy 65, 631–643.
The Royal Society of Chemistry, 2015. Still Only One Earth—Progress in the 40 Years Since the First UN Conference on the Environment. Cambridge, United Kingdom.
The United Nations, 1992. United Nations Framework Convention on Climate Change.
The United Nations, 1998. Kyoto Protocol to the United Nations Framework Convention on Climate Change.
The United Nations, 2015a. Available at: http://www.un.org/en/sections/about-un/overview/index.html (accessed September 2017).
The United Nations, 2015b. Paris Agreement.
The United Nations, 2016a. The Sustainable Development Agenda. Available at: http://www.un.org/sustainabledevelopment/development-agenda/ (accessed September 2017).
The United Nations, 2016b. Transforming our World: The 2030 Agenda for Sustainable Development.
World Resources Institute, 2016. WRI. Available at: http://www.wri.org/indc-definition (accessed 2017).

FURTHER READING

Ghana, 2015. Intended Nationally Determined Contributions (iNDC).
United Nations Climate Change, 2015. UNFCCC. Available at: http://unfccc.int/focus/items/10240.php (accessed November 2017).

7.4

Bioenergy and Food Security: Synergies and Trade-offs

Siwa Msangi
Environment & Production Technology Division, International Food Policy Research Institute, Washington DC, United States

7.4.1 INTRODUCTION

In the phase of increasing socioeconomic growth and the resulting demand for goods and services, the demand for energy has steadily grown in the last decades, along with other demands on the natural resource base. Aside from the demand for transportation fuel, energy is becoming increasingly demanded for productive uses in industry as well as at the household level, where increasingly wealthy households acquire energy-using products and desire better quality heating and lighting (among other energy services). In the desire to reduce dependence on imported fossil-based energy and enhance domestic energy security, many countries have begun to look more closely at the possibilities that bioenergy from renewable resources could offer in supplementing their domestic energy portfolio. Countries pursuing bioenergy development in both the developed and developing world often state their desire to achieve multiple objectives such as the enhancement of domestic energy security, the promotion of environmental sustainability, and the stimulation of rural economic development. Depending on how the supporting policies are designed, however, not all of these goals can be achieved through the same national program.

While a number of developing countries find the prospect of bioenergy development attractive, the degree to which they are able to invest (or attract investment) in building up capacity for their own domestic production varies widely, given the uncertain price of fossil-based energy, the varying level of long-term commitment that national governments are willing to give to support fledgling bioenergy industries (through subsidies or incentives), and other key factors (Food and Agriculture

Organization of the United Nations, 2008). A number of interested countries in sub-Saharan Africa, South and Southeast Asia, and Latin America have favorable conditions for growing the large volumes of biomass as feedstock for bioenergy production, both in terms of suitable climates and agro-ecological conditions as well as sufficient land and water resource availability (Langeveld et al., 2014). While these conditions might appear favorable, the actual commercial feasibility of large-scale bioenergy production for either domestic consumption or export may be hindered. In terms of bioenergy production in Africa, the degree of infrastructure development across the continent varies widely, posing a severe impediment to scaling up bioenergy production in landlocked countries, especially, whereas other countries with more developed value chains in the necessary commodities and better transportation networks find themselves better connected to global export markets.

In terms of food security, some of the same countries in which bioenergy is most desired and needed find it equally difficult to competitively sell agricultural exports on world markets and face unfavorable terms-of-trade when trying to import needed food commodities. Here we use the definition of food security first put forward at the World Food Summit of 1996 (Food and Agricultural Organization of the United Nations (FAO), 1996), and frequently used in the literature (Kline et al., 2017)—which states that food security is the condition that exists when all people, at all times, have physical and economic access to sufficient, safe, and nutritious food that meets their dietary needs and food preferences for a healthy and active life. The food price spike of 2007–08 brought up a vigorous debate and discussion over the role that key global drivers of socioeconomic change might have played in causing it—including the rising price of crude oil and the demand for biofuels and other alternative renewable energy sources. This touched off a debate over the sustainability of biofuels production and the larger role that bioenergy has to play in meeting future energy demands while also meeting important goals in the reduction of global carbon emissions. The "food versus fuel" aspect of that debate has largely been answered with the hindsight of empirical evidence that has allowed researchers to test hypotheses concerning which determinants might have played the biggest part in driving up food prices. Nonetheless, concerns remain about the importance of safeguarding food security (i.e., dietary energy access) while pursuing other energy security goals.

Given our focus on food security, we do not address the context of OECD countries like the United States, where much of the policy debate tends to focus on improving the environmental performance of existing biofuel and domestic bioenergy programs—such as through "low carbon" standards introduced in western states like California (Khanna et al., 2017). Within the Latin American context, most bioenergy schemes involve relatively large-scale, commercially oriented enterprises, which tend not to have the same socioeconomic impacts as the small holder–based schemes that have been proposed for Africa—and raise concerns more related to the environmental impacts of land conversion rather than food security (Ludena et al., 2007; HLPE, 2013). In the Asian context, countries like China have mitigated potential food security effects of bioenergy developments through strict limits on use of food-based feedstocks (Huang et al., 2012).

In this chapter, we review some recent literature that looks at the linkage between food security and bioenergy and put forward a framework that is useful for understanding the way in which they interact positively or negatively, within the context of developing economies. We mostly focus on the African context, where issues of food security remain the most acute and where the need to reduce "energy poverty" with additional exploitation of both fossil and renewable sources is still urgent—making bioenergy and food security linkages highly relevant. We discuss both the micro- and macro-level linkages that connect energy to agricultural markets and propose a typology of countries that could provide useful guidance in determining where bioenergy ventures might be best placed to thrive, whether for supplying domestic energy needs or for creating exportable energy products to the world market. This framework will enable us to better appreciate some of the trade-offs inherent in bioenergy production systems with other types of agricultural enterprises, and where potential synergies could arise that could actually strengthen food growing capacity. This framework can allow both policy makers and researchers to better understand how programs which expand biofuel production can synergize with investment and development strategies aimed at strengthening the function of food systems.

7.4.2 THE "FOOD VERSUS FUEL" DEBATE: A FALSE TRADE-OFF?

In this section, we provide a brief overview of the current literature that addresses the food-versus-fuel issue for biofuels and summarize the key insights that have been gained, since the 2007–08 food price spike, into the nature of the potential trade-offs and synergies between bioenergy and food-based agriculture. This overview will help to orient the discussion on bioenergy and food security linkages that will be discussed later in the chapter.

Following the 2007–08 food price spike, a large body of literature began to look at connections between various global drivers of socioeconomic and environmental change, and food prices as seen in global and local markets. The close coincidence of energy price rises, biofuel production increases, and the increases in food prices within the 2005–07 time period made the job of analysts somewhat complicated, in terms of trying to attribute the effects of one phenomenon upon the other and disentangling the causal linkages. An example of an attempt to sort through the various drivers of agricultural commodity price change on world markets is an analysis by the USDA's Economic Research Service (Trostle, 2008), Chatham House (Evans, 2008), and by the International Food Policy Research Institute (IFPRI) (Headey and Fan, 2008). In this literature, immediately following the price spikes, the direct connection to food security was made largely through the inference that higher food prices would be hurtful to households that are overall net buyers of food, compared to those who might be net sellers and actually benefit from higher market prices for their marketed output. A number of papers tried to use detailed micro-level data to tie food price increases to direct poverty impacts observed or inferred for various types of households (Ivanic and Martin, 2008; de Hoyos and Medvedev, 2011).

Among the papers that drew very strong connections between biofuels growth and a deterioration in food security within the developing world are those from Oxfam International (Bailey, 2008) and the Center for Global Development (Elliott, 2008)—which put doubt on the overall sustainability of biofuels in the strongest terms—as well two articles in foreign policy journals (Runge and Senauer, 2007, 2008) and the congressional testimony given by IFPRI (von Braun, 2008). For the purposes of focusing this paper on food security dimensions, we leave aside those studies that mostly look at the environmental dimensions of biofuels growth (Fargione et al., 2008; Searchinger et al., 2008). We can point to a large body of literature that tried to use regional or global-level market equilibrium model-based analysis to infer the contribution of biofuels growth on agricultural commodity prices and determine their potential welfare effects. Key among those are the papers of IFPRI (Rosegrant et al., 2008), the OECD (von Lampe, 2006), LEI-Wageningen (Banse et al., 2008), and that of the World Bank (which inferred the highest impact on world prices to biofuels, and which drew much criticism, subsequently) (Mitchell, 2008).

The wide range of model-based estimates also generated meta-analyses which tried to determine the causes for divergences in these estimates (Zhang et al., 2013; Crago, 2014; Condon et al., 2015), and their implications for assessing the biofuels–food security linkage. Some differences were attributed to the underlying assumptions, whereas others related more to model structure and design, such as whether or not the coproducts from biofuels are directly accounted for in terms of its contribution to feed supply (thereby mitigating some commodity price effects). Some key papers have pointed to the importance of this coproduct effect (Taheripour et al., 2008, 2010), while others point to the importance of accounting for inventory demand (Hochman et al., 2014).

The basic aim of much of this literature was to focus on the impacts of increased biofuel production on agricultural commodity prices (and agricultural land use) and to attempt separating out the effect of crop-based biofuel growth from other factors affecting food markets. Many of the studies acknowledged the close connection between energy and agricultural markets—both in terms of the direct effects of energy prices on agricultural production costs and prices, as well as with respect to the effect of added demand for agricultural products to produce energy products, such as crop-based biofuels. In much of the literature, however, agricultural commodity prices were treated synonymously to food prices and assumed that an increase in one would inevitably lead to a similar (even if smaller) increase in the other. One of the more recent papers on bioenergy and food security linkages (Kline et al., 2017) has tried to comprehensively address some of the underlying misunderstandings and inaccuracies in the literature and to point out some important definitions and concepts that are key to capturing the potential trade-offs or synergies between them. Among the points raised by the authors is that using indicators of food prices to indicate changes in food security status may not be accurate or even valid, especially where local markets are relatively isolated from outside or international markets, or where they are heavily regulated. Given the separation between macro-level market dynamics and more micro-level food security outcomes—which must be measured at the household or individual level—the challenges in using regional equilibrium models to infer food security status become apparent.

In contrast to global-level market studies, more country-focused studies have also used equilibrium models to connect the growth in biofuels production to household-specific welfare effects but with more of an economy-wide focus. An example of such studies are those which utilize computable general equilibrium models to determine if certain aspects of biofuel production (and export) could be pro-growth and pro-poor—such as the 2010 study on Mozambique (Arndt et al., 2010a,b) and the 2012 study on Tanzania (Arndt et al., 2012). These studies ended up having a more positive view of the potential for national biofuels programs to promote economic growth and reduce poverty levels for specific household types, depending on the path of biofuels development taken by country policies. In general, additional economic activity around biofuel feedstock production, conversion processing, and export resulted in economic gains—mostly through the kind of labor demand that they generate and the subsequent impact on household incomes and purchasing power (which disseminates itself through the rest of the economy). In situations where labor is scarce, however, the competition between sectors for unskilled laborers might cause growth in the biofuels feedstock and processing sector to detract from another export-generating agricultural sector, resulting in some economic losses. These studies tended to find that schemes which encourage greater smallholder participation through out-grower schemes tend to produce more substantial benefits for rural households—partly through the fact that more households are drawn into the economic activity (both landless laborers as well as smallholder farmers), as well as through the "spillovers" created by having larger growers and processors disseminate better agricultural methods and technologies to the smaller growers, thereby raising their on-farm productivity and earnings (Arndt et al., 2010a,b). The critical issue of land availability and access, however, remains a key driver of what these studies are able to say about the implications of large-scale schemes on the welfare of smallholders, and overall economic well-being. In cases where the growth of biofuel production does not impede the availability of land for other agricultural activities, the effects are positive. But in situations where the area under an important staple or exportable could be reduced by expansion of dedicated energy crop production, the welfare effects become negative for the rural poor (Arndt et al., 2010b).

Based on these studies, therefore, we are left to conclude that the effects of biofuels and bioenergy (more generally) on economic well-being and food security do not need to be negative. A more recent paper looking at bioenergy and food security in Malawi (Schuenemann et al., 2018) drew a closer connection to bioenergy for domestic use and household-level food security than typical market-level equilibrium studies do, given that it relates direct food consumption to the need for cooking energy from improved sources (relative to the crude biomass that is often used by the rural poor). This type of household energy security perspective is one of the key aspects of bioenergy and food security that tends to be overlooked in studies that only consider the part of bioenergy that relates to liquid fuels for transportation. The household-level implications for having more reliable, cleaner, and convenient sources of energy for cooking, heating, and lighting relate directly to the time-use that rural women have to spend on collecting raw biomass that is scattered over the landscape and which directly detracts from their ability to engage in other income-earning (or welfare-increasing) activities (Ewing and Msangi, 2009).

This is among the key aspects of bioenergy and food security that we will address more closely in the following sections of this chapter.

7.4.3 RETHINKING FOOD SECURITY AND BIOENERGY LINKAGES

In this section, we present a simplified analytical framework to show the linkages between bioenergy and household-level food security in a way that more directly relates to the key dimensions of welfare relevant to rural households within developing countries. The reason why we limit ourselves to developing-country households in this discussion is due, partly, to the fact that rural households in these countries tend to spend a much larger share of their income on food, compared to households in wealthier countries, and are therefore more vulnerable to food price shocks. These households also source their food less from retail chains, whose processed and final food products tend to exhibit less fluctuations compared to basic (and unprocessed) food commodity markets to which the poor are more closely connected to. While we have not operationalized this framework explicitly within a computational model that can be simulated for a specific empirical case, we use it to discuss the important and relevant linkages that determine whether households derive positive or negative impacts from bioenergy development.

This approach will help us to illustrate more clearly how bioenergy (writ broadly) can lead to important synergies and trade-offs with food availability, access, and household welfare.

The pathways through which energy affects human welfare within a rural, developing country setting mostly follows through the linkages that households have to critical (and often limited) resources they depend upon for production and consumption activities, as well as the ties that connect households to local or regional markets for food and energy. In less-developed countries where many rural (or even urban) households are disconnected from modern electricity grids (or natural gas networks), such households must rely upon their own labor and limited resources to procure energy for cooking, heating, and lighting either through marketed products of inferior quality (like kerosene for lamps or wax for candles) or from freely collecting crude biomass that can be directly used as fuel. In order to understand the nature of these linkages and how they relate to human well-being, we take a broader view of the household as a productive unit that uses its own labor to produce goods for consumption and sale. Such households rely on markets for purchasing inputs that they either use in their own production or that they consume after some process of transformation. This transformation could be in the form of cooking or other type of food preparation (e.g., grinding grain) and can be thought of as its own type of household-level production process, which requires the input of unpaid labor. These social dimensions—especially those relating to the unpaid (and often under-appreciated) domestic work within the household—are highly relevant to the gender aspects of household-level welfare, where women tend to bear the brunt of responsibility for providing care and nurture to other household members.

Fig. 7.4.1 conceptualizes the household as a productive unit that draws from both natural and human resources to produce goods that it uses for its own consumption or for generating income, through the mediation of markets and is adapted from an earlier conceptual paper (Msangi, 2012).

The labor used for productive activities at the household level can be sourced internally or can be obtained through market-mediated wages (where such markets exist). On the consumption side, households make decisions which are based on their own production possibilities as well as on the goods that are available through exchange or market interactions. Household-level savings and investment decisions (or remittances from others) play a role in allowing the household to accumulate wealth and to enhance its own welfare, while also building up buffers against potential shocks to production or income that enable it to smooth its own consumption over time. This schematic, in a very compact form, summarizes the linkages and feedbacks that are most relevant to how energy affects household-level production and consumption decisions, and the subsequent linkages to health and well-being outcomes.

From this diagram we also see that energy is both an input into production as well as a good (or service) that can be consumed by the household. If the household is engaged in bioenergy crop production, it is actually a producer of energy (in feedstock form) that can then be sold to the market. Some energy products used in household level production might be marketed (as in the case of liquid fuels, gas, or electricity for powering machinery or in energy-intensive fertilizers) or else they might be own-produced, as in the case of own-labor or farm animals used for traction. The energy used in the transformation of raw food commodities to consumable food can be obtained through refined energy products or in the form of raw biomass (whether purchased through informal markets or own-collected).

One can distinguish between the various types of households that consume energy or are involved in energy feedstock production. They might be landless laborers who allocate their effort to the labor market for a wage, or households with land that either produces food or other nonfood cash crops that can be used as feedstocks into bioenergy production processes. Each of these household types is nested within the graphic of Fig. 7.4.1 (with some of the boxes removed). We summarize these cases in Table 7.4.1 and discuss their relevance to the question of bioenergy and welfare impacts.

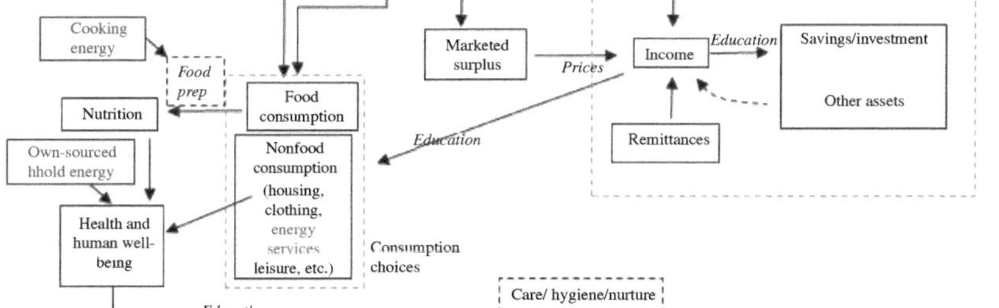

Figure 7.4.1 Schematic of key linkages relating to energy and food security at household level. *From Author.*

Table 7.4.1 Key household types and their linkages to energy availability and use

Type	Key household characteristics	Contextual relevance	Key trade-offs	Effect of food price changes
1	Households that sell their own labor to the market and buy all of the food they consume	Applies to most urban households	Trade-off in time use between leisure and paid work	Affects purchasing power of labor wages
		Applies to landless rural households		
2	Households produce food which is consumed and/or marketed	Applies mostly to rural households with land	Trade-off between leisure and the returns from two competing economic activities—one of which is defined by return per unit wage and the other by returns per unit land	Affects net return (if household markets food) or compensating power of off-farm wage (if household is net food buyer)
	The same food commodity is used as a feedstock for bioenergy production and is either sourced from the household or drawn from other marketed surplus	Price effects that are driven by biofuel production affect the production and consumption decisions of the household (whether they are net sellers or buyers of the good)		
3	Households that produce crops that are for food and for (potential) bioenergy feedstock use separately and only consume the food crop, while both can be sold to the market	Applies mostly to rural households with land	Trade-off in returns per unit land of two activities which also compete for limited labor time. Therefore, productivity per unit labor is key	Determines the degree to which the household is exposed to food price risk and should compensate with feedstock production
		The key resource constraints are that of land (which must be divided between both types of crops) and labor (for which the household might be constrained, if labor markets are thin or missing)		

Source: Adapted from Msangi, S., 2012. Biofuels and Africa: impacts and linkages at the household level. In: Gasparatos, A., Stromberg, P. (Eds.), Socioeconomic and Environmental Impacts of Biofuels. Cambridge, United Kingdom: Cambridge University Press, pp. 231–253. https://doi.org/10.1017/CBO9780511920899.017.

From the table we can see that an increase in either food prices (due to domestic or international impacts from bioenergy production) affects the various household types differently and that the implications for increased labor demand in bioenergy production processes (whether in feedstock production or in the conversion and processing stage) are also varied. For a household which only consumes food (and energy) products, the effect of market-level changes in food and energy availability directly impact the household budget and income possibilities. Since these households sell their labor to the market, any increase in labor demand generated by a bioenergy production process will have a positive effect on them.

In the case of households that both produce and consume food commodities and products (labeled as "Type 2" in Table 7.1), the impacts of higher food prices that might come as a result of bioenergy-related impacts (or not) will have a positive or negative impact, depending on their overall position as a net seller or buyer of the food commodity. This is the case that the World Bank authors looking at the poverty impacts of food price changes in the 2007–08 period were considering in their papers (Ivanic and Martin, 2008; de Hoyos and Medvedev, 2011). The third type of household (where some of the agricultural production activity can relate directly to bioenergy feedstocks) embodies the cases considered in the computable general equilibrium model studies of Tanzania and Mozambique (Arndt et al., 2010a,b; Arndt et al., 2012), where the production possibilities for bioenergy feedstock producers might interact with their production of food commodities and create impacts on food availability and food security.

The impacts of substituting crude biomass-based energy sources for cleaner and more efficient energy products (whether from bioenergy or not) relate to all of these household types and have implications for cost and expenditure (in the case where energy products are purchased), or for domestic time use (if cleaner bioenergy products reduce the need to source energy by one's own labor). Both urban and rural households in developing countries use raw biomass to meet domestic energy needs, although to varying degrees. In more urbanized regions, where households may be connected to the grid, the unreliable nature of power supply (in the form of scheduled or unscheduled outages) may cause households to have "backup" sources of energy for cooking, lighting, and heating, in the form of solid or liquid fuel cookstoves or other appliances (Ali, 2016). Whereas urban households may not spend as much time collecting the raw biomass they use, compared to their rural counterparts, the cost implications of substituting purchased biomass with better-quality fuels is still there (Peters and Sievert, 2016).

The illustrations drawn from this overall framework of household-level consumption and production—tied to energy use and availability—point to the diversity of welfare impacts that may be possible when countries engage in bioenergy production ventures of sufficient scale as to affect product and labor markets. The way in which analysts should measure food security impacts coming from bioenergy development should be aware of the different household types and the ways in which their production and consumption choices are linked to energy and food markets. This perspective will be used to inform the discussion, in the following section of this chapter, on how bioenergy development can create synergies or trade-offs with food security, when considered in the case of specific country cases.

7.4.4 WHERE BIOENERGY CAN HELP (OR HURT) FOOD SECURITY

Given our conceptual orientation towards household-level linkages between food security and energy, we can now discuss the potential positive and negative interactions between bioenergy and food security with greater clarity. The micro-level perspective on the household enables us to appreciate the issue of "energy poverty" that some authors pose as being the most relevant connection between bioenergy development and poverty reduction (Clancy, 2013). If increased (or cheaper) bioenergy availability helps to reduce the constraints that households face in meeting their basic needs for domestic energy, then an important aspect of household welfare (closely tied to food security) can be addressed. A badly designed bioenergy scheme, however, can also serve to undermine socioeconomic benefits and desired environmental goals if the implementation causes unintended impacts on the production or consumption possibilities of households. These impacts could come about as a consequence of adverse price impacts, labor market shortages, or the reduced access to key productive resources (such as land). In this section, we discuss some of the ways in which policy design and implementation can help or hurt the food security outcomes related to bioenergy development programs.

7.4.4.1 Viable value chains and feedstock choice

The great interest in *jatropha curcas* as a nonedible biofuel feedstock that was there in the mid to late 2000s has now waned, in light of the practical realities that producing and processing this type of bioenergy feedstock presents. Although jatropha—a perennial, oil-bearing shrub—is not a food commodity whose direct use can reduce food availability, some authors found that the incentives to grow it on primary agricultural land as a cash crop (rather than on the marginal lands that its proponents intended its cultivation for) can end up crowding out food-based agricultural production in terms of land area and input usage (von Maltitz et al., 2012). The crowding-out effect of jatropha, as well as other edible though dedicated biofuel feedstock crops, was cited in another recent paper as also having a potentially negative impact on rural smallholders in sub-Saharan Africa, where large-scale projects are being pursued (Kiggundu et al., 2017).

Other authors have argued against the large-scale development of bioenergy programs around feedstocks like jatropha due to the fact that it generates coproducts that are low in value and are not amenable for use as animal feed (as is the case with dried distillers grains and solubles from grain ethanol processing), and that its productivity is rather low, compared to other oil-bearing crops (Hazell and Evans, 2011; Msangi and Evans, 2013). Hazell and Evans go on to point to the fact that very little yield-enhancing research and development has gone into making jatropha the kind of high productivity feedstock that would be needed to keep a large-scale processing operation continuously and profitably operating. The fact that the most successful bioenergy crop in Africa, currently, is sugarcane comes from the fact that the value chains for sugar are much further developed, compared to little-tried crops like jatropha, and that the vertical integration of these value chains (from the feedstock production to the extraction and conversion processes) enables operators to maintain a year-round level of production that can satisfy both domestic and international demand. In the detailed country analysis for Tanzania, the FAO team looking at bioenergy and food security linkages (Maltsoglou and Khwaja, 2010) found that among the candidate feedstock crops for bioenergy production and conversion, sugarcane was the most viable, given its proven productivity and marketability in the country in the face of considerable challenges to transportation and marketing infrastructure.

The fact that the production and marketing of food products in Africa face some of the same barriers that bioenergy products would face argues for an approach that focuses on addressing the constraints on food-based agriculture first, so that the production, processing, and marketing possibilities of nonfood products (including bioenergy products) can be strengthened (Msangi and Evans, 2013). In other words, the type of infrastructure investments and technological research and development that enhances productivity can improve the performance, and delivery of food systems can also benefit bioenergy systems and make developing country ventures in these areas much more viable and sustainable (Hazell and Evans, 2011).

7.4.4.2 Prioritizing bioenergy development pathways and policies

When discussing the challenges facing the achievement of inclusive green growth in Africa, one must keep in mind that there is a broad diversity of countries on the continent and, therefore, a broad diversity of challenges. If we construct a broad typology of countries—borrowing from the analysis of Thorbecke, in which he proposes a growth typology of African countries (Thorbecke, 2009)—we can group a range of African countries, according to several key characteristics that are relevant to their growth potential with that of agro-ecological suitability and agricultural potential, the degree to which countries are "resource-rich" or "resource-poor," and, finally, whether the countries are landlocked or not. A summary of the classifications is given in Fig. 7.4.2.

By looking at the three basic dimensions or criteria embedded in this typology, we can focus on some of the key challenges that African countries who wish to scale up production of agriculture-based products for export would face in achieving market competitiveness and long-term, sustainable growth in that sector—whether it is focused on food, feed, or bioenergy uses. The landlocked countries face a special set of challenges in terms of access to world markets and keeping open channels of communication and commerce. The country infrastructure diagnostic for Africa (AICD) noted, in particular, that the key challenge of maintaining the competitiveness of landlocked African regions on world markets is in improving their infrastructure, as well as that of the neighboring countries on which they depend (Foster and Briceno-Garmendia, 2009).

The typological characteristics in Thorbecke's framework play a role in determining the degree of viability and competitiveness of potential domestic or export-oriented bioenergy sectors in these countries, given that resource- (especially oil-) rich countries would like to subsidize

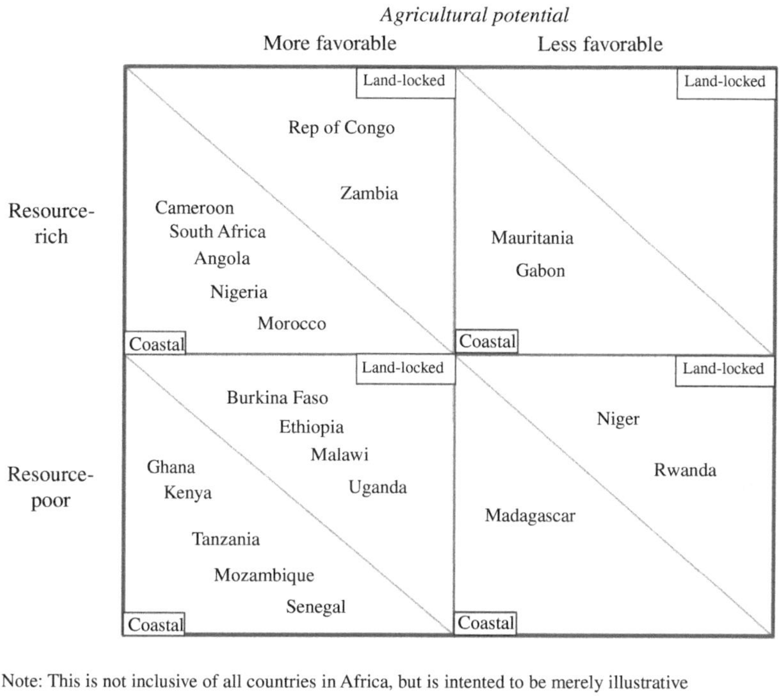

Figure 7.4.2 Growth typologies of African countries.
Note: This is not inclusive of all countries in Africa but is intended to be merely illustrative. *From Author.*

fossil-based fuels in a way that would make bioenergy products uncompetitive (as we see in the case of Nigeria, Angola, and Morocco). The degree of agricultural potential would, of course, determine the likely productivity of potential bioenergy feedstock crops (either food or non-food based), and the coastal or landlocked geographical disposition of the country is key in determining the transactions costs and logistical barriers to trade that these countries would face in exporting their agricultural (or even nonagricultural) goods to market.

In addition to this, if we also try and capture the degree to which a country is "energy poor"—in terms of its reliance upon low-quality, low-efficiency sources of energy (such as raw biomass in place for more energy-intense liquid and gel-based fuel products)—we might also uncover an important determinant of the kind of bioenergy strategy that should be pursued. Fig. 7.4.3 is shown here to illustrate a crude measure of "energy poverty," and comes from a paper looking at welfare, energy,

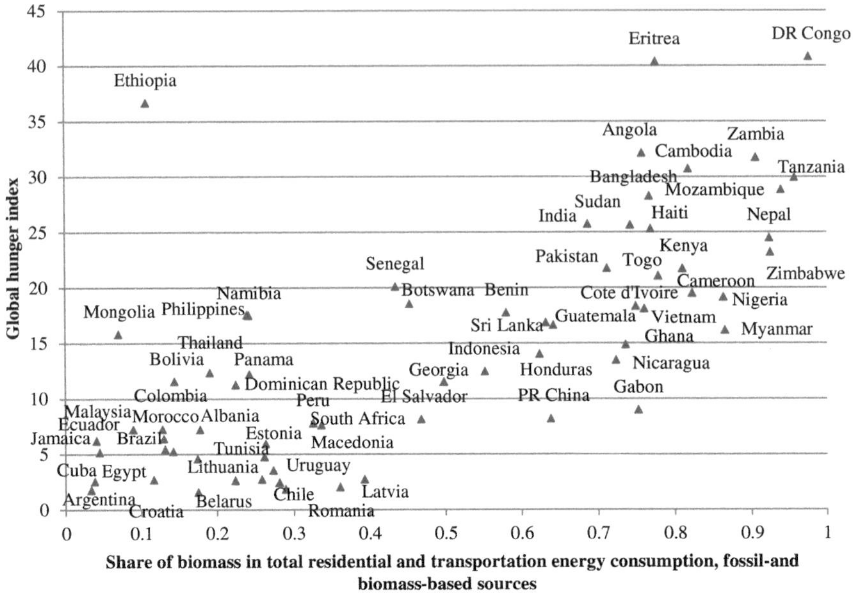

Figure 7.4.3 Energy demand and hunger in non-OECD, non-petroleum exporting countries. *From Ewing and Msangi (for illustrative purposes).*

and food security in a developing country context (Ewing and Msangi, 2009).

From their paper, they describe the strong correlation between those countries experiencing high levels of hunger and food insecurity (as captured in an earlier version of the continually updated Global Hunger Index (GHI) of IFPRI (see http://www.globalhungerindex.org/)—and their dependence upon crude biomass for meeting their energy needs. Given that we would expect that countries with high levels of insecurity—as captured by the GHI—would also have high levels of poverty, then we can see that this diagram gives a rough correspondence between energy poverty and income poverty.

Combining the elements of energy poverty and export potential from the previous two graphics, we can now propose a typology of countries that indicates where the policy emphasis on bioenergy development versus food production should be placed, and whether the production of bioenergy should be more domestic or export-oriented. Table 7.4.2 shows how the export potential of bioenergy products and the relative energy poverty of the rural and urban population combine to determine the kind of bioenergy development strategy that a country should undertake to enhance its food (and energy) security goals.

Table 7.4.2 Mapping of countries onto a prioritization matrix of bioenergy development, according to export potential and domestic energy poverty

	Low levels of domestic energy poverty	High levels of domestic energy poverty
Low export potential for liquid biofuels or other bioenergy products	If low export potential is due to low productivity of bioenergy feedstock and uncompetitive processing costs, then better to opt out of bioenergy production If domestic demand for transport biofuels is sizeable (with the help of mandates) and sufficiently competitive, then liquid biofuels development for the domestic market may be beneficial	If low export potential is due to poor productivity of bioenergy feedstock, R&D investments might be warranted alongside imports of needed bioenergy products **(Rwanda, Niger)** Focus on non-transport bioenergy and clean alternative fuels for cooking and other household uses if feedstock productivity is sufficiently high **(Zambia, Republic of Congo)**
High export potential for liquid biofuels and other bioenergy products	Focus on export markets if transportation costs are favorable and if the domestic value chains can deliver (consistently and competitively) **(Morocco, South Africa)**	Biofuels with cogeneration potential for domestic energy (such as sugarcane) if sufficient quality and quantity of biomass can be generated for processing into exportable and domestically consumable bioenergy products **(Nigeria, Angola, Cameroon)**

Source: Author.

From this table, we see that in situations where export potential is low—either because of high transportation costs and market barriers and/or because of low productivity and profitability of possible bioenergy export products (relative to market competitors)—there is greater incentives for domestic-focused production. If the productivity of the feedstock is the key constraint, the country should either opt out of doing bioenergy altogether (as in the case where household energy poverty is relatively low) or else engage in long-term research and development focused investments to boost overall feedstock productivity in the longer term, while considering importable bioenergy substitutes (or other means of expanding domestic energy provision). In the case where transportation costs and transactional barriers are at the root cause of low export potential (but where feedstock productivity and processing is relatively good), then there is a case for either instituting blending mandates in transportation fuels to boost domestic consumption of biofuels or else focusing on non-transport bioenergy products that have sufficient demand at the household level to sustain bioenergy production and profitability. This relates to the second row of criteria listed under the "low export potential" case.

In the case where high export potential exists for bioenergy products—whether due to productivity and/or cost-competitive exportability of the feedstock or processed product— a country would have sufficient justification for scaling up bioenergy production for either domestic use or export market. If the most favorable crop is an energy crop like sugarcane, which has combustible coproducts that can be used for the cogeneration of electricity, then that might provide further justification of the program to meet household energy needs.

By taking the country growth typology derived from Thorbecke's discussion (in Fig. 7.2), we can use the "landlocked" versus "coastal" status to indicate relative transportation and marketing costs for food or nonfood products on export markets. Similarly, we can take the ranking of countries shown in Fig. 7.3—in terms of their degree of overall dependence on raw biomass for energy consumption—and use it to indicate the degree of energy poverty that is captured in Table 7.2. By doing so, we can tentatively map the countries that overlap between Figs. 7.2 and 7.3 onto the matrix in Table 7.2, in the bold text that is shown.

From this mapping, we can see that countries like South Africa, which already has a robust sugar and ethanol sector, maps well into the prioritization matrix. The countries with high export potential as well as high levels of energy poverty, like Nigeria, Angola, and Cameroon, are also the resource-rich countries (mapped in Fig. 7.2) that are likely to favor their oil sectors over agriculture, and perhaps even subsidize fossil-based fuels to the point where bioenergy alternatives become noncompetitive on the domestic market. Nonetheless, their tremendous agricultural potential could be the basis for organizing competitive and productive value chains around agricultural products in general and around bioenergy products, in particular (Cushion et al., 2010).

In the right-hand column of Table 7.2, on the higher rows, we can see the countries which would merit a non-transportation fuel-focused bioenergy strategy—such as Zambia and the Republic of Congo—but who would find it difficult to achieve export competitiveness due to their landlocked nature. By contrast, the countries who would be advised to invest more R&D in crop feedstock technologies that boost productivity (i.e., Niger and Rwanda) before attempting to develop a bioenergy sector focused on domestic energy uses.

We have not placed some of the countries in Table 7.2, due to the fact that not all African countries were mapped in Fig. 7.3, and that many of the countries lying closer to the origin of the axes in Fig. 7.2 are non-African, leaving fewer candidate countries to map onto Table 7.2 in the same way. In further work, we hope to expand this mapping exercise to more countries, so that we can make a comparison with the policies that are actually being pursued in those countries. Nonetheless, we can see from this exercise that countries which face some of the greatest barriers to achieving export competitiveness (and domestic market viability) in agricultural food markets also face the same challenges when trying to expand to bioenergy products. Therefore, prioritization of crop technology investments and R&D to boost productivity and marketing cost reductions through marketing infrastructure improvements and lowering of transactions costs within the distribution network can serve to improve both food security outcomes as well as energy security outcomes, related to bioenergy production.

7.4.5 CONCLUSIONS

In this chapter, we have laid out some ideas to better organize the thinking of policy analysts and decision-makers around bioenergy development, and how the growth of such a sector in the pursuit of energy security can either generate synergies or cause unintended trade-offs with food security. We have reviewed some of the key literature that underpinned the "food versus fuel" debate and have shown that its almost exclusive focus on liquid biofuels for transportation may have overlooked some important ways in which bioenergy development—more broadly—might provide socioeconomic benefits. Some of these points have come out in more recent literature on bioenergy and food security linkages, which we point out to the reader.

By drawing upon a conceptual framework that links household-level well-being more tightly to energy use and availability, we highlighted some of the important gendered dimensions of bioenergy and food security that policy makers need to be aware of when accounting for the human welfare gains generated by policies that expand and improve energy access. These gender-specific benefits come from the time use savings that are generated if women's unpaid time burdens were lightened by access to improved energy sources, thereby allowing them more time to pursue other income-generating or welfare-enhancing activities. We also used this framework to explain the difference in welfare impacts that various household types might experience when faced with a food or energy price shock and related it to empirical results found in the literature.

The prioritization of bioenergy development that we propose in the chapter comes from considering the overlay of energy poverty levels with productivity and market potential of bioenergy products, in terms of both export market competitiveness, as well as its viability within domestic markets. Our typology draws a distinction between those countries whose focus is better spent on promoting rural energy access and security within its own borders, rather than trying to scale up bioenergy production and processing activities for export markets. From this exercise we found a close connection between countries that already struggle with low agricultural productivity and those with limited potential to develop large-scale bioenergy production schemes, thereby suggesting that the improvements in the functioning of markets and performance of crop technologies could lead to improvements in both food and energy security.

In further work, we hope to expand this typology to include a larger array of countries—both within Africa as well as in Asia and Latin America—so that a more complete global picture of bioenergy development potential and the risks or rewards it poses to food security. This work departs from more market-level, equilibrium model-based analyses that we (among many other researchers) have done on bioenergy growth and food security and offers a useful perspective on the underlying socioeconomic characteristics of households that determine the nature and magnitude of welfare impacts coming from bioenergy production growth and development. We advocate for continued study and examination of this issue at both the micro- and macro-levels so that economic decision-making and policy design can be better informed by the full range of evidence on bioenergy and food security impacts that have emerged in the decade following the 2007–08 food price crisis.

REFERENCES

Ali, A., 2016. The Impact of Electricity Outages on Households. University of Toronto. Available from: https://tspace.library.utoronto.ca/bitstream/1807/76179/3/Ali_Ayesha_201611_PhD_thesis.pdf.

Arndt, C., Benfica, R., Tarp, F., Thurlow, J., Uaiene, R., 2010a. Biofuels, poverty, and growth: a computable general equilibrium analysis of mozambique. Environ. Dev. Econ. 15 (1), 81–105. Available from: https://doi.org/10.1017/S1355770X09990027.

Arndt, C., Msangi, S., Thurlow, C.J., 2010b. Are Biofuels Good for African Development? An Analytical Framework with Evidence from Mozambique and Tanzania. 110/2010. WIDER Working Paper Series. WIDER Working Paper. World Institute for Development Economic Research (UNU-WIDER). Available at: https://www.wider.unu.edu/sites/default/files/wp2010-110.pdf.

Arndt, C., Pauw, K., Thurlow, J., 2012. Biofuels and economic development: a computable general equilibrium analysis for Tanzania. Energy Econ. 34 (6), 1922–1930. Available from: https://doi.org/10.1016/j.eneco.2012.07.020.

Bailey, R., 2008. Another Inconvenient Truth: How Biofuels Policies Are Deepening Poverty and Accelerating Climate Change. 114. Oxfam Briefing Paper. Available at: https://d1tn3vj7xz9fdh.cloudfront.net/s3fs-public/file_attachments/bp114-inconvenient-truth-biofuels-0806_3.pdf.

Banse, M., van Meijl, H., Tabeau, A., Woltjer, G., 2008. Will EU biofuel policies affect global agricultural markets? Eur. Rev. Agric. Econ. 35 (2), 117–141. Available from: https://doi.org/10.1093/erae/jbn023.

von Braun, J., 2008. Biofuels, International Food Prices, and the Poor. Washington, DC. Available at: http://ebrary.ifpri.org/cdm/ref/collection/p15738coll2/id/10408.

Clancy, J., 2013. Biofuels and Rural Poverty. Routledge. Available at: https://www.routledge.com/Biofuels-and-Rural-Poverty/Clancy/p/book/9781844077199.

Condon, N., Klemick, H., Wolverton, A., 2015. Impacts of ethanol policy on corn prices: a review and meta-analysis of recent evidence. Food Policy 51 (February), 63–73. Available from: https://doi.org/10.1016/j.foodpol.2014.12.007.

Crago, C.L., 2014. "Measuring the welfare impact of biofuel policies: a review of methods and findings from numerical models.". In: Pinto, A.A., Zilberman, D. (Eds.), Modeling, Dynamics, Optimization and Bioeconomics. Springer, Cham, pp. 221–236. Available from: https://doi.org/10.1007/978-3-319-04849-9_13.

Cushion, E., Whiteman, A., Dieterle, G., 2010. Bioenergy Development: Issues and Impacts for Poverty and Natural Resource Management. The World Bank, Washington, DC. Available from: https://doi.org/10.1596/978-0-8213-7629-4.

Elliott, K., 2008. Biofuels and the Food Price Crisis: A Survey of the Issues. Washington, DC. Available at: https://www.cgdev.org/sites/default/files/16499_file_Elliott_Biofuels_Food_Crisis.pdf.

Evans, A., 2008. Rising Food Prices Drivers and Implications for Development. 08/01. Chatham House Briefing Papers. Available at: https://reliefweb.int/sites/reliefweb.int/files/resources/7FBA41E4E121A8284925743B000C5C6E-Full_Report.pdf.

Ewing, M., Msangi, S., 2009. Biofuels production in developing countries: assessing tradeoffs in welfare and food security. Environ. Sci. Policy 12 (4), 520–528. Available from: https://doi.org/10.1016/j.envsci.2008.10.002.

Fargione, J., Hill, J., Tilman, D., Polasky, S., Hawthorne, P., 2008. Land clearing and the biofuel carbon debt. Science (New York, N.Y.) 319 (5867), 1235–1238. Available from: https://doi.org/10.1126/science.1152747.

Food and Agricultural Organization of the United Nations (FAO), 1996. Rome Declaration on World Food Security and World Food Summit Plan of Action. World Food Summit 13–17 November 1996. Rome: FAO. Available at: http://www.fao.org/docrep/003/w3613e/w3613e00.HTM.

Food and Agriculture Organization of the United Nations, 2008. The State of Food and Agriculture 2008. Food and Agriculture Organization of the United Nations. Available at: http://www.fao.org/docrep/011/i0100e/i0100e00.htm.

Foster, V., Briceno-Garmendia, C., 2009. Africa's Infrastructure. The World Bank, Washington, DC. Available from: https://doi.org/10.1596/978-0-8213-8041-3.

Hazell, P.B.R., Evans, M., 2011. Environmental, economic and policy aspects of biofuels. In: Gararraga, I., Gonzalez-Eguino, M., Markandya, A. (Eds.), Handbook of Sustainable Energy. Edward Elgar, pp. 375–392. Available at: https://www.researchgate.net/profile/Peter_Hazell2/publication/295585924_Biofuels_and_food_Mutually_exclusive/links/56f8004a08ae95e8b6d35a9c.pdf.

Headey, D., and S. Fan. 2008. Anatomy of a crisis: the causes and consequences of surging food prices. Agric. Econ. 39 (Suppl 1): 375–391. https://doi.org/10.1111/j.1574-0862.2008.00345.x.

HLPE, 2013. Biofuels and Food Security: A Report by the High Level Panel of Experts on Food Security and Nutrition of the Committee on World Food Security. Rome, Italy. Available at: http://www.fao.org/3/a-i2952e.pdf.

Hochman, G., Rajagopal, D., Timilsina, G.R., Zilberman, D., 2014. Impacts of biofuels on food prices. The Impacts of Biofuels on the Economy, Environment, and Poverty. Springer New York, New York, NY, pp. 47–64. Available from: https://doi.org/10.1007/978-1-4939-0518-8_4.

de Hoyos, R.E., Medvedev, D., 2011. Poverty effects of higher food prices: a global perspective. Rev. Dev. Econ. 15 (3), 387–402. Available from: https://doi.org/10.1111/j.1467-9361.2011.00615.x.

Huang, J., Yang, J., Msangi, S., Rozelle, S., Weersink, A., 2012. Global biofuel production and poverty in China. Appl. Energy 98, 246–255. Available from: https://doi.org/10.1016/j.apenergy.2012.03.031.

Ivanic, M., Martin, W., 2008. Implications of Higher Global Food Prices for Poverty in Low-Income Countries. 4594. Policy Research Working Paper. Washington, DC. Available at: http://documents.worldbank.org/curated/en/253001468150585777/Implications-of-higher-global-food-prices-for-poverty-in-low-income-countries.

Khanna, M., Wang, W., Hudiburg, T.W., DeLucia, E.H., 2017. The social inefficiency of regulating indirect land use change due to biofuels. Nat. Commun. 8 (June), 15513. Available from: https://doi.org/10.1038/ncomms15513.

Kiggundu, N., Kabenge, I., Arhin, S.G., Banadda, N., 2017. Impacts of biofuel policies on welfare and food security: assessing the socioeconomic and environmental tradeoffs

in sub-Saharan Africa. Int. J. Renew. Energy Res. 7 (4), 2162–2171. Available at: http://www.ijrer.org/ijrer/index.php/ijrer/article/view/6272.
Kline, K.L., Msangi, S., Dale, V.H., Woods, J., Souza, G.M., Osseweijer, P., et al., 2017. Reconciling food security and bioenergy: priorities for action. GCB Bioenergy 9 (3), 557–576. Available from: https://doi.org/10.1111/gcbb.12366.
von Lampe, M., 2006. Agricultural Market Impacts of Future Growth in the Production of Biofuels. Paris. Available at: http://www.oecd.org/tad/agricultural-trade/36074135.pdf.
Langeveld, H., J. Dixon, and H. van Keulen. 2014. Biofuel Cropping Systems : Carbon, Land, and Food. Routledge. Available at: https://www.researchgate.net/publication/260405563_Biofuel_cropping_systems_Carbon_land_and_food.
Ludena, C.E., Razo, C., Saucedo, A., 2007. Biofuels Potential in Latin America and the Caribbean: Quantitative Considerations and Policy Implications for the Agricultural Sector. Santiago de Chile. Available at: https://ageconsearch.umn.edu/bitstream/9986/1/sp07lu01.pdf.
Maltitz, G.P.V., Sugrue, A., Gush, M.B., Everson, C., Borman, G.D., Blanchard, R., 2012. Environmental and socioeconomic considerations for Jatropha growing in Southern Africa. In: Gasparatos, A., Stromberg, P. (Eds.), Socioeconomic and Environmental Impacts of Biofuels. Cambridge University Press, Cambridge, pp. 278–306. Available from: https://doi.org/10.1017/CBO9780511920899.019.
Maltsoglou, I., Khwaja, Y., 2010. Bioenergy and Food Security The BEFS Analysis for Tanzania. 35. Environment and Natural Resources Management Working Paper. Rome. Available at: http://www.fao.org/docrep/012/i1544e/i1544e.pdf.
Mitchell, D., 2008. A Note on Rising Food Prices. 4682. Policy Research Working Paper. Available at: http://documents.worldbank.org/curated/en/229961468140943023/A-note-on-rising-food- prices.
Msangi, S., 2012. Biofuels and Africa: impacts and linkages at the household level. In: Gasparatos, A., Stromberg, P. (Eds.), Socioeconomic and Environmental Impacts of Biofuels. Cambridge University Press, Cambridge, pp. 231–253. Available from: https://doi.org/10.1017/CBO9780511920899.017.
Msangi, S., Evans, M., 2013. Biofuels and developing economies: is the timing right? Agric. Econ. 44 (4–5), 501–510. Available from: https://doi.org/10.1111/agec.12033.
Peters, J., Sievert, M., 2016. Impacts of rural electrification revisited – the African context. J. Dev. Effect. 8 (3), 327–345. Available from: https://doi.org/10.1080/19439342.2016.1178320.
Rosegrant, M.W., Zhu, T., Msangi, S., Sulser, T., 2008. Global scenarios for biofuels: impacts and implications. Rev. Agric. Econ. 30 (3), 495–505. Available from: https://doi.org/10.2307/30225893.
Runge, C.F., Senauer, B., 2007. How biofuels could starve the poor. Foreign Aff. 86 (3), 41–53. Available at: https://www.foreignaffairs.com/articles/2007-05-01/how-biofuels-could-starve-poor.
Runge, C.F., Senauer, B., 2008. How ethanol fuels the food crisis. Foreign Aff. May, . Available at: https://www.foreignaffairs.com/articles/global-commons/2008-05-28/how-ethanol-fuels-food-crisis.
Schuenemann, F., Msangi, S., Zeller, M., 2018. Policies for a sustainable biomass energy sector in Malawi: enhancing energy and food security simultaneously. World Dev. 103 (March), 14–26. Available from: https://doi.org/10.1016/J.WORLDDEV.2017.10.011.
Searchinger, T., Heimlich, R., Houghton, R.A., Dong, F., Elobeid, A., Fabiosa, J., et al., 2008. Use of U.S. croplands for biofuels increases greenhouse gases through emissions from land-use change. Science (New York, N.Y.) 319 (5867), 1238–1240. Available from: https://doi.org/10.1126/science.1151861.

Taheripour, F., Thomas, W.H., Tyner, W.E., Beckman, J.F., Birur, D.K., 2010. Biofuels and their by-products: global economic and environmental implications. Biomass Bioenergy 34 (3), 278–289. Available from: https://doi.org/10.1016/J.BIOMBIOE.2009.10.017.

Taheripour, F., Thomas, W.H., Tyner, W.E., Beckman, J.F., Birur, D.K., 2008. Biofuels and Their By-products: Global Economic and Environmental Implications. West Lafayette, IN. Available at: https://www.gtap.agecon.purdue.edu/resources/download/3974.pdf.

Thorbecke, E., 2009. The Anatomy of Growth and Development in Sub-Saharan Africa. Economic Growth and Poverty Reduction in Sub-Saharan Africa. Framework Paper Prepared for African Economic Research Consortium (AERC). doi:10.1093/acprof:oso/9780198728450.003.0002.

Trostle, R., 2008. Global Agricultural Supply and Demand: Factors Contributing to the Recent Increase in Food Commodity Prices. Washington, DC. Available at: https://www.ers.usda.gov/webdocs/publications/40463/12274_wrs0801_1_.pdf?v = 41057.

Zhang, W., Elaine, A.Y., Rozelle, S., Yang, J., Msangi, S., 2013. The impact of biofuel growth on agriculture: why is the range of estimates so wide? Food Policy 38 (February), 227–239. Available from: https://doi.org/10.1016/j.foodpol.2012.12.002.

CHAPTER EIGHT

Land–Water–Energy Nexus of Biofuels Development in Emerging Economies: A Case Study of Bioethanol Policy in Thailand

Thapat Silalertruksa[1,2] and Shabbir H. Gheewala[1,2]
[1]The Joint Graduate School of Energy and Environment, King Mongkut's University of Technology Thonburi, Bangkok, Thailand
[2]Center for Energy Technology and Environment, Ministry of Education, Bangkok, Thailand

Contents

8.1 Emerging Economies and Biofuels Development 380
8.2 Land–Water–Energy Nexus and Sustainability of Food and Biofuels Production 381
8.3 Life Cycle–Based Indicators for Land–Water–Energy Nexus Assessment 384
8.3.1 WF assessment 384
8.3.2 CF assessment 386
8.3.3 Ecological footprint 386
8.3.4 Relationships of the footprints and land–water–energy nexus 387
8.4 A Case Study: Land–Water–Energy Nexus of Biofuels Policy Mandate in Thailand 388
8.4.1 About Thailand and agriculture 389
8.4.2 The government policy on biofuels promotion 389
8.4.3 Water scarcity and the bioethanol policy target 391
8.4.4 Land-GHG emissions nexus of the bioethanol policy target 394
8.4.5 Feedstock balance for food and bioethanol 395
8.4.6 Land–water–energy nexus of irrigation system for improving biofuel crop productivity 396
8.4.7 Nexus management for sustainable bioethanol production in Thailand 397
8.5 Conclusions 398
References 399

The Role of Bioenergy in the Emerging Bioeconomy.
DOI: https://doi.org/10.1016/B978-0-12-813056-8.00008-X

8.1 EMERGING ECONOMIES AND BIOFUELS DEVELOPMENT

Emerging economies have grown fast over the past decades due to their high share of the world population, and consequently, high production as well as consumption. The trends of world energy consumption for electricity as well other forms of final energy indicate a shift toward greater consumption among emerging economies that are not members of the Organisation for Economic Cooperation and Development (OECD) (OECD/IEA, 2013). A notable example is that China has surpassed the United States as the largest net importer of petroleum and other oil-based liquid fuels since 2013 by accounting for more than a quarter of the world's growth in oil consumption (IEA Bioenergy, 2016). Nevertheless, the rapid development has raised the sustainability challenges common to all emerging economies, i.e., energy security, energy affordability, and environmental quality protection (OECD/IEA, 2013).

Biofuels have been recognized as one of the promising alternative energy carriers to be promoted to counter the energy security and climate change challenges, especially for the countries in Asia, where the economies are still highly reliant on the agro-industry and there is rich variety of biomass resources available (Yan and Lin, 2009). The expected benefits from the biofuels promotion are to maximize the use of the indigenous feedstocks for bioenergy which in turn would reduce dependency on oil import and mitigate the climate change impact. In addition, the new biofuel industry can bring about rural development through employment and income generation for farmers along with associated workers. The farmers' income from selling their crops can be stabilized because the biofuel industry generally has a constant demand for feedstocks.

Currently, the biofuels industry can be broadly classified into two distinct sectors: bioethanol and biodiesel. The bioethanol production in 2016 was approximately 100 billion liters, the major producers being the United States (58%), Brazil (28%), Europe (5%), China (3%), Canada (2%), and Thailand (1%) (Statista, 2017). The global bioethanol production almost tripled from 17.1 billion liters in 2000 and 49.5 billion liters in 2007 (Statista, 2017). Meanwhile, for biodiesel production, the total production was approximately 24 billion liters in 2016 with the leading producers being the United States (23%), Brazil (16%), Germany (13%),

Indonesia (3%), Argentina (13%), and Thailand (6%) (Statista, 2017). This global biodiesel production increased more than 10 times from 0.9 billion liters in 2000 and 10.2 billion liters in 2007 (WBGU, 2010). The vast majority of feedstocks for commercial bioethanol production currently are crops such as corn, sugarcane, molasses, and cassava. Meanwhile, the major feedstocks for biodiesel are the vegetable oils derived from oil palm, soybean, rapeseed, and sunflower.

The widespread use of biomass for biofuels is facing the challenge on resource scarcity because the arable land, freshwater, as well as nutrient availability essential to its production, are finite. This is especially for the freshwater resource whose availability is limited spatially and temporally. Increased use of land and water for fulfilling the biofuel feedstock demand has posed concerns on food and fuel competition. In addition, expansion of cropland for biofuel might result in biodiversity and habitat losses, soil erosion, and increased GHG emissions due to land-use change (LUC) and land degradation. All these potentially limit the ability of the biofuels produced from food crops to achieve the target set in the developing countries as alternative fuels for transport to substitute fossil fuels. Higher concern is attributed to the developing countries because most of them generally have lower efficiency and productivity as compared to the developed regions; they also have the widened gaps between policy and implementation (Gheewala et al., 2017a). To counter the sustainability challenges of biofuels development, in recent years, the terms "Nexus" and "Nexus assessment" have been widely discussed among researchers as well as policy makers because it is necessary to have the integrated research approaches and policy formulations to address often complicated interactions, such as the water–food, water–energy, water–food–energy, land–water–energy, and climate-related nexus (WWAP, 2014; Endo et al., 2017).

8.2 LAND–WATER–ENERGY NEXUS AND SUSTAINABILITY OF FOOD AND BIOFUELS PRODUCTION

The global challenges of food and energy production due to the freshwater and arable land scarcity can be revealed by the statistical records. For example, agriculture accounts for around 70% of freshwater withdrawals from rivers and aquifers globally and up to more than 90% in some developing countries (UNESCO, 2009; Gheewala et al., 2011). In addition, agriculture also occupies more than 50% of the vegetated land (Foley et al., 2005). On the other hand, the world population is forecasted to increase to 9.6 billion by 2050 (UNDESA, 2012); this implies the need for increasing global food production by an estimated 60% to meet demand (FAO, 2012) which in turn will lead to more requirements for land and water for agriculture. Meanwhile, more than 1.2 billion of the world's population has already been living in water-scarce areas (CA, 2007). It is estimated that the increasing demand for irrigation coupled with growing water use in the domestic and industrial sectors will increase the number of people at risk from water stress to one-third of the world's population by 2050 (De Fraiture et al., 2010). For energy consumption, food production is also significant as it has accounted for 30% of the global energy demand (FAO, 2011). The GHG emissions from food production contribute 19%–29% of the global anthropogenic emissions, 80% of which is from agriculture (Vermeulen et al., 2012). Hence, if business as usual continues, this will create unprecedented pressures which will exacerbate current sustainability challenges associated with provision of food, energy, and water (Azapagic, 2015).

Nowadays, there is an increasing concern about the Land–Water–Energy Nexus because these resources are in a close relationship and essential to economic and social development. The nexus approach has been proposed to enhance efficiency and balance between the different uses of ecosystems resources like water, land, and energy by various stakeholders in a particular region (Flammini et al., 2014; Azapagic, 2015; Sanders and Masri, 2016; Smajgl et al., 2016). Fig. 8.1 shows the simplified linkage and trade-offs between land, water, and energy in agriculture as well as food and bioenergy production (Silalertruksa and Gheewala, 2018). According to the figure, the "energy" can be represented as either energy for economic activities such as fuels, electricity, or energy for human (i.e., food). Hence, the increase in the production of energy carriers like biofuels will require the land for biomass feedstock cultivation and water for crop's irrigation. In the meantime, energy is used to distribute the water for agriculture and other uses. Also, energy is used for agricultural machinery since the land preparation, crop plantation until harvesting. The land–water–energy trade-off is such that the use of irrigated water might be able to increase crop yield which leads to improved land productivity; however, this has to trade off

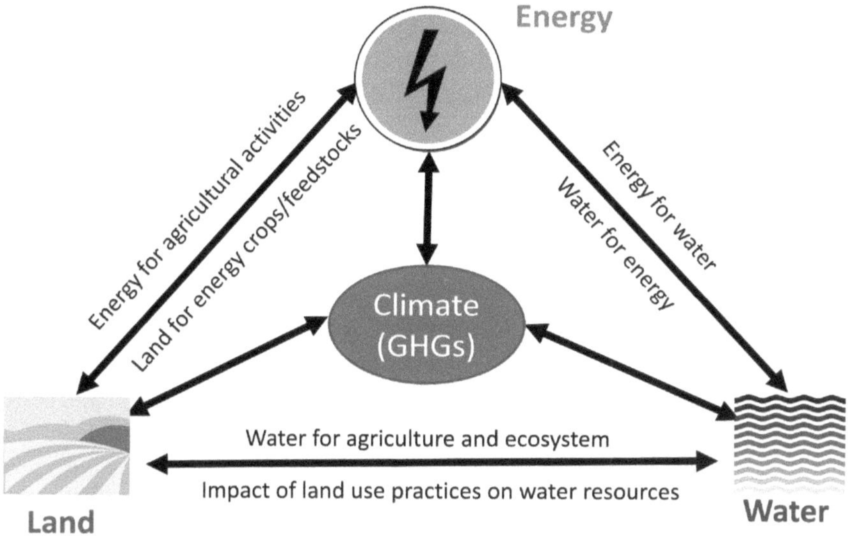

Figure 8.1 Land–water–energy–climate nexus in agriculture.

with more energy required for irrigation and more potential water competition with other water users in the region. Recognizing these synergies and balancing, these trade-offs are therefore essential to ensure the land, water, fuel, and food security. Thus, it is gaining attraction as the key issue that needs to be understood in biofuel policy-making to move the implementation of policy towards the sustainable biofuel production.

Land use is in general concerned with the sustainable standards of food and bioenergy production. The key issue of discussion is the LUC consequences from expansion of agriculture such as forest land conversion to cropland which will deteriorate biodiversity-abundant ecosystems and cause a substantial amount of greenhouse gas (GHG) emission (Tilman et al., 2002, 2006). In addition, utilization of land for biofuels can compete with food production when the agricultural expansion of energy crops takes place on the prior cropland. Nevertheless, water is often implicitly considered in conventional biofuel potential studies since the amount of precipitation and soil-water characteristics are two of the parameters determining suitability for biofuel production. But so far the studies provide little insight into local and regional water consequences of actually realizing the assessed biofuel potentials by shifting land use to provide feedstock for biofuels (Gheewala et al., 2013, 2017b). It is of utmost importance for the future state of the freshwater resource of how—and where—biomass is produced for energy.

8.3 LIFE CYCLE–BASED INDICATORS FOR LAND–WATER–ENERGY NEXUS ASSESSMENT

To understand the interaction of land, water, and energy for food and biofuels production, life cycle–based assessment is recognized as a crucial concept used to bind the whole interrelationship of those all resources in the food and fuel supply chain. The aims of the nexus assessment generally are to improve the resource use efficiency along with economic efficiency and livelihood options (Bazilian et al., 2011). A variety of approaches have been introduced for nexus assessment including life cycle assessment (LCA) and water footprint (WF) (Vanham, 2016; Scholz et al., 2015; Jeswani et al., 2015; Brancoli et al., 2017), simulation and optimization (Garcia and You, 2016), and numerical modeling and economic modeling (Jalilov et al., 2016; Pacetti et al., 2015). However, in this section, three life cycle–based indicators that are used as the stand-alone indicators for evaluating the environmental impacts on water, land, and energy along with the climate change are selected as the nexus assessment tools viz., WF, ecological footprint (EF), and carbon footprint (CF).

8.3.1 WF assessment

The International Organization for Standardization (ISO) has recently announced the principles, requirements, and guidelines related to WF assessment in the ISO 14046. The ISO-WF is developed based on the LCA concept by defining the term of "water footprint" as a metric that quantifies the potential environmental impacts related to water (ISO 14046, 2014; Pfister et al., 2017). There are different types of WF defined in ISO-WF such as "water scarcity footprint" and "water availability footprint," a particular type of WF to be used depending upon the goal and scope of the WF study (ISO/TR 14073, 2017). The water scarcity footprint is used to assess the contribution of the product or production systems to water scarcity by considering the quantity of water consumption and the water stress index (WSI) of the region where the water is extracted, to determine the impact of freshwater consumption in view of the water deprivation potential (Pfister et al., 2009; Pfister and Bayer, 2014). Meanwhile, the water availability footprint is used to present the impacts from the lower water availability due to water consumption and degradation (ISO/TR 14073, 2017).

The WF assessment is widely used as a tool to enhance water use efficiency, sustainability of water use within the watersheds, and mitigate impact of water use by improving water resource management (Pfister and Bayer 2014; Lovarelli et al., 2016). This is especially for the agricultural products as well as biofuels for which generally a substantial amount of freshwater is required during the cultivation of crops (Mekonnen and Hoekstra 2011; 2014). To determine the WF, firstly, the amount of water evapotranspired by the crops in a specific climate regime and adequate soil water is maintained by rainfall and/or irrigation will be estimated by the general formula (Eq. (8.1)) (Allen et al., 1998). ET_C represents the crop evapotranspiration. K_C represents the crop coefficient of Penman–Monteith, and ET_0 represents to the reference crop evapotranspiration of Penman–Monteith. The crop evapotranspiration (ET_C) and the effective rainfall can be calculated for the given set of data on ET_0, monthly rainfall, K_C, and the crop calendar.

$$ET_C = K_C \times ET_0 \tag{8.1}$$

To determine the water scarcity footprint of biofuels, Eq. (8.2) shows the general formula for the "water privation potential" or the so-called as "water scarcity footprint."

$$\text{Water scarcity footprint}_{\text{product},i} = \text{Water consumption}_{\text{product},i} \times \text{WSI}_i \tag{8.2}$$

where, water consumption$_{\text{product},i}$ represents the amount of freshwater consumed for the production of product since the biofuel crop cultivation, feedstock processing until the biofuel conversion process (Gheewala et al., 2013); WSI_i represents the water stress index of watershed (i) where the water is consumed. WSI or the so-called "water scarcity index" commonly refers to the indices that are developed to quantitatively evaluate water resources vulnerability. Although many water stress indices have been proposed, the general indicator is defined as the ratio of total annual freshwater withdrawals to hydrological availability (Brown and Matlock, 2011). WSI developed by Pfister et al. (2009) for assessing the environmental impacts of freshwater use is a well-established method applied as the characterization factor for the midpoint category, the so-called water deprivation (Pfister et al., 2011; Ridoutt and Pfister, 2013; Gheewala et al. 2013; 2014; 2017b). To maintain the same structure of unit based on the LCA approach, a nonconventional unit "m^3 H_2Oeq" is defined as the unit of water deprivation in m^3 water equivalents (Gheewala et al., 2014).

8.3.2 CF assessment

CF is an indicator used to indicate the total amount of GHGs emitted over the life cycle of products, i.e., from the extraction of raw materials, production, transportation, use, to waste disposal. The CF is therefore known as the life cycle impact category indicator "global warming potential (GWP)" in LCA. For land–water–energy nexus of biofuels, the assessment of CF is essential to explain the impacts related to water, land, and energy use for agriculture as well as for biofuel production on GHG emission which in turn relates to the climate change impact potential as shown in Fig. 8.1. The CF assessment will account for the emissions of all GHGs including CO_2, N_2O, and CH_4, and families of gases such as hydrofluorocarbons and perfluorocarbons, and each gas is converted into a CO_2 equivalent value or expressed in terms of "kg CO_2eq." The basic formula to evaluate the total GHG emissions of biofuel is as shown in Eq. (8.3) (Silalertruksa and Gheewala, 2011).

$$E_{\mathrm{total}} = E_{\mathrm{luc}} + E_{\mathrm{cul}} + E_{\mathrm{fp}} + E_{\mathrm{ecv}} + E_{\mathrm{u}} + E_{\mathrm{td}} + E_{\mathrm{ext}} - E_{\mathrm{crd}} \quad (8.3)$$

where E_{total} is the total GHG emissions from the entire the life cycle of biofuel (kg CO_2 eq/unit of biofuels); E_{luc} is the GHG emissions from carbon stock (C-stock) changes caused by LUC and management; E_{cul} represents the GHG emissions from cultivation and harvesting of feedstocks; E_{fp} is the GHG emissions from feedstock processing; E_{ecv} represents GHG emissions from biofuel conversion; E_{u} represents the GHG emissions from the use of biofuel in vehicle; E_{td} is the GHG emissions from transport and distribution over the entire life cycle; E_{ext} is the GHG emissions from the extraction of input material over the entire life cycle; and E_{crd} represents GHG credits obtained from the biofuel system (e.g., credits for biogas recovery, excess electricity from cogeneration, etc.).

8.3.3 Ecological footprint

The EF is a measure of the area of biologically productive land and water that is required for an individual or an activity to produce all the resources it consumes and to absorb the waste it generates, using prevailing technology and resource management practices (Wackernagel and Rees, 1997). The impact of resource use can be estimated by the level of stress in terms of the EF by comparing with the carrying capacity of the planet in terms of "global hectare (gha)." However, in the concept of LCA, the EF of a product is defined as the sum of time-integrated direct land occupation (EF_{direct}) and indirect land occupation, related to nuclear energy use and to CO_2 emissions ($EF_{CO_2} + EF_{nuclear}$) from fossil energy use and cement burning (Huijbregts et al., 2008). Eq. (8.4) shows the general formula of EF-LCA of biofuel that can be scoped as the sum of direct land occupation (EF_{direct}) and indirect land occupation (m^2a) related to CO_2 emissions from fossil energy use (EF_{CO_2}) (Frischknecht et al., 2007).

$$EF = EF_{direct} + EF_{CO_2} + EF_{nuclear} \tag{8.4}$$

The direct land occupation is divided into five types of lands: built-up area, forest, cropland, pasture, and hydropower area; the EF_{direct} can be calculated by Eq. (8.5).

$$EF_{direct} = \sum_a A_a \times EqF_a \tag{8.5}$$

where A_a is the occupation of area classified by land use type a (m^2a), EqF_a is the equivalence factor of land use type a (dimensionless). The equivalence factor is used to adjust each type of land in terms of bioproductivity (Huijbregts et al., 2008). The high-productivity lands, such as cropland, have a higher EqF than the low-productivity lands such as pastures. The equivalence factors for forest, built-up area, cropland, hydropower, and pasture are about 1.4, 2.2, 2.2, 1.0, and 0.5, respectively, as referred from Wackernagel et al. (2005). However, in the life cycle impact assessment, the actual yield is used for crops to obtain the direct land occupation A (Huijbregts et al., 2008), instead of global average yields as referred from Wackernagel et al. (2005). This is to understand the efficiency of land use which differs widely based on the agricultural practices, climate, and productivity. The indirect EF as the land occupation by fossil CO_2 (EF_{CO_2}) is estimated as the additional biologically productive area required to sequester atmospheric fossil CO_2 and calcination CO_2 from cement burning through afforestation. The EF factor is about 2.7 m^2a/kg CO_2 emitted (Huijbregts et al., 2008). The EF factor of indirect land occupation by the use of nuclear energy is about 0.2 m^2a/MJ consumed nuclear primary energy (Huijbregts et al., 2008).

8.3.4 Relationships of the footprints and land–water–energy nexus

To assess the nexus on land, water, energy, and GHGs, which occurs from biofuels production and use, it is necessary to understand the scope

Table 8.1 Relationships of the footprints and land–water–energy nexus

	EF (land occupation)	Water scarcity footprint (water use and scarcity)	CF (GHGs)
Land 	Land occupation for biofuel crop cultivation		GHGs emission from land use/LUC for biofuel crop cultivation
Water 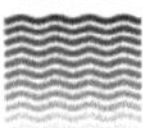		Impact of irrigation water use in view of water deprivation potential	GHG emission from water treatment, irrigation water use, and water distribution
Energy 	Land occupation for energy production	Impact of water use for energy production and use in view of water deprivation potential	GHG emissions from energy production and use in processes
GHGs 	Land occupation for capturing CO_2 emissions		Total GHG emissions of biofuel

of the each footprint because each footprint indicator generally has its own focus and way of interpretation. Table 8.1 shows the overlapping of the three footprint indicators on impact assessment of land, water, and energy resources' use for biofuels production systems. EF overlaps to some extent with the CF, i.e., the impact of CO_2 emission will be taken into account in terms of land equivalent for capturing it. It can be seen that there is no one footprint indicator that can capture and explain the whole land–water–energy nexus of biofuel production. CF is the indicator that can reveal the whole GHG emissions associated with the use of land, water, and energy for biofuel production but the results cannot reveal the impact in view of resource use and scarcity which is a key issue of the nexus assessment. Thus, there is still a gap for future improvement of the footprint indicators for capturing the land–water–energy nexus impact.

8.4 A CASE STUDY: LAND–WATER–ENERGY NEXUS OF BIOFUELS POLICY MANDATE IN THAILAND

This section shows a case study of biofuels policy promotion target in Thailand and its implication on land–water–energy nexus using several life cycle–based assessment tools including the water scarcity footprint, CF, along with feedstock balance assessment. The case study addresses a concern on environmental impact of the biofuel promotion policy target of the government which is set at 9 ML/day of bioethanol production in the year 2021–22. The data presented for nexus analysis in the chapter are from previous studies related to the sustainability assessment of biofuel in Thailand (Gheewala et al., 2013, 2014; Silalertruksa and Gheewala, 2010, 2011, 2018).

8.4.1 About Thailand and agriculture

Thailand is located in the south eastern region of Asia, between 5°N and 20°N and 97°E and 105°E. The climate region is tropical where a variety of crops, fruits, and plants grow well. Thailand is therefore known as one of the world's leading countries in agricultural production and exports. Of the total country's land area of about 51.3 million hectares, 46% is agricultural land, followed by 32% of forest land and 22% of other lands (OAE, 2016). Rice (paddy) fields occupy the highest at around 47% of the total agricultural land followed by perennial crops and fruit orchards, cropland, vegetable and flowers, and others at about 23%, 21%, 1%, and 8%, respectively. This has led Thailand to be the sixth-largest rice producer and the world's leading country for rice export; to be the world's largest cassava producer and exporter contributing about 70% of the world market share; and to be recognized as the world's second leading sugar exporter. Moreover, Thailand is also a key producer of other agricultural commodities such as palm oil, natural rubber, maize, beans, fruits, and vegetables (OAE, 2016). Fig. 8.2 shows the map of Thailand classified by geographical and hydrological boundaries. Based on geographical boundaries, the country can be divided into five regions: North, Northeast, Central, East, and South covering all the 76 provinces. Based on hydrological boundaries, the country can be divided into 25 major watersheds.

8.4.2 The government policy on biofuels promotion

The Thai government has promoted the use of alternative energy as a national agenda since 2004, especially bioethanol derived from the indigenous feedstocks such as molasses, cassava, and sugarcane. Thailand's first 15 years renewable development plan (REDP 2008–22) was implemented in 2008 and the goals of bioethanol production were set at 9.0 ML/day by 2022. Promotion was focused on the ethanol derived

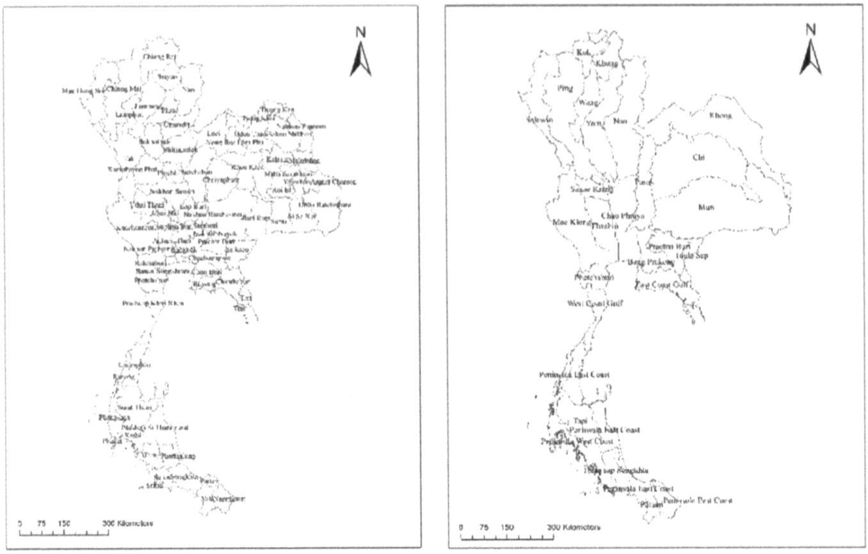

Figure 8.2 Map of Thailand classified by geographical and hydrological boundaries.

from cane molasses, cassava, and sugarcane. At the beginning, promotion strategies started from blending 10% ethanol into octane 95 gasoline (also called E10) to replace methyl tertiary butyl ether as well as to replace 10% of the oil content in octane 91. In 2008, as E10 was already well established in the market, a 20% ethanol blend (E20) was introduced. Later in the same year, E85 gasohol was launched. The biofuel target has been revisited again in 2012 to be the Alternative Energy Development Policy (AEDP 2012–21) which still maintains the target of producing 9 ML/day of bioethanol set for the year 2021 (DEDE, 2012).

The production of ethanol has continuously been increasing, tripling in only 6 years from 1.2 ML/day in 2010 to 3.7 ML/day in 2016 (DEDE, 2017). About 59% of the total production of ethanol produced is from molasses, followed by cassava (37%) and sugarcane juice (4%). The total ethanol production capacity of the 21 existing ethanol plants amounts to 4.19 million liters/day with 64% from molasses, 31% from cassava, and 5% from sugarcane juice (BOT, 2017). However, this growth is inclined to continue as per the ambitious goal of "AEDP (2012–21)" which set to produce 9 ML ethanol/day by 2021. The proliferation of bioethanol production promises to increase stress on land, water, as well as food competition, especially the water scarcity, because the agriculture currently consumes 73% of active freshwater storage in the country. This is of particular concern because Thailand has a large agricultural base for both food for local consumption and export as well as for feed and fuel. The examples of quantitative assessment of the environmental impacts related to land–water–energy nexus of bioethanol production are summarized in the following sections.

8.4.3 Water scarcity and the bioethanol policy target

The implications of the bioethanol policy mandate of producing 9 ML ethanol/day by 2021 on water use and water deprivation can be revealed by the water scarcity footprint (Gheewala et al., 2013). The assessment has been done by determining the water requirement for the bioethanol in Thailand, i.e., molasses ethanol and cassava ethanol by separating into rainwater, irrigation water, and process water required from biofuel crop production to ethanol conversion process. Table 8.2 summarizes the WF of bioethanol in Thailand, which is seen to vary between 2372 and 2838 L water/L cassava ethanol and 1976–3105 L water/L molasses ethanol. The molasses ethanol production uses more surface and groundwater, with the feedstock cultivation stage sharing around 97%–98% of the total surface/groundwater use. The industrial stages such as sugar milling and ethanol conversion contribute just about 2%–3% of the total surface/groundwater consumption. However, the same amount of water use may result in the different magnitude of impacts depending on the water scarcity level that exists in each region. The impact of water use is therefore evaluated in terms of water deprivation potential or so-called "water scarcity footprint" of bioethanol. The WSIs for 25 major watersheds of Thailand have been referred as the characterization factor (Gheewala et al., 2014). The water scarcity footprint is measured in "m^3 H_2Oe" and represents a surrogate indicate for the amount of water deficiency to downstream human users and ecosystems. For example, the water scarcity footprint of molasses ethanol in the Mun and Chao Phraya watersheds are ranging between about 468–591 m^3 H_2Oe/L ethanol and 224–235 m^3 H_2Oe/L ethanol. This higher water scarcity footprint of molasses ethanol in Mun implies that the production of molasses ethanol in Mun potentially results in the higher impact on water competition as compared to molasses ethanol production in Chao Phraya. Table 8.2 indicated that Mun and Chi, the two important watersheds in the northeast of Thailand, would potentially face serious water stress if the water resources are not properly managed.

Table 8.2 Water requirement for cassava and molasses bioethanol production in Thailand

		Water requirement (L water/L bioethanol)			Water scarcity footprint (m^3 H_2Oe/L bioethanol)	
		Rain water	**Surface/groundwater**	**Total WF**	**Key watersheds**	**Water scarcity footprint**
Cassava ethanol	Average range	2051 (1806–2389)	528 (449–566)	2582 (2372–2838)	Mun	442–548
					Chi	267–320
					Chao Phraya	266
					Prachinburi	6–7
Molasses ethanol	Average range	1711 (1147–1885)	829 (699–1220)	2410 (1976–3105)	Mun	468–591
					Chi	279–337
					Chao Phraya	162–367
					Tha Chin	224–235

Table 8.3 Estimated water requirement for achieving the policy mandate of 9 ML/day of bioethanol production in Thailand (Classified by type of ethanol plants)

	Water requirements (Million m^3/year)					
	Dry season		Wet season		Total	
	Rain water	Surface/ ground water	Rain water	Surface/ groundwater	Rain water	Surface/ groundwater
Cassava ethanol plants	754	933	3974	25	4728	958
Sugarcane ethanol plants	8	17	75	1	83	18
Molasses ethanol plants	102	295	1039	159	1141	454
Multi-feedstock plants	74	146	534	49	608	195
Total	938	1391	5622	234	6560	1625

The potential impacts on water use with respect to the policy mandate of bioethanol production in Thailand are summarized in Table 8.3. The assessment has been done by assuming that the AEDP's target of producing 9 ML ethanol/day by 2021 will be satisfied by the 48 licensed ethanol plants. The scenarios assume that only molasses, cassava, and sugarcane juice are the feedstocks. The ratios of feedstocks used for the 8 multi-feedstock bioethanol plants are assumed on a 50/50 basis by sharing between the molasses and cassava. It was estimated that, to satisfy the AEDP target in 2021, around 8185 million m^3/year would be required with 6560 million m^3 rain water (80%) and 1625 million m^3 of surface water/groundwater (for irrigation). The highest amount of the additional irrigation water requirement is for cassava ethanol production, i.e., around 958 million m^3 of surface/groundwater, followed by for molasses ethanol production and multi-feedstock ethanol production plants, i.e., 454 and 195 million m^3, respectively (Table 8.3). This total additional amount of water required in the future is equivalent to about 3% of the current active water storage of Thailand. Mun, Chi, and Prachinburi are the key watersheds that would have the significant increase in irrigation water demand. Considering the water scarcity footprint (m^3 eq/year), Mun and Chi would be the two main watersheds that have high potential to confront the pressures on water stress and competition with other users if the water resources were not properly managed in the future.

Importantly, both watersheds are in the Northeastern region of Thailand which has the largest crops plantation areas in Thailand. By provinces, Nakhonratchasima, Ubon Ratchathani, and Chaiyaphum are the three provinces under the Mun and Chi watersheds that potentially have the high impact on water use due to the ethanol policy mandate.

8.4.4 Land-GHG emissions nexus of the bioethanol policy target

To satisfy bioethanol production as target in 2021, five scenarios for the future bioethanol production systems are developed using the key assumption that the 48 registered bioethanol plants will be operated and the additional feedstocks like cassava and sugarcane will be produced from different scenarios as follows (Silalertruksa and Gheewala, 2011):

- *Base case*: Cassava and sugarcane yields are projected to increase to reach the genetic potentials of the current varieties (i.e., 50 and 125 t/ha, respectively). Hence, expansion of new cultivated areas will not be required.
- *Scenario 1*: New plantation areas for cassava and sugarcane take place on grassland.
- *Scenario 2*: The new plantation areas for both cassava and sugarcane take place on grassland. However, the ethanol systems widely adopt the sustainability measures (e.g., waste utilization and replacing coal by biomass in ethanol plants).
- *Scenario 3*: New plantation areas for cassava and sugarcane take place on forest land.
- *Scenario 4*: The new plantation areas for cassava and sugarcane take place on forest land. However, the ethanol systems widely adopt the sustainability measures (e.g., waste utilization and replacing coal by biomass in ethanol plants).

Land use and LUC for ethanol feedstock production were found to be the key linkage to the GHG performance of biofuels. The average GHG emissions of the overall bioethanol system in Thailand would possibly be 0.49−3.7 kg CO_2-eq/L ethanol (Table 8.4). Nevertheless, the GHG performances of the Thai bioethanol system are likely to decrease in the long term due to the effects of expansion of plantation areas to satisfy the deficits of cassava and molasses. For example, Scenarios 3 and 4 are worse for the future development of ethanol production in Thailand, i.e., the forest land is trespassed by farmers for cassava and sugarcane cultivation. This would bring about the carbon stock loss which in turn is taken into account in the GHG emissions. Hence, the net avoided GHG emissions

Table 8.4 Estimation of the net GHG emissions and the average GHG emission factors of future bioethanol production system in Thailand*

	Average GHG emission factors of bioethanol (kg CO_2-eq/L ethanol)	Net avoided GHGs emission compared to gasoline (M tons CO_2-eq/year)	GHGs emissions reduction compared to gasoline (%)
Base case	0.49	4.63	74
Scenario 1	1.84	0.20	3
Scenario 2	0.85	3.45	55
Scenario 3	3.70	− 5.91	−95
Scenario 4	2.71	− 2.68	−43

*Negative values (−) means bioethanol does not provide GHG emission benefit as compared to gasoline.

of bioethanol as compared to gasoline for Scenarios 3 and 4 are negative (Table 8.4). This implies that using ethanol where feedstocks are from plantation areas on forest land (Scenarios 3 and 4) to substitute gasoline will increase the GHG emissions by about 5.91 and 2.68 M tons CO_2-eq/year, respectively. The best result is found for the base case, i.e., yield improvement for both cassava and sugarcane which brings about the highest GHG emission reduction as compared to gasoline. Thus, improvements of sugarcane and cassava yields as per the expectation by the government through the high-yield varieties' development and adopting good and appropriate varieties nationwide would be a vital measure to avoid direct LUC and its consequential impacts on GHG emissions. Deforestation is actually illegal and prohibited by the government; however, the strict enforcement must be required to ensure that the biofuel policy promotion will not cause the drawback to the environment.

8.4.5 Feedstock balance for food and bioethanol

Silalertruksa and Gheewala (2010) have addressed the concern about long-term security of feedstock supply, i.e., molasses and cassava for achieving bioethanol target and the competition with the food, feed, and other sectors. The net feedstock balance was used as the indicator to describe the physical availability of feedstocks to supply for bioethanol production. The resources potential for sustainable bioethanol production in Thailand is defined as the feedstock surplus available after satisfying the demand for food and the other existing industries by considering on both domestic and export demand. This is for policy makers to avoid the impacts to other current industries. Three scenarios for yields

Table 8.5 The net feedstock balances after accounting for the projected bioethanol demand in 2021–22

Net balances (Mt feedstocks/year)	Molasses	Cassava	Sugarcane
Scenario 1: Low yields improvement	− 0.17	− 20.95	6.24
Scenario 2: Moderate yields improvement	− 0.08	− 20.63	8.23
Scenario 3: High yields improvement	1.44	− 0.48	41.18

improvement have been set. Scenario 1: Low yields improvement (Crop yields are projected to continue growing as usual); Scenario 2: Moderate yields improvement (Crop yields are anticipated to be improve as per the short-term government's policy targets as mentioned in the renewable development plan); and Scenario 3: High yields improvement (Crop yields are projected to increase to reach the genetic potential of cassava and sugarcane varieties). It was found that only the high yield scenario could ensure that the country has enough cassava for catering to the export market and still have surplus feedstock for bioethanol production in the year 2021–22 (Table 8.5). The deficits of cassava supply for bioethanol in the future could possibly be fulfilled by decreasing the export of cassava products. However, this would affect the existing dried chips and starch industry in the country and the global cropping systems due to the loss of cassava in the global market. On the other hand, sugarcane juice would play an important role as feedstock for the medium- and long-term bioethanol production in Thailand due to its surplus availability and high net balances as compared to cassava and molasses.

8.4.6 Land–water–energy nexus of irrigation system for improving biofuel crop productivity

From the case studies mentioned earlier, yields improvement was found to be the key for sustainable biofuel production in the long run. Irrigation water is recognized as one of the key for improving the productivity. Table 8.6 shows an example of the trade-off between the increase of water and energy use for sugarcane irrigation and its productivity improvement using the EF, WF, and CF (Silalertruksa and Gheewala, 2018). The results shown that for the same area of sugarcane cultivation, i.e., Nakhon Sawan province which is located in Chao Phraya watershed, the irrigation system can help spur the yield of sugarcane as compared to the nonirrigated system. The use of irrigation will increase the consumption of energy, i.e., diesel, which in turn also induces the increase of water scarcity footprint and the additional GHG

Table 8.6 EF, water scarcity footprint, and CF of sugarcane production in different conditions

	Indicators	Nakhon Sawan (Chao Phraya watershed)	
		Irrigated	Nonirrigated
Crop yield	Sugarcane (t cane/ha)	111	72
Water use	Rain water (m^3/t cane)	68	105
	Irrigation water requirement (m^3/t cane)	34	52
	Actual irrigation water used (m^3/t cane)	17	
	Total water scarcity footprint (m^3 H_2O eq/t cane)	11	5
Climate change	Life cycle GHG emissions (kg CO_2eq/t cane)	30	47
Land	EF (m^2a/t cane)	242	377
	Direct land occupation (m^2a/t cane)	197	303
	Indirect land occupation from raw materials (m^2a/t cane)	1	2
	Carbon dioxide (m^2a/t cane)	44	72

emissions per hectare. However, accounting for the yields improvement due to irrigation, the CF and EF of the sugarcane product are decreased. Nevertheless, the increased freshwater resources used for irrigation bring about an increase in the water scarcity footprint of irrigated sugarcane.

8.4.7 Nexus management for sustainable bioethanol production in Thailand

The assessment results so far have revealed the significant implication of bioethanol target on the potential water scarcity, LUC, feedstock balance for food, feed, and fuel in the future, and their related GHGs emissions. There was a clear linkage of land, water, energy, and GHGs impacts from the changes of the biofuel production system to satisfy the increased demand for feedstock and biofuels as targeted by the government. To satisfy the government plan of bioethanol production in 2021 but without the drawback to the ecosystem as well as food versus fuel competition, the key measures for improving the productivity of land, water, and energy resources in the country are as follows:

- Crop yield improvement is the most important factor for ensuring the long-term security of feedstock supply for future bioethanol production. Crop varieties improvement and good agricultural practices promotion need to be continuously and widely encouraged by the government. Improvement of the irrigation infrastructure and promotion of the efficient irrigation system such as the subsurface drip irrigation to expand the crop's irrigated areas is also important.

- The use of suitable land for crop along with the consideration of the water stress level in the region should also be considered and promoted by the policy makers. This is to improve the land productivity and avoid the water competition among other users.
- Feedstock balances between the demands of cassava and sugarcane for biofuel as well as the other industries should be forecasted in advance with more precision.
- The biorefinery concept to maximize the use of waste generated over the supply chain of biofuels and producing the value-added products such as bioplastics and biochemicals needs to be encouraged. The biorefinery can be initiated from the conventional system like the vinasse utilization for biogas, fertilizer, as well as animal feed; biomass utilization for steam and power generation; and further on to advanced systems for producing biomaterials and biochemicals which are more value added.

8.5 CONCLUSIONS

The sustainable biofuel development is essential to the emerging economies. However, the biofuel promotion creates the unique linkage between food, water, land, and energy. Growing crops for biofuels can jeopardize food security and increase water demand. The rapid increase and widespread use of biomass for biofuels production is facing challenges on resource scarcity because the arable land, freshwater, energy, and nutrient availability are limited. The magnitude of impacts such as water depletion and water system deterioration along with water deprivation potential among water users will vary significantly across regions and countries. The variations are depending on the size of the biofuel targets adopted, the key technologies, biomass feedstocks identified, and especially the water availability and scarcity level of that particular region of biofuel promotion. The land–water–energy nexus is introduced in this chapter because those resources have a close relationship and are essential to economic and social development. The chapter has provided a comprehensive look on the relationship of land–water–energy and the unique nexus challenges of the biofuels development. The life cycle–based indicators and assessment methodology including water scarcity footprint, CF, and EF are introduced for the nexus assessment. The case study of bioethanol target of Thailand on land, water, energy, and GHG impacts has been revealed and discussed. The recommendations for further sustainable biofuel production based on the nexus assessment have also been provided.

It is recommended that for policy makers, the implementation of the nexus to enhance efficiency and balance the different uses of ecosystems resources like water, land, and energy by various stakeholders in a particular region is necessary. The key main steps of the nexus assessment for biofuel development can be summarized as follows: (1) setting the goal and interest of biofuels development by considering the land–water–energy resource availability as well as the interest of the stakeholders in management of those resources in the specific area; (2) quantitative assessment of the impacts of the biofuels development on the land, water, and energy, along with the GHG emissions and other socioeconomic impacts. The combination of life cycle–based indicators such as water scarcity footprint, CF, EF, as well as the feedstock balance can provide useful information for policy decisions. However, the assessment should be more comprehensive looking at the geographical and hydrological boundaries; (3) scenario analysis to identify the most suitable ways of achieving target; and (4) decision on the appropriate policy goal and measures through the participatory approaches with the relevant stakeholders. Currently, although the individual assessment of WF, CF, and EF can provide the relevant information to address the land–water–energy nexus perspective, nevertheless, the development of the land–water–energy nexus assessment methodology and a single indicator that can capture the whole interaction of the land–water–energy and the related GHG emissions is still a challenge for future research.

REFERENCES

Allen, R.G., Pereira, L.S., Raes, D., Smith, M., 1998. Crop Evapotranspiration: Guidelines for Computing Crop Water Requirements. FAO Drainage and Irrigation Paper 56. Food and Agriculture Organization, Rome.

Azapagic, A., 2015. Special issue: sustainability issues in the food-energy-water nexus. Sustainable Prod Consumption 2, 1–2.

Bazilian, M., Rogner, H., Howells, M., Hermann, S., Arent, D., Gielen, D., et al., 2011. Considering the energy, water and food nexus: towards an integrated modelling approach. Energy Policy 39, 7896–7906.

BOT, 2017. Ethanol situation report year 2016 [in Thai]. Bank of Thailand, Bangkok. Available from: https://www.bot.or.th/Thai/MonetaryPolicy/NorthEastern/Pages/commodities.aspx.

Brancoli, P., Rousta, K., Bolton, K., 2017. Life cycle assessment of supermarket food waste. Resour. Conserv. Recycl. 118, 39–46.

Brown, A., Matlock, M.D., 2011. A Review of Water Scarcity Indices and Methodologies. White Paper 106.

CA (Comprehensive Assessment of Water Management in Agriculture), 2007. Water for Food, Water for Life: A Comprehensive Assessment of Water Management in Agriculture. IWMI, London, UK: Earthscan; and Colombo, Sri Lanka.

DEDE, 2012. The Renewable and Alternative Energy Development Plan for 25 Percent in 10 Years (AEDP2012–2021). Department of Alternative Energy Development and Efficiency, Bangkok.

DEDE, 2017. Ethanol Production Statistics. Department of Alternative Energy Development and Efficiency. http://www.dede.go.th/more_news.php?cid = 81&filename = index (Accessed May 1, 2017).

De Fraiture, C., Moden, D., Wichelns, D., 2010. Investing in water for food, ecosystems, and livelihoods: an overview of the comprehensive assessment of water management in agriculture. Agric. Water Manag. 97, 495–501.

Endo, A., Tsurita, I., Burnett, K., Orencio, P.M., 2017. A review of the current state of research on the water, energy, and food nexus. J. Hydrol: Regional Stud. 11, 20–30.

FAO, 2011. Energy-Smart for People and Climate—Issue Paper. Food and Agriculture Organization of the United Nations, Rome.

FAO, 2012. World Agriculture Towards 2030/2050: The 2012 Revision. Food and Agriculture Organization of the United Nations, Rome.

Flammini, A., Puri, M., Pluschke, L., Dubois, O., 2014. Walking the Nexus Talk: Assessing the Water-Energy-Food Nexus. Climate, Energy and Tenure Division (NRC), Food and Agriculture Organization of the United Nations (FAO), Rome.

Foley, J.A., DeFries, R.S., Asner, G.P., Barford, C., et al., 2005. Global consequences of land use. Science 309, 570–574.

Frischknecht, R., Jungbluth, N., Althaus., H.J., Doka, G., Dones, R., Hischier, R., et al., 2007. Implementation of Life Cycle Impact Assessment Methods: Data v2.0. Ecoinvent Report No. 3. Swiss Centre for Life Cycle Inventories, Dübendorf, Switzerland.

Garcia, D.J., You, F., 2016. The water-energy-food nexus and process systems engineering: a new focus. Comput. Chem. Eng. 91, 49–67.

Gheewala, S.H., Berndes, G., Jewitt, G., 2011. The bioenergy and water nexus. Biofuels Bioprod. Bioref. 5, 353–360.

Gheewala, S.H., Silalertruksa, T., Nilsalab, P., Mungkung, R., Perret, S.R., Chaiyawannakarn, N., 2013. Implications of the biofuels policy mandate in Thailand on water: the case of bioethanol. Bioresour. Technol. 150, 457–465.

Gheewala, S.H., Silalertusksa, T., Nilsalab, P., Mungkung, R., Perret, S.R., Chaiyawannakarn, N., 2014. Water footprint and impact of water consumption for food, feed, fuel crops production in Thailand. Water 6, 1698–1718.

Gheewala, S.H., Silalertruksa, T., Malakul, P., Prapaspongsa, T., 2017a. Preface: Promoting sustainability in emerging economies via life cycle thinking. Int. J. Life. Cycle. Assess. 22, 1641–1643.

Gheewala, S.H., Silalertruksa, T., Nilsalab, P., Lecksiwilai, N., Sawaengsak, W., Mungkung, R., et al., 2017b. Water stress index and its implication for agricultural land-use policy in Thailand. Int. J. Environ. Sci. Technol. 15 (4), 833–846. Available from: https://doi.org/10.1007/s13762-017-1444-6.

Huijbregts, M.A.J., Hellweg, S., Frischknecht, R., et al., 2008. Ecological footprint accounting in the life cycle assessment of products. Ecol. Econ. 64, 798–807.
IEA Bioenergy, 2016. The Potential of Biofuels in China. IEA Bioenergy.
ISO 14046, 2014. Environmental Management—Water Footprint—Principles, Requirements and Guidelines. International Organization for Standardization, Geneva, Switzerland.
ISO/TR 14073, 2017. Environmental Management—Water Footprint—Illustrative Examples on How to Apply ISO 14046. International Organization for Standardization, Geneva, Switzerland.
Jalilov, S.M., Keskinen, M., Varis, O., Amer, S., Ward, F.A., 2016. Managing the water-energy-food nexus: Gains and losses from new water development in Amu Darya River Basin. J. Hydrol. 539, 648–661.
Jeswani, H.K., Burkinshaw, R., Azapagic, A., 2015. Environmental sustainability issues in the food-energy-water nexus: breakfast cereals and snacks. Sustainable Prod. Consumption 2, 7–28.
Lovarelli, D., Bacenetti, J., Fiala, M., 2016. Water footprint of crop productions: a review. Sci. Total Environ. 548–549, 236–251.
Mekonnen, M.M., Hoekstra, A.Y., 2011. The green, blue and grey water footprint of crops and derived crop products. Hydrol. Earth Syst. Sci. 15, 1577–1600.
Mekonnen, M.M., Hoekstra, A.Y., 2014. Water footprint benchmarks for crop production: a first global assessment. Ecol. Indic. 46, 214–223.
OAE, 2016. Basic Information of Agricultural Economics. Office of Agricultural Economics, Bangkok.
OECD/IEA, 2013 Energy Technology R&D Needs of Emerging Economies: Experts' Group on R&D Priority Setting and Evaluation Workshop, Beijing, China, 28–29 November 2012. OECD/IEA.
Pacetti, T., Lombardi, L., Federici, G., 2015. Water-energy nexus: a case of biogas production from energy crops evaluated by water footprint and life cycle assessment (LCA) methods. J. Clean. Prod. 101, 278–291.
Pfister, S., Bayer, P., 2014. Monthly water stress: spatially and temporally explicit consumptive water footprint of global crop production. J. Clean. Prod. 73, 52–62.
Pfister, S., Koehler, A., Hellweg, S., 2009. Assessing the environmental impacts of freshwater consumption in LCA. Environ. Sci. Technol. 43, 4098–4104.
Pfister, S., Bayer, P., Koehler, A., Hellweg, S., 2011. Environmental impacts of water use in global crop production: hotspots and trade-offs with land use. Environ. Sci. Technol. 45, 5761–5768.
Pfister, S., et al., 2017. Letter to the Editor: Understanding the LCA and ISO water footprint: a response to Hoekstra (2016) "A critique on the water-scarcity weighted water footprint in LCA", Ecol. Indic., 72. pp. 352–359.
Ridoutt, B.G., Pfister, S., 2013. A new water footprint calculation method integrating consumptive and degradative water use into a single stand-alone weighted indicator. Int. J. Life Cycle Assess. 18 (1), 204–207.
Sanders, K.T., Masri, S.F., 2016. The energy-water agriculture nexus: the past, present and future of holistic resource management via remote sensing technologies. J. Clean. Prod. 117, 73–88.
Scholz, K., Eriksson, M., Strid, I., 2015. Carbon footprint of supermarket food waste. Resour. Conserv. Recycl. 94, 56–65.
Silalertruksa, T., Gheewala, S.H., 2010. Security of feedstocks supply for future bio-ethanol production in Thailand. Energy Policy 38, 7476–7486.
Silalertruksa, T., Gheewala, S.H., 2011. Long-term bio-ethanol system and its implications on GHG emissions: a case study of Thailand. Environ. Sci. Technol. 45, 4920–4928.

Silalertruksa, T., Gheewala, S.H., 2018. Land-water-energy nexus of sugarcane production in Thailand. J. Clean. Prod. 182, 521–528.
Smajgl, A., Ward, J., Pluschke, L., 2016. The water-food-energy nexus—realising a new paradigm. J. Hydrol. 533, 533–540.
Statista, 2017. https://www.statista.com/statistics/281606/ethanol-production-in-selected-countries/.
Tilman, D., Cassman, K.G., Matson, P.A., Naylor, R., Polasky, S., 2002. Agricultural sustainability and intensive production practices. Nature 418, 671–677.
Tilman, D., Hill, J., Lehman, C., 2006. Carbon-negative biofuels from low-input high-diversity grassland biomass. Science 314, 1598–1600.
UNESCO, 2009. The United Nations World Water Development Report 3: Water in a Changing World. Earthscan, Paris: UNESCO, and London.
UNDESA, 2012. World Urbanization Prospects: The 2012 Revision. United Nations Department of Economic and Social Affairs, New York.
Vanham, D., 2016. Does the water footprint concept provide relevant information to address the water-food-energy-ecosystem nexus?. Ecosyst. Serv. 17, 298–307.
Vermeulen, S.J., Campbell, B., Ingram, J., 2012. Climate change and food systems. Annu. Rev. Environ. Resour. 37, 195–222.
Wackernagel, M., Rees, W., 1997. Perceptual and structural barriers to investing in natural capital: economics from an ecological footprint perspective. Ecol. Econ. 20, 3–24.
Wackernagel, M., Monfreda, C., Moran, D., Wermer, P., Goldfinger, S., Deumling, D., et al., 2005. National footprint and biocapacity accounts 2005: the underlying calculation method. Global Footprint Network, Oakland.
WBGU (German Advisory Council on Global Change), 2010. Future bioenergy and sustainable land use. Earthscan, London, UK.
WWAP, 2014. The United Nations World Water Development Report 2014: Water and Energy. United Nations World Water Assessment Programme, UNESCO, Paris.
Yan, J., Lin, T., 2009. Biofuels in Asia: an introduction to the special issue. Appl. Energy 86 (Suppl), S1–S10.

PART III

Future Trends and Policy

CHAPTER NINE

Innovation on Bioenergy

Francisco Gírio
LNEG, Bioenergy Unit, Lisboa, Portugal

Contents

9.1 Introduction 405
9.2 Emerging Biofuels From Biochemical-Based Technologies 406
9.2.1 Bio-solar cell factories 407
9.2.2 Long-chain fatty alcohols 409
9.2.3 Long-chain fatty acids 411
9.2.4 Advanced consolidated bioprocessing 411
9.3 Emerging Biofuels From Thermochemical-Based Technologies 413
9.3.1 Solar fuels obtained from gasification and pyrolysis 413
9.3.2 Novel fuels from hydrothermal liquefaction of biomass 414
9.3.3 Next generation of aviation and maritime biofuels 415
9.4 Advanced Renewable Fuels 418
9.4.1 Fuels from CO_2 and renewable energy source—producing power 418
9.4.2 Combined biomass-to-gas and PtG 420
9.4.3 Waste-to-energy 420
9.5 Future Trends on Heat, Cooling, and Power Technologies 421
9.5.1 Farms and village scale 422
9.5.2 Smart urban cities 423
9.5.3 Fuel cells 424
9.6 Conclusions 427
Acknowledgments 429
References 429

9.1 INTRODUCTION

Nowadays bioenergy is being challenged about its future role in a clean energy-based society for all Europeans. More than ever, bioenergy needs to reinvent its role under the environment of continuous changes in the energy integration mix based on renewables sources both for transportation purposes and for stationary purposes (heat, cooling, and power). Longer term vision for bioenergy as well as looking at through the

The Role of Bioenergy in the Emerging Bioeconomy.
DOI: https://doi.org/10.1016/B978-0-12-813056-8.00009-1

potential game-changing biotechnologies currently still at laboratory level is the purpose of this chapter.

Bioenergy is based on the biomass conversion into biofuels, biomaterials, power, and heat. Contrarily, to other renewable sources (e.g., wind, solar), biomass suffers the competition of other value chains (e.g., pulp and paper industry, furniture manufacturing, food and feed, etc.), and for bioenergy applications, two main sustainability criteria are in place in EU countries for biofuels and bioliquids production: (1) greenhouse gas emissions and (2) land use (Directive (EU)2015/1513)).

For transportation sector, current biofuels in the market does not meet enough sustainability standards for standing in the market for long term in competition with the rapid evolution of cleaner and energy-efficient technologies, e.g., electric mobility and/or renewable hydrogen from water electrolysis for fuel cell vehicles (FCV). On the other hand, these latter technologies still need a long-term deployment to meet both enough well-to-wheel sustainability and cost-competitive and, consequently, emerging biofuels at very low carbon-printing or even neutral carbon emissions as well as meeting the strict technical quality requirements of some transportation modes, e.g., long-distance aviation sector, are needed until 2050.

Not only for many power and heat applications but also for many biofuel production pathways, biomass upgrading into high-quality bioenergy carriers is crucial to make biomass value chains more efficient and cost-effective. Challenges connected to biomass use must be dealt with at the biomass source, and the resulting bioenergy carriers should meet the requirements of logistics and end-use. For medium and long term, technologies like torrefaction (solid product), pyrolysis (bio-oils), and gasification (syngas) are likely to substitute the more pollutant combustion-based power plants playing a dominant role for the power and heat sector and at the same time contributing for the stability of electrical grids due to its dispatchable and nonintermittent features.

9.2 EMERGING BIOFUELS FROM BIOCHEMICAL-BASED TECHNOLOGIES

Current biofuels in the market are largely dominated by ethanol in the United States and Brazil for petrol-driven vehicles and biodiesel for diesel-driven vehicles in Europe. In Europe, these so-called first-generation biofuels can only be blended on limited amounts according to the EN228 and EN590 for petrol and diesel quality requirements, respectively. Since they are mainly produced from food crops, they hardly meet the strict sustainability criteria currently in-place by the Directive (EU)2015/1513 for the European road market.

Additionally, the strict fuel specifications for maritime and, particularly, aviation engines are not met by current first-generation biofuels, which require biofuels with higher energy density, lower freezing point, and a hydrocarbon range between C8 and C16 of linear and aromatics mixture, in the case of aviation biofuels (Zschoke and Scheuermann, 2012).

The next generation of biofuels, also called advanced biofuels, must meet the composition required as drop-in biofuels to avoid the costs of new powertrain infrastructures. In the next sections, some emerging new biofuels obtained from biochemical-based pathways are discussed.

9.2.1 Bio-solar cell factories

Many technological options for novel biofuels are nowadays emerging from the bench-scale laboratories. One of the most interesting and challenging technologies is to harness solar energy by bio-solar cell factories (BSCF), in which phototrophic microorganisms (e.g., cyanobacteria, eukaryotic algae) directly catalyze the conversion of CO_2 and H_2O into oxygen and chemical energy (e.g., fuel) in a CO_2-neutral way and bypassing the biomass production phase (Fig. 9.1). These fuels, produced either from biogenic CO_2 or from flue gas released from fossil-based plants, and sunlight, might have a major role beyond 2030 for production of clean energy for transportation sector.

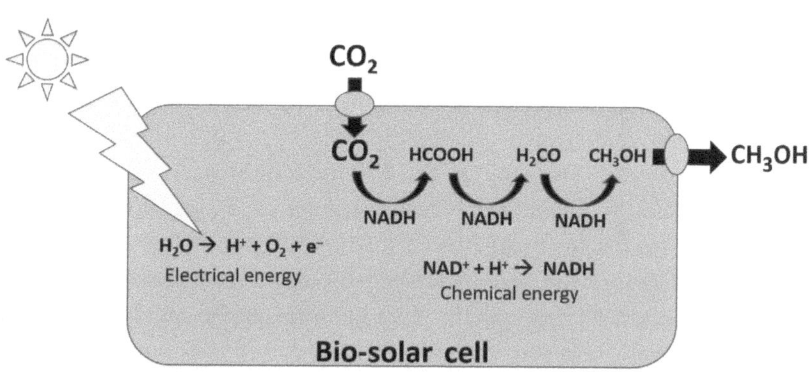

Figure 9.1 Bio-solar cell that uses photosynthesis to directly convert sunlight into useable energy for methanol production.

The main limitation for biofuel production from plant biomass is its low photosynthetic efficiency. Plants already convert solar into chemical energy under the form of adenosine triphosphate, using the photosynthesis but this natural process is very inefficient. Less than 1% of the energy in the sunlight is stored in the biomass of the plant which means that the earth produces far too little biomass to sustain our energy demands (de Vriend and Purchase, 2013). The maximum photosynthetic efficiency is 6.7% (Hall et al., 1992). Researchers are currently trying to mimic plants by building artificial photosynthesis systems by placing tiny solar panels on microbes using synthetic biology tools (Kim and Logan, 2011). Indeed, sustainable energy production from phototrophic biomass only seems feasible if we can dramatically increase the conversion efficiency of the photosynthesis process itself.

The use of synthetic biology tools is therefore essential for designing BSCF with improved sunlight-absorbing machinery of the thylakoid membrane. This should constitute a breakthrough for getting novel drop-in fuels. Currently, such designer BSCF use less than 5% of the incoming CO_2 and solar energy to growth and the remainder 95% CO_2 is used for the formation of biofuels (liquid energy carriers) and/or chemical feedstock like methanol, ethanol, butanol, ethylene, and lactate. The current overall conversion rates are about 5%–10% photosynthesis for algae or cyanobacteria. The theoretical limit for artificial photosynthesis is around 40%, and conversion rates of 18%–20% seem achievable (de Vriend and Purchase, 2013). Calculations show that it is likely that in the long run with these BSCF an energetic efficiency (relative to the total energy from the sun that reaches the surface of the earth) of up to 15% can be obtained (Gírio et al., 2013).

BSCF biotechnology could also play another important role in the future by storing or "peak-shaving" electricity from the grid. The theoretical energy conversion efficiency is even higher, up to 35%, when photobiological organisms are grown solely with 700-nm light which can be generated by electricity. Photo-bioreactors equipped with infrared light-emitting diodes, emitting at ca. 690 nm could be connected to the power grid, which would thereby allow storage of electrical energy in the form of liquid fuel at an estimated efficiency of over 20%, under periods of excess electricity in the grid. Such photo-bioreactors can also be linked directly to large-scale photovoltaic (PV) facilities and result in "PV fields" that produce liquid fuels (Gírio et al., 2013).

As already said, the current main barrier to commercialization is the overall efficiency of conversion from solar into chemical energy, which is still far from that required for commercialization. Other barriers lie on the cost of photo-bioreactor systems, the high sensitivity of BSCF catalysts to most biofuels (e.g., toxicity), and the risk of contamination in both open and closed cultivation systems. Extremophiles microorganisms can be in future engineered as BSCF since their ability to grow at extreme conditions (high temperature, high saline, high/low pH values) led for minimizing the risk of microbial contaminations at open ponds as well as in closed unsterile photobioreactors.

9.2.2 Long-chain fatty alcohols

The pathways for production of hydrocarbons-based molecules by microorganisms are well established both for bacteria and yeasts (Fig. 9.2). Yeasts are usually preferred over bacteria because of their demonstrated tolerance to low pH, high robustness, and inhibitor tolerance (Demeke et al., 2013). There are several reports of engineering microbial strains for production of long-chain fatty alcohols (LCFAls) (e.g., butanol, iso-butanol, and n-hexadecanol; Guo et al., 2016; Zhou et al., 2016; Fillet et al., 2015). Butanol has advantages over ethanol. The energy content of butanol is higher than that of ethanol and closer to that of gasoline and, more importantly, it has no compatibility, miscibility, or materials problems. However, the cost of producing synthetic butanol from fossil fuels is extremely sensitive to the price of crude oil, and for bio-butanol (n-butanol or iso-butanol) there are substantial R&D needs regarding microorganism strain selection and development and bioprocess design, with an emphasis on continuous operation and recovery technologies (Gírio et al., 2013). The US company, Gevo Inc., was the first investor to build a pilot plant for iso-butanol plant at Luverne, Minnesota, The United States, using a recombinant yeast strain. Another US company, Amyris is trying to get a cost-effective biochemical-based process for advanced biofuels through engineering *Saccharomyces cerevisiae*. The engineered strain produces large amounts of isoprenoid beta-farnesene. This C_{15} molecule can further be chemically derivatized into a variety of

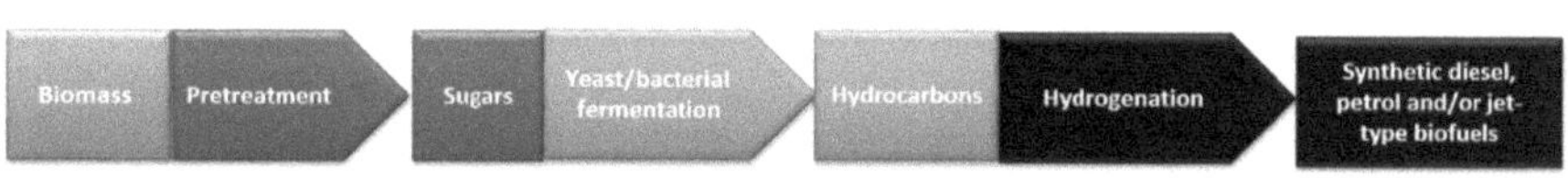

Figure 9.2 General scheme of biochemical and catalytic conversions for integral conversion of lignocellulosic sugars into diesel, petrol, and jet-fuel substitutes.

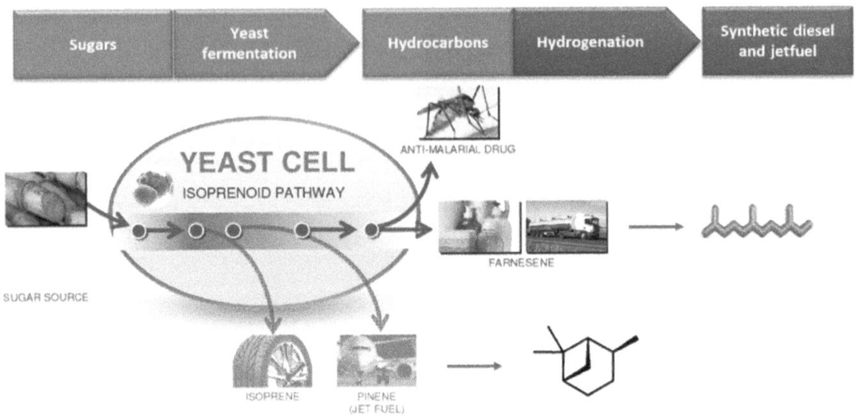

Figure 9.3 Hydrocarbon production (farnesene) from sugars through a biochemical-based process. Farnesene can further be converted in to farnesane, a molecule used as biojet-fuel.

products, including diesel substitute, bio-based chemicals (e.g., a surfactant, lubricants, or other useful chemicals), and bio-jet fuel. ASTM did already accept farnesane as Synthesized Iso-Paraffinic fuel (DSCH-SIP) in the ASTM D7566 standard (Fig. 9.3).

Other higher alcohols and hydrocarbons still require great technological developments when compared to lignocellulosic ethanol, namely with respect to yields and productivities. Moreover, these advanced biofuels are, in general, much more toxic to microbial cell factories than ethanol, and, therefore, the development of cost-effective technologies for continuous removal of advanced biofuels (or their precursors) from the fermentation broth is mandatory. Feng et al. (2015) reported a strain of *S. cerevisiae* MDS130 that have excellent performance for n-hexadecanol production in concentrated undetoxified lignocellulose sugar hydrolysates. However, the reported yield was only 0.03 g/g glucose which is still far below the theoretical yield, 0.34 g/g. Further improvements in the fatty alcohol production shall be focused on suitable enzyme overexpression, on the inactivation of inhibitory regulators, by abolishing ethanol production, and by downregulating other competing pathways. Both metabolic engineering and even random engineering studies are expecting to increase the LCFAl yields close to the maximal theoretical one. The theoretical yield of fatty alcohol production from sugar by yeast is 33.6%, and theoretical heat of combustion is −43 MJ/kg, which is in the range of the specific energy of jet fuels.

9.2.3 Long-chain fatty acids

Similar to LCFAls, the production of long-chain fatty acids (LCFAc) poses identical metabolic problems for their biochemical production from sugars. The goal is to carry out suitable metabolic engineering of oleaginous yeasts that already exhibit a natural capacity for lipid accumulation, e.g., *Yarrowia lipolytica*, *Trichosporon* sp., and *Rhodosporidium* sp. Oleaginous yeasts can produce up to 70% w/w dry cell weight of lipids, and studies on their use as production hosts for various oleochemicals are rapidly increasing (Adrio, 2017). Recently *Y. lipolytica* was engineered to reach yield of 0.27 g/g glucose (Qiao et al., 2017). This value is near the target (0.28 g fatty acids/g lignocellulosic sugars) determined by the National Renewable Energy Laboratory for the US Department of Energy's 2017 cost goal of 5$/gallon gasoline equivalent (Davis et al., 2013). Besides production yields, oleaginous strains need to show a considerable tolerance behavior over harsh industrial media and not all yeasts are suitable to attain an acceptable industrial performance. Yeast species, for example, from the genera *Trichosporon* sp. and *Rhodosporidium* sp. are able to efficiently utilize C5 sugars originating from hemicellulose. They are also more tolerant to inhibitors. Koivuranta and co-workers did modify *Trichosporon oleaginosus* for LCFAc production and reported good productivity on both glucose and xylose (Koivuranta et al., 2011).

9.2.4 Advanced consolidated bioprocessing

Current processes for biorefining lignocellulosic biomass, especially for biofuels, are built upon several unit operations, including pretreatment, biomass hydrolysis, and bioconversion, and completed by downstream processing. The enzymatic hydrolysis is usually achieved by adding commercial enzymes supplied by manufacturers or by in situ enzyme cocktails production from fungi and later added to the hydrolytic pot. This operation unit is often combined with the fermentation reactor (e.g., simultaneous saccharification and fermentation) for better process integration. Moreover, to further reduce cost and improve integration, it will be desirable to devise new low-temperature and low energy-demanding for biomass pretreatments. Consolidated bioprocessing (CBP), featuring microbial enzyme production and concomitant microbial conversion of suitable feedstock into value-added products in a single step, offers great potential for cost-effective lignocellulosic bioconversion (Guo et al., 2017).

Saccharomyces cerevisiae is the most known inhibitors-tolerant and simultaneously the most powerful hexoses fermenting yeast used in the main industrial bioprocesses, such as bakery's yeast, brewing, and ethanol production, for fuel grade. However, several constraints limit the use of *S. cerevisiae* for CBP, namely, neither ferment naturally pentoses nor grow well in cellulose and hemicellulose which are the main component of plant cell wall. In contrast to *S. cerevisiae*, many cellulolytic fungi including *Neurospora crassa* naturally grow on cellulose and hemicellulose (Tian et al., 2009). Other enzymatic paradigms might include cell-bound enzyme systems, e.g., cellulosomes and secreted multicatalytic domain enzymes, both of which owe their efficiency to an increased synergy between complementary enzymatic activities.

For a successful biorefining process, metabolic engineering and synthetic biology are tools needed for constructing suitable robust and improved cell factories for CBP. In a practical way, we should start designing one microorganism, or two or more compatible microorganisms that possess both hydrolysis capability and the target molecule production capacity to be intended to be used for biorefining. Ultimately, we should reach a one-step fully integrated process or CBP, which directly converts pretreated biomass into target molecules in a single reactor (den Haan et al., 2015; Olson et al., 2012). To date only a few naturally occurring species of bacteria within the genus *Clostridium* have been described with some of those capabilities. For instance, *Clostridium lentocellum* can convert cellulose to acetic acid (Tammali et al., 2003), and *Clostridium thermocellum* can rapidly solubilize and ferment cellulosic biomass for the production of ethanol (Tian et al., 2016). An excellent review about *C. thermocellum* as a candidate for CBP applications is available elsewhere (Akinosho et al., 2014). Despite CBP is an extremely attractive concept, there are still important challenges to overcome before this came into a commercial bioprocess. Any of the engineered strains reported so far have convincingly hydrolyzed the cellulose feedstock, and currently no commercially viable CBP organism has been reported (Guo et al., 2017). Moreover, advanced CBP (*a*CBP) concept would go even further by using low-energy, low-input pretreatment, thus providing an almost seamless overall process that involved minimum heat exchange and recycling (e.g., water, minerals, and catalysts; Gírio et al., 2013) (Fig. 9.4).

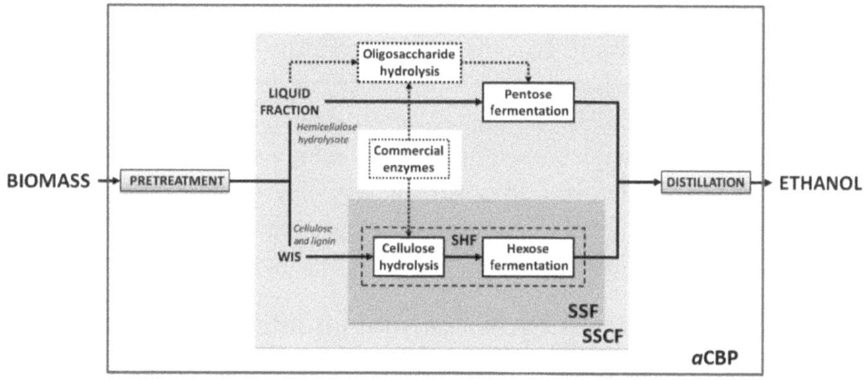

Figure 9.4 Ideal boundaries of a one-step *a*CBP for bioethanol production. WIS, water insoluble solids; SHF, separate hydrolysis and fermentation; SSF, simultaneous saccharification and fermentation; SSCF, simultaneous saccharification and combined fermentation.

9.3 EMERGING BIOFUELS FROM THERMOCHEMICAL-BASED TECHNOLOGIES

9.3.1 Solar fuels obtained from gasification and pyrolysis

Direct solar energy can be utilized in many alternative ways to provide the required heat for thermochemical conversion processes. These systems were studied and evaluated in detail already in the 1980s, but only now interesting applications are reported in development of renewable fuels from solar reaching an energy conversion efficiency of 18%, defined as the ratio of the heating value of the syngas produced to the solar radiative energy input and the heating value of the feedstock (Muller et al., 2018). Conventional autothermal gasification has been reported from a huge number of crop residues (e.g., rice husk, jatropha shell, sunflower and soya residues, hazelnut shells, and cotton stalk; Zhao et al., 2009; Maiti et al., 2014; Zabaniotou et al., 2010; Karatas et al., 2013). However, in most cases the electrical conversion efficiencies are quite low. In contrast solar-driven allothermal gasification process at elevated temperatures has a number of advantages: (1) it delivers higher syngas output per unit of feedstock, as no portion of the feedstock is combusted for process heat, (2) it produces a syngas free of combustion by-products and with little tar or soot content, (3) it produces syngas with higher calorific value and lower CO_2 concentration, and (4) it eliminates the need for an upstream air separation unit, as steam is the only gasifying agent (Piatkowski et al., 2011; Gírio et al., 2013). Ultimately, solar-driven gasification offers an efficient means of storing intermittent solar energy in a transportable and dispatchable chemical form. In a similar way, concentrated solar energy can be utilized for pyrolysis processes and for secondary gasification of pyrolysis oils or reforming of gasification-derived hydrocarbons.

9.3.2 Novel fuels from hydrothermal liquefaction of biomass

In addition to conventional gasification and pyrolysis technologies, other types of thermochemical conversion systems have been suggested and studied to produce high-quality liquid and gaseous fuels (Lyckeskog et al., 2016). These processes often operate at very high pressures but relatively mild temperatures and in the presence of catalysts. These high-pressure conversion systems are, in general, suitable for biomass which is in the liquid state or which can be fed in the form of slurries. Hydrothermal liquefaction (HTL) of biomass is becoming popular as one the most promising thermochemical-based technologies for many applications. Contrarily to fast pyrolysis, HTL can use wet biomass (e.g., algae), as biomass is fed to the process as slurry, and the primary product may be more easily upgraded towards transportation fuels. The current challenges of this technology are the improvement of the major process variables (liquid media, catalysts, reaction gas, and process conditions) to enhance the overall performance of this technology. Fig. 9.5 compares the conventional processes and sub/super critical processes, including HTL, hydrothermic carbonization, (HTC) and supercritical water/steam gasification (SCWG) as more effective technologies to directly convert wet algae into bioliquids or biogases without the need of an intermediate operation unit of drying. Black liquors from pulping kraft process also offers an excellent medium for HTL processing as it already includes alkali salts, which act as a catalyst in biomass liquefaction. Total oil yield of 11% from kraft lignin has been reported (Zhou, 2014). Another potential interesting technology to produce bio-oil from kraft lignin has been developed at both Chalmers University of Technology and Innventia, Sweden (LignoBoost technology). Lignin is depolymerized in water at subcritical conditions using a catalyst system, $ZrO_2/K_2CO_3/KOH$, and phenol as capping agent (Lyckeskog, 2016). This HTL technique increases the bio-oil yield and simultaneously decreases the char production.

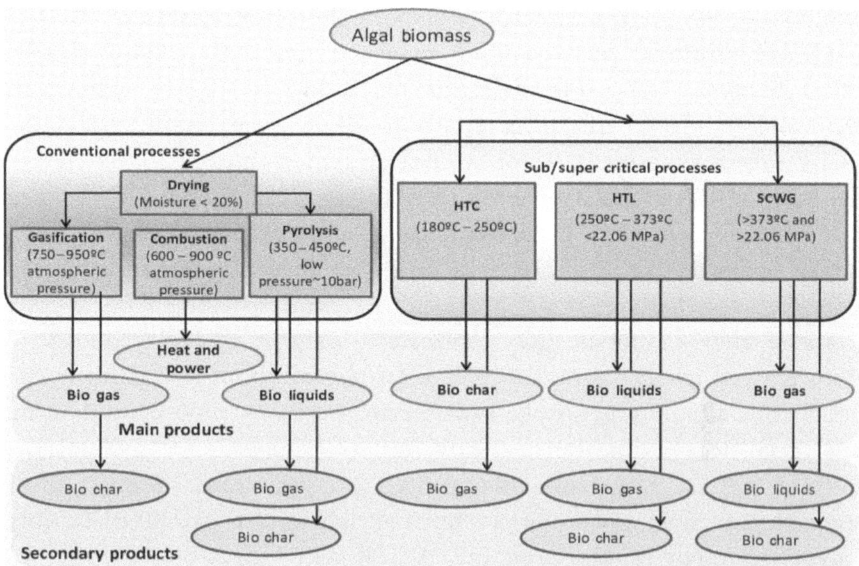

Figure 9.5 Comparison among conventional and sub/super critical processes for converting algae biomass into biofuels and bio-based products (Pinto et al., 2017).

Coprocessing of other biomass feeds with black liquor offers a process route, which has the potential to increase biomass processing capacity by integrating HTL into existing pulping industries. Some other high-pressure conversion technologies are being studied rather intensively, e.g., supercritical water gasification, but which has not yet reached industrial demonstration (Sikarwar et al., 2017). This so-called mild gasification process run at supercritical state (374°C, 220 bar) with biomass as feedstock produces the higher conversion yields of biohydrogen, and it has the great advantage to use wet biomass. The oxygen atoms from supercritical water interact with C atoms of biomass to generate CO, which in turn reacts with steam to produce biohydrogen and CO_2 via water gas shift (Parthasarathy and Narayanan, 2014). The disadvantage is the biohydrogen production cost via this route which is currently much higher than the conventional method of steam methane reforming due to a high energy demand (Sikarwar et al., 2017).

9.3.3 Next generation of aviation and maritime biofuels

The aviation sector is responsible for 2%–3% of global anthropogenic GHG emissions (Hamelinck et al., 2013; Kousoulidou and Lonza, 2016).

Worldwide, the maritime sector is responsible for an additional 3%–4% of GHG emissions, corresponding to more than 1000 million tons of CO_2 (Ecofys, 2012). Electrification of long-haul flights, which are responsible for 80% of GHG emissions of aviation sector, and long-distance freight maritime sector are not an option until 2030. Therefore, the most obvious alternative for substituting fossil fuels is low-carbon drop-in biofuels used in the current aircraft and shipping engines.

For aviation sector, as stated before, due to the requirements of jet fuels (e.g., very high energy content, lubricity, and specific chemical composition), they cannot be fully replaced by existing biofuels. Indeed, jet biofuels need to accommodate C8–C16 hydrocarbons to fulfill physical and chemical requirements like low freezing point ($< -40°C$) and, at the same time, to provide the necessary heat of combustion (> 42 MJ/kg). Current semi-commercial sustainable jet fuels are derived from vegetable and waste oils (comprising triacylglycerides) produced through a hydrotreatment process (i.e., HEFA-SPK and HEFA-SKA) (Fig. 9.6). Several other technologies are emerging at demo scale, for example, Biomass and MSW (municipal solid waste)-based Fischer–Tropsch (FT-SPK/SKA) synthesis (not shown); alcohol to jet (ATJ-SPK/SKA), and isoparaffins synthesized by direct sugar conversion to hydrocarbons (DSHC-SIP) (Fig. 9.7). Additional biochemical-based conversion pathways are under review at ASTM. These biofuels can be blended with fossil Jet-A1 in different blend ratios (usually ranging from 10% to 50%).

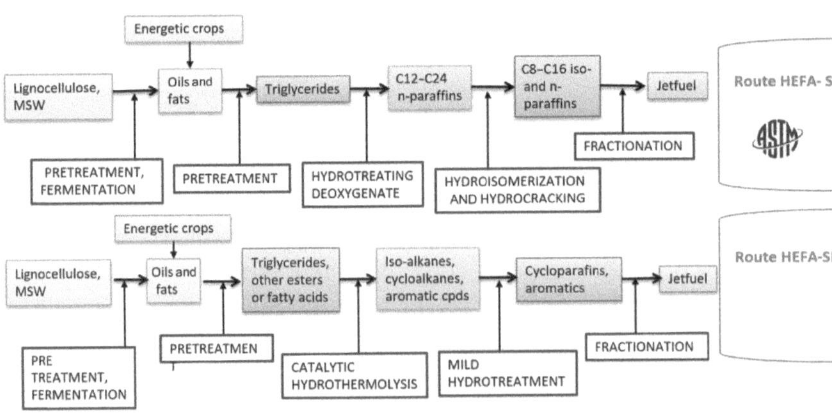

Figure 9.6 Main technological biochemical platforms from oils and fats for bio-jet fuels. HEFA-SPK, hydroprocessed esters and fatty acids-synthetic paraffinic kerosene; HEFA-SKA, hydroprocessed esters and fatty acids-synthetic paraffinic kerosene with aromatics.

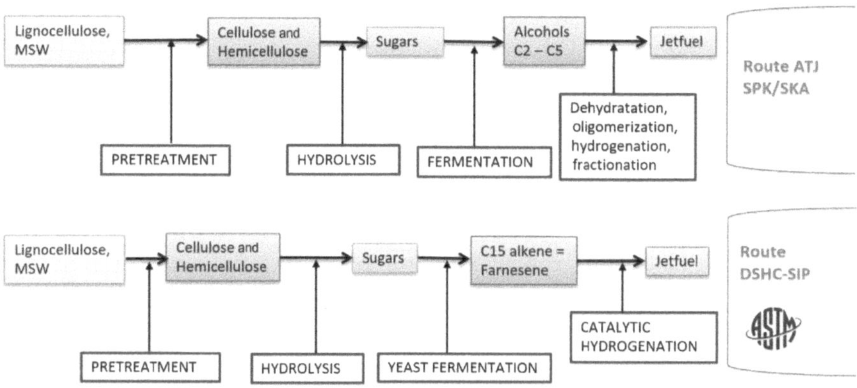

Figure 9.7 Main technological biochemical platforms from sugars for bio-jet fuels. ATJ-SPK/SKA, alcohol-to-jet synthetic paraffinic kerosene/synthetic paraffinic kerosene with aromatics; DSHC-SIP, direct sugars to hydrocarbons-synthesized isoparaffinic.

Although all current conversion pathways are technically capable of producing bio-jet fuel within set specifications, they suffer from challenges in sustainability, scalability, and cost. For example, HEFA production has been certified from vegetable oils. In the future, alternative feedstocks, such as lignocellulosic sugar hydrolysates, need to be deployed as soon as they reach a cost-effective comparison with fossil kerosene.

For maritime sector, the international shipping sector has not formulated any explicit CO_2 emission reduction target neither any orientation towards biofuels. However, the next EU Directive RED2 that shall be in place on January 1, 2021, did include maritime sector under the sustainability criteria for all transportation modes in Europe. Currently, marine fuels are produced from crude oil at refineries or in case of liquefied natural gas (LNG) from natural gas (NG) and can reach up to 50% of operational costs for long-distance shipping (Ecofys, 2012). They must respect the international standard ISO 8217 regarding bunker oil fuel quality. Therefore, a transition from bunker oil to LNG and/or compressed natural gas (CNG) is expected for short-sea and inland shipping. Biogas should be the main feedstock for bio-LNG/bio-CNG along with power-to-gas (PtG) methane at longer term. The storage characteristics of PtG methane are similar to those of bio-CNG. The availability of biofuels worldwide does not seem to be a bottleneck for possible introduction of biofuels in ships. All major ports or bunker stations in Europe and North America also have biofuel production facilities nearby (Ecofys, 2012).

From a list of potential biofuels to be ready for use in shipping sector, besides bio-LNG, small percentage of biodiesel blends (up to 20%) with marine diesel oil/marine gas oil seems to be the most promising fuel for shipping aside the replacement of heavy fuel oils by straight vegetable oils (SVO) due to the best compatibility with current engines and supply chains (Ecofys, 2012). From a sustainability point of view, it is necessary to secure that vegetable oils are nonedible oils which do not use land competing with food crops land. Microbial oils from yeast/bacterial cultivation are the best alternative at medium–long term for unlimited production of oils at large-scale facilities fulfilling the strict sustainable criteria of reducing GHG emissions and avoiding competition for land use.

Moreover, the cost-economic benefits of using biofuels in shipping could be more favorable than their current use for road transportation, if sulfur restrictions are increasing in maritime fuels. This is particularly true for the SVO and biodiesel due to their no-sulfur content.

9.4 ADVANCED RENEWABLE FUELS

Steinberg was the first that envisioned the closed-loop CO_2-recycled synthetic fuels, including CO_2 capture from the atmosphere (Steinberg, 1977) The state of the art of enabling technologies from CO_2 capture, H_2O and CO_2 dissociation, and fuel synthesis can be found elsewhere (Goede, 2015; Pearson et al., 2012; Graves, 2010).

9.4.1 Fuels from CO_2 and renewable energy source–producing power

Renewable energy sources (RES), like wind and PV, due to their intermittent activity, originate a mismatching between offer-demand of power, and during several periods of the day/night an electricity surplus is generated. Although other useful applications can also be foreseen for this surplus (e.g., pumping water upwards in a hydroelectric plant), water electrolysis is another way to fulfill the energy requirements for splitting water into hydrogen and oxygen.

Hydrogen itself is a solar fuel and can be used as transport fuel on fuel cell–driven vehicles. Hydrogen can also be used in fuel cells for electricity generation in stationery applications (e.g., buildings) or used for many

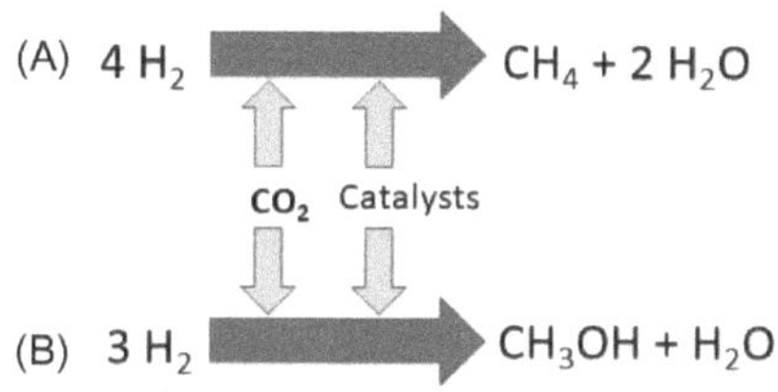

Figure 9.8 (A) PtG and (B) PtL technologies. The hydrogen is obtained from water electrolysis using renewable electricity.

existing manufacturing technologies, including the use on oil refineries. Solar fuels are also easily transported using existing distribution networks by pipeline, road, rail, or sea (de Vriend and Purchase, 2013).

In another way, a reaction can occur between the produced hydrogen with CO_2 (either from biogenic sources or flue gas from industrial sites) using suitable catalysts and pressure to produce gaseous biofuels (e.g., methane) or liquid biofuels (e.g., methanol, dimethylether). Fig. 9.8A exemplifies the PtG technology. This methane formed from CO_2 and H_2 is also known as synthetic NG and can be liquefied or concentrated to LNG or CNG, respectively, for end use as engine fuel.

PtG can also act as a powerful way of power "storage" via methane directly injected into the NG grid or any other grid compatible gas, via a two-step process: H_2 production by water electrolysis and H_2 conversion with an external CO or CO_2 source to CH_4 via methanation reaction (Gotz et al., 2016). The injection of the methanation products into the NG grid must follow the national regulations since country-specific standards applies for H_2 to be injected in the NG grid. In general, it is limited to a maximum of 0%–12% (Newton, 2014).

The main drawbacks for PtG are a relatively low efficiency and still high costs. An excellent review about the systemic analysis of PtG, including energy efficiency, a comparison between different CO and CO_2 sources for methanation reaction and cost analysis can be found in Gotz et al. (2016).

An alternative to PtG is the so-called power-to-liquids (PtL; Fig. 9.8B) that instead of the production of CH_4, the final products of the methanation reaction are liquid energy carriers such as methanol, dimethyl ether, and FT products or the production of chemicals (Budzianowski, 2012; Centi and Perathoner, 2014; Iglesias et al., 2015; Schaub et al., 2013). The current challenge is how to produce those liquid renewable fuels at large-scale in a cost-effective way to compete with current biofuels in the market.

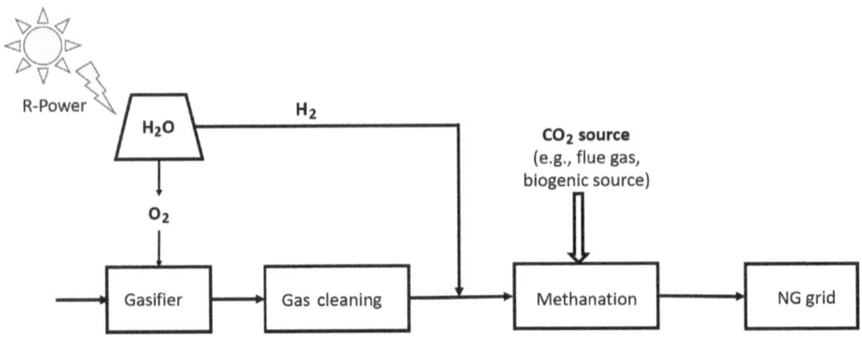

Figure 9.9 Integration of PtG with BtG.

9.4.2 Combined biomass-to-gas and PtG

Emerging technologies can also be foreseen by combining biomass-to-gas (BtG) with PtG, firstly proposed by Gassner and Maréchal (2008). Fig. 9.9 exemplifies such combination that offers some advantages, if H_2 production is obtained from water electrolysis using renewables energy (Gassner and Maréchal, 2008; Bajohr et al., 2013, 2014): (1) total carbon exploitation is more than doubled, (2) higher overall energy efficiency; (3) O_2 from electrolysis can be added to the biomass gasifier, and (4) water-gas shift reaction can be avoided by adding H_2 from electrolysis to syngas. Furthermore, BtG and PtG integration shall avoid large H_2 storage since H_2 is added to the syngas in the gasification plant. On the other hand, when the intermittent RES (e.g., wind or solar) are not in surplus production, water electrolysis is interrupted and the methanation can be operated with syngas only (Gotz et al., 2016).

9.4.3 Waste-to-energy

It is estimated that global waste generation will double by 2025 to over 6 million tons of waste per day, and the rates are not expected to peak by the end of this century (WEC, 2016). This is the consequence of population growth and economic development. The amount of MSWs per capita shall increase drastically during this century. At the same time, food and energy demands also increase. MSWs in developing countries are composed by a big amount of biomass materials, such as food, paper, and wood wastes (Omari et al., 2014). The non-recyclable combustible materials from MSWs can be used for energy recovery respecting the principles of the waste hierarchy regulations, as they can reduce the utilization of fossil fuels assisting in minimizing global warming. Besides the organic

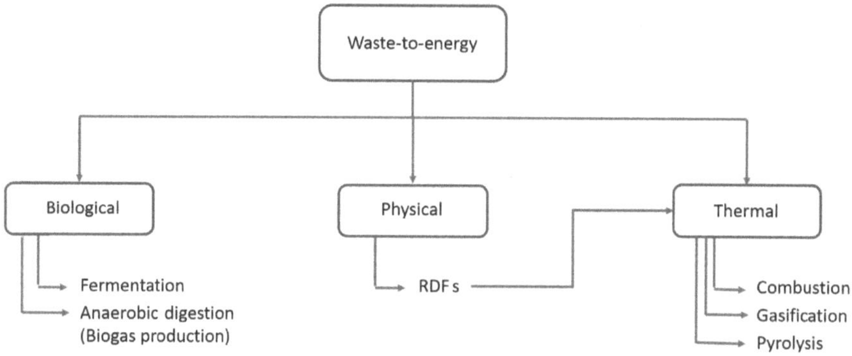

Figure 9.10 Technological methods on waste disposal for energy generation. RDFs, refused derived fuels.

fraction, nonorganic wastes, such as plastics, tires, metal and tin, glass, textiles, and ash are also present. The technologies available of waste disposal for energy recovery can be biological, physical, and thermal (Fig. 9.10). The most common thermal treatment technology is a conventional combustion on a moving grate with over 500 plants in operation worldwide. Advanced thermal treatment technologies are generally less developed than more conventional systems although progress is being made. Some fluidized bed gasification and plasma gasification plants are planning to enter in operation, particularly in the United Kingdom. However, the operational experience of these plants is not yet satisfactory.

9.5 FUTURE TRENDS ON HEAT, COOLING, AND POWER TECHNOLOGIES

Bioenergy is the major RES in the world for heat, cooling, and power applications, mostly based on wood fire for domestic/residential purposes and combustion-based power and heat plants in the industrial sector. At present, globally 50% of the final energy is used in the form of heat. Although theoretically this demand could be reduced at medium term due to energy efficiency measures, like energy-saving buildings including the near-zero energy buildings, the realistic potential to reduce heat demand until 2035 is limited. The cities are already built, and not much more than 20%–25% of the houses will be rebuilt until 2035 (Kopetz, 2015).

9.5.1 Farms and village scale

In rural areas which are less dense-populated regions, wood pellet stoves and boilers are a better option than district heating for heat supply, while the larger consumers (e.g., agro-industrials farms) prefer wood chips. Some countries had consistent and stable policies to push biomass-based heat supply, like Sweden and Italy. In the former, about three decades ago environmental taxes on fossil fuels were implemented, making heat from biomass cheaper than using oil or coal (Kopetz, 2015).

For electricity production, nowadays there is a possibility to economically produce power from biomass at a small scale (e.g., <5 MWe) using gasifiers. Conversely, ordinary combustion and steam cycles are not economically viable as such scale unless using cogeneration and produced heat can be sold. High total efficiency, however, requires that heat generated can also be utilized or that state of the art gas engines should be replaced with fuels cells or some other innovation (Table 9.1). Integration of biomass gasification with biogas or solar and wind power (RES-Hybrids) offers interesting alternatives for farm- and village-scale production of renewable energy, in both rural and agricultural areas of Europe (Gírio et al., 2013).

Gasification followed by hot-gas filtration can also be applied to generate clean gaseous fuels from many low-grade fuels such as different wastes and straw. The removal of gas contaminants prior to combustion makes it possible to minimize corrosion and deposition in boilers. Recovery of valuable metals from waste gasification and different nutrients for fertilizing purposes are interesting new possibilities for gasification-based technologies. In direct

Table 9.1 Domestic micro-CHP technologies as function of ICE, SE, and FC

	ICE	SE	FC
Capacity (electrical)	1 kW +	1 kW +	1 kW +
Electrical efficiency (%)	20	10	35–60
Overall efficiency (%)	85	90	85
Heat to power ratio	3:1	10:1	1:1
Able to vary output	No	No	Yes
Fuel used	Gas, biogas, liquid fuels	Gas, biogas, butane	Hydrocarbon fuel
Maturity	High	Fair	Low

ICE, internal combustion engine; SE, Stirling engine; FC, fuel cell.
Source: Adapted from Elmer, T., Riffat, S., 2012. State of the art review: fuel cell technologies in the domestic built environment. Conference paper, pp. 247–271. doi:10.1007/978-3-319-07977-6_17.

cofiring the biomass/waste-derived ashes are mixed with coal ash, which makes it difficult to recover valuable components.

For bigger consumers, large-scale high-efficiency power production by means of innovative and optimized power cycles, such as IGCC (integrated gasification combined cycles) or combined cycles with gas engines, has the potential for over 45% electric efficiency together with over 80% total efficiency in CHP applications. Innovative combined power and cooling cycles are also interesting alternatives.

The commercialization of gasification systems has been rather slow, and many promising technologies, such as biomass IGCC, have not yet been demonstrated on an industrial scale. Often, the main barriers to take-up of new high-efficiency technologies have been economic. Especially the economics of the first demonstration and flagship plants are very challenging, and these often require long-term incentives. The target should be to maximize the biomass conversion efficiency, targeting zero-GHG emissions and creating remarkable income for the stakeholders of the value chain. Both thermochemical and biotechnical (e.g., biogas) routes are enabling technologies that should be developed.

9.5.2 Smart urban cities

In the last 70 years, urban cities in the world are becoming larger metropoles, some of them having more than 10 million inhabitants which bring new challenges; among them the increase of GHG emissions is one of the major problems. In a century, from 1950 to 2050 (estimation), the increase of urban world's population is from 30% to about 66%, respectively (Rosa, 2017). Moreover, cities account for 75% of global energy consumption and 80% of global GHG emissions (UN Habitat, 2011). Targeted policies based on low-carbon renewables use are essential to fight climate mitigation by decreasing the use of fossil fuels in cities. Bioenergy is a proven option to replace fossil fuels in the heat supply and partially in the transport sector and generation of electricity (WBA, 2016). The transition for a greener environment and green growth in urban cities is crucial to fight the climate changes but also to ensure better quality of life to the people who live in these urban cities.

In Graz, Austria, about 70% of the city's heat is generated in a coal-fired CHP plant situated in Mellach. Paris's district heating is based on geothermal and biomass. In Copenhagen more than 98% of households are connected to district heat, mainly based on renewable energy, such as pellets, wood chips, straw, waste, and geothermal, as well as fossil fuels for peak demand (WBA, 2016). Although Europe is on the front-runner on using bioenergy in urban cities, this experience can be transferred and adapted to cities elsewhere, particularly in Asia, North America, and Africa, regardless their different climates.

Although the supply biomass availability is somehow questioned, European forests are growing. According to WBA (2016), 290 million m^3 of wood could be harvested annually from forest for wood supply without decreasing the live wood volume below its replacement. Moreover, from 2004 to 2014, global pellet production grew 21% annually from 4 million metric tons to 27.1 million metric tons. In the next 15 years, it is estimated that world pellet production will surpass the 50 million metric tons (Fletcher, 2016).

The role of bioenergy on electricity production in cities is also expected to increase significantly due to their back-up capacity in the periods when intermittent renewable electricity (IRE) output is low. Bioenergy is both dispatchable and carbon neutral and can hence be a promising option to back up IRE (Jiang et al., 2017). According to Kondziella and Bruckner (2016), the increasing penetration of IRE will lead to an increasing demand for electrical grid balance. An open question is, however, how much bioenergy will be required as back-up capacity in the future electricity systems of cities and megacities (Jiang et al., 2017).

In the transportation sector, renewable electricity from biomass plants and other renewable sources is steadily wide-spreading the electric mobility, in particular on urban cities. We should keep in mind that huge amounts of biomass feedstock are generated in the urban cities, e.g., MSWs, municipal liquid sewage, organic waste from food industry and waste food in Horeca (Hotels, Restaurants, and Cafes) sector, and solid biomass from urban waste or landscaping wood from gardens and parks inside the cities. Furthermore, the use of waste biomass streams from cities can contribute to a more effective urban waste management.

Besides the electric mobility, other low-carbon technologies are becoming more important in the urban mobility. Public buses run on biomethane, obtained from biogas upgrading are operating in Goteborg, Sweden.

9.5.3 Fuel cells

Fuel cells are considered nowadays the most efficient means of converting various fuels, especially hydrogen, to clean, efficient, and reliable power and heat for a wide range of energy-related applications (Niakolas et al., 2016). And, hydrogen is being considered the next generation clean energy carrier both for stationery applications (e.g., including portable devices, cogeneration of power and heat) and for road and non-road transport.

Indeed, FCV based on hydrogen did attract the car manufacturers since 2013. After Hyundai (model ix35), several others, such as BMW (model Hydrogen 7) and Honda (model FCX Clarity), have released limited editions of cars with hydrogen fuel cells to produce water and electricity or with engines that burn hydrogen. General Motors, Daimler, Ford, Nissan, and Toyota are also in this field, while others, such as VW, announced in 2017 to stop research for development FCV (de Vriend and Purchase, 2013).

Fuel cells directly convert chemical energy to electricity. They have low sulfur, low NOx emissions, low noise of operation, and have higher efficiency compared to the conventional power sources as their efficiencies are not limited by the Carnot cycle (Yarguddi and Dharme, 2014). Irrespective of the type of fuel cells, they all consist of an anode (negative side), a cathode (positive side), and an electrolyte that allows charges to move between the two sides of the fuel cell. Electrons are drawn from the anode to the cathode through an external circuit or load, producing direct current. In a proton exchange membrane fuel cell (PEMFC), air is fed into the cathode, whereas hydrogen is fed into the anode, separated by the membrane that functions as electrolyte and reactant separation. Protons and electrons are obtained from the hydrogen when it diffuses through the membrane and dissociates in the cathodic side. Because membrane is electrically insulating, electrons are forced to travel through an external circuit, supplying power. In the cathode catalyst, oxygen molecules react with the electrons, which have traveled through the external circuit, and protons to form water. The production cost of hydrogen is one of the main disadvantages associated with the high manufacturing cost of PEMFC and catalyst.

Of interest for industry is the carbon molten carbonated fuel cell (MCFC) and the solid oxide fuel cell (SOFC), both belongs to the so-called direct carbon fuel cell systems (Rady et al., 2012). MCFC operates at 650°C and prototypes of 100 kW are being tested, the SOFC or third-generation fuel cell operates at close to 1000°C but is more efficient and more versatile as power generation system (Yarguddi and Dharme, 2014).

One of the major bottlenecks for these fuel cells are the limited type of biofuels used on it. In general, they exhibit poor performance with biomass, low rank coals, and many other alternative fuels. Other weaknesses are concerned with stability, useful lifetime, and production cost use of (rare-noble) expensive raw elements and materials (Niakolas, et al., 2016).

An alternative to chemical-based fuel cells are the microbial fuel cells (MFC) that are emerging as a versatile renewable energy technology. This is particularly because of the multidimensional applications of this eco-friendly technology. Fig. 9.11 shows an MFC constituted by a phototrophic bacterium fed by sunlight and CO_2. The MFC converts the energy available from the available CO_2 directly into electricity. In an alternative way, a heterotrophic microorganism actively catabolizes any other biodegradable substrate and electricity is generated. Nearly all MFCs consist of anode and cathode chambers, physically separated by a proton exchange membrane. Since oxygen inhibits the production of electricity, microbial cells are only fed into the O_2-free environment of anode and separated from the aerated cathode compartment. Mixed substrates, e.g., sewage sludge, are inexpensive substrates and are among the best performing MFCs in the literature reaching 6000 mW/m^2 of maximum power density (Franks and Nevin, 2010; Schroder et al., 2003).

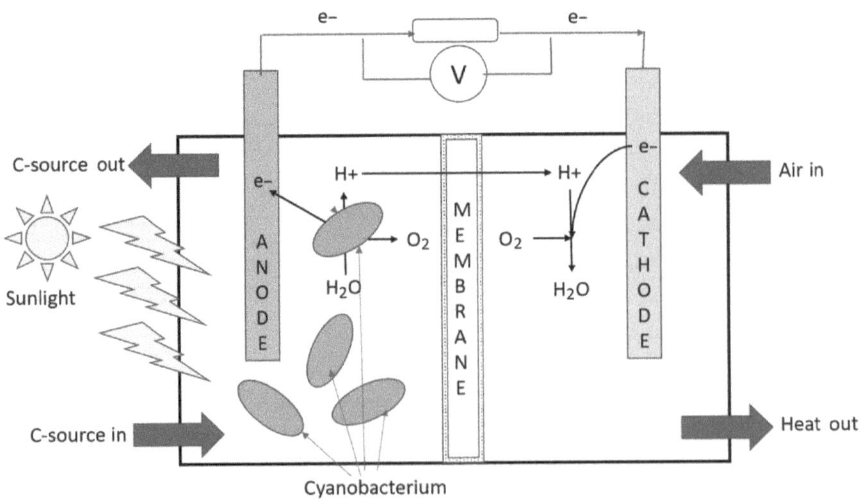

Figure 9.11 An MFC converts the energy available from a biodegradable substrate directly into electricity (in volts). In this example, it was considered a phototrophic cyanobacterium using CO_2 as substrate.

MFCs could also be utilized as power generator in small devices such as biosensor (Rahimnejad et al., 2015).

MFCs can be readily adjusted to the harvest of biohydrogen instead of producing electricity; therefore, MFCs can also be a part of the solution for clean energy in the transport sector using MFCs vehicles based on hydrogen.

Besides the technological issues and costs associated to FCV, the current major constraint is the lack of road infrastructure to supply them. For instance, the "Hydrogen Mobility Europe" project announced, in September 2015, 29 operational stations in 10 countries by 2019 (E4tech, 2015), which is just a starting point for a widespread biohydrogen-based road infrastructure in Europe.

In medium term, beyond 2030, it is expected that fuel cells (either microbial or chemical-based) might also contribute for different final energy applications, e.g., road transportation (as clean fuel), low power portable devices (battery replacement), and/or distributed power generation. Indeed, they are already commercial available but at high costs.

9.6 CONCLUSIONS

Based on evidences compiled in this paper, bioenergy shall represent beyond 2020 an important role for new value chains based on the conversion of biomass into next generation of sustainable liquid and gaseous biofuels, on electric mobility either from renewable power generation in the mix of renewable sources for 100% power supply before 2100 or on for FCV based on hydrogen and for stationery applications, including domestic and small-urban micro- and mini-CHP units.

In terms of biofuel uses, the medium- to long-term picture may not be drastically different from today, due to the lag time necessary to install new technologies for engines and for the distribution of biofuels. So, the medium-term market will be mainly based on gasoline- or diesel-compatible solutions (the so called "2 G" biofuels). However, in 2020–25 the situation will not be the same as today, since some tendencies are already emerging, such as (Gírio et al., 2013):

- the use of electricity for short mobility day-to-day transportation (from home to work, also short-distance delivery and craftsman tours) in close integration with other uses of electricity in terms of optimization of the production ("smart grid" evolution);

- higher flexibility in the use of liquid biofuels in engines ("dual-fuel" evolution, incorporation of alcohols in diesel-like engines, better yield by using the oxygen atoms from alcohols in gasoline-like engines, etc.);
- new ways to produce diesel-like biofuels and jet fuels by biological routes, such as production of hydrocarbons by yeasts, which is already being tested in pilot scale; and
- evolution of long-distance road transportation (trucks) towards a more integrated system with railways (increase the overall energy efficiency in transports), and a change of the majority of trucks fleet for direct use NG and biomethane with gas-turbine engines (better yield, less pollutant), or as alternative at short term the use of CNG and LNG from NG or bio-CNG and bio-LNG from biogas/biomethane.

More widely, it is expected that a fast evolution from the second-generation (2G) processes to next generation (2G+) processes along with a significant and competitive market increase of sustainable biomass utilization can be reached by making use of innovative biorefinery concepts that using biochemical or thermochemical conversion routes or by combining these two conversion technologies. Biorefineries will offer market-tailored production of high-value products and conversion of low-cost by-products and residues to fast growing green-energy markets (Gírio et al., 2017). New solutions can be designed by concept development assisted by technology, new chemical-based catalysts, and/or biocatalysts (enzymes and cell factories).

In the longer term, beyond 2030 sugar-based biochemical processes offer many possibilities for disruptive technologies. The following longer term technological strategies should be developed to enhance the use of biomass for market fuels for transportation in the most efficient way:

- *Biofuels by innovative sugar-based biochemical and chemical processes.* The number of new biofuel molecules produced from biomass by biological processes (e.g., new jet fuel molecules obtained from hydrocarbons produced by microorganisms) is expected to become much larger than at present. Other technologies to produce jet biofuels from sugars include chemical condensation processes such as aqueous-phase processing of the dehydration and condensation of alcohols produced by fermentation. Lignocellulosic ethanol can therefore play an important (intermediary) role in jet biofuel production.
- *Biofuels by innovative co-production at existing oil refineries.* The existing oil refineries offer cost-efficient solutions to decarbonizing the transport sector, including the road, marine, and aviation markets. These new

solutions need innovative technology development for dedicated biomass carriers and conversion technologies in this specific oil-refinery environment that combine top-level biomass conversion and oil refining competencies.

- *Biofuels by innovative photosynthetic processes from CO_2 and light.* The direct production of electrons by microorganisms and/or enzymes in so-called "MFCs" (H_2 route) or "artificial leaves," respectively. Biosolar microorganisms are another game-changing technology for direct production of biofuels and chemicals from CO_2 and sunlight at very high (above 60%) light conversion efficiency.
- *Innovative technologies for using algae and other wet biomass streams by using nontraditional bioenergy solutions.* Cultivation of algae that exhibit a long depth of penetration of photons in water is a major breakthrough that opens the way for novel concepts for the fast development of biofuels from algal stored carbon, either in the form of lipids or carbohydrates.

ACKNOWLEDGMENTS

Francisco Gírio would like to acknowledge all participants of the Workshop EERA-EIBI, on "*Longer-term R&D needs and priorities on Bioenergy: Bioenergy beyond 2020*" held in Brussels on May 7–8, 2013, which contributes to some of the ideas expressed in this chapter that are still updated nowadays.

REFERENCES

Adrio, J.L., 2017. Oleaginous yeasts: promising platforms for the production of oleochemicals and biofuels. Biotechnol. Bioeng. 114 (9), 1915–1920. Available from: https://doi.org/10.1002/bit.26337.

Akinosho, H., Yee, K., Close, D., Ragauskas, A., 2014. The emergence of *Clostridium thermocellum* as a high utility candidate for consolidated bioprocessing applications. Front. Chem. 2, 66. Available from: https://doi.org/10.3389/fchem.2014.00066.

Bajohr, S., Graf, F., Gotz, M., 2013. Bewertung der Kopplung von PtG-konzepten mit Biomassevergasung. gwf—Gas/Erdgas 154, 222–227.

Bajohr, S., Schollenberger, D., Buchholz, D., Weinfurtner, T., Gotz, M., 2014. Kopplung der PtG-Technologie mit thermochemischer Biomassevergasung. Das KIC prokect "DemoSNG". gwf—Gas/Erdgas 155, 470–475.

Budzianowski, W.M., 2012. Negative carbon intensity of renewable energy technologies involving biomass or carbon dioxide as inputs. Renew. Sustain. Energy Rev. 16, 6507–6521.

Centi, G., Perathoner, S., 2014. Perspectives and state of the art in producing solar fuels and chemicals from CO2. Green Carbon Dioxide. John Wiley & Sons, Inc, New York, pp. 1–24.

Davis, R., Tao, L., Tan, E.C.D., Biddy, M.J., Beckham, G.T., Scarlata, C., et al., 2013. Process design and economics for the conversion of lignocellulosic biomass to hydrocarbons: dilute-acid and enzymatic deconstruction of biomass to sugars and biological conversion of sugars to hydrocarbons. Technical Report, NREL/TP-5100-602232013.

Demeke, M.M., Dietz, H., Li, Y., Foulquié-Moreno, M.R., Mutturi, S., Deprez, S., et al., 2013. Development of a D-xylose fermenting and inhibitor tolerant industrial *Saccharomyces cerevisiae* strain with high performance in lignocellulose hydrolysates using metabolic and evolutionary engineering. Biotechnol. Biofuels 6 (1), 89. Available from: https://doi.org/10.1186/1754-6834-6-89.

den Haan, R., van Rensburg, E., Rose, S.H., Görgens, J.F., van Zyl, W.H., 2015. Progress and challenges in the engineering of non-cellulolytic microorganisms for consolidated bioprocessing. Curr. Opin. Biotechnol. 33, 32–38. Available from: https://doi.org/10.1016/j.copbio.2014.10.003.

de Vriend, H., Purchase, R., 2013 Solar Fuels and Artificial Photosynthesis: Science and Innovation to Change Our Future Energy Option. In: de Vriend, H., Groot, H., Lankhorst, R.K., (Eds.), Biosolar Cells, Wageningen, The Netherlands.

Ecofys, 2012. Potential of Biofuel for shipping. Final Report for EMSA-European Maritime Safety Agency, contract no BIONL11332.

Elmer, T., Riffat, S., 2012. State of the art review: fuel cell technologies in the domestic built environment. Conference paper, pp. 247–271. doi:10.1007/978-3-319-07977-6_17.

E4tech, 2015. Report "The fuel cell industry review", E4tech. www.FuelCellIndustryReview.com

Feng, X., Lian, J., Zhao, H., 2015. Metabolic engineering of *Saccharomyces cerevisiae* to improve 1-hexadecanol production. Metab. Eng. 27, 10–19. Available from: https://doi.org/10.1016/j.ymben.2014.10.001.

Fillet, S., Gibert, J., Suárez, B., Lara, A., Ronchel, C., Adrio, J.L., 2015. Fatty alcohols production by oleaginous yeast. J. Ind. Microbiol. Biotechnol. 42, 1463–1472. Available from: https://doi.org/10.1007/s10295-015-1674-x.

Fletcher, K., 2016. WBA issues report on the importance of bioenergy in Europe, Biomass Magazine online, July 25. http://biomassmagazine.com/articles/13521/wba-issues-report-on-the-importance-of-bioenergy-in-europe

Franks, A.E., Nevin, K., 2010. Microbial fuel cells, a current review. Energies 3, 899–919.

Gassner, M., Maréchal, F., 2008. Thermo-economic optimization of the integration of electrolysis in synthetic natural gas production from wood. Energy 33, 189–198. Available from: https://doi.org/10.1016/j.energy.2007.09.010.

Gírio, F., Kurkela, E., Kiel, J., Lankhorst, R.K., 2013. EERA-EIBI Workshop report on "Longer term R&D needs and priorities on bioenergy: bioenergy beyond 2020". https://setis.ec.europa.eu/system/files/EERA_EIBI_WORKSHOP_Report.pdf.

Gírio, F., Marques, S., Pinto, F., Oliveira, A.C., Costa, P., Reis, A., et al., 2017. Biorefineries in the world. In: Rabaçal, M., Ferreira, A.F., Silva, C.A.M., Silva, M.C. (Eds.), Biorefineries: Targeting Energy, High Value Products and Waste Valorisation. Springer International Publishing, Cham, Switzerland, 978-3-319-482866. pp. 227–281.

Goede, A.P.H., 2015. CO2-neutral fuels. EPJ Web Conference 98, 07002.

Gotz, M., Lefebvre, J., Mors, F., Koch, A.M., Graf, F., Bajohr, S., et al., 2016. Renewable power-to-gas: a technological and economic review. Renew. Energy 85, 1371–1390.

Graves, C.R., 2010. Recycling CO2 into sustainable hydrocarbon fuels: electrolysis of CO2 and H2O. PhD Thesis, New York, NY: Columbia University.

Guo, W., Sheng, J., Zhao, H., Feng, X., 2016. Metabolic engineering of *Saccharomyces cerevisiae* to produce 1-hexadecanol from xylose. Microb. Cell Fact. 15, 24. Available from: https://doi.org/10.1186/s12934-016-0423-9.

Guo, Z.-P., Duquesne, S., Bozonnet, S., Cioci, G., Nicaud, J.-N., Marty, A., et al., 2017. Conferring cellulose-degrading ability to *Yarrowia lipolytica* to facilitate a consolidated bioprocessing approach. Biotechnol. Biofuels 10, 132. Available from: https://doi.org/10.1186/s13068-017-0819-8.

Hall, D.O., Casillo-Calle, F., Williams, R., Woods, J., 1992. Biomass for energy: supply prospects. In: Johansson, T.B., et al. (Eds.), Renewable Energy Sources for Fuels and Electricity, 593. Island Press, Washington, DC1-55963-138-4, Prepared for the UN Rio Conference in 1992.

Hamelinck, C., Cuijpers, M., Spoettle, M., van den Bos, A., 2013. Biofuels for Aviation. Ecofys, Utrecht, NL.

Iglesias, M., de Vires, G.C., Claeys, M., Schaub, G., 2015. Chemical energy storage in gaseous hydrocarbons via iron Fischer-Tropsch synthesis from H2/CO2—kinetics, selectivity and process considerations. Catal. Today 242, 184–192.

Jiang, Y., Werf, E., van der, Ierland, E.C., Keesman, K.J., 2017. The potential role of waste biomass in the future urban electricity system. Biomass Bioenergy 107, 182–190.

Karatas, H., Olgun, H., Akgun, F., 2013. Experimental results of gasification of cotton stalk and hazelnut shell in a bubbling fluidized bed gasifier under air and steam atmospheres. Fuel 112, 494–501. Available from: https://doi.org/10.1016/j.fuel.2013.04.025.

Kim, Y., Logan, B., 2011. Hydrogen production from inexhaustible supplies of fresh and salt water using microbial reverse-electrodialysis electrolysis cells. Proc. Nat. Acad. Sci. USA 108, 16176–16181. Available from: https://doi.org/10.1073/pnas.1106335108.

Koivuranta, K., Ruohonen, L., Penttilä, M., 2011. Genetically modified fungi and their use in lipid production. Patent Number(s): US2017088842.

Kondziella, H., Bruckner, T., 2016. Flexibility requirements of renewable energy based electricity systems—a review of research results and methodologies. Renew. Sustain. Energy Rev. 53, 10–22.

Kopetz, H., 2015. The future role of bioenergy in the global energy system. In: Swaaij, W., van Kersten, S., Palz, W. (Eds.), Biomass Power for the World. Pan Stanford Series on Renewable Energy, volume 6, Pan Stanford Publishing, Singapore, pp. 141–161. Chapter 3.

Kousoulidou, M., Lonza, L., 2016. Biofuels in aviation: fuel demand and CO2 emissions evolution in Europe toward 2030. Transp. Res. D Trans. Environ. 46, 166–181.

Lyckeskog, H.N., 2016. Hydrothermal liquefaction of lignin into bio-oil. Influence of the reaction conditions and stability of the bio-oil produced. PhD Thesis, Chalmers University of Technology. 978-91-7597-503-0.

Lyckeskog, H.N., Mattsson, C., Åmand, L.-E., Olausson, L., Andersson, S.I., Vamling, L., et al., 2016. Storage stability of bio-oils derived from the catalytic conversion of softwood Kraft lignin in subcritical water. Energy Fuels 30, 3097–3106.

Maiti, S., Bapat, P., Das, P., Ghosh, P.K., 2014. Feasibility study of jatropha shell gasification for captive power generation in biodiesel production process from whole dry fruits. Fuel 121, 126–132. Available from: https://doi.org/10.1016/j.fuel.2013.12.048.

Muller, F., Patel, H., Blumenthal, D., Pozivil, P., Das, P., Wieckert, C., et al., 2018. Co-production of syngas and potassium-based fertilizer by solar-driven thermochemical conversion of crop residues. Fuel Process. Technol. 171, 89–98. Available from: https://doi.org/10.1016/j.fuproc.2017.08.006.

Newton, J., 2014. Power-to-gas and methanation—pathways to a "hydrogen economy". In: 14th Annual APGTF Workshop London. http://www.apgf-uk.com/files/workshops/14thWorkshop2014/212/JohnNewton.pdf.

Niakolas, D., Daletou, M., Neophytides, S.G., Vayenas, C., 2016. Fuel cells are a commercially viable alternative for the production of "clean" energy. Ambio 45, S32–S37. Available from: https://doi.org/10.1007/s13280-015-0731-z.

Olson, D.G., McBride, J.E., Shaw, A.J., Lynd, L.R., 2012. Recent progress in consolidated bioprocessing. Curr. Opin. Biotechnol. 23, 396–405.

Omari, A.M., Kichonge, B.N., John, G.R., Njau, K.N., Mtui, P.L., 2014. Potential of municipal solid waste, as renewable energy source—a case study of Arusha, Tanzania. Int. J. Renew. Energy Technol. Res. 3, 1–9.

Parthasarathy, P., Narayanan, K.S., 2014. Hydrogen production from steam gasification of biomass: influence of process parameters on hydrogen yield—a review. Renew. Energy 66, 570–579.

Pearson, R.J., Eisaman, M.D., Turner, J.W.G., Edwards, P.P., Jiang, Z., Kuznetsov, V.L., et al., 2012. Energy storage via carbon-neutral fuels made from CO2 water, and renewable energy. IEEE Proc. 100, 440–460.

Piatkowski, N., Wieckert, C., Weimer, A.W., Steinfeld, A., 2011. Solar-driven gasification of carbonaceous feedstock-a review. Energy Environ. Sci. 4, 73–82.

Pinto, F., Costa, P., Miranda, M., 2017. Thermochemical treatment (Part II: chapter 5). In: Pereira, Leonel (Ed.), Algal Biofuels. CRC Press/Taylor & Francis Group, Boca Raton, FL, ISBN 978-1-4987-5231-2.

Qiao, K., Wasylenko, T.M., Zhou, K., Xu, P., Stephanopoulos, G., 2017. Lipid production in *Yarrowia lipolytica* is maximized by engineering cytosolic redox metabolism. Nat. Biotechnol. 35, 173–177. Available from: https://doi.org/10.1038/nbt.3763.

Rady, A.C., Giddey, S., Badwal, S.P.S., Ladewig, B.P., Bhattacharya, S., 2012. Review of fuels for direct carbon fuel cells. Energy Fuels 26, 1471–1488.

Rahimnejad, M., Adhami, A., Darvarii, S., Zirepour, A., Oh, S.-E., 2015. Microbial fuel cell as new technology for bioelectricity generation: a review. Alexandria Eng. J. 54, 745–756. Available from: https://doi.org/10.1016/j.aej.2015.03.031.

Rosa, R.N., 2017. The role of synthetic fuels for a carbon neutral economy. J. Carbon Res. 3, 11. Available from: https://doi.org/10.3390/c3020011.

Schaub, G., Eilers, H., González, M.I., 2013. Chemical storage of renewable electricity via hydrogen—principles and hydrocarbon fuels as an example. Transition to Renewable Energy Systems. Wiley-VCH Verlag GmbH & Co. KGaA, Weinheim, Germany, pp. 619–628.

Schroder, U., Nissen, J., Scholz, F., 2003. A generation of microbial fuel cells with current outputs boosted by more than one order of magnitude. Angew. Chem. Int. Edit. 42, 2880–2883.

Sikarwar, V.S., Ming, Z., Fennell, P.S., Shah, N., Anthony, E.J., 2017. Progress in biofuel production from gasification. Progr. Energy Combust. Sci. 61, 189–248. Available from: https://doi.org/10.1016/j.pecs.2017.04.001.

Steinberg, M., 1977. Synthetic carbonaceous fuels and feedstocks from oxides of carbon and nuclear power. Brookhaven National Laboratory BNL-22785. In: Proceedings of the Twelfth Intersociety Energy Conversion Engineering Conference, Washington, DC, August 28–September 2, 1977. Available at: http://www.osti.gov/scitech/servlets/purl/7220354 (accessed September 12, 2016).

Tammali, R., Seenayya, G., Reddy, G., 2003. Fermentation of cellulose to acetic acid by *Clostridium lentocellum* SG6: induction of sporulation and effect of buffering agent on acetic acid production. Lett. Appl. Microbiol. 37, 304–308.

Tian, C., Beeson, W.T., Lavarone, A.T., Sun, J., Marletta, M.A., Cate, J.H.D., et al., 2009. Systems analysis of plant cell wall degradation by the model filamentous fungus *Neurospora crassa*. Proc. Natl. Acad. Sci. USA 106 (52), 22157–22162. Available from: https://doi.org/10.1073/pnas.0906810106.

Tian, L., Papanek, B., Olson, D.G., Rydzak, T., Holwerda, E.K., Zheng, T., et al., 2016. Simultaneous achievement of high ethanol yield and titer in *Clostridium thermocellum*. Biotechnol. Biofuels 9, 116. Available from: https://doi.org/10.1186/s13068-016-0528-8.

UN Habitat, 2011. Cities and Climate Change: Global Report on Human Settlements. Earthscan, London.

WBA, 2016. Report on "The role of bioenergy in European cities". World Bioenergy Association.

WEC, 2016. Report on "Waste to energy 2016", World Energy Council.

Yarguddi, O., Dharme, A.A., 2014. Fuel cell technology: a review. Int. J. Innov. Res. Eng. Technol. 3, 14668–14673.

Zabaniotou, A., Kantarelis, E., Skoulou, V., Chatziavgoustis, Th, 2010. Bioenergy production for CO2-mitigation and rural development via valorization of low crop residues and their upgrade into energy carriers: a challenge for sunflower and soya residues. Bioresour. Technol. 101, 619–623. Available from: https://doi.org/10.1016/j.biortech.2009.08.004.

Zhao, Y., Sun, S., Tian, H., Qjian, J., Su, F., Ling, F., 2009. Characteristics of rice husk gasification in an entrained flow reactor. Bioresour. Technol. 100, 6040–6044. Available from: https://doi.org/10.1016/j.biortech.2009.06.030.

Zhou, X.-F., 2014. Conversion of kraft lignin under hydrothermal conditions. Bioresour. Technol. 170, 583–586.

Zhou, Y.J., Buijs, N.A., Zhu, Z., Qin, J., Siewers, V., Nielsen, J., 2016. Production of fatty acid-derived oleochemicals and biofuels by synthetic yeast cell factories. Nat. Commun. 7. Available from: https://doi.org/10.1038/ncomms11709.

Zschoke, A. and Scheuermann, S., 2012. High biofuel blends in aviation (HBBA), Final report ENER/C2/2012/420-1. Germany: Lufthansa.

CHAPTER TEN

Future Role of Bioenergy

Nicolae Scarlat and Jean-Francois Dallemand
Directorate for Energy, Transport and Climate, European Commission Joint Research Centre, Ispra (VA), Italy

Contents

10.1 Bioenergy Perspectives 435
10.1.1 Biomass resources 435
10.1.2 Bioenergy in future low-carbon energy systems 450
10.1.3 Perspectives for aviation biofuels 459
10.1.4 Perspectives of algae for bioenergy 464
10.1.5 Waste and residues for sustainable bioenergy 470
10.1.6 Role of trade in bioenergy 476
10.2 Policies and Measures to Support Sustainable Bioenergy 489
10.2.1 Bioenergy policy context and bioeconomy: evolving framework 489
10.2.2 Sustainability certification and standards 498
10.2.3 GHG emissions and carbon accounting for bioenergy 507
10.2.4 Support schemes for bioenergy 520
10.2.5 Bioenergy for sustainable development 529
Disclaimer 537
References 537
Further Reading 546

10.1 BIOENERGY PERSPECTIVES

10.1.1 Biomass resources

10.1.1.1 Assessing biomass potentials

Biomass availability for energy use is a key issue for bioenergy deployment. Various feedstocks can contribute to meet bioenergy demand, including energy crops and wooden biomass, residues from agriculture and forestry, organic waste from households and industry, as well as algae and aquatic biomass. Biomass availability depends closely on the biophysical capacities of ecosystems, the various constraints related to land, water

The Role of Bioenergy in the Emerging Bioeconomy.
DOI: https://doi.org/10.1016/B978-0-12-813056-8.00010-8

resources, biodiversity protection, and several competitive uses of biomass (food, feed, fiber, biomaterials, and biochemicals). There are several types of biomass potentials that have different scope and based on different approaches and methodologies, which include theoretical, technical, environmental, economic, and sustainable potentials (BEE, 2010, WBGU, 2009).

Theoretical potential represents the overall maximum amount of biomass or the biophysical limit that can be available using current resources available (e.g., land, water, soil, etc.) without additional restrictions for energy production.

Technical potential is a fraction of the theoretical potential that is available with the current technological level (e.g., infrastructure and accessibility, harvesting techniques, and processing technologies), and spatial constraints (e.g., terrain, altitude, slope, etc.).

Environmental potential describes the fraction of the theoretical potential that meets certain ecologic criteria related to soil, water, and air protection, biodiversity protection, etc.

Economic potential describes how much of the technical potential is economically usable within the given economic framework conditions, in competition with other uses (e.g., food, feed, fuels, biomaterials, biochemicals, etc.).

Sustainable potential is the potential that meets the various technical, economic, and environmental constraints (sustainability criteria).

The estimations of biomass potentials are basically resource-focused, demand-driven, and integrated assessments. Different approaches employ different methodologies, from simple statistical to more complex spatially explicit methods and integrated modeling assessment. Resource-focused assessments typically estimate the theoretical, technical, or environmental potentials, focusing on the supply side. Certain limitations are included into the assessments, comprising biophysical and environmental limitations for the production of biomass and the competition between different uses of resources. The resource-focused approach studies employ statistical and spatially explicit methods. Statistical analyses of bioenergy potentials follow a bottom-up approach to determine the availability of biomass for energy, making use of data from statistics on land use, agricultural production, crop yields, etc., based on expert judgment, field studies, or literature review. Several assumptions are made on different scenarios and technical, environmental, or social constraints. It is not always possible to account for site-specific information and local constraints, and only average values can be included in the assessments. Spatially explicit methods combine spatially explicit data and statistical data to assess biomass availability and provide insight on the distribution of the biomass potentials across regions or countries. Spatially explicit analyses are more suitable, compared to statistical analyses, to consider the local conditions on land use, climate, soil type, vegetation, and detailed local technical, environmental, and social constraints for biomass use. The yields of energy crops can be estimated based on crop growth models that use spatially explicit data on climate, soil type, and crop management (BEE, 2010). Spatially explicit methods allow the performance of spatial analyses and suitability analyses that consider the spatial distribution of biomass, infrastructure, and various constraints allowing technical–economic analyses for bioenergy production.

The demand-driven assessments analyze the competitiveness of biomass-based energy systems with other energy systems or estimate the amount of biomass required to meet certain targets set for bioenergy in certain framework conditions, focusing on the demand side. These studies typically access the economic and implementation potentials and include cost–supply analysis and energy system modeling. Cost–supply analyses combine bottom-up bioenergy technical estimates with cost evaluation of biomass production, transportation, and conversion. The results are normally expressed as cost–supply curves. The analyses are based on assumptions on the availability of land for energy crops, yields, as well as the demand of land and biomass for other purposes and technical, environmental, and social limitations. Energy system models simulate energy markets dynamics and evaluate the competitiveness of various energy options through cost optimization of energy systems and are suitable to evaluate the costs and effectiveness of energy and climate policy options. The models include key market and economic factors influencing energy demand: energy supply options and relevant supply and demand sectors (food, feed, fiber, fuel, and materials). Economic correlations used in the models are often based on expert judgment and assumptions complementing historical data on energy use and prices.

Integrated approaches combine resource-focused and demand-driven methodologies and include integrated assessment models (IAMs) to assess various policy options including the possible contribution of biomass to energy and climate strategies. IAMs integrate key elements of economic and biophysical system in a comprehensive modeling framework, based on explicit assumptions about how the modeled system behaves. The integrated modeling frameworks include economic models on different sectors (economic, energy, agriculture, forestry, and land use) to assess trade-offs and synergies of policy options across economic, social, and environmental dimensions and estimate the impacts on agricultural markets, land use, and greenhouse gas (GHG) emissions (Batidzirai et al., 2012; BEE, 2010; Wicke et al., 2014a).

10.1.1.2 Global biomass potentials

The global potential of bioenergy is limited. Land resources have multiple uses and provide multiple ecosystem services for food, feed, timber, fiber production, biomaterials, biochemicals, nature conservation, etc. Biomass could make an important contribution to the future global energy supply through a combination of residues, wastes, and energy crops grown on different land types (Table 10.1). Several studies provide a wide range of estimates of the global biomass potentials, describing a variety of potentials (theoretical, technical, and economic) that are based on different definitions and thus not directly comparable. The studies provide estimates over different time frames: current, short term (forecasts up to 2030), mid-term (2050), and long term (2100).

The available biomass potential could amount from almost zero up to 1500 EJ theoretical potential, in various studies (Fig. 10.1), as compared to about 576 EJ global primary energy demand in 2016 (IEA, 2017a). The medium range of estimates is between 300 and 500 EJ. In comparison, the global primary energy supply from biomass, in traditional biomass use and modern bioenergy technologies, has reached about 56 EJ in 2015, representing a share of the total global primary energy consumption of about 10% (IEA, 2017a). The annual amount of biomass produced by plant growth, i.e., the net amount of carbon assimilated by vegetation or the net primary production is estimated at about 118 billion tons of dry matter per year (Gt) with a heating value of 2200 EJ, of which about 1240 EJ are allocated to above-ground biomass. The total annual amount of biomass harvest is about 20 Gt, which represents about 373 EJ when expressed in equivalent heat content, of which 12 Gt biomass harvested (219 EJ) is used for food, fodder, fiber, and forest products (GEA, 2012; Popp et al., 2014; IPCC, 2011; Haberl et al., 2010).

The large variability in biomass potentials reflects large views on scenarios and assumptions, such as market and policy conditions, trends in population growth, food demand, and developments in agriculture. The differences in the estimates of bioenergy resources are due to a broad

Table 10.1 Classification of biomass resources

Resource	Category	Source
Energy crops	Conventional crops	Annual crops: grains, sugar, and oil crops
	Perennial energy crops	Short rotation coppice/short rotation forestry: willow, poplar, eucalyptus
		Energy grasses: miscanthus, switchgrass
Primary residues	Forestry and forestry residues	Logging residues: branches, twigs, tops, thinning residues
		Low-quality stem wood (fuelwood)
		Landscape care residues
	Agricultural residues	Straw from cereals, oil seed rape, and other crops
		Trimming residues from vineyards or orchards
		Manures and slurries from cattle, pigs, poultry, sheep and goats
Secondary residues	Wood processing	Sawmill coproducts
		Wood residues, shavings, sawdust and bark from wood processing
	Agricultural residues	Food industry residues
		Residues from processing of agricultural products (e.g., rice husks, peels)
		Slaughterhouse residues
Waste	Urban wood residues	Construction and demolition wood
		Clean and contaminated waste wood
		Consumer durables
	Organic waste	Paper/card, food/kitchen, garden/plant, textiles wastes, packaging wastes
	Sewage sludge	Wastewater treatment plants
	Landfill gas	Captured gases from decomposing biodegradable waste in landfill sites

Source: From BEE, 2010. Status of Biomass Resource Assessments Version 3 Deliverable No: D 3.2; Slade R., Saunders R., Gross R., Bauen A., 2011. Energy From Biomass: The Size of the Global Resource. London: Imperial College Centre for Energy Policy and Technology and UK Energy Research Centre; Batidzirai B., Smeets E.M.W., Faaij A.P.C., 2012. Harmonising bioenergy resource potentials—methodological lessons from review of state of the art bioenergy potential assessments. Renew. Sustain. Energy Rev. 16 (9), 6598–6630.

variety of approaches, methodologies, assumptions, and data sets. The estimates also differ in terms of the range of feedstocks included in the analysis: agricultural and forestry residues, energy crops, waste, etc., and on the constraints to biomass supply. A wide range of estimates are due to the assumptions on population growth and food demand, diet, or improvement in agriculture production and yield improvement and the

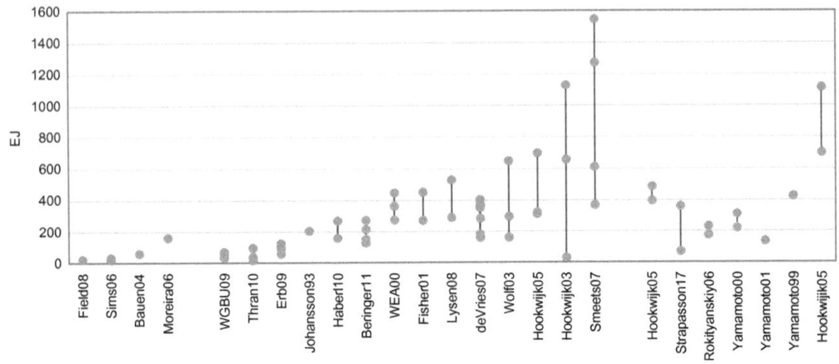

Figure 10.1 Global biomass potentials from various studies. *From Batidzirai B., Smeets E.M.W., Faaij A.P.C., 2012, Harmonising bioenergy resource potentials—methodological lessons from review of state of the art bioenergy potential assessments. Renew. Sustain. Energy Rev. 16 (9), 6598–6630; Bauen, A., Woods, J., Hailes, R., 2004. Bioelectricity vision: achieving 15% of electricity from biomass in OECD countries by 2020. E4tech (UK) Ltd.; Beringer, T., Lucht, W., Schaphoff, S., 2011. Bioenergy production potential of global biomass plantations under environmental and agricultural constraints. GCB Bioenergy 3, 299-312; de Vries, B.J.M., van Vuuren, D.P., Hoogwijk, M.M., 2007. Renewable energy sources: their global potential for the first-half of the 21st century at a global level: an integrated approach. Energy Policy, 35 2590–2610; Erb, K.H., Haberl, H., Kraussmann, F., Lauk, C., Plutzar, C., Steinberger, J.K., et al., 2009. Eating the Planet: Feeding and Fuelling the World Sustainably, Fairly and Humanely—A Scoping Study. Vienna: Institute of Social Ecology and PIK Potsdam; Field, C.B., Campbell, J.E., Lobell, D.B., 2008. Biomass energy: the scale of the potential resource. Trends Ecol. Evol., 23:65–72; Fischer, G., Schrattenholzer, L., 2001. Global bioenergy potentials through 2050. Biomass Bioenergy 20 (3), 151–159; Hoogwijk, M., Faaij, A., van den Broeka, R., Berndes, G., Gielen, D., Turkenburg, W., 2003. Exploration of the ranges of the global potential of biomass for energy. Biomass Bioenergy 25, 119–133; Hoogwijk, M., Faaij, A., Eickhout, B., 2005. Potential of biomass energy out to 2100, for four IPCC SRES land-use scenarios. Biomass Bioenergy 29, 225–257; Haberl H., Beringer T., Bhattacharya S.C., Erb K.-H., Hoogwijk M., 2010. The global technical potential of bio-energy in 2050 considering sustainability constraints. Curr. Opin. Environ. Sustain. 2 (5–6), 394–403; IEA, 2017a. Energy Technology Perspectives. International Energy Agency; IFEU, 2007. Nachwachsende Rohstoffe für die chemische Industrie: Optionen und Potenziale für die Zukunft. Heidelberg: Institute for Energy and Environmental Research (IFEU); Johansson, T.B., Kelly, H., Reddy, A.K.N., Williams, R.H., 1993. A renewables-intensive global energy scenario (RIDGES). In: Johansson, T.B., et al., (Eds.), Renewable Energy: Sources for Fuels and Electricity. Washington, DC: Island Press; Lysen, E., van Egmond, S., et al., 2008. Biomass assessment—assessment of global biomass potentials and their links to food, water, biodiversity, energy demand and economy—inventory and analysis of existing studies. Netherlands Research Program on Scientific Assessment and Policy Analysis for Climate Change, WAB report no. 500102014; Moreira, J.R., 2006. Global biomass energy potential. Mitig. Adapt. Strat. Global Change 11, 313–342; Rokityanskiy, D., Benítez, P.C., Kraxner, F., McCallum, I., Obersteiner, M., Rametsteiner, E., et al., 2007. Geographically*

(Continued)

available land for energy crops. As a consequence, the estimates of the global biomass supply potentials vary widely (Batidzirai et al., 2012; BEE, 2010; Wicke et al., 2014a).

Several studies, providing very low biomass potentials (below 100 EJ), follow a conservative approach considering high population growth and high demand for food, extensive agriculture, and low improvements in agriculture in the future. These studies also consider little or no expansion of agriculture land. The studies assume that no land or little land area (<0.15 Gha) or somewhat larger area of marginal land (<0.4 Gha), with low yields, is available for energy crops due to high demand for food and strict sustainability criteria. As a reference, the global land area is about 13 Gha, of which 1.5 Gha of arable land, 3.5 Gha pasture land, 4 Gha forest land, and 4 Gha other land. The contribution from waste and residues to bioenergy is generally not considered, or their contribution is low in these studies.

Some other studies provide moderate values for biomass potentials, between 100 and 300 EJ, and assume that food crop yields grow at the same rate with population growth and increasing food demand. Marginal, degraded, and deforested land, with increased yields, may be

◀ *explicit global modeling of land-use change, carbon sequestration, and biomass supply. Technol. Forecast. Soc. Change 74, 1057–1082; Popp J., Lakner Z., Harangi-Rákos M., Fári M., 2014, The effect of bioenergy expansion: food, energy, and environment. Renew. Sustain. Energy Rev. 32, 559–578; Sims, R., Hastings, A., Schlamadinger, B., 2006. Energy crops: current status and future prospects. Global Change Biol. 12, 2054–2076; Slade R., Saunders R., Gross R., Bauen A., 2011. Energy From Biomass: The Size of the Global Resource. London: Imperial College Centre for Energy Policy and Technology and UK Energy Research Centre; Strapasson, A., Woods, J., Chum, H., Kalas, N., Shah, N., Rosillo-Calle, F., 2017. On the global limits of bioenergy and land use for climate change mitigation, GCB Bioenergy 9, 1721–1735 doi:10.1111/gcbb.12456; Yamamoto, H., Yamaji, K., Fujino, J., 1999. Evaluation of bioenergy resources with a global land use and energy model formulated with SD technique. Appl. Energy 63, 101–113; Yamamoto, H., Yamaji, K., Fujino, J., 2000. Scenario analysis of bioenergy resources and CO2 emissions with a global land use and energy model. Appl. Energy 66, 325–337; Yamamoto, H., Fujino, J., Yamaji, K., 2001. Evaluation of bioenergy potential with a multi-regional global-land-use-and-energy model. Biomass Bioenergy 21, 185–203; WBGU, 2009. Future Bioenergy and Sustainable Land Use. German Advisory Council on Global Change (WBGU); Wicke B., van der Hilst F., et al., 2014a. Model collaboration for the improved assessment of biomass supply, demand, and impacts. GCB Bioenergy 7 (3), 422–437; Wolf, J., Bindraban, P.S., Luijten, J.C., Vleeshouwers, L.M., 2003. Exploratory study on the land area required for global food supply and the potential global production of bioenergy. Agric. Syst. 76 (3), 841–861; Doornbosch, 2008.*

available to some extent for energy crops (0.1–0.5 Gha), but no surplus agricultural land is available for energy crops. An expansion of global agricultural area is expected in case of high demand for food and materials, leading to some decrease in the pasture and forest area. The various residues and wastes can have a significant contribution to biomass supply, if taken into account. The higher estimates of biomass potentials, between 300 and 600 EJ, follow an optimistic approach and assume an intensive agriculture, with high yields. The increase in food crop yields would be higher than the increasing demand for food, and large areas of good agricultural land would be available for energy crops (>1 Gha), with marginal and degraded land but also some pasture lands converted to energy crops (Slade et al., 2011; WBGU, 2009; Lysen et al., 2008; IEA, 2007).

Few studies envisage biomass potentials in excess of 600 EJ, assuming extreme scenarios with high agricultural inputs and yield increases and less land needed for food production. This is linked to low food demand and a largely vegetarian global diet. High biomass potential can be achieved with intensified agriculture, high crop yields that outpace demand, leading to lower area being needed for food production and/or with the expansion of agricultural area, conversion of grassland and deforestation. The primary aim of the highest estimates is to provide a theoretical maximum biomass potential (Slade et al., 2011; WBGU, 2009; Lysen et al., 2008; IEA, 2007). More recent studies have tried to provide an improved view on the real biomass potentials by performing a deeper analysis of the assumptions and reasons for the wide range of biomass potentials in an attempt to reduce the ranges (Haberl et al., 2010; Lysen et al., 2008; WBGU, 2009). The results show that the potential deployment levels of biomass for energy by 2050 could be in the range of 150–300 EJ (IEA, 2007; IPCC, 2011).

The biomass potentials, consists of three main categories of biomass: energy crops, residues and waste, and forestry (Fig. 10.2). The greatest contribution comes in most studies from energy crops, grown on marginal, degraded, fallow land and surplus agricultural land. The widest range of potentials also relate to energy crops, ranging from very small (27 EJ) to well beyond current global primary energy supply (1272 EJ), which is mainly due to the differences in the assumptions related to the land area available and the expected yields. Studies trying to narrow down the estimates of energy crops resulted in a range of 100–150 EJ that could be available by 2050 (Haberl et al., 2010; Lysen et al., 2008; GEA, 2012).

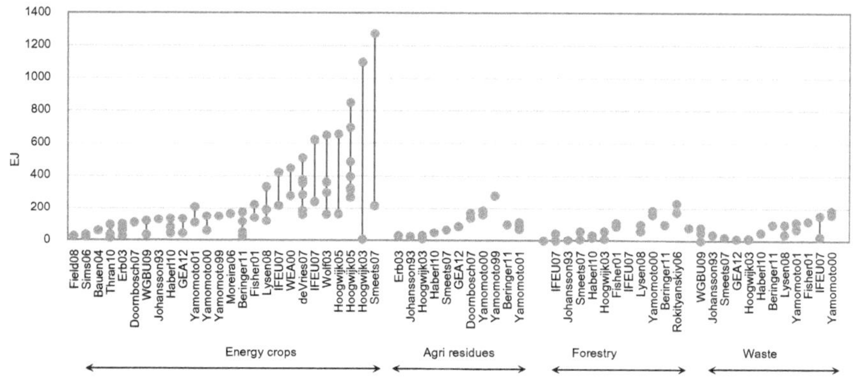

Figure 10.2 Contribution of different biomass sources to global biomass potentials. *From Batidzirai B., Smeets E.M.W., Faaij A.P.C., 2012, Harmonising bioenergy resource potentials—methodological lessons from review of state of the art bioenergy potential assessments. Renew. Sustain. Energy Rev. 16 (9), 6598–6630; Bauen, A., Woods, J., Hailes, R., 2004. Bioelectricity vision: achieving 15% of electricity from biomass in OECD countries by 2020. E4tech (UK) Ltd.; Beringer, T., Lucht, W., Schaphoff, S., 2011. Bioenergy production potential of global biomass plantations under environmental and agricultural constraints. GCB Bioenergy 3, 299-312; de Vries, B.J.M., van Vuuren, D.P., Hoogwijk, M.M., 2007. Renewable energy sources: their global potential for the first-half of the 21st century at a global level: an integrated approach. Energy Policy, 35 2590–2610; Erb, K.H., Haberl, H., Kraussmann, F., Lauk, C., Plutzar, C., Steinberger, J.K., et al., 2009. Eating the Planet: Feeding and Fuelling the World Sustainably, Fairly and Humanely—A Scoping Study. Vienna: Institute of Social Ecology and PIK Potsdam; Field, C.B., Campbell, J.E., Lobell, D.B., 2008. Biomass energy: the scale of the potential resource. Trends Ecol. Evol., 23:65–72; Fischer, G., Schrattenholzer, L., 2001. Global bioenergy potentials through 2050. Biomass Bioenergy 20 (3), 151–159; Hoogwijk, M., Faaij, A., van den Broeka, R., Berndes, G., Gielen, D., Turkenburg, W., 2003. Exploration of the ranges of the global potential of biomass for energy. Biomass Bioenergy 25, 119–133; Hoogwijk, M., Faaij, A., Eickhout, B., 2005. Potential of biomass energy out to 2100, for four IPCC SRES land-use scenarios. Biomass Bioenergy 29, 225–257; Haberl H., Beringer T., Bhattacharya S.C., Erb K.-H., Hoogwijk M., 2010. The global technical potential of bioenergy in 2050 considering sustainability constraints. Curr. Opin. Environ. Sustain. 2 (5–6), 394–403; IEA, 2017a. Energy Technology Perspectives. International Energy Agency; IFEU, 2007. Nachwachsende Rohstoffe für die chemische Industrie: Optionen und Potenziale für die Zukunft. Heidelberg: Institute for Energy and Environmental Research (IFEU); Johansson, T.B., Kelly, H., Reddy, A.K.N., Williams, R.H., 1993. A renewables-intensive global energy scenario (RIDGES). In: Johansson, T.B., et al., (Eds.), Renewable Energy: Sources for Fuels and Electricity. Washington, DC: Island Press; Lysen, E., van Egmond, S., et al., 2008. Biomass assessment—assessment of global biomass potentials and their links to food, water, biodiversity, energy demand and economy—inventory and analysis of existing studies. Netherlands Research Program on Scientific Assessment and Policy Analysis for Climate Change, WAB report no. 500102014; Moreira, J.R., 2006. Global biomass energy potential. Mitig. Adapt. Strat. Global Change 11,*

(Continued)

The residues and waste can be important sources for expansion of biomass production for energy, both in the short and longer term. The use of residues and waste materials is considered to be more sustainable compared to the use of energy crops, for example, because they entail no direct impacts on land use and no competition with food and materials. While the potential contribution from residues and wastes are less than the estimates for energy crops, these potentials appear significant compared to total global energy consumption. Overall, forestry, agricultural residues, and other organic wastes would provide between 50 and 150 EJ/year: agricultural residues might provide between 50 and 100 EJ, forestry residues might provide between 20 and 50 EJ, and organic wastes could provide between 10 and 30 EJ. However, the use of some crop residues might have negative impacts on soil productivity if removed in an unsustainable way and there might be some other current competitive uses (for

313–342; Rokityanskiy, D., Benítez, P.C., Kraxner, F., McCallum, I., Obersteiner, M., Rametsteiner, E., et al., 2007. Geographically explicit global modeling of land-use change, carbon sequestration, and biomass supply. Technol. Forecast. Soc. Change 74, 1057–1082; Popp J., Lakner Z., Harangi-Rákos M., Fári M., 2014, The effect of bioenergy expansion: food, energy, and environment. Renew. Sustain. Energy Rev. 32, 559–578; Sims, R., Hastings, A., Schlamadinger, B., 2006. Energy crops: current status and future prospects. Global Change Biol. 12, 2054–2076; Slade R., Saunders R., Gross R., Bauen A., 2011. Energy From Biomass: The Size of the Global Resource. London: Imperial College Centre for Energy Policy and Technology and UK Energy Research Centre; Strapasson, A., Woods, J., Chum, H., Kalas, N., Shah, N., Rosillo-Calle, F., 2017. On the global limits of bioenergy and land use for climate change mitigation, GCB Bioenergy 9, 1721–1735 doi:10.1111/gcbb.12456; Smeets, E., Faaij, A., Lewandowski, I., Turkenburg, W., 2007. A bottom-up assessment and review of global bio-energy potentials to 2050. Prog. Energy Combust. Sci. 33, 56–106; Yamamoto, H., Yamaji, K., Fujino, J., 1999. Evaluation of bioenergy resources with a global land use and energy model formulated with SD technique. Appl. Energy 63, 101–113; Yamamoto, H., Yamaji, K., Fujino, J., 2000. Scenario analysis of bioenergy resources and CO2 emissions with a global land use and energy model. Appl. Energy 66, 325–337; Yamamoto, H., Fujino, J., Yamaji, K., 2001. Evaluation of bioenergy potential with a multi-regional global-land-use-and-energy model. Biomass Bioenergy 21, 185–203; WBGU, 2009. Future Bioenergy and Sustainable Land Use. German Advisory Council on Global Change (WBGU); Wicke B., van der Hilst F., et al., 2014a. Model collaboration for the improved assessment of biomass supply, demand, and impacts. GCB Bioenergy 7 (3), 422–437; Wolf, J., Bindraban, P.S., Luijten, J.C., Vleeshouwers, L.M., 2003. Exploratory study on the land area required for global food supply and the potential global production of bioenergy. Agric. Syst. 76 (3), 841–861; Doornbosch, 2008; Thrän, D., Seidenberger, T., Zeddies, J., Offremann, R., 2010. Global biomass potentials—resources, drivers and scenario results. Energy Sustain. Dev. 14, 200–205.

feed and fodder, horticulture, mushroom production, etc.) (Slade et al., 2011; IEA, 2017a; Popp et al., 2014; WBGU, 2009).

10.1.1.3 European biomass potentials

Several studies provide estimates of the biomass potentials available in the European Union (EU) (Fig. 10.3). This is important to understand the limits of the bioenergy deployment as result of European policies supporting the development of renewable energy toward the goal of achieving of a low-carbon energy system. As for the global studies, the different studies differ largely with regard to the biomass categories considered (e.g., energy crops, agriculture or forest residues, waste, or total potentials), the geographical scale (e.g., EU15, EU25, EU27, etc.), the time frame of the analysis, and different definitions of the potentials considered. The analysis of biomass resource assessments shows that the reported total potentials differ to a considerable degree, mostly related to the potential of energy crops. These uncertainties can mainly be explained by varying methods and assumptions of estimating land available or yields as well varying assumptions on multiple factors that influence potentials, including the consideration of sustainability issues. The Biomass Energy Europe project emphasized the need for a coherent and harmonized approach for the estimation of biomass types and potentials (BEE, 2010; JRC, 2015).

The estimates of total biomass potentials range between 4 and 20 EJ on short term (2020) to 8.5–24 EJ on mid-term (2030) and between 18–59 EJ on long term (2050). While the estimates of energy crop potentials vary between 1–10 EJ on short term, the estimates increase to 2–13 EJ on mid-term and reach between 15 and 20 EJ on long term. The trends for agricultural and forestry residues and forestry are decreasing over time. The potentials of agricultural residues vary between 1.8 and 3.8 EJ on short term with 1.8 and 2.8 EJ on longer term. The potentials for forestry and forest residues vary between 1.8 and 3.3 EJ on short term, 1.4 and 4.0 EJ on medium term, and 1.8 and 2.3 EJ on long term. However, the highest estimates should be regarded as the maximum (theoretical) potentials for bioenergy for the EU (Fig. 10.4). The final amount of available biomass potential will depend on the capacity to mobilize further unexploited potential (BEE, 2010).

Latest studies looked into the quantification of the available biomass that could contribute to achieving the EU energy and climate targets for 2030 and 2050. Biomass Future project provided a comprehensive

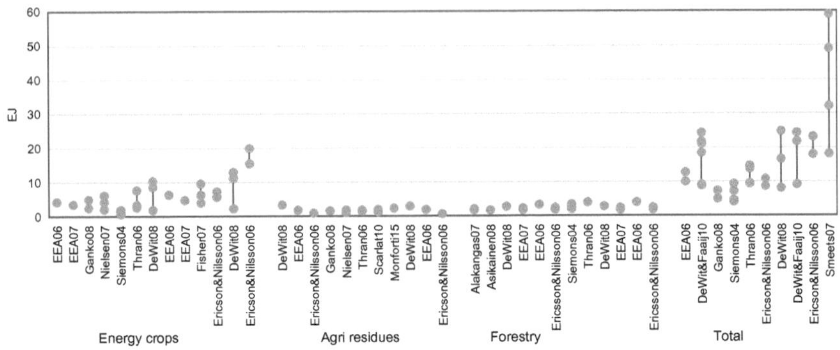

Figure 10.3 Assessments of biomass potential in the EU. *From Alakangas, E., Heikkinen, A., Lensu, T., Vesterinen, P., 2007. Biomass Fuel Trade in Europe. Summary Report VTTR0350807. Jyväskylä, Finland: VTT, 57; Asikainen, A., Liiri, H., Peltola, S., Karjalainen, T., Laitila, J., 2008. Forest energy potential in Europe (EU27). Working Papers of the Finnish Forest Research Institute. Joensuu: Finnish Forest Research Institute. 69, p. 33; BEE, 2010. Status of Biomass Resource Assessments Version 3 Deliverable No: D 3.2; EEA, 2006. How Much Bioenergy Can Europe Produce Without Harming the Environment? European Environment Agency, Report No 7/2006 ISBN 92–9167–849-X2006; EEA, 2007. Environmentally Compatible Bio-Energy Potential From European Forests. Copenhagen, European Environmental Agency, 54; Elbersen, B., Startisky, I., Hengeveld, G., Schelhaas, M.-J., Naeff, H., Böttcher, H., 2012. Atlas of EU Biomass Potentials. Deliverable 3.3: Spatially Detailed and Quantified Overview of EU Biomass Potential Taking Into Account the Main Criteria Determining Biomass Availability From Different Sources; Ericsson, K., Nilsson, L.J., 2006. Assessment of the potential biomass supply in Europe using a resource-focused approach. Biomass Bioenergy 30(1): 1–15; Ganko, E., Kunikowski, G., Pisarek, M., Rutkowska Filipczak, M., Gumeniuk, A., Wróbel, A., 2008. Biomass Resources and Potential Assessment. Final Report WP 5.1 of RENEW Project, Contract no SES-CT-2003-502705. Project Final Report. Warsaw: EC BREC/IPiEO Institute for Fuels and Renergable Energy. Available at: www.renew-fuel.com; Fischer, G., Schrattenholzer, L., 2001. Global bioenergy potentials through 2050. Biomass Bioenergy 20 (3), 151–159; Nielsen J.B.H., Oleskowicz-Popiel P., Al Seadi T., 2007. Energy crop potentials for bioenergy in EU-27; Scarlat N., Martinov M., Dallemand J.F., 2010. Assessment of the availability of agricultural crop residues in the European Union: potential and limitations for bioenergy use. Waste Manage. 30 (10) 1889–1897; Siemons, R. M. Vis, van den Berg, D., Mc Chesney, I., Whiteley, M., Nikolaou, N., 2004. Bioenergy's role in the EU Energy market, a view of developments until 2020; de Wit, M. P., Faaij, A.P.C., 2008. Biomass Resources Potential and Related Costs. Assessment of the EU-27, Switzerland, Norway and Ukraine. Report. Utrecht, The Netherlands: Utrecht University, Copernicus Institute. Available at: www.refuel.eu/fileadmin/refuel/user/docs/REFUEL_D9.pdf; Thrän, D., Weber, M., Scheuermann, A., Fröhlich, et al., 2006. Sustainable Strategies for Biomass Use in the European Context. Institute for Energy and Environment, Leipzig, p. 387.*

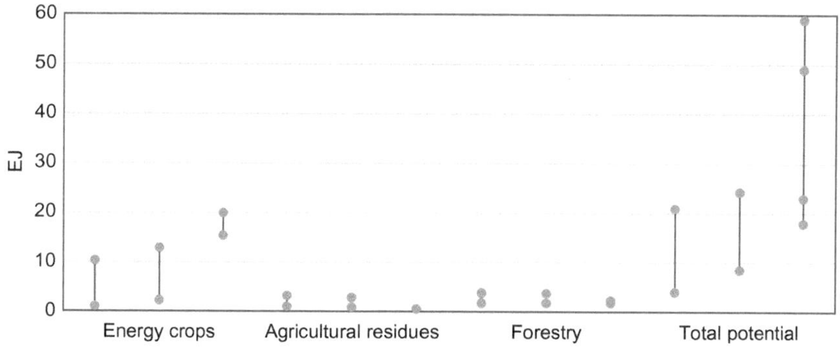

Figure 10.4 Contribution of different biomass sources to European biomass potential. *From Alakangas, E., Heikkinen, A., Lensu, T., Vesterinen, P., 2007. Biomass Fuel Trade in Europe. Summary Report VTTR0350807. Jyväskylä, Finland: VTT, 57; Asikainen, A., Liiri, H., Peltola, S., Karjalainen, T., Laitila, J., 2008. Forest energy potential in Europe (EU27). Working Papers of the Finnish Forest Research Institute. Joensuu: Finnish Forest Research Institute. 69, p. 33; BEE, 2010. Status of Biomass Resource Assessments Version 3 Deliverable No: D 3.2; EEA, 2006. How Much Bioenergy Can Europe Produce Without Harming the Environment? European Environment Agency, Report No 7/2006 ISBN 92–9167–849-X2006; EEA, 2007. Environmentally Compatible Bio-Energy Potential From European Forests. Copenhagen, European Environmental Agency, 54; Elbersen, B., Startisky, I., Hengeveld, G., Schelhaas, M.-J., Naeff, H., Böttcher, H., 2012. Atlas of EU Biomass Potentials. Deliverable 3.3: Spatially Detailed and Quantified Overview of EU Biomass Potential Taking Into Account the Main Criteria Determining Biomass Availability From Different Sources; Ericsson, K., Nilsson, L.J., 2006. Assessment of the potential biomass supply in Europe using a resource-focused approach. Biomass Bioenergy 30(1): 1–15; Ganko, E., Kunikowski, G., Pisarek, M., Rutkowska Filipczak, M., Gumeniuk, A., Wróbel, A., 2008. Biomass Resources and Potential Assessment. Final Report WP 5.1 of RENEW Project, Contract no SES-CT-2003-502705. Project Final Report. Warsaw: EC BREC/IPiEO Institute for Fuels and Renergable Energy. Available at: www.renew-fuel.com; Fischer, G., Schrattenholzer, L., 2001. Global bioenergy potentials through 2050. Biomass Bioenergy 20 (3), 151–159; Scarlat N., Martinov M., Dallemand J.F., 2010. Assessment of the availability of agricultural crop residues in the European Union: potential and limitations for bioenergy use. Waste Manage. 30 (10) 1889–1897; de Wit, M.P., Faaij, A.P.C., 2008. Biomass Resources Potential and Related Costs. Assessment of the EU-27, Switzerland, Norway and Ukraine. Report. Utrecht, The Netherlands: Utrecht University, Copernicus Institute. Available at: www.refuel.eu/filead-min/refuel/user/docs/REFUEL_D9.pdf.*

analysis of biomass supply options and their availability taking into account different sustainability constraints to evaluate the biomass available for energy uses in the medium term. Biomass Future project provided sustainable potentials in the reference and sustainability scenarios for 2020 and 2030 with the sustainability scenario applying more stringent criteria intended to limit the environmental impact of bioenergy (Elbersen et al., 2012).

The Joint Research Centre (JRC) used the JRC-EU-TIMES partial equilibrium (PE) energy system model to analyze the role of energy technologies development and their potential contribution to decarbonization pathways of the energy system until 2050 (JRC, 2015). The JRC has performed the quantification of current and future biomass potentials for energy systems based on the agricultural PE CAPRI model, combined with the JRC LUISA model, EFISCEN forestry model, and waste statistics. JRC provided sustainable potentials for the low biomass availability scenario (biomass use for energy is not a key priority, sustainability, and resource efficiency highly considered), the reference scenario (reference case, continuation of current trends, sustainability, and resource efficiency), and the high biomass availability scenario (high demand, stimulation measures). These studies concluded that the total biomass potential is between 14.8 and 21.1 EJ: agriculture potential ranges between 6.0 and 9.6 EJ, forestry potential ranges between 7.0 and 9.9 EJ, and waste potentials range between 1 and 1.8 EJ.

10.1.1.4 Outlook—making use of the full biomass potential

Most estimates show that biomass is likely to be sufficient to play a significant role in the global energy supply system until 2050. Given the complexity of factors that affect the future biomass resource availability for energy and materials, it is not possible to present the future biomass potential in one simple figure. The real biomass contribution to energy supply will ultimately depend on the capacity to mobilize sustainable biomass potentials. Despite the uncertainties about existing potentials already mentioned and concerns about sustainability and mobilization, a level of biomass supply of around 145 EJ, as required by the 2DS (2 Degree Scenario) and B2DS (Beyond 2 Degree Scenario), is challenging but achievable. The biomass supply for these scenarios is unlikely to be achieved just through the use of waste and agricultural and forestry residues and requires further contribution from energy crops. Reaching the upper range of potentials level would, however, require long-term policy vision and adequate policy support, targeting improvements in the agricultural sector and in the efficiency of biomass use. Without policy support, bioenergy development might be largely constrained to the use of residues and waste and some possible cultivation of bioenergy crops on marginal and degraded lands, whenever biomass is a cheaper option compared to the main reference options (IPCC, 2011; Lysen et al., 2008; Strapasson et al., 2017; IEA, 2017a).

The main challenge in realizing the potential of bioenergy is in the large-scale deployment. The biomass mobilization of increased biomass amounts is a critical issue, and the need to increase agricultural productivity and efficiency mostly in low income countries is essential. The increased competition between food, feed and fiber, wood products or new bio-based materials, and bioenergy needs to be properly addressed allowing the prioritization of biomass use according to the societal needs. The risks of delivery of unsustainable feedstock increase with the increase of biomass use thus requiring good governance and land-use planning (IPCC, 2011; Lysen et al., 2008). Biomass potential for energy would benefit from the improved sustainable food production that makes better use of limited resources including land, water, and fertilizers. The large variability in the estimates needs a detailed and local analysis of biomass feedstock production and various uses that requires due consideration of specific local sustainability issues, that include water constraints, biodiversity protection, and food security in certain areas.

The major issue related to the use of energy crops for energy supply is that they compete for water, land, and nutrients crops with food and feed crops, and that they could cause land-use changes (LUCs), ecosystems damage, and loss of habitats. Water is a critical issue that constrains food production and energy crops and needs to be further investigated in more refined studies at regional level. In case of use of perennial crops (either grasses or trees), they can have a positive impact on biodiversity and carbon stock, especially grown on low quality land as well as additional benefits such as soil protection, improved water retention and water purification, and ecosystem services (Slade et al., 2011, Lysen et al., 2008). Good quality land and improvements in energy crop cultivation can lead to some large yields (>12 dry ton ha^{-1}), while marginal and degraded land could have only modest yields (<5 dry ton ha^{-1}). However, the large uncertainties related to the availability and suitability of land for energy crops, the potential of yield increases, future area demand for food, and trade-offs with biodiversity and water availability require further investigation to better understand the real potential of energy crops.

10.1.2 Bioenergy in future low-carbon energy systems

10.1.2.1 Future low-carbon energy system

Several attempts tried to reveal the future configuration of global energy system that would be able to deliver the carbon emission reductions necessary to achieve the long-time goal of limiting climate change. The IEA prepared a Bioenergy Roadmap based on the Energy Technology Perspectives modeling framework that covers energy supply, buildings, industry, and transport sectors (IEA, 2017b). The Roadmap has been developed for three scenarios with different energy technologies in a low-carbon energy system. The Roadmap (IEA, 2017c) identified in each scenario the role of a portfolio of technologies in a future sustainable global energy system that allows the achievement of the long-term goal to limit the temperature increase. The following scenarios have been analyzed:

- **Reference Technology Scenario (RTS)**—The baseline scenario that takes into account existing and planned energy and climate-related commitments, following the global climate agreement reached during the 21st Conference of the Parties (COP21) of the United Nations Framework Convention on Climate Change (UNFCCC).
- **2DS**—Scenario of an energy system that allows limiting the global average temperature increase to 2°C by 2100.
- **B2DS scenario**—Accelerating clean energy technology deployment reaching more ambitious climate goals of limiting the global average temperature increase to 1.75°C by 2100.

The RTS scenario is built on the assumption of the implementation of the Nationally Determined Contributions (NDCs) proposed by the signatories of the Paris Agreement that requires significant changes in energy system and policies. This would anyway result in an average temperature increase of 2.7°C by 2060. The 2DS requires major improvements in energy efficiency across all sectors, the widespread deployment of renewable energies, fuel switching, and the application of carbon capture and storage (CCS) technologies. The B2DS scenario looks into a more ambitious decarbonization pathway that relies on an important contribution from bioenergy and a stronger role for CCS, to deliver the additional emission reductions. Thus, both scenarios require a challenging and ambitious transformation of the energy sector, with additional technical and political challenges for the B2DS scenario.

Global total CO_2 emissions from the energy sector were around 34.3 Gt in 2014 with industrial process emissions included. In the 2DS scenario, annual energy sector CO_2 emissions are reduced by 70% from current levels by 2060, with cumulative emissions of around 1170 Gt CO_2 between 2015 and 2100, including industrial process emissions. In the 2DS, CO_2 emissions continue to decline after 2060 to reach carbon neutrality in the energy system by 2100. The 2DS allows a cumulative carbon budget over the period to 2060 about 40% lower compared to the RTS, requiring the abatement of an additional 760 Gt CO_2 over this period. Renewable energy technologies are deployed mostly in the power sector, driven by the need for rapid decarbonization in the 2DS, and further applications in transport (biofuels), heating in buildings, and industry (Fig. 10.5).

The B2DS results in cumulative emissions from the energy sector of around 750 Gt CO_2 between 2015 and 2100, and carbon neutrality of the energy system by 2060, supported by negative emissions through deployment of bioenergy with CCS. The deployment of Bioenergy with Carbon Capture and Storage (BECCS) is essential in the B2DS to reach net-zero emissions in 2060. The negative emissions from BECCS compensate for the emissions in industry and transport that are very difficult or very costly to abate (IEA, 2017b). The B2DS scenario should limit cumulative energy sector CO_2 emissions to around 750 Gt in the period 2015–60 which requires the reduction of cumulative emissions by almost 60% by 2060 compared with the RTS or about 1000 Gt CO_2. In the B2DS the power sector is already virtually decarbonized by 2060. In the

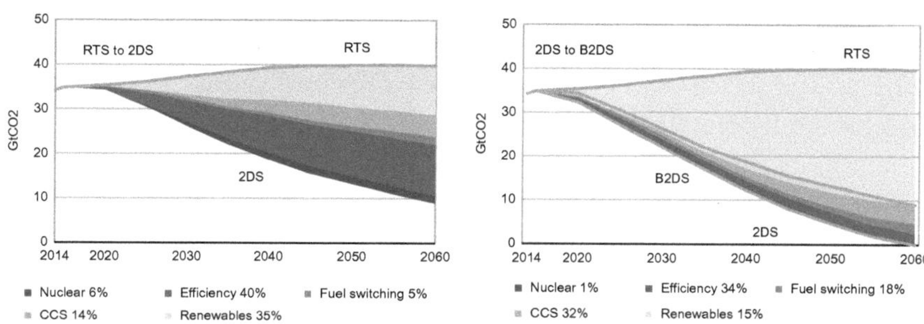

Figure 10.5 Cumulative global CO_2 reductions in different scenarios until 2060. *From IEA, 2017b. Technology Roadmap. Delivering Sustainable Bioenergy. International Bioenergy Agency.*

B2DS, CO_2 emissions rapidly decline to reach net-zero emissions in 2060 following a with a much faster decarbonization pathway compared to the 2DS (Fig. 10.5). Energy efficiency is crucial for the energy sector transformation and accounts for 40% cumulative emissions reductions needed for the shift from the RTS to the 2DS, and additional 34% emissions reductions needed for the shift from the 2DS to B2DS (IEA, 2017b).

In the RTS scenario, the global primary energy demand increases from 576 EJ in 2015 to reach 843 EJ in 2060. Fossil fuels continue to dominate primary energy supply, falling from 82% in 2014 to 67% in 2060, with the remaining share in 2060 coming from biomass and waste (12%), other renewables (14%), and nuclear (7%). In the 2DS, growth in primary energy demand is limited and is around 20% (180 EJ) lower than in the RTS. The role of fossil fuels declines substantially from 82% in 2014 to just 35% of the mix in 2060, while renewables dominate the primary energy mix, with 52% (348 EJ). The share of biomass and waste doubles by 2060 to reach 144 EJ and to account for 22% of the energy mix. The power sector approaches carbon neutrality toward 2100. In the B2DS, the growth in primary energy demand is limited to only 10% (or 55 EJ) until 2060, due to the energy efficiency measures. The share of fossil fuels decreases from 82% in 2014 to 26% in 2060 and total renewables increase from 13% to 60%. Bioenergy experiences only a limited growth compared with the 2DS, reaching 24% of the energy mix in 2060, reflecting the constraints on the availability of sustainable biomass. According to the different scenarios, carbon intensity of power generation decreases from the current level of around 520–254 gCO_2/kWh in the RTS and below zero to -10 gCO_2/kWh in 2060 in the B2DS. In the 2DS, 98% of electricity generation is from low-carbon sources by 2060, with renewables delivering around two-thirds of the emissions reductions in the power sector, with CCS providing 18% and nuclear 16% of CO_2 emission reductions (IEA, 2017b).

10.1.2.2 Biomass in future global low-carbon energy system

IEA modeling indicates that modern bioenergy is an essential component of the future low-carbon global energy system in both the 2DS and the B2DS (IEA, 2016c). Nowadays, bioenergy is the main source of renewable energy worldwide and plays an important role as a modern and efficient source of energy to generate electricity, heat for buildings or for industrial processes, and biofuels for transport (IEA, 2017b). Traditional use of biomass plays a significant role for cooking and heating in low-income countries with about 2.7 billion people relying on traditional use of biomass for energy (IEA, 2016c).

Compared to the current level of deployment, the contribution of bioenergy across all sectors should increase significantly if global climate change commitments are to be met. According to projections, biomass use is expected to grow from 56 EJ in 2015 (IEA, 2017b) to almost 100 EJ in 2060 in the RTS and to around 145 EJ in both 2DS and B2DS (Fig. 10.6). As mentioned above, the contribution of bioenergy was limited at around 145 EJ due to the constraints on biomass availability. In the 2DS, the contribution of bioenergy to electricity production is higher than in the RTS. A decline in traditional use of biomass from 30 to 17 EJ is expected by 2060 in the RTS, driven by the progress on improved access to clean energy and better economic circumstances in a number of low-income countries in sub-Saharan Africa and Asia (IEA, 2017b). Traditional biomass use follows the same pattern in the 2DS and B2DS as in the RTS, reducing by around 40% between 2015 and 2060, still playing a major role in supplying energy in residential sector in low income countries.

Bioenergy can play an important role in decarbonizing certain sectors for which there are no other options or they are very limited, such as in the transport sector (freight road transport, aviation, and maritime transport). Considering the limited available resources, however, bioenergy contribution should be targeting areas where CO_2 emission reduction is higher, such as in power generation with BECCS, employing bioenergy pathways that deliver maximum carbon benefits. Biomass use for electricity generation can play an enhanced role in the electricity grid by providing flexible dispatchable power and thus allowing high levels of variable

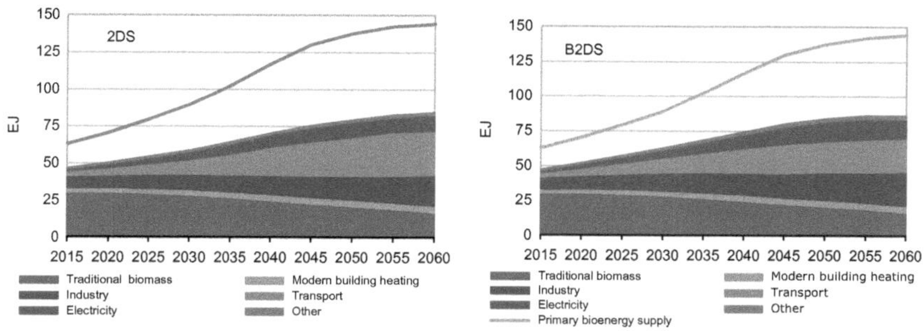

Figure 10.6 The global contribution of bioenergy to energy supply in 2DS and B2DS. *From IEA, 2017b. Technology Roadmap. Delivering Sustainable Bioenergy. International Bioenergy Agency.*

electricity from wind and solar into the power grid. The IEA estimates show that the biomass use for electricity generation is much higher in the B2DS in comparison to 2DS, due to a strong shift to the increased use of electricity, coupled to BECCS to generate negative emissions. Major expansion is expected for the use of biomass in the transport sector under the 2DS, reaching nearly 30 EJ in 2060. In the B2DS, the contribution of biofuels to transport is lower in the than in the 2DS (24 EJ), due to the reduction of energy demand in transport and a higher role of biomass and BECCS to generate negative emissions.

10.1.2.3 Future EU low-carbon energy system

In the EU, the *Energy Roadmap 2050* investigated possible pathways for a transition towards the decarbonization of the energy system until 2050 and the associated impacts, challenges, and opportunities. A number of scenarios to achieve 80% reduction in GHG and about 85% reduction of energy-related CO_2 emissions have been examined on the basis on a modeling framework including PRIMES, PROMETHEUS, GAINS, and GEM-E3 models. According to the Roadmap, different energy options can contribute to the achievement of the 2050 decarbonization goals, with energy efficiency and renewable energy playing a major role. The highest share of energy supply in 2050 is expected to come from renewables. The reference scenario 2050 includes the existing trends and long-term projections on population growth, economic development, fossil fuel prices, and technological developments in the framework of the EU policies and measures implemented by March 2010. The scenarios analyzed include the following:

- Current trend scenarios
  - Reference scenario: current trends and long-term projections on economic development, policies adopted by March 2010, including the 2020 targets for RES and GHG reductions as well as the Emissions Trading Scheme (ETS) Directive.
  - Current Policy Initiatives (CPI): updated measures adopted and proposed in the Energy 2020 strategy and proposed actions concerning the energy efficiency plan and the new energy taxation directive.
- Decarbonization scenarios
  - High energy efficiency: very high energy savings and stringent requirements for appliances, new buildings and energy utilities.

- Diversified supply technologies: all energy sources, including nuclear and CCS, compete on a market basis with no specific support measures while decarbonization is driven by carbon pricing.
- High renewable energy sources (RES): strong support measures for RES leading to a very high share of RES in gross final energy consumption (75% in 2050) and a share of RES in electricity generation reaching 86%.
- Delayed CCS: all energy sources, but assuming that CCS is delayed, leading to higher shares for nuclear energy with decarbonization driven by carbon prices rather than technology push.
- Low nuclear: all energy sources, but no new nuclear plants in addition to those currently under construction are being built, resulting in a higher penetration of CCS (around 32% in power generation).

Energy efficiency is an essential element in all decarbonization scenarios and thus the Roadmap projects a decrease of primary energy consumption between 11% and 20% by 2030 and between 30% and 41% by 2050, as compared to the Reference scenario, with the maximum values reached in the energy efficiency scenario. The Roadmap shows that, driven by a strong support, RES are expected to increase considerably their share in primary energy supply in all decarbonization scenarios to reach between 22% and 26% by 2030 and between 41% and 60% by 2050, with the maximum values reached in the high RES scenario (Fig. 10.7). The Reference scenario assumes that the RES target (20%

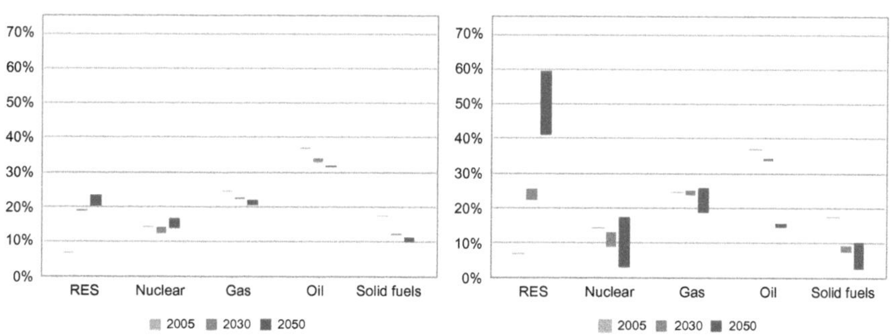

Figure 10.7 RES shares in primary energy supply in the EU in CPI scenarios (left) decarbonization scenarios (right) in 2030 and 2050. *From Energy Roadmap 2050 (COM, 2011a. 885 Final. Energy Roadmap 2050 Communication from the Commission to the European Parliament, the Council, the European Economic and Social Committee and the Committee of the Regions, 2011; COM, 2011b. 112 Final. A Roadmap for Moving to a Competitive Low Carbon Economy in 2050 Communication from the Commission to the European Parliament, the Council, the European Economic and Social Committee and the Committee of the Regions).*

RES in gross final energy supply) is reached in 2020 and the RES share in gross final energy consumption rises to at least 24% in 2030 and 26% in 2050. The RES share is projected to rise substantially in all decarbonization scenarios, reaching between 28% and 31% in 2030 and between 55% and 75% in 2050 (Fig. 10.8) (SEC(2011) 1565).

There is a strong increase of power generation in all scenarios compared to 2005 levels. Power generation will be lower in all decarbonization scenarios by 2050 compared with Reference and CPI scenarios. The share of renewables in electricity generation is expected to reach between 51% and 60% in 2030 and between 60% and 86% in 2050 in the high RES scenario. The share in power consumption could reach even 97% in 2050 (when calculating the RES electricity share in line with the RES directive, i.e., excluding energy losses to pumped storage, etc.) compared to almost 30% in 2016. Wind is expected to be the main RES source of power generation in most decarbonization scenarios, with about one third, except the high RES case, with a wind share of almost 50% in 2050. The high RES scenario is the most challenging scenario regarding the restructuring of the energy system, including major investments in power generation with RES capacity in 2050 reaching 1740 GW, from about 400 GW in 2016. Biomass is expected to have a share of about 10% of electricity supply in all decarbonization scenarios. The share of renewable energy in transport rises to 19%–20% in 2030 and to

Figure 10.8 RES share in a) gross final energy supply and b) in power generation in the EU. *From Energy Roadmap 2050 (COM, 2011a. 885 Final. Energy Roadmap 2050 Communication from the Commission to the European Parliament, the Council, the European Economic and Social Committee and the Committee of the Regions, 2011; COM, 2011b. 112 Final. A Roadmap for Moving to a Competitive Low Carbon Economy in 2050 Communication from the Commission to the European Parliament, the Council, the European Economic and Social Committee and the Committee of the Regions).*

62%–73% in 2050, using biofuels and renewable electricity in electric vehicles (SEC(2011) 1565.

10.1.2.4 Biomass in future EU low-carbon energy system

Biomass is expected to play a significant role in the EU in all decarbonization scenarios. In order to fulfill this role, the total use of biomass for heat, electricity, and transport is expected to rise significantly until 2050 in all scenarios. The biomass use has already increased from 3.6 EJ in 2005 to about 5.2 EJ in 2016. The biomass use in the reference scenario was projected to reach about 7.5 EJ in 2030 and 7.8 EJ in 2050. In the decarbonization scenarios, biomass consumption was projected to reach between 6.8 and 8.0 EJ in 2030 and between 10.1 and 12.6 EJ in 2050 in the high RES scenarios. This implies an increase in biomass supply by about 4.9–7.4 EJ by 2050, in addition to the present use of 5.2 EJ. The biomass supply for energy in different scenarios analyzed is presented in Fig. 10.9, showing a significant increase in certain scenarios (especially high RES scenario) until 2050. The key issue for bioenergy development is related to the availability of reliable and affordable sustainable biomass supply (SEC(2011) 1565).

In the EU, the installed bioenergy power capacity in 2015 was 30 GW and the installed bioenergy power capacity is expected to reach 43 GW in 2020. The installed biomass capacity would increase significantly in all scenarios until 2050 to reach 87 GW in the reference scenario. The

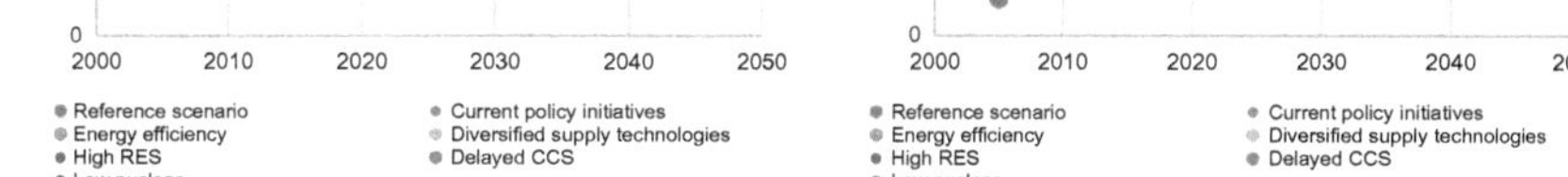

Figure 10.9 Expected biomass use in the EU (left) and bioenergy installed plant capacity. *From Energy Roadmap 2050 (COM, 2011a. 885 Final. Energy Roadmap 2050 Communication from the Commission to the European Parliament, the Council, the European Economic and Social Committee and the Committee of the Regions, 2011; COM, 2011b. 112 Final. A Roadmap for Moving to a Competitive Low Carbon Economy in 2050 Communication from the Commission to the European Parliament, the Council, the European Economic and Social Committee and the Committee of the Regions).*

growth in biomass installed capacity is much higher in different decarbonization scenarios, which should reach between 106 and 163 GW in 2050 (Fig. 10.9). Biomass electricity generation in the EU increased from 69 in 2005 to 190 TWh in 2016 and is expected to reach 232 TWh in 2020. The biomass electricity production is projected to further grow to 360 TWh in 2050 in the reference scenario and to 460–494 TWh in 2050 in different decarbonization scenarios. Biomass electricity contribution could rise from 2.6% share in power generation in 2005 and 5.5% in 2015 to 7.3% in 2050 in the reference scenario and between 9.3%–10.9% in decarbonization scenarios SEC(2011) 1565.

The RES share in gross final consumption of heating and cooling is expected to increase from just more than 10% in 2005 and 19% in 2016 to 21 % in 2020. Under different scenarios until 2020, RES share in heating and cooling would double, to reach at least 44% by 2050 under various decarbonization scenarios and up to 54% in the high RES scenario. In absolute terms, this would require an increase in the RES use for heating and cooling of 20%–60%, depending on the accompanying measures for reducing energy use and improvements in energy efficiency (Scarlat et al., 2015). Currently, heating is the main bioenergy market, accounting for 3.3 EJ in the EU, and about 90% of renewable heating and 15% of total heat generation in the EU in 2016. Direct use of biomass for heating is expected to rise to approximately 33% in 2050 in the high RES scenario (SEC(2011) 1565).

In the long term, the share of renewables in transport (biofuels and renewable electricity) is expected to reach 19%–20% in 2030 and up to 62%–73% in 2050 in all decarbonization scenarios (SEC(2011) 1565). The use of biofuels in transport sector in decarbonization scenarios was projected to increase to 1.0–1.5 EJ in 2030 and 2.8–3.0 EJ in 2050, in different scenarios. In the light of the last changes in the EU vision for the first-generation biofuels, future contribution of biomass to the transport sector could be different. Biofuels could probably be the main option for aviation, freight road transport, and maritime transport that cannot be electrified. The future technology developments in the advanced biofuels will be critical in the achievement of the decarbonization targets in transport, in relation to the policy changes and the cap set for food-based biofuels in the EU.

The EU Reference Scenario (REF2016) has been updated in 2016, projecting the impact of current EU policies on energy and transport trends until 2050 as well as the amount of GHG emissions. The projections are based on updated assumptions on population growth, economic development and fossil fuel price trends, technology improvements, and updated policies. The REF2016 shows that current policies and market conditions will not allow reaching the 2030 targets and the long-term 2050 objective of 80%–95% GHG emission reductions. In this scenario, the RES share in gross final energy consumption reaches 21% in 2020, 24% in 2030 and 31% in 2050. The energy savings, relative to the baseline will reach 18% in 2020, below the 2020 energy efficiency target, and 24% in 2030. The overall GHG emissions decrease by 26% in 2020, 35% in 2030, and 48% in 2050 (EC, 2016).

The IEA Bioenergy Roadmap data for the EU shows that the primary energy supply is expected to grow in the EU from 5.0 EJ in 2014, the reference year, to 9.1 EJ in the RTS, with a higher growth expected in the D2S and B2DS, with 11.5 EJ and 11.8 EJ in 2050, respectively (IEA, 2017b). This is well in accordance with the results of the high RES scenario of the EU Renewable Energy Roadmap for 2050. The IEA Roadmap shows that the bioenergy capacity, with and without BCCS, will increase in the RTS to 81 GW in the EU in 2050, while the expected installed capacity would be between 103 and 137 GW in 2050, which are also within the ranges projected by the different decarbonization in the EU. The results confirm that the high RES scenario would be in line with the global effort aiming at the achievement of the global long-term goal to limit the temperature increase to maximum 2°C.

10.1.3 Perspectives for aviation biofuels

10.1.3.1 Market for biofuel use in aviation

Although a relatively small contributor to global anthropogenic CO_2 emissions (about 2.6% annually), commercial aviation activity is expected to grow at about 5% per year in the coming decades. As a result, aviation's contribution to global CO_2 emissions could grow significantly to 4.6%–20.2% by mid-century. States, industry, and international organizations are taking measures to address the climate impact of aviation.

At the beginning of 2018—10 years after the first flight operated by a commercial airline in 2008 between London and Amsterdam—several airline companies have performed commercial flights using aviation biofuel. Since 2009 the international standard ASTM d7566 is in place to ensure alternative jet fuels' compliance with jet fuel specifications suitable for aircrafts which are in operation today. As of March 2018, there are five conversion processes approved with different blending levels ranging from

10% to 50% and several others are currently being considered (Kostova, 2017).

Aviation biofuels are defined here as fuels obtained from biomass and having the potential to generate lower life cycle GHG emissions than conventional petroleum-derived jet fuel. Drop-in aviation biofuels are here defined as fully interchangeable and compatible substitutes for petroleum-derived jet fuel: in other words, not requiring adaptation of the aircraft/engine fuel system or the fuel distribution network, and usable "as is" on currently flying aircrafts. Aviation biofuels can reduce aviation GHG emissions through savings achieved in the phase of production of renewable, biological material (biomass feedstock) and its subsequent conversion into fuels. Drop-in biofuels blended with conventional petroleum-based jet fuel do not reduce emissions in the actual combustion phase.

Potential emissions savings from the use of biofuels may be as large as 80% but depend highly on the pathway considered, i.e., the combination of feedstock type/origin and conversion processes. For biofuels produced from agricultural crops, as in the case of road transport, the potential direct and indirect effects—including the conversion and use of arable land—are of key importance. Due to competition of bioenergy with other potential economic uses of biomass (e.g., food, feed, fiber, biomaterials, and green chemistry) along with possible reduced waste management burdens, alternative aviation fuels produced from waste streams are of special interest.

Although aviation biofuels are technically operational, their production at commercial scale is still an unaccomplished task. Large expectations exist on aviation biofuels' potential to reduce the rapidly growing GHG emissions generated by aviation in Europe and worldwide. As a result, numerous initiatives are in place by promoting the production and uptake of biofuels. This is the case—for example—of the European Advanced Biofuels Flightpath geared towards achieving 2 million tons per year of aviation biofuels by 2020, or the Farm to Fly initiative in the United States aimed at producing one billion gallons of sustainable jet fuel by 2018. It is worth noting that the definitions of sustainable biofuels differ in the EU and US regulatory frameworks.

To support the development of aviation biofuels in Europe, the EU Renewable Energy Directive 2009a,b/28/EC (RED) sets a 10% target for the use of renewable energy in transport by 2020. It is measured against the use of fuel in road transport but can be fulfilled by any renewable energy in any form of transport, thus including air transport. The EU Fuel Quality Directive 98/70/EC (FQD) sets a 6% target of GHG emission reduction from all energy used in transport for 2020 compared with 2010. The RED and the FQD have harmonized requirements regarding biofuel sustainability, which act as exclusion criteria for counting towards the regulatory target. Minimum emissions reduction thresholds becoming increasingly stringent over time are designed to steer investment to the best performing technologies in terms of GHG reduction potential.

In the United States, the Congress established the Renewable Fuel Standard (RFS) in 2005 which set annual targets for biofuel blending into gasoline up to 2008. In 2007 the RFS was amended (RFS2) to include other biofuel types and set annual volume targets up to 2022, as well as minimum GHG savings compared to conventional fuels. The United States uses tradable certificates called Renewable Identification Numbers (RINs) to facilitate demonstrating compliance with the RFS2. RINs can also be traded with obligated parties. Although jet fuel is not mandated under the RFS2, RINs can be generated for biojet fuel provided the fuel meets the correct definition of renewable fuel.

Aviation was brought into the EU's ETS in 2012 covering all flights to and from EU airports. The scope was reduced to cover intra-EU flights and to give time for the UN agency which regulates aviation, International Civil Aviation Organization (ICAO), in charge of proposing a global measure. Despite biofuels being zero emissions rated in the EU ETS, they have anyhow to demonstrate that they meet these RED/FQD sustainability criteria.

At the international level, ICAO is thus preparing the global Carbon Offsetting and Reduction Scheme for International Aviation (CORSIA) scheme as adopted by the ICAO Assembly in September 2016 and developing ICAO Standard and Recommended Practices (SARPs) dedicated to CO_2 aircraft emissions. This activity is supported by the ICAO Alternative Fuels Task Force, within the framework of ICAO Committee on Aviation Environmental Protection (CAEP). These activities have induced the European Commission to propose a further extension of the EU ETS.

Commercial facilities exist in Europe for hydroprocessing of vegetable oils and animal fats, currently producing road transport biodiesel and not specifically designed for aviation biofuels. Investment is under way for conversion of municipal solid waste (MSW) into jet fuel.

Production costs for aviation biofuels are certainly the main barrier to their availability and uptake by airlines. Cost curves can be reasonably expected to decrease over time as a result of technology maturity, scaling of production, and market dynamism. Low-cost biomass is also a necessary requirement for cost competitiveness with conventional jet fuel, with production/harvesting mechanization playing an important role in countries with high labor cost such as the United States and Europe.

10.1.3.2 Potential for emissions savings in aviation

Staples et al. (MIT Boston, 2017) made an attempt at estimating the potential savings of aviation CO_2 emissions reductions attributable to aviation biofuels. Their scenario-based analysis assessed the availability of feedstocks suitable for aviation biofuel, the resulting volumes of aviation biofuel, and an estimate of their emissions' profiles compared to petroleum-based jet fuel. The authors also provided quantitative estimates for the number of biorefineries and capital investment required to achieve the aspirational emissions reductions targets announced by industry, namely carbon-neutral growth by 2020, and halving of emissions by 2050. The authors concluded that significant policy measures would be required to incentivize significantly the production of aviation biofuels giving this use priority over other potential uses of the same resources. More specifically, according to the authors of the study, emission reductions of 15% by 2050 would require the construction of about 60 new biorefineries every year between 2015 and 2050 (similar to the growth in global biofuel production capacity in the early 2000s), and capital investment of approximately 12 billion USD 2015 per year (approximately 20% of annual capital investment in petroleum refining). The maximum emissions reduction potential by the use of aviation biofuels achieved in one of the scenarios considered in the MIT analysis is 68.1% in 2050. All six 2050 scenarios assume the availability of large quantities of primary bioenergy and waste feedstock as a result of environmental and society constraints with policy measures—and price signals—incentivizing the production of aviation biofuels relative to other potential competitive uses of biomass. Biorefining capacity is also a fundamental constraint to achieve aspirational targets of aviation emissions' reductions by the use of biofuels.

Finally, the analysis by Staples et al. shows that even 100% replacement of petroleum-derived jet fuel with biofuels in 2050 may result in an absolute increase in aviation life cycle GHG emissions compared to a 2005 baseline. The aviation biofuel scenario in the MIT study presenting the most optimistic results in terms of emissions reductions estimates 2050 aviation GHG emissions at 1101 Mt/year, whereas ICAO estimates of GHG emissions from aviation petroleum-derived fuel burn globally in 2005 was substantially lower at 711 Mt. The authors conclude by stating that this consideration "is especially relevant in the policy context of CORSIA, and for IATA and ICAO's emissions reduction goals, which are unlikely to be possible without significant use of CO_2 offsets from other sectors."

10.1.3.3 Outlook

As a conclusion about the perspectives related to the future use of biofuels in aviation, air transport is a dynamic economic sector with rapidly increasing rates in transport activity, loads, and fuel demand. Aircraft emissions are impacting GHG emissions and thus inducing climate change. Since the reduction of carbon emissions is recognized by UNFCCC as essential at global level in relation to climate change, aviation biofuels can contribute to this goal in the medium term. However, alternative fuel production requires the establishment of a new industrial sector worldwide and trade-offs for the use of limited natural resources. There is substantial interest in this development path in many parts of the world, such as North America, Europe, Indonesia, and Malaysia (world leaders in palm oil production), and Brazil (world leader in the use of bioethanol from sugar cane in road transport). So far, Indonesia is the only country to have introduced a biojet fuel mandate (2% in 2016 increasing to 5% in 2025). Although the GHG emission performance of some biofuel pathways might improve in the future as well as their economic attractiveness, there is the need to continue exploring new opportunities for the production of renewable jet fuel.

Reaching an international agreement on certifying the sustainability of aviation biofuels is also an essential milestone for the development of this sector. Such certification encompassing the entire biofuel supply chain should address environmental and social criteria of sustainability it could draw from the experience matured in the field of sustainability certification of biofuels for road transport, bioenergy, but it could also be enriched by the operational experiences gained in other sectors, such as food and agriculture, forestry, or textile. A strong sustainability certification system (with environmental criteria not limited to GHG emissions reductions but spanning to include air quality, land use, and conservation and sustainable management of water resources) could, for example, help to minimize controversies about mechanisms of carbon offsetting and the link with deforestation in tropical or subtropical ecosystems.

10.1.4 Perspectives of algae for bioenergy

10.1.4.1 Algae systems

Algae receive increasing interest as potential source for the production of biomass for multiple uses. Between 40,000 and 100,000 species of algae have been identified so far, having different morphological, structural, and chemical features and with different content of lipids, proteins, and carbohydrates. There are different types of algae: photoautotrophs, being able to fix inorganic carbon from atmospheric CO_2 and to convert sunlight into chemical energy via photosynthesis; heterotrophs, utilizing an organic carbon substrate (glucose, fructose, glycerol, etc.) as the only carbon and energy source; mixotrophic, using both phototrophic and heterotrophic processes, to accumulate energy for growth, while consuming both inorganic CO_2 and organic carbon substrates (Rocca et al., 2015).

Overall, algae can be classified into two major groups based on their size. Macroalgae (or seaweeds) are multicellular organisms growing from 50 cm up to 60 m in length. Microalgae are unicellular organisms varying from few micrometers to a few hundred micrometers. The use of algae for energy production is expected to offer several advantages compared to land-based biomass crops, including high photosynthetic efficiency and high yield, as well as the possibility to grow on nonfertile land in a variety of water sources (i.e., fresh, brackish, and saline) and additional CO_2 capture potential. A wide range of marketable coproducts can be extracted from algae, e.g., chemicals and nutrients, along with the production of biofuels, within a biorefinery concept (FAO, 2009; van der Velde et al., 2017).

Seaweed or marine macroalgae, such as green (chlorophyta), red (rhodophyta), and brown (ochrophyta) species, are multicellular plants growing in salt or fresh water and are highly photosynthetically efficient, growing primarily in near-shore marine coastal waters, where they grow attached to rocks or suitable substrates. Depending on the species, macroalgae contain different proportions of lipids, proteins, and carbohydrates. Macroalgae can be exploited for the production of chemicals with high economic value and for the production of biomethane, bioethanol, and biobutanol via microbiological conversion processes (Jiang et al., 2016). The majority of seaweeds are cultivated in Asia (99% of the global production of about 28 million tons) especially for food and additives for food, pharmaceutical, cosmetics, and chemical industry. In Europe, the commercial farming of seaweed is at an early stage of development, with the seaweed used in industry supplied almost exclusivity from harvesting (FAO, 2016).

Macroalgae are typically grown in marine water offshore attached to dedicated growth structures, like anchored lines/netting, although different farming systems have also been employed onshore and land-based facilities. Land-based ponds, both as free standing farms or in combination with land-based aquaculture systems, could reach higher productivities (up to 50 tons of dry matter per ha per year) (FAO, 2009; van der Velde et al., 2017). Harvesting algae requires different steps and approaches, including dewatering and drying to reduce the water content of seaweeds from 80%–85% to 20%–30%. Macroalgae are currently cultivated in large open ponds or lagoons in Asia, to produce food and additives for food, pharmaceuticals, cosmetics, and chemical industry (Vigani et al., 2015; Rocca et al., 2015; Scarlat and Dallemand, 2016).

Microalgae are unicellular organisms, including green algae (chlorophyte), blue-green algae or cyanobacteria (cyanophyta), golden brown algae (chrysophyta), and diatoms (bacillariophyta). Currently there are commercial cultures of microalgae for high-value, low-volume food, feed, and nutraceuticals in Asia, United States, Israel, and Australia since the 1980s (Vigani et al., 2015; Rocca et al., 2015; Scarlat and Dallemand, 2016). The cultivation of microalgae is limited to large open ponds or lagoon while commercial production in photobioreactors (PBR) is limited. Microalgae cultures can be established using three approaches: phototrophic (using of light as energy source and CO_2 from air or from flue gas from power plants as carbon source); heterotrophic (using organic substrates, such as nutrients from wastewater, in a dark environment); and mixotrophic conditions (based on phototrophic and heterotrophic processes to extract the energy for growth, depending on the concentration of organic carbon sources and light intensity). Mixotrophic algal growth is a combination of phototrophic and heterotrophic growth and may have advantages in low light and low nutrients environment. Heterotrophic cultures can attain high lipids yield and biomass productivity but require organic carbon feedstock and significant energy consumption and present the risks of contamination by other microorganisms (Mehrabadi et al., 2015; Judd et al., 2015).

Microalgae could be cultivated on land in open or closed reactors, allowing accurate and continuous monitoring and control of most relevant parameters. In this design, algae absorb the sunlight and CO_2 from air or from flue gas and use fertilizers (N, P, and K) added to the water. Open Raceway Ponds (ORP) are inexpensive and easy to operate and maintain, but there are several drawbacks including having lower productivity, poor light utilization, high water evaporation losses, and high risk of contamination. The PBR are closed systems with controlled conditions, which allow the culture of single species of microalgae with low risks of contamination, can provide greater productivity and reduced contamination but rely on complex design and require high investment and maintenance cost. Harvesting microalgae requires different steps and approaches, depending on the features of the selected strains and desired concentration. Harvesting includes thickening (flocculation), separation, and dewatering (filtration through a membrane and centrifugation) that increase the algal mass concentration from less than 0.1% to 10%–25%, followed by drying. The existing designs of ORP and PBR have been mainly investigated at small/experimental scale only; algae production have not been implemented at the large scale and they are still far from commercialization (FAO, 2009; Vigani et al., 2015; Milledge and Heaven, 2013; Rocca et al., 2015).

10.1.4.2 Energy options for algae cultivation

Possible bioenergy pathways from algae include their conversion to biogas, bioalcohols, bio-oil, biodiesel, and bio-hydrogen. These include various processes such as oil extraction, biochemical (anaerobic digestion, fermentation, etc.), and thermochemical conversion (gasification, pyrolysis, and hydrothermal liquefaction (HTL)) technologies. Algae can produce high amounts of lipids, hydrocarbons, and other complex oils. The extraction of oil can be performed through chemical solvent extraction (dry biomass, 60%–98 %) and supercritical fluid extraction (wet biomass 10%–25%).

The high content of moisture and carbohydrates in macroalgae make them suitable for wet conversion methods, including anaerobic digestion and fermentation. Anaerobic digestion of algae for biogas production is one of the most viable technologies considering the high moisture (85%–90%), high carbohydrate content, and low amount of lignin of algae. There are still some technical issues to be addressed, such as the high salinity and sand accumulation over time (IEA, 2017d). Algae exhibits high biomethane potential, depending on the different species, and the biogas yields can be increased through co-digestion of some species of algae with nitrogen rich substrates (such as manure) and manipulation of the microbial composition of the inoculums. Algae can be a suitable substrate for bioethanol production via hydrolysis followed by fermentation. Algae can also be a suitable feedstock for bio-hydrogen production via photo fermentation or dark fermentation by means of a pure or mixed culture of hydrogen-producing bacteria or via a combination of dark, photo fermentation, and anaerobic digestion in three-stage processes (Redwood et al., 2009; Rocca et al., 2015; Scarlat and Dallemand, 2016).

Algae can be suitable feedstocks for bio-oil via thermochemical conversion pathways, such as pyrolysis and HTL. Pyrolysis produces bio-oil (or biocrude), bio-char, vapors, and an aqueous phase upon condensation. A major limitation to pyrolysis is the high moisture content (70%–80%) of algae, requiring significant energy for drying. The bio-oil yields may significantly vary depending on the macroalgae composition and operating conditions. HTL allows algae to be processed without drying, using supercritical water at high temperature and high pressure in the presence of a catalyst (zeolites or alkali salts) (Milledge and Heaven, 2014; López Barreiro et al., 2013; Rocca et al., 2015).

The rapid growth and high photosynthetic efficiency of algae potentially allows higher energy yields to be achieved compared to terrestrial crops. In this context, algae have a clear potential to be used for the production of bioenergy to address future energy and sustainability challenges. There are significant barriers currently impeding commercialization and economic production of algae for relatively low-value energy and fuel that range from incomplete knowledge of algae biology to the challenges associated with the integration of technologies at large scale. One important prerequisite to grow algae commercially for energy production is the need for large-scale systems so that economy of scale could reduce production costs. Systems have to be optimized for particular bioenergy pathway, integrating cultivation, harvesting, and conversion into the final product for improved energy output and economic performance. There are good market opportunities for algal biomass, e.g., for food and feed applications, that will contribute to the advancement in the algae production systems with potential benefits for longer term bioenergy production. The use of algae in biorefineries can help to produce higher value food, feed, nutraceutical, and biochemical products in integrated processes, together with algae-based biofuels. As such, a biorefinery could contribute decisively to improve the algae economic viability and drive the development of bioenergy production (IEA, 2017d; Scarlat and Dallemand, 2016).

10.1.4.3 Outlook

Given the large number of algae with different characteristics it is not yet clear which specific algae species would be best suited for bioenergy applications. The identification and development of energy efficient and cost-effective biofuels pathways based on algae is a challenge, given the large variety of algal strains, their growth conditions, yields, and chemical compositions. There is a need for species identification, the analysis of microalgae strains, and selection of those with advantageous traits and/or genetic engineering to enhance the biomass productivity and yield of target components. Numerous parameters, including the method of cultivation, species of algae, yield per hectare, method of harvest, suitability of seaweed to ensiling, carbon balance, as well as costs of the algae and the final cost of the produced bioenergy or biofuel, have not yet been adequately assessed (IEA, 2017d).

Algae cultivation for bioenergy production requires a combination of technical breakthroughs on cultivation under different location-specific conditions, but also on harvesting, dewatering, drying, and conversion to the final product. In particular, considering the microscopic size and properties of microalgae strains, the development of harvesting and dewatering technologies represents a major challenge and a critical issue with respect to energy requirements and costs. Algae-based bioenergy production is not foreseen to be economically viable in the near to medium term. The relatively high cost of producing algal biomass remains the most critical barrier to commercial viability of algae-based production (IEA, 2017d). There is also a need for demonstration at full scale of pretreatment/hydrolysis processes (e.g., ultrasound and use of enzymes), oil extraction and biochemical (anaerobic digestion, fermentation), and thermochemical (pyrolysis, HTL) conversion technologies (Rocca et al., 2015; Scarlat and Dallemand, 2016).

There is a need for developing large-scale cultivation systems, including the development of cost-effective methodologies for off-shore and farms/land-based ponds cultivation, harvesting and conversion, improving yields, and proving economical production. Appropriate sea farming techniques and infrastructure would have to be developed for macroalgae cultivation, based on existing experience in macroalgae cultivation for food and additives for food, pharmaceutical, cosmetics, and chemical industry. One of the major challenges to large-scale sustainable cultivation of algae includes high demands of water and nutrients for algal growth.

Water use requirements for algal biomass and biofuel production vary depending on growth conditions, but effective wastewater recycling is essential to minimize freshwater and nutrients consumption. The integration of algae production with wastewater treatment (WWT) provides opportunities for the treatment of waste streams and the use of organic substrate such as nutrients (N, P) from wastewater to produce algae. This solution might be economically feasible in the near term, allowing algae production for bioenergy to achieve a better economic and environmental sustainability. Carbon dioxide can be used for cultivating algae to facilitate optimal algae growth but procuring CO_2 and pumping it to the algae growing facility is one costly and energy demanding and the algae production facility would have to be located near a suitable gas source (such as a power plant, anaerobic digestion plant). However, flue gases can contain high amounts of NOx and SOx, which can change the pH of the algae cultivation medium and inhibit algae growth and lead to accumulation of heavy metals in biomass (Judd et al., 2015; Scarlat and Dallemand, 2016; IEA, 2017d).

The available potential of algae for the production of energy has not been sufficiently assessed. The amount of harvestable natural seaweed resources was not yet quantified, although for offshore seaweeds system, the combination of nutrients and light can be a measure of the capacity of the ecosystems to ensure algal growth. The ecological impact of harvesting natural resources needs to be understood and properly addressed. However, it is also unlikely that seaweed can be harvested at the scale sufficient to provide significant quantities necessary for bioenergy production. Therefore the large-scale supply of macroalgae for bioenergy applications might be based on aquaculture. When selecting a location for seaweed cultivation, temperature, nutrient, and light need to be taken into consideration as well as the competition with other functions of the sea, such as conservation areas, recreation, harbors, shipping routes, etc. Considering all these aspects, the potential area for macroalgae in temperate oceanic climates, in coastal areas, might be significant. For land-based systems, the cultivation systems have much lower land requirement than terrestrial cropping systems. Anyway, the land needed for algae production could be low quality, low fertility land, such as desert land, saline land, and other land with low economic and ecologic value. Due to the ability of growing algae on non-arable land, the competition for land with existing food and feed supply is thus avoided.

10.1.5 Waste and residues for sustainable bioenergy

10.1.5.1 Opportunities for waste and residues for energy

Biomass availability, competition between the alternative uses of biomass, as well as the sustainability issues are major concerns for bioenergy deployment. Biomass potentials appear to be large enough to allow bioenergy to play a significant role in the future energy systems that require wide range of feedstock categories to be mobilized. Sustainability constraints for the use of biomass for energy purposes can reduce biomass availability for energy production, in addition to a range of competitive uses for food, feed, fiber, and materials. In this context, the use of waste or residual streams of biological materials from agriculture and forestry could have a significant contribution to bioenergy production, with no impact of land use/LUC, food security, or with no or reduced competition for raw materials with other sectors. In particular, the valorization of various waste streams for energy recovery is essential for the improvement of the economic viability of various processes (Scarlat et al., 2018a). However, there may also be still some negative environmental effects if unsustainable collection rates will be applied to agricultural and forest residues or if some competition develops for existing uses in different sectors. There is also a danger that increased amounts of by-products and residues are produced in different processes to provide a feedstock for bioenergy production. However, the use of wastes and residues offer potential options to avoid the use of food crops or good quality wood for energy. In particular, as food production needs to grow to feed an increasing population, more organic residues and waste are produced providing additional amounts of feedstocks for energy.

10.1.5.2 Circular economy and waste

A long-term vision in the EU is to establish a circular economy (COM (2014a,b) 398 final) in which material use and waste generation are minimized, waste is recycled and reused in the production of new materials. The unavoidable waste is then treated in the least harmful way to the environment and human health or is used for energy recovered and the remaining waste is landfilled. Turning waste into a resource is an essential part of increasing resource efficiency and closing the loop in a circular economy, through greater recycling and re-use. The circular economy package aims to stimulate the transition from a linear economy where resources are extracted, used, and thrown away, towards a concept of circular economy, involving multiple uses of materials in various sectors in the economy (Scarlat et al., 2018a). Energy from waste can play an important role in a circular economy provided that prevention, reuse, and recycling are prioritized in the waste management cycle.

The more efficient way of using raw materials is the main concept of the resource-efficient initiative (COM(2010a,b,c) 2020 final) that aims to increase resource productivity and decouple economic growth from resource use and its environmental impact, and to support the shift towards sustainable growth via a resource-efficient and low-carbon economy. The resource-efficient use of biomass is essential given the significant anticipated increase of biomass use for energy and bio-based products along with the growing demand for food and feed. The cascading use principle is the guiding principle to improve resource efficiency and limit unsustainable pressure on natural resources.

When discussing the multiple uses of biomass (food, feed, fiber, biochemical, biomaterials, and bioenergy), this entails a combination of several biomass applications in a cascade of uses, based on the prioritization of biomass use. A number of factors could be considered in the prioritization of biomass use, such as the economic or social value of biomass products, the conversion efficiency of biomass, and GHG emission reduction performances. This allows to differentiate between energy and non-energy uses and finally between different energy pathways as well. The principle of cascading use of biomass resources also applies to the use of agricultural and forest wastes and residues, allowing the reuse and recycling of products and raw materials first, with the last option being energy use (Scarlat et al., 2018a).

Waste management has been developed in the EU on the concept of a hierarchy of waste management options, based on a legally binding prioritization of waste management activities. Waste management can open up new economic opportunities, improve raw materials supply to industry, and provide green energy. Energy recovery from waste as part of the waste hierarchy is expected to enable circular economy to play a significant role to reach the decarbonization objectives of the energy system (COM(2017a,b) 34 final). In essence, waste prevention is the most desirable option, followed by material recovery and recycling (metal, glass, paper recycling, or organic waste), energy recovery from waste (through incineration or anaerobic digestion of biodegradable waste), and finally disposal (landfilling) with no recovery of either materials and/or energy as the least desirable option. Within this framework, the EU Member States have increasingly changed focus from disposal of waste to prevention and recycling. The amount of waste recycled almost tripled and landfilled reduced by half in the last decade in the EU. However, a large share (more than 30%) of the EU's municipal waste is still being landfilled, with significant variations between the EU Member States (Scarlat et al., 2018a).

10.1.5.3 Agricultural and forest residues

Agricultural residues, associated with crop production, are potential feedstock categories for bioenergy production or for bio-based materials. Agricultural residues include crop residues, such as straw, maize stover, residues from sugar beet, oilseeds, pruning and cutting materials from permanent crops and landscape management, and processing residues in the form of olive pits, husks, nut shells, etc. (Elbersen et al., 2012). Agricultural residues constitute a large biomass resource and provide an enlargement of the biomass feedstock base.

The amount of crop residues that can be utilized for energy purposes is limited due to environmental constraints for straw removal from land, technical restrictions for harvesting, economics of biomass collection, and mobilization, as well as various competitive uses. Crop residues are an important source of organic carbon and play a critical role for the maintenance of the organic matter balance and nutrient cycles in the soil. A share of crop residues is required for soil management, depending on local circumstances (soil, temperature, precipitation patterns, etc.) crop type, and yield. Therefore, the sustainable extraction rates need to be considered in relation to local conditions. Excessive residue removal from the field can reduce the soil carbon pool, affecting soil fertility and soil productivity. Some competitive uses need to be considered for feed or animal bedding, horticulture, mushroom production, construction material, etc.

Forestry residues include primary and secondary residues. Primary residues include sawdust, small branches, twigs, or tops from harvesting or logging activities. The removal of forest residues must consider the potential negative impacts on forest ecosystems, e.g., biodiversity and ecosystem functions, carbon stock in forest, etc. Secondary residues include residues from the wood processing industry, such as sawdust, woodchips, wood remains, and black liquor. Most of the by-products from the processing industry are already utilized in the wood industry itself in a variety of uses such as particleboard and panels, while the remaining part is already used for energy generation mainly to cover the internal energy needs for processing. Significant volumes of forest wood that currently have no industrial use, such as low quality wood, can be used for bioenergy.

Although the volume of forest residues appears to be significant, the major barrier to their mobilization is the need to establish appropriate extraction levels, depending on local conditions and forest management. As with crop residues, excessive forest residue extraction might lead to loss of biodiversity through the removal of deadwood, reduction of soil organic matter, nutrient availability, and increased erosion risks. Stump removal can have particular negative effects on carbon stocks. The application of sustainable forest management practices, together with guidelines for sustainable extraction rates is critical. The limited accessibility, for example, due to slope and distance, for harvesting of forest biomass is an additional constraint for full mobilization. Various forest residues could have different competitive uses, for particleboard manufacture and for pulp and paper production, but also for the production of new bio-based materials and bio-based chemicals. The high costs of extraction of forest harvesting residues over large areas, with difficult terrain might be limiting the economic potential.

One important challenge for the mobilization of agriculture and forestry residues is related to the high bulk, low energy density, and low value. Crop and forestry residues are highly dispersed requiring high harvesting, collection, and transport costs. This limits the range over which they can be used in absence of densification through pelletization that makes it merely a local feedstock.

10.1.5.4 Biogas production

A range of agricultural residues (livestock manure, crop residues), industrial residues (sewage sludge, food industry waste, and slaughterhouse residues), energy crops, organic fraction of municipal waste, etc., could be a suitable feedstock for biogas production in anaerobic digestion plants. In the last decade, there has been a significant development in Europe towards the use energy crops (silage maize, grasses), industrial and municipal waste for anaerobic digestion.

Biogas production has increased enormously in the last years from the treatment of wet-waste biomass, WWT, and landfill gas recovery. Biogas production is used for generating energy, such as electricity, heat and fuel, displacing the use of fossil fuels. Biogas production brings a number of economic, environmental, and climate benefits, such as reduction in air, water, and soil pollution, and GHG emission reduction. In particular, biogas production from various waste and residues has no indirect effects, such as land use/LUC, indirect land-use change (ILUC), and negative impact on food security, and provides significant GHG emission reductions.

In low-income countries, biogas is mainly produced in small, domestic-scale digesters to provide a fuel for cooking or even lighting, in comparison to high income countries, where biogas developments focused on larger scale, farm-based, and commercial biogas plants for electricity and heat generation. Modern anaerobic digesters provide electricity and heat in electricity-only plants, heat-only, or combined heat and power (CHP) plants with electrical capacity ranging from tens of kWe up to a few MWe (Scarlat et al., 2018b). The heat generated can also be used to meet the local heat demand on farm or delivered to external users. Energy from biogas can help to substitute traditional bioenergy that has negative environmental effects and indoor pollution. Traditionally, manure has been used as fertilizer in agriculture, which can cause environmental problems, including soil and water contamination and pollution. Natural degradation of manure leads to emissions of methane and carbon dioxide during storage. Anaerobic digestion of manure brings the highest GHG emissions reduction among many bioenergy pathways, in comparison to the reference fossil fuel system, due to the avoidance of methane emissions from this natural decomposition. Solid residues from anaerobic digestion (digestate) can still be used as fertilizer, just like manure, having the same content of nutrients as manure. This brings additional economic benefits by reducing the use of chemical fertilizers in farms and reducing nutrient runoff.

Biogas can be upgraded to biomethane and injected into natural gas network or used it in transport vehicles, after proper purification to remove trace gases, such as H_2S, water, and CO_2. Biogas upgrading and biomethane production offer new opportunities for the use of biogas and for the substitution of fossil fuels in transport sector, overcoming the limitations for the use of heat and for improving the economics of biogas plants. The use of upgraded biogas (biomethane) in transport has emerged as a good alternative to the use of food-based crops biofuels to replace fossil fuels for transport, with very low GHG emissions if compared to food-based crops biofuels or advanced biofuels. Biomethane could even result in negative GHG emissions when produced from feedstocks which otherwise would emit methane during decomposition, such as manure (Scarlat et al., 2018b).

10.1.5.5 Outlook for the use of waste

The generation of waste increases with population growth, and there is also a correlation between waste generation per capita and growth in gross domestic product leading to more waste expected to be produced. In this context, energy recovery from waste, including various waste treatment processes generating energy, can play a significant role in the circular economy (COM(2017a,b) 34), provided that the EU waste hierarchy is used as a guiding principle. The use of the energy potential of waste within the waste management strategy has brought additional value for energy supply and for reaching renewable energy targets. The recovery of energy from waste has a potential to become an important player in the renewable energy sector in Europe. Currently, this potential is still far from being fully exploited as only a minor fraction of waste available is actually sent for incineration with energy recovery. The advantages of recovering energy from waste include the generation of local renewable energy and the decrease the volume of solid waste dumped in landfills, which in turn may have positive effects on carbon emissions by avoiding methane emissions from landfills and carbon dioxide emissions from fossil fuel use (Smith et al., 2001). Energy recovery, as a source of renewable energy, is expected to play an increasingly important role in sustainable management of municipal waste at global level. It is estimated that a reduction of about 10%–15% in the global GHG emissions could be achieved through improved solid waste management (recycling, waste diversion from landfill, and energy recovery from waste) (UNEP, 2015). Various processes could contribute and create synergies with EU energy and climate policy: co-incineration of waste in combustion plants and in cement production; waste incineration in dedicated facilities; anaerobic digestion of biodegradable waste; production of waste-derived fuels (solid, liquid or gaseous); and other processes including pyrolysis or gasification (Scarlat et al., 2018a).

The EU waste policies and targets include minimum requirements for managing certain waste types. The most relevant targets for waste are the landfill diversion targets for biodegradable municipal waste, recycling targets, and the target on recycling and preparing for reuse. The Waste Framework Directive 2008/98/EC (WFD) has established legal requirements for the waste treatment within the EU, aiming to protect the environment and human health from the harmful effects of waste generation. The WFD has set a 50% target for preparing for reuse and recycling of household and similar waste and a 70% target for preparing for reuse, recycling, and other material recovery of nonhazardous construction and demolition waste by 2020 (Scarlat et al., 2018a). Reducing the landfilling of biodegradable waste is particularly important from a climate perspective to reduce methane emissions. The landfill gas recovery from existing managed landfills is an established technology and represents a good opportunity for the capturing and energy use of gas generated through natural decomposition of organic waste.

With the new legislative package on waste, the European Commission proposed new targets for municipal waste for waste recycling, landfilling, and other waste-related targets, to reach 65% recycling and preparing for reuse by 2030. The proposed targets include (1) 65% reuse and recycling target for MSW by 2030; (2) 75% reuse and recycling target for packaging waste by 2030; (3) minimum targets for reuse and recycling for specific materials contained in packaging waste: 75% of wood, 85% of ferrous metal, 85% of aluminium, 85% of glass, 85% of paper and cardboard; (4) maximum 10% municipal waste landfilled of the total amount of generated MSW and a ban was set on landfilling of separately collected waste.

The reduction of landfilling of biodegradable waste, in combination with the requirements to ensure separate collection of bio-waste, could result in the reduction of waste available for landfilling but increase in the feedstock available for biogas production, thus opening up new opportunities for biogas production. Also, the rules on separate collection and ambitious recycling rates covering wood, paper, plastic, and biodegradable waste are expected to reduce the amount of waste potentially available for energy recovery and thus limiting the contribution of waste to energy generation. The development of new waste treatment capacity needs to be therefore framed in a long-term circular economy perspective and to be consistent with the EU waste hierarchy. To promote the efficient use of energy generated from waste, the WFD includes an energy efficiency criterion (R1, referred to as the *R1 criterion* or the *R1 formula*). This allows a distinction between waste incineration with energy recovery (R1) and without energy recovery or incineration on land for the purposes of disposal (D10). The R1 criterion takes into account the energy recovered from waste and its effective uses as electricity or as heat for heating and cooling or as processing steam in industry.

10.1.6 Role of trade in bioenergy

10.1.6.1 Bioenergy markets and trade

Bioenergy provides about 56 EJ representing almost 10% of world primary energy supply, out of which traditional bioenergy still makes up more than 60% (IEA, 2017a). Traditional use of biomass has played a significant role for heating and cooking for long time. Traditional bioenergy relies on the use of agricultural by-products (animal dung and crop residues), fuelwood, and charcoal that are mostly used locally. Modern use of biomass for electricity and heat generation and provision of high temperature heat for industry and for biofuels production has started recently as a more efficient option of providing energy. Modern bioenergy relies on energy carriers such as bioethanol, biodiesel, wood pellets, and wood chips, through a range of modern technologies (Sharmina et al., 2017). Bioenergy markets, despite significant development over the last decade, are still immature and face complex interactions between agriculture, forestry, and energy sectors. The development of bioenergy and the increasing demand for biomass have been key drivers for international trade to exploit available biomass resources and local market potentials, which are currently underutilized in many world regions. International trade flows in bioenergy markets are still relatively low compared to fossil fuel markets and have intricate connections with food and wood commodities trade flows that make them difficult to estimate (Popp et al., 2014).

10.1.6.2 International bioenergy trade

International trade is being driven by excess biomass availability on the supply side and increasing demand in other markets. Support policies for bioenergy, such as biofuels mandates and renewable energy obligations together with various support schemes, have significantly stimulated biomass demand and thus international trade over the past decade. The biomass supply originates from producing and exporting regions with considerable biomass resources, high agriculture production and forest areas with strong forestry, wood processing, and/or pulp and paper industries. The availability of no/low-cost raw material, such as forest and agricultural residues and waste, has been key drivers for bioenergy development using locally sourced feedstock. In particular, the availability of forest residues from a competitive forest sector, with proper infrastructure for processing, handling, and transport in the wood processing and pulp and paper industry, has stimulated bioenergy development. Support policies for renewables have been a key driver for the development of modern bioenergy markets. In contrast, bioenergy markets are small or nonexistent in countries with high fossil fuel reserves and no bioenergy support policies, despite high biomass resources (Lamers et al., 2012; Popp et al., 2014).

As the economic viability of bioenergy production depends on the existence of cheap feedstock, the availability of such cheap feedstock in other regions of the world has promoted international trade. In addition, the cost of biomass production and transport determines whether a feedstock is worth transporting over long distances. In order to make international bioenergy trade economically viable, appropriate, well-developed logistics systems are crucial. International trade thus took advantage of the availability of handling, storage, and transport infrastructure and other factors, despite the additional costs of long-distance transport (Lamers et al., 2012).

Bioenergy trade is constrained by the wide and poor biomass characteristics in comparison to fossil fuels, which pose large difficulties for transportation, handling, and storage. Biomass feedstocks have generally variable properties, high bulk volume, high moisture, low energy density, and low economic value that made them difficult to trade. Energy density is low for various biomass feedstocks, which means that large volumes need to be transported for a certain energy value. The improvement of biomass feedstock to increase bulk density and energy density through pretreatment (e.g., drying, pelletizing, torrefying, etc.) before transport facilitated biomass trade. Low value, high bulk volume products, such as fuelwood, charcoal, forestry, and agricultural residues, are usually used and traded locally, whereas refined, homogeneous liquid biofuels and biomass with high energy density and high economic value, such as wood pellets and wood chips, are traded globally (Junginger et al., 2014; Popp et al., 2014; Lamers et al., 2012).

Despite the increasing importance of international trade of biomass for energy, there is no complete database for various biomass trade streams. The global trade statistics include data for various commodities, such as wood products (such as roundwood and wood chips) that are used in wood, pulp, and paper industry. Trade statistics also include detailed info on agricultural commodities (cereals, sugar crops, oilseeds, etc.) used for food production. Bioenergy markets are closely related to the agricultural commodities, and wood product markets and various biomass commodities could be used directly for bioenergy production although they might have some other uses as well. For example, data related to the ethanol trade flows include ethanol that can be used for fuel, beverage production, or multiple industrial applications. Palm oil and soybean oil are both used mostly as raw material in the food sector, and a small fraction is used for biodiesel production (Popp et al., 2014; USDA, 2017). In addition, a share of wood from processing ends up as by-products and residues, for example bark, sawdust, wood remains, and black liquor that are used for energy. Thus, the international trade in agricultural products and wood for the manufacture of wood products and for pulp and paper could also involve indirect trade of biomass for energy. In order to quantify this indirect trade, biomass flows have to be broken down into different end-uses for material and energy uses. There are some uncertainties in the estimations of this indirect trade of biomass for energy related to the conversion efficiency of raw material, the production processes, the level of technology, and the integration of the production processes, which affect the conversion into various materials (Junginger et al., 2014; Proskurina et al., 2018).

10.1.6.3 International trade in biofuels

The liquid biofuel markets are rather mature markets, highly dynamic and complex and are closely related to agriculture commodities. Biofuels (bioethanol, biodiesel) are traded globally as such, and additional large amounts of feedstocks (starch, sugar, and oil crops) are traded and processed into biofuels in the importing country (USDA, 2017). The United States and Brazil are leaders in the ethanol market with 70% of the global ethanol production in 2016, respectively, followed by China and the EU. Ethanol trade statistics do not differentiate between ethanol used for biofuels and for other purposes, but the share of ethanol traded for energy can be estimated from the share of bioethanol used in the importing country.

Bioethanol production and use in transport has increased significantly in the last years from 15 billion liters in 2000 and 30 billion liters in 2005 to reach about 99 billion liters in 2016 (Fig. 10.10). The world leaders in the use of bioethanol in transport in 2015 were the United States and Brazil, with more than 80% of global bioethanol market, followed by the EU and China. The EU bioethanol market has decreased, however in the last years, in relation to the sustainability debate and uncertainties about future policies. The statistics show that bioethanol is mostly used locally, with the share of bioethanol international trade relative to the total global bioethanol market of only about 7.5% in 2016. The US Renewable Fuels Standard, requiring a specific share of advanced bioethanol (which includes sugarcane ethanol), has led to an increase in the export of ethanol from Brazilian to the United States and also an increase in the export of US corn ethanol to Brazil and to the EU (Proskurina et al., 2018; OECD, 2017).

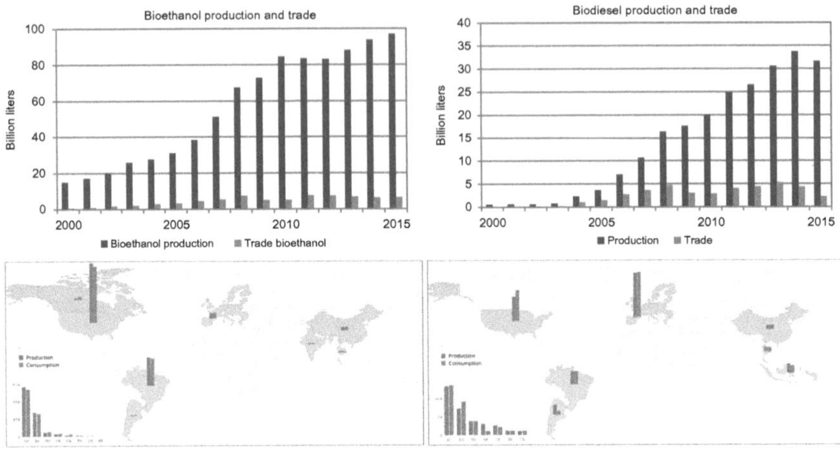

Figure 10.10 International trade and major producers and users of bioethanol (left) and biodiesel (right) in 2016. *From FAO, 2018. Food and Agriculture Organization. Available at: http://www.fao.org/faostat/en/#home.*

Global biodiesel production has increased significantly over the last years, from 0.5 billion liters in 2000 to 3.6 billion liters in 2005 and 32 billion liters in 2016 (Fig. 10.10). The biodiesel market is dominated by the EU with about 38% of the global use of biodiesel in transport, due to the high penetration of diesel-powered automobiles, followed by the United States and Brazil, all accounting for almost 75% of the global market. The share of the EU market for biodiesel has stagnated during the last years, while other markets had a strong growth, mostly in the United States, but also in Brazil. Biodiesel trade represents only about 10% of global biodiesel production. Biodiesel trade has changed substantially in the last 5 years, driven by the uncertainties in the biofuel policies, especially in the EU. These changes made the United States the largest importer of biodiesel, with almost 80% of global trade, as opposed to the EU trade that dropped from almost 60% in 2010 to about 7% in 2016. Argentina has become the largest exporter of biodiesel with 48% of the global trade, followed by the United States and the EU. In addition, there is a large international trade in raw materials that are used for biodiesel production in the form of vegetable oils (palm oil, rapeseed oil) and oil crops (rapeseed, soy) (OECD, 2017; USDA, 2017).

10.1.6.4 Global solid biomass trade

Global solid biomass trade for energy includes direct trade of wood pellets, fuelwood, wood chips, wood residues, and charcoal, as well as

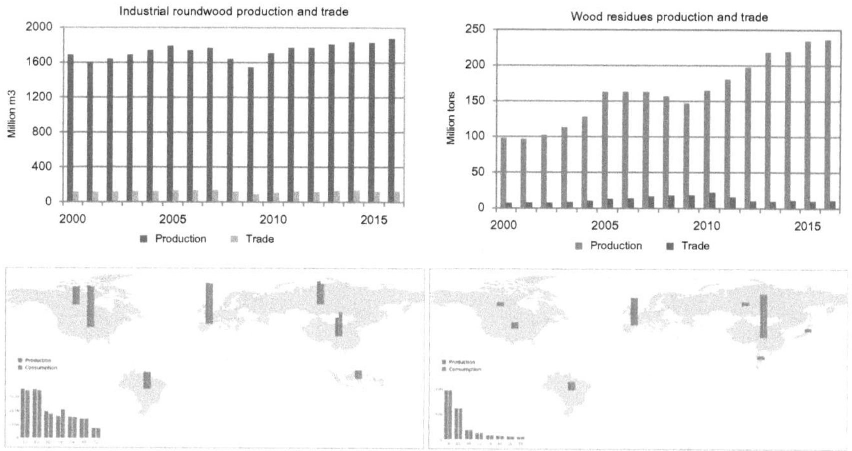

Figure 10.11 International trade, major producers and users of roundwood (left) and wood residues (right) in 2016. *From FAO, 2018. Food and Agriculture Organization. Available at: http://www.fao.org/faostat/en/#home.*

indirect trade from industrial roundwood, wood chips, wood residues, and charcoal. Industrial roundwood is the largest solid biomass trade stream, followed by fuelwood, wood chips, charcoal, sawdust, and wood waste. The industrial roundwood trade is almost constant since 2000, with a growth from 115 Mm^3 in 2000 to 125 Mm^3 in 2016, in comparison to a global production of 1874 Mm^3 in 2016 (Fig. 10.11). Leaders in the industrial roundwood production are the United States (357 Mm^3), EU (355 Mm^3), Russia (198 Mm^3), China (163 Mm^3), and Canada (158 Mm^3). Relative to production, almost 7% of globally produced industrial roundwood is been traded internationally. This indicates that industrial roundwood is being primarily used locally. Major exporting countries include Russia (20 Mm^3), New Zealand (16 Mm^3), the United States (11 Mm^3), Canada (7 Mm^3), and Australia (4 Mm^3), while major importing countries include China (49 Mm^3), India (6 Mm^3), Canada (4 Mm^3), Korea (4 Mm^3), and Japan (4 Mm^3) (FAO, 2018). Roundwood, whether used local or traded globally, is destined to wood processing into a range of wood products and not directly linked to energy production. However, there is a large amount of indirect trade in the form of bark and residues from wood processing. After processing, some by-products such as wood residues and waste are used as for bioenergy generation. Depending on the material use, technology and degree or process integration, approximately 30%–45% of by-products from roundwood usage can end up being used for energy purposes (Lamers et al., 2012; Proskurina et al., 2018).

Wood residues include a variety of streams from forest harvesting and wood processing residues, small diameter/low-quality roundwood, recycling and recovered wood. Statistical data show a significant production of wood residues. Global production of wood residues has grown from 97 Mm^3 in 2000 to 236 Mm^3 in 2016 (Fig. 10.11). The largest part of this volume is made up of forestry residues (e.g., tops and branches) and processing (e.g., sawdust) residues. Key producers include China (98 Mm^3), Brazil (20 Mm^3), the United States (14 Mm^3), Canada (9 Mm^3), and Russia (8 Mm^3). Between 2000 and 2016, however, the trade in wood residues has increased from 7 to 11 Mm^3, but still having a marginal share of total global production, of 7% and 5%, respectively (FAO, 2018).

Fuelwood plays a major role in the global bioenergy production, primarily in the form of traditional bioenergy, used for heating and cooking. The global fuelwood production has grown from 1772 Mm^3 in 2000 to 1863 Mm^3 in 2016. World trade in fuelwood has grown from 2.8 to 7.7 Mm^3 between 2000 and 2016. Statistical data show that fuelwood trade represents only about 0.4% of fuelwood production (Fig. 10.12). This shows that fuelwood is used rather locally or traded locally, over short distances. However, fuelwood trade occurs as informal cross-border trade and in small volumes, which are anyway difficult to estimate and

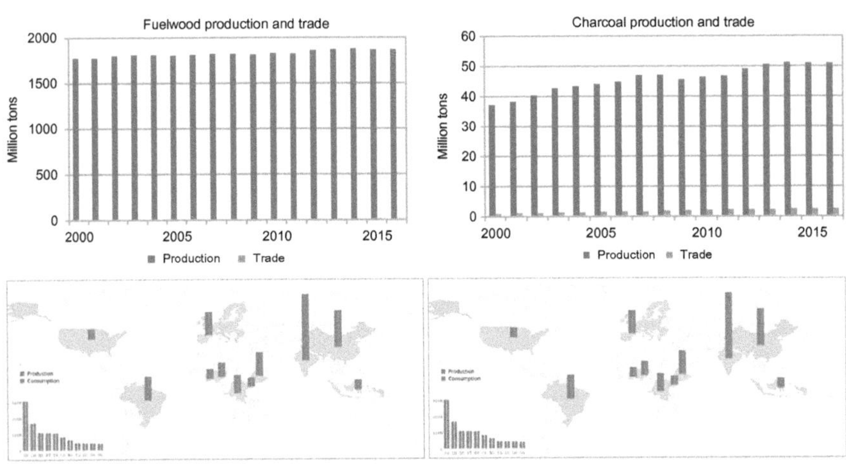

Figure 10.12 International trade and major producers and users of fuelwood (left) and charcoal (right) in 2016. *From FAO, 2018. Food and Agriculture Organization. Available at: http://www.fao.org/faostat/en/#home.*

thus excluded from statistics. The EU is a large market for fuelwood use, with about 109 Mm^3 in 2016, for residential heating, driven by low cost and certain support schemes. The EU has been the key driver of fuelwood trade and importer of fuelwood with an increase in fuelwood use from 0.15 to 1.5 Mm^3 between 2000 and 2016. Thus, the share of EU net import of global trade increased to 20% in 2016 from only 5% in 2000. Statistical data reveals high uncertainties, shown in the differences in the trade flows between importing and exporting countries. Leading countries in fuelwood production and use include India (306 Mm^3), China (169 Mm^3), Brazil (112 Mm^3), as well as several African countries—Ethiopia (109 Mm^3), Dr Congo (83 Mm^3), Nigeria (65 Mm^3), etc. (FAO, 2018). The share of biomass (e.g., fuelwood) in sub-Saharan Africa represents up to 80% of total energy supply. Fuelwood is thus used mostly in low-income countries for heating and cooking, as traditional bioenergy, in appliances with low efficiency and high emission, with severe impact on indoor air quality and health. This low efficiency of biomass use leads to high consumption of biomass, with huge impact on deforestation, loss of habitats, biodiversity, and other ecosystem services.

Charcoal is an important source of energy in many parts of the world, in particular, in low-income countries. World charcoal production has grown from 37 Mt in 2000 to around 51 Mt in 2016. Leading producing countries include Brazil (5.5 Mt), Nigeria (4.4 Mt), Ethiopia (4.3 Mt), India (2.9 Mt), and the Democratic Republic of Congo (2.5 Mt) (Fig. 10.12). Charcoal is used in traditional heating and cooking as well as in large-scale industrial applications in the chemical (as active coal) and in the iron and steel industry, as a reducing agent and energy source. Charcoal is largely used in sub-Saharan Africa for domestic cooking purposes, due to the lack of access to other sources of energy. With population growth and rapid urbanization, charcoal use is likely to continue to increase. Trade flows in charcoal are quite marginal, with 2.4 Mt in 2016, representing only about 5% in global charcoal production, which shows that, similar to fuelwood, charcoal is used locally with informal or little cross-border trade (FAO, 2018).

The wood pellet market has become the most dynamic solid biomass market, with the production of wood pellets increasing from 1.7 Mt in 2000 to more than 29 Mt in 2016 (Fig. 10.13). Key factors for this growth include homogeneity of the product properties due to standardization, high heating value, and high bulk density. International standards (ISO 17225-2), defining product requirements, i.e., moisture, energy

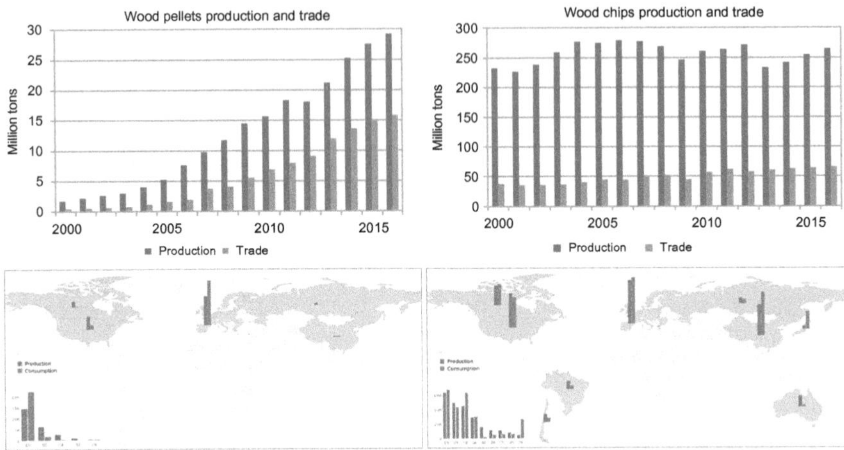

Figure 10.13 International trade and major producers and users of wood pellets (left) and (right) in 2016. *From FAO, 2018. Food and Agriculture Organization. Available at: http://www.fao.org/faostat/en/#home.*

density, particle size, and shape for wood pellets, have turned this feedstock into a commodity that can be easily traded globally (Thrän et al., 2017). Wood pellets also benefit for large flexibility in operation, from small scale, residential stoves, to small-scale boilers and to large-scale CHP power plants. The leading wood pellet producers are the EU (15 Mt), the United States (6.4 Mt), Canada (2.8 Mt), Viet Nam (1.4 Mt), and Russia (1.0 Mt), which all produced in total over 26 Mt of wood pellets in 2015. The EU is the major market for wood pellets, with 50% of global production (15 Mt) and more than 75% of wood pellet consumption (23 Mt). A large intra-EU trade of wood pellets also occurs mostly for residential wood pellets use. Wood pellets are been used in the EU in residential heating, district heating, and large-scale power production (Belgium, Denmark and the United Kingdom). In addition to the EU, major consumer countries include the United States (1.9 Mt), Korea (1.7 Mt), China (0.5 Mt), Japan (0.5 Mt), and Canada (0.4 Mt). New countries and regions have entered the market for both, pellet production (such as Canada, South-East Asia, China, and Russia) and pellet consumption (such as East Asia—Korea, Japan, and China) (FAO, 2018).

Wood chips represent the second largest single biomass trade stream. Wood chips, produced from roundwood, residues, and waste wood, mainly consist of high-quality chips for pulp and paper production or other uses, such as fiber and particle boards. The global wood chip production has grown to 235 Mm^3 in 2000 to 265 m^3 in 2016, while the traded annual production volume has increased faster over time from 37 Mm^3 in 2000 to 66 Mm^3 in 2016, with the share of trade increasing from 16% to 25% between 2000 and 2016 time (Fig. 10.13). Key producers include the EU (64 Mm^3), the United States (50 Mm^3), China (46 Mm^3), Canada (29 Mm^3), and Australia (16 Mm^3). Major consumers include EU (68 Mm^3), China (64 Mm^3), the United States (44 Mm3), Canada (30 Mm^3), and Japan (26 Mm^3) (FAO, 2018). The extent of the trade flow of wood chips used for energy is difficult to assess. Wood chips do not constitute a homogeneous commodity, showing large variations in particle size, moisture content, and energy density, depending on the source material. Wood chips for energy purposes are mainly derived from forestry or processing residues (branches, tree tops, thinnings, bark, etc.), and other low-quality wood that is not suitable for material or pulp and paper production. In contrast, high-quality chips derived from roundwood are a high value feedstock in pulp and paper production. Statistical data show that that less than 10% of wood chips trade volumes are used as feedstock for energy generation. In addition, wood chips used for energy could originate from wood that is chipped at the site of combustion. Generally, wood chips for energy purposes is being transported over shorter distances than wood pellets due to the high moisture content, low heating value, and low and bulk density that make their transportation not economically viable over long distance (Lamers et al., 2012; Junginger et al., 2014). The increasing trend in wood chips is likely to continue, driven mainly by the increasing demand in pulp and paper and materials, with energy having a marginal role.

10.1.6.5 Outlook

The picture of biomass trade flows is very intricate, as it includes direct biomass trade for energy, as fuelwood, wood residues, wood pellets, and biodiesel but also significant indirect trade flows, as industrial roundwood, wood chips, charcoal, and ethanol that are used to some extent for energy production. Our calculation shows that the global biomass trade for energy has grown from 568 PJ (57.2 Mt) to 1278 PJ (95.6 Mt) between 2000 and 2016 (Table 10.2). Of this large trade, solid biofuel trade has grown from 651 PJ (66.8 Mt) to 1092 PJ (97.2 Mt) between 2000 and 2016. Solid biofuel trade developments of the past decade show a highly heterogeneous picture: wood pellet, wood chip, wood waste, fuelwood, and residue trade streams. Wood pellets have become a globally traded

Table 10.2 Global trade in bioenergy carriers

	PJ					Mt				
	2000	**2005**	**2010**	**2015**	**2016**	**2000**	**2005**	**2010**	**2015**	**2016**
Industrial roundwood	427	413	321	350	464	45.5	43.9	34.2	37.2	49.4
Wood residue	46	86	144	66	74	4.8	9.2	15.3	7.0	7.9
Wood chips	28	34	42	48	50	3.0	3.6	4.5	5.1	5.3
Fuelwood	26	35	55	71	71	2.0	2.7	4.2	5.5	5.4
Wood pellets	8.5	28.7	120.4	261	277	0.5	1.6	6.9	14.9	15.8
Charcoal	20	33	45	54	52	0.9	1.5	2.0	2.5	2.4
Bioethanol	11	66	104	136	158	0.4	2.5	3.9	5.1	5.9
Biodiesel	1	48	94	75	132	0.0	1.3	2.5	2.0	3.5
Total	**568**	**743**	**926**	**1060**	**1278**	**57.2**	**66.2**	**73.5**	**79.2**	**95.6**
Solid biomass	556	629	728	850	989	56.7	62.4	67.1	72.2	86.2
Biofuels	13	114	199	210	289	0.5	3.7	6.4	7.0	9.4
Direct solid biomass trade[a]	80	150	319	398	422	7.3	13.5	26.4	27.4	29.2

[a]Including wood residues fuelwood and wood pellets.
Source: Our own calculation based on data from FAO, 2018. Food and Agriculture Organization. Available at: http://www.fao.org/faostat/en/#home.

commodity, having the strongest development over the past decade with world trade growing from 8.5 PJ (0.5 Mt) to 277 PJ (15.8 Mt). In contrast, fuelwood is mostly used locally, and the trade streams in fuelwood represent a very low share of total fuelwood use for energy. Despite this, global trade in fuelwood has risen notably, from 26 PJ (2.0 Mt) in 2000 to 71 PJ (5.4 Mt) in 2016. The global liquid biofuel trade also had a significant increase, from 13 PJ (0.5 Mt) to 289 PJ (9.4 Mt) between 2000 and 2016. The estimated values for biomass trade have a large degree of uncertainty, due to differences in the databases of FAO, UN, and Eurostat and in reporting in exporting and importing countries. As global energy related trade data does not allow to distinguish bioenergy from other trade flows, the estimations of biomass trade depends on the assumptions made on the degree of use in various sectors. The markets are highly dynamic and technologies are improving. The differences in biomass feedstock properties, in energy and moisture content, make the results even more uncertain.

In the context of the expected increase of global bioenergy production to about 145 EJ in 2050, the traditional use of biomass might decrease and be replaced by modern bioenergy, with higher conversion efficiency and lower environmental impacts. The future role of bioenergy in the future energy system depends on the capacity to mobilize large amounts of biomass. The huge demand for biomass in all future energy scenarios will have an impact on the trade flows and changes in trade patterns, given the global distribution of biomass resources and expansion of bioenergy deployment in key global markets. Without significant bioenergy trade between world regions, less contribution of bioenergy to the energy and climate goals is achievable.

Many studies show a huge increase of bioenergy trade in the coming decades in most scenarios, in particular in the more ambitious bioenergy scenarios. In moderate scenarios, bioenergy trade represents 0%–20% and 7%–26% of global bioenergy demand in 2030 and 2050, respectively. In ambitious low-carbon scenarios, bioenergy trade increases to 14%–26% and 14%–30% of global bioenergy demand in 2030 and 2050, respectively (Matzenberger et al., 2015). This implies an annual bioenergy trade between 765 and 2860 Mt and up to 840 and 4720 Mt in 2030 and 2050, respectively, in low-carbon scenarios, compared to a level of 23 Mt in bioenergy trade today. Other study (Sharmina et al., 2017) estimate that bioenergy trade would be limited to lower levels, to 91–641 Mt in 2030 and 173–1255 Mt in 2050 in low-carbon energy scenarios.

Long-term trade depends on a number of factors, such as biomass availability and cost, demand, trade barriers, logistics, and sustainability considerations. Critical issues for further market and for biomass trade developments are the policy framework conditions, targeting to mitigate climate change, and the future sustainability requirements for biomass feedstock. The latest developments in biofuels markets show clearly that the uncertainties and ongoing changes in policies and the debates on sustainability requirements and iLUC risks for biofuels cause markets to stagnate. Long-distance trade of biomass requires optimized logistics and organized supply chains to facilitate trade. Developing the infrastructure and logistics for increasing levels of trade will be challenging. Other trade barriers (such as import tariffs) will need to be removed if in place (Matzenberger et al., 2015; Thrän et al., 2017).

The international bioenergy market is expected to increase and include a large number of suppliers and consumers from different world regions. The last years' experience shows that, in contrast to the slowdown in the EU bioenergy market, the world bioenergy markets are rapidly growing (China, Korea, Japan, etc.). New trade routes might develop, involving, for example, solid biomass from Latin America, Africa, and Russia toward new emerging markets. Key potential future biomass export regions are Russia, Canada, South America, Central Africa, with key import regions including EU, India, and China (Matzenberger et al., 2015; Thrän et al., 2017). Some trade streams have been largely sourced regionally (e.g., fuelwood, charcoal, and wood residues), and this is expected to continue. Trade flows of biofuels and of biofuel feedstocks toward the EU are likely to decrease in the future, due to the cap set on the use of first-generation biofuels in transport. Despite large uncertainties about the commercial availability of advanced biofuels, biomass trade is likely to grow for advanced biofuel production in the longer term, while advanced biofuels are likely to be used locally, driven by the mandates set in the EU and the United States.

Wood pellet market and trade could continue to increase using the existing infrastructure, such as storage, loading, and handling capacities in production areas and harbors. However, future developments in the wood pellet trade are uncertain, as depending on additional sustainability requirements for solid biomass that might hinder future bioenergy development. Increasing demands are expected for small-scale applications in residential heating systems and in large-scale power plants. Although wood pellets are mostly derived from either forestry or processing residues, increasing use of wood pellets may increase competition for woody biomass resources, with concerns that high quality, stem wood might be used for pellets production (Junginger et al., 2014). Similar feedstocks could benefit from the existing infrastructure and logistics, driven by biomass demand. New trade streams of biomass for energy can emerge, such as torrefied biomass and pyrolysis oil, with improved characteristics and enhanced energy density (Thrän et al., 2017). Pelletization, pyrolysis, and torrefaction increase the energy density and thus tradability of lignocellulosic feedstocks (such as wood residues and agricultural residues), making them more homogeneous and thus more suitable for long-distance transport (Popp et al., 2014). Pyrolysis oil could become a global commodity and could be traded using the existing infrastructure as oil or biofuels.

Standardization is essential to support the creation of new markets and trade opportunities for different commodities. In order to become a commodity that could be traded globally, the standardization of biomass feedstock is needed, providing a guarantee that the feedstock is delivered with certain characteristics. The standards can help removing trade barriers, increase market transparency and increase public acceptance. Harmonization of technical characteristics and quality specifications through standardization were crucial in the establishment of wood pellets as global commodities. The future trade in agricultural residues will depend on the application of pretreatment processes for quality improvement and densification (through pelletization, torrefaction, or pyrolysis), as well as through standardization, although they are likely to be used mostly locally.

Voluntary sustainability schemes for the use of solid biomass have been established and implemented in a number of countries in the EU (Belgium, Denmark, the Netherlands, and the United Kingdom) driven by industry. Additional sustainability requirements (such as the new stronger sustainability requirements proposed in the EU) can have multiple effects on existing markets and trade. Sustainability certification can facilitate trade, bringing confidence about the achievement of environmental and socioeconomic benefits of bioenergy. On the other side, sustainability certification could also become a trade barrier in the future. Some sourcing areas or feedstock categories could be excluded from bioenergy trade or at least could be influenced by sustainability considerations.

10.2 POLICIES AND MEASURES TO SUPPORT SUSTAINABLE BIOENERGY

10.2.1 Bioenergy policy context and bioeconomy: evolving framework

10.2.1.1 Towards a low-carbon economy

The EU has established the ambitious goal of building a competitive low-carbon economy in 2050 and set the objective of reducing GHG emissions by 80% by 2050, compared to 1990 levels to keep climate change below 2°C (COM(2011a,b) 112 final). An increase of 2°C compared to the temperature in preindustrial times is seen by scientists as the threshold beyond which there is a high risk that dangerous and possibly catastrophic changes in the global environment will occur. This requires a domestic emission reduction of at least 80% by 2050. The European Commission has produced a Low-Carbon Roadmap (COM(2011a,b)112 final) and an Energy Roadmap (COM(2011a,b) 885) to detail the perspective to 2050.

The Roadmap for moving to a competitive low-carbon economy in 2050 (COM(2011a,b)112 final) sets out the key elements for the EU's climate action helping the EU become a competitive low-carbon economy by 2050. The Roadmap showed that, as part of a global effort to meet the 2°C target, the EU should cut its emissions to 80% below 1990 levels by 2050 through domestic reductions alone (i.e., rather than relying on international offset credits). The analysis of different scenarios showed that a cost-effective pathway requires a 40% domestic reduction of GHG emissions for 2030 compared to 1990 levels, 60% by 2040, and 80% for 2050. The roadmap sets intermediate milestones for a cost-efficient pathway and GHG emission reductions, policy challenges, investment needs, and opportunities in different sectors. All sectors need to contribute to the low-carbon transition depending on their technological and economic potential—energy sector, industry, transport, buildings, construction, and agriculture.

The Energy Roadmap 2050 (COM(2011a,b) 885), investigated possible pathways for a transition towards a decarbonization of the energy system, while ensuring energy security and competitiveness, and the impacts, challenges, and opportunities for modernizing the energy system. A number of energy scenarios to achieve an 80% reduction in GHG emissions and about 85% reduction of energy-related CO_2 emissions have been examined. According to the decarbonization scenarios of the Energy Roadmap 2050, it needs to achieve significant reductions in GHG emissions already in 2030 (57%–65%) and to reach near-complete decarbonization by 2050 (96%–99%). RES are the key in any decarbonization scenarios. The share of renewable energy rises substantially in all scenarios to 28%–31% in 2030 and 55%–75% in the high RES scenario in 2050. The share of renewables in transport is expected to increase to 19%–20% in 2030 and 62%–73% in 2050. Bioenergy is expected to have an important role within the long-term goal to become a competitive low-carbon economy according.

10.2.1.2 Energy and climate change

In 1997, the *White Paper for a Community Strategy and Action Plan Energy for the future: Renewable sources of energy* (COM(97)599 final) sets the basis for the EU policy on renewable energy. This proposed doubling the share of renewable energy in the EU gross energy consumption from 6% to 12% by 2010. Several technology-specific targets were set for 2010 (135 Mtoe of bioenergy; 40 GW capacity for wind; 3 GWp for solar PV; 5 GWth for geothermal heat; 1 GW for geothermal electricity; and 105 GW for hydro). The reality has shown significant progress especially for some renewables (solar and wind), with the share of renewable energy in gross inland energy consumption reaching almost 10% in 2010, falling short of the indicative target. Toward developing the renewable energy in the EU, two Directives set targets for the renewable electricity and biofuels. The Renewable Electricity Directive 2001/77/EC set in 2001 a target of 21% of total electricity to be produced from renewable sources by 2010. In 1997, the share of renewable electricity in the EU27 was 12.9%. A national indicative target was defined for electricity generation from renewable sources. The share of green electricity has grown continuously, reaching almost 20% in 2010. In 2003, the Biofuels Directive 2003/30/EC set a target for 2010 for the biofuels and other renewable fuels replacing petrol and diesel of 5.75% of all petrol and diesel used in transport, again in the form of indicative targets. The data showed that the 2010 target set was not met, despite certain progress. The biofuel consumption in transport has increased from 1% to 4.4% biofuels in 2010, below the target (Eurostat, 2018).

The Green Paper *A European Strategy for Sustainable, Competitive and Secure Energy* (COM(2006) 105) sets another milestone in the EU energy policy following the request of the European Council to develop a long-term and coherent energy policy. The Commission has proposed an integrated *Energy and Climate Change package* in 2007 that includes both the energy and climate goals: *Energy policy for Europe* (COM(2007) 1 final) and *Limiting Global Climate Change to 2 degrees Celsius-The way ahead for 2020 and beyond* (COM(2007) 2 final). This package included the climate and energy targets aiming to achieve (1) 20% reduction of GHG emissions by 2020 compared to 1990 levels, (2) 20% renewable energy in the EU's energy mix, including a 10% target for renewable energy for 2020, and (3) 20% energy efficiency improvements by 2020.

The RED2009a,b/29/EC (RED) aims in particular, to promote renewable energy sources, including bioenergy, and to deliver GHG emissions reductions as part of the EU's policy to tackle climate change. The RED translated energy targets for 2020 into legally binding requirements: (1) a 20% target for renewable energy of gross final energy consumption at EU level and (2) a 10% target for renewable energy as a share of energy used in the transport sector. The RED specifies national objectives and legally binding targets for the share of renewable energy. The RED includes provisions to facilitate the development of renewable energy and to prepare National Renewable Energy Action Plans with detailed roadmaps and measures taken to reach the RES targets.

FQD2009a,b/30/EC sets a target of a 6% GHG reduction for the fuels used in transport in 2020 compared to a 2010 baseline. Biofuels used in transport are expected to deliver most of this reduction. The RED and FQD include criteria for sustainable biofuel production and procedures for verifying that these criteria are met. The calculation of the GHG intensity of fuels is based on a life cycle analysis that includes all emissions from extraction, processing, and distribution of fuels. In order to reduce the risk of indirect ILUC and to prepare the transition towards advanced biofuels, the Directive 2015/1513 introduced new rules to amending the RED and the FQD: (1) a limit of 7% for the share of biofuels from crops grown on agricultural land; (2) an indicative 0.5% target for biofuels produced from feedstocks that are not in competition with food crops and wastes, residues, nonfood cellulosic material or lignocellulosic material; and (3) GHG emission saving of at least 60% from the use of biofuels and bioliquids in new installations.

The 2020 energy and climate goals have been incorporated into the *Europe 2020 Strategy* (COM(2010a,b,c) 2020) *for smart, sustainable and inclusive growth* and into its flagship initiative *Resource efficient Europe.* The *Europe 2020 Strategy* sets out a vision to achieve smart growth (research and in novation), sustainable growth (resource efficient and low-carbon economy) and inclusive growth (employment, productivity, social, and territorial cohesion). The objectives of the strategy are supported by seven "*flagship initiatives.*" Five headline targets have been set for the EU for 2020 on employment; research and development; climate and energy; education; and social inclusion and poverty reduction. The *Energy 2020-A strategy for competitive, sustainable and secure energy* (COM(2010a, b,c) 639) sets the actions of an energy strategy until 2020 along the five energy priorities: reduce energy consumption; build internal market and develop infrastructure; extend technological leadership; ensure secure, safe, and affordable energy; and reinforce external dimension. These objectives are part of the Europe 2020 strategy and the "*Resource Efficient Europe*" initiative, with one specific goal to support the development of innovative new low-carbon technologies, including through Strategic Energy Technology Plan (SET-Plan). Resource Efficient Europe flagship initiative supports the shift towards a resource-efficient, low-carbon economy, and to achieve sustainable growth. The aim is to "decouple economic growth from resource and energy use, reduce CO_2 emissions, enhance competitiveness and promote greater energy security." The *Roadmap for a resource-efficient Europe* (COM(2011a,b) 571) sets a framework for the actions to develop a resource efficient, sustainable economy by 2050 and proposes ways to increase resource efficiency and decouple economic growth from resource use. It sets out a vision for the structural and technological change needed to achieve by 2050, with milestones to be reached by 2020.

10.2.1.3 Bioenergy and bioeconomy

Europe has a number of well-established traditional bio-based industries, ranging from agriculture, food, feed, fiber, forest-based industries, including pulp and paper and wood products, to biotechnology, chemical, biofuels, and bioenergy industries. Bioeconomy is already one of the biggest and most important components of the EU economy, estimated at €2.4 trillion (Scarlat et al., 2015). Several EU policies and initiatives have an impact on the bio-based economy: agriculture, forestry, industry, energy, environment, and climate change.

The first steps toward a bioeconomy have been made in 2002 when the *Life Science and Biotechnology Strategy* (COM(2002) 27), was proposed to develop and apply life sciences and biotechnology, setting out several actions for the development of biotechnology. The Bioeconomy Strategy *Innovating for Sustainable Growth: A Bioeconomy for Europe* (COM(2012) 60) was set to develop an "innovative, resource efficient and competitive society that reconciles food security with the sustainable use of renewable resources for industrial purposes." The strategy proposes a comprehensive approach to address five societal challenges through the introduction of a bioeconomy: (1) ensuring food security; (2) managing natural resources sustainably; (3) reducing dependence on nonrenewable resources; (4) mitigating and adapting to climate change; (5) creating jobs and maintaining European competitiveness. The bio-based economy plays a key role, as part of a green economy, to replace fossil fuels on a large scale, not only for energy applications but also for chemicals and materials applications. Several EU Member States have designed national bioeconomy strategies. A key component of the strategy is the production of food, feed, bio-based products and bioenergy, and the sustainable use of renewable sources.

Agriculture is part of the bioeconomy with multiple interactions with the bioenergy sector. In the EU, the Common Agricultural Policy (CAP) provides the overall EU framework for food production, with the aim to increase agricultural productivity to ensure a fair standard of living for farmers, secure the food supply, stabilize markets, and provide affordable food. Farmers have to comply with environmental requirements promote the development of agricultural practices preserving the environment and safeguarding the countryside. The rural development was introduced as a second pillar of the CAP in under the *Agenda 2000* reform to improve the competitiveness of farming and forestry, to protect the environment and the countryside, to diversify the rural economy, and to support rural development. The EU rural development policy includes measures aimed at encouraging the production and use of renewable energy. The European Commission presented on June 1, 2018, the legislative proposals on the future CAP for the 2021–27 period, following the Communication on the future of food and farming (COM(2017a,b) 713 final). The Communication COM(2017a,b) 713 final) outlined the vision for the future of food and farming in the EU focusing on (1) fostering a smart and resilient agricultural sector, (2) reinforcing environmental protection and climate action, (3) strengthening the socioeconomic fabric of rural areas.

A Circular Economy Package (COM(2015) 614) includes measures that will help stimulate the transition towards a circular economy where resources are used in a more sustainable way. The bioeconomy, providing a range of products (food, feed, fibers, construction wood, furniture, pulp and paper, chemicals, etc.) and energy, provides alternatives to fossil-based products and energy and can contribute to a sustainable circular economy. A circular economy relies on a cascading use of renewable resources, with several reuse and recycling cycles including of the bio-based materials, which can be used in multiple ways, thus bringing benefits for both the environment and the economy, fostering energy savings and reducing GHG emissions.

10.2.1.4 Outlook—future developments on bioenergy

The Paris Agreement, concluded at the 21st Conference of the Parties (COP21) of the UNFCCC, established on December 12, 2015, a long-term goal and set out a plan for limiting the increase of global average temperature to well below 2°C above preindustrial levels and to pursue efforts to keep it to 1.5°C above preindustrial levels. The Paris Agreement replaces the 1997 Kyoto Protocol that will not continue beyond 2020. The countries submitted comprehensive national climate action plans with their NDCs mainly through renewable energy and energy efficiency measures to keep global warming below 2°C. High-income countries intend to make available USD 100 billion per year by 2020 to support low-income countries. The Paris Agreement entered into force on November 4, 2016, after at least 55 Parties representing at least 55% of the total global emissions have ratified the Agreement.

The EU has adopted *A policy framework for climate and energy in the period from 2020 to 2030* (COM(2014a,b) 15 final) building up on the 2020 climate and energy package, in line with the longer term perspective set out in the Roadmap for moving to a competitive low-carbon economy in 2050 and the Energy Roadmap 2050. It sets EU-wide targets and policy objectives for 2030 to drive progress towards a low-carbon economy: (1) a target to reduce the EU GHG emissions by 40% relative to emissions in 1990, (2) a renewable energy binding target of at least 27% at Union level, and (3) an indicative target for energy efficiency of at least 27% at EU level compared to the business-as-usual scenario. The 40% GHG emission target will be achieved by reducing ETS sector emissions by 43% and those of non-ETS sectors by 30% compared to the 2005 baseline level. The ETS is to be reformed and strengthened and the emissions cut will be translated into individual binding targets for Member States. This target will enable the EU to take cost-effective steps towards its long-term objective of cutting emissions by 80%–95% by 2050 in line with the Energy Roadmap 2050.

The European Commission presented in 2016 a new package of measures, *Clean Energy for All Europeans* (COM(2016) 860 final) to facilitate the clean energy transition and the creation of the Energy Union to help the EU energy sector become more stable, more competitive, and more sustainable. Aimed at enabling the EU to deliver on its Paris Agreement commitments, the proposals are intended to help the energy sector become more stable, more competitive, and more sustainable. The Clean Energy package includes eight different legislative proposals covering: Energy Efficiency, Energy Performance in Buildings, Renewable Energy, Governance, and so on. The package pursues three main goals: (1) putting energy efficiency first, (2) achieving global leadership in renewable energies, and (3) providing a fair deal for consumers.

As part of the package *Clean Energy for all Europeans*, the European Commission has published in 2016 a proposal for a revised Renewable Energy Directive (RED II) to make the EU a global leader in renewable energy and to ensure that the 27% target for the share of renewable energy consumed in the EU in 2030 is met (COM(2016a,b,c,d) 767 final). The Commission proposal introduces some key changes to the promotion of renewables in the EU. The 27% renewable energy target in the final consumption is binding at the EU level but will not be translated into nationally binding targets. Therefore, one of the key challenges is to ensure that the 2030 target is collectively met in a cost efficient-way while avoiding a disproportionate burden on some Member States. For this goal, the 2020 national targets will be set as baseline to build on the progress achieved and Member States will not be allowed to go below their 2020 targets from 2021 onwards. Member States will prepare Integrated National Energy and Climate Plans that is part of the governance proposal to collectively achieve the EU target.

Specific measures to close any possible gaps to ensure the achievement of the renewables target will be introduced by the Energy Union Governance. By 2030, half of European electricity should be renewable. Around 40% of the renewable energy consumption by 2030 should come from renewable heating and cooling. An annual increase of 1 % in the share of renewable energy in heating and cooling is introduced. In order to foster the decarbonization of the EU transport sector, the new RED introduces an obligation on European transport fuel suppliers to provide an increasing share of renewable and low-carbon fuels: advanced biofuels, renewable fuels of nonbiological origin (e.g., hydrogen), waste-based fuels, and renewable electricity. The level of this obligation progressively increases from 1.5% in 2021 to 6.8% in 2030, including at least 3.6% of advanced biofuels. To address potential ILUC effects, a cap applies on the contribution of food-based biofuels towards the EU renewable energy target, starting at 7% in 2021 and decreasing progressively to 3.8% in 2030. Guarantees of Origin (GO) shall be issued for electricity, heating, and cooling produced from renewable energy for a standard size of 1 MWh. Member States shall issue such GO and transfer them to the market through auctioning, with revenues raised to be used to offset the costs of renewable support. RED II requires support schemes to be granted in an open, transparent, competitive, nondiscriminatory, and cost-effective, and aligned with state-aid rules. Support schemes may be opened to cross-border participation through, opened tenders, joint tenders, opened certificate schemes, joint support schemes, etc.

The proposed RES directive is under debate at the European Parliament and at the Council. The Council supports the binding 27% EU target for 2030. It proposes to supplement this with a 14% target for the transport sector (including a 3% share of advanced biofuels), and cap at 7% the share of first generation biofuels. The report of the Industry, Research and Energy (ITRE) committee of the European Parliament was adopted in January 2018, together with a mandate to start inter-institutional negotiations. The plenary resolution endorses the proposed binding 35% target for renewables and the 12% target for renewables in transport. Furthermore, biofuels from feed and crop would be capped at 2017 levels, with a maximum of 7% in road and rail transport. Palm oil would also be banned as a source of biofuel from 2021.

A core element of the EU climate and energy policy has been putting a price on CO_2 emissions. The EU ETS, a so-called cap-and-trade system, creates such a price, to ensure the emission reduction target is met. The EU ETS covers half of the EU's CO_2 emissions, mainly from the power sector and manufacturing industry, with almost 2 billion tons of CO_2 in 2017 and this amount is declining gradually. The Commission presented in July 2015 a legislative proposal on the revision of the EU ETS for Phase 4 (2021–30), in line with the 2030 climate and energy policy framework and as part of its contribution to the Paris Agreement. This proposal aims to reduce EU ETS emissions by 43% compared to 2005 and ensure the energy sector and energy intensive industries deliver the emissions reductions needed and accelerate the low-carbon transition. The proposal introduces some changes. A single, EU-wide cap on emissions declining at an annual rate of 2.2% from 2021 onwards, compared to 1.74 applies in place of the previous system of national caps. A progressive shift towards auctioning of allowances is introduced as default method in place of free allocation and harmonized allocation rules apply to the allowances still given away for free (6.3 billion allowances will still be allocated for free). The EU ETS creates incentives to invest in technologies that cut emissions by capping overall GHG emissions. Currently, NER 300 is a funding program for supporting innovative low-carbon energy demonstration projects for CCS and innovative renewable energy technologies on a commercial scale. Under the new EU ETS for the period after 2020, several support mechanisms will be established to help the industry and the power sectors meet the challenges of the transition to a low-carbon economy. For this goal, two new funds will be set: Innovation Fund, extending existing support for the demonstration of innovative technologies and a Modernization Fund, facilitating investments in modernizing the power sector and boosting energy efficiency in lower income Member States.

Sectors of the economy not covered by the EU ETS must reduce emissions by 30% by 2030 compared to 2005. These sectors, including transport, buildings, agriculture, and waste management, account for almost 60% of total EU emissions. The European Commission presented in July

2016 a legislative proposal, the *Effort Sharing Regulation* (COM(2016a,b,c,d) 482 final), setting out binding annual GHG emission targets in the non-ETS sectors for Member States for the period 2021–30. The proposed Regulation established binding annual GHG emission targets for each Member State for the period 2021–30, based on the principles of fairness, cost-effectiveness, and environmental integrity. In line with the guidance of the European Council, the resulting 2030 targets range from 0% to 40% compared to 2005 levels. The Member States have the flexibility to achieve their national targets by covering some emissions in the non-ETS sectors with EU ETS allowances which would normally have been auctioned (maximum 100 million tons CO_2 EU-wide over the period 2021–2030). Ms can use up to 280 million credits over the entire period 2021–30 from certain land use categories to comply with their national targets.

In July 2016, the Commission published a proposal for a regulation on the inclusion of GHG emissions and removals from Land Use, LUC and Forestry (LULUCF) in the 2030 climate and energy framework (COM (2016a,b,c,d)/479 final. The proposed LULUCF regulation introduces binding commitment for all Ms for GHG emission reduction in forestry and land use for all Member States, as well as related compliance rules. A no-debit rule is also proposed to ensure that accounted emissions from land use are entirely compensated by an equivalent removal of CO_2 from the atmosphere through action in the sector (e.g., afforestation or improving sustainable management of existing forests). Emissions of biomass used in energy will be recorded and counted towards each Member State's 2030 climate commitments, addressing the criticism that emissions from bioenergy production are not currently accounted for under EU law.

10.2.2 Sustainability certification and standards

10.2.2.1 Introduction

Certification schemes have been developed long before biofuel sustainability certification started, addressing a wide range of products from agriculture, forestry, and other sectors. Certification schemes have been developed as a result of various concerns or specific purposes (fair-trade, environmentally sound cultivation, organic agriculture, etc.) to gain market access, or to develop a green business profile. In the forest sector, forestry standards (such as the Forest Stewardship Council (FSC) and the Program for the Endorsement of Forest Certification (PEFC)) have been set to ensure sustainable management of forests. FSC was setting international principles for sustainable forest management, and local stakeholders developed region-specific standards. PEFC acts as an umbrella organization for certification by endorsing national-level regimes in compliance with a harmonized set of standards. No forest certification systems have yet included carbon accounting, due to the different scope of certification or disagreements about methodologies (Scarlat and Dallemand, 2011; Pelkmans et al., 2014).

Agricultural certification schemes (such as International Federation of Organic Agriculture Movements, Worldwide Standard for Good Agricultural Practices—Global GAP, Rainforest Alliance Sustainable Agriculture Standard, and FAIR Fairtrade Labelling Organizations International) were primarily developed to ensure sustainable farming practices, health and safety, and food traceability or for the certification of organic products, addressing mainly environmental aspects. These agricultural schemes include environmental, economic, and social aspects and cover to some extent soil conservation and air quality issues. Some schemes focused on a specific crop, like the Round Table on Responsible Soy (soy), Roundtable of Sustainable Palm Oil (palm oil), and Better Sugar Cane Initiative (Bonsucro) (sugarcane) to promote sustainable production, with reduced social, economic, and environmental impacts of these crops. Due to the fact that these systems have been developed with different interests and priorities (i.e., by governments, NGOs, companies), their scope and requirements vary from scheme to scheme (Scarlat and Dallemand, 2011).

10.2.2.2 Biofuels sustainability certification

The sustainability of biofuels has been mostly questioned based on the concerns over the real GHG emission reduction of biofuels, potential environmental, social, and economic impacts as well as the indirect effects, such as impact on food security or iLUCs. As result of these concerns on biofuels, the need for sustainability certification has emerged to make sure that biofuels are produced in a sustainable way, while negative effects are reduced or eliminated. These certification schemes tackle some environmental, economic, and social aspects and address specifically GHG emission performance.

Along with the proposed targets for renewable energy, the RED2009a,b/28/EC of the European Union (EU-RED) proposed a set of mandatory sustainability criteria as part of an EU sustainability scheme that include monitoring and reporting requirements for biofuels and bioliquids. Similar sustainability requirements were set in the FQD2009a,b/

30/EC on the specification of petrol, diesel, and gas-oil and introduced a mechanism to monitor and reduce GHG emissions. Biofuels are required to fulfill all sustainability criteria to count towards EU targets and to be eligible for financial support. The EU Member States are responsible for checking compliance with the sustainability criteria, while the European Commission recognizes the compliant voluntary sustainability certification schemes. The EU-RED requires a Single Harmonized Scheme in the EU, and, therefore, Member States may not lay down legal requirements that go further. The European Commission has recognized 16 voluntary schemes, as of early 2018, that can be used to prove compliance with the sustainability requirements for biofuels for the EU-RED.

In the United States, the RFS sets a target of 36 billion gallons of different biofuels that have to be blended with conventional fuel by 2022, with 21 billion gallons of advanced biofuels, according to the Energy Independence and Security Act of 2007. The RFS2 requires a reduction of life cycle GHG emissions depending on the renewable fuel category: 20% for first generation biofuels (corn ethanol), 50% for advanced biofuels (biodiesel and sugarcane ethanol), and 60% GHG emission reductions for cellulosic biofuels (lignocellulosic ethanol). The GHG emissions methodology should include all life cycle GHG emissions, including the emissions from both direct land-use change (dLUC) and iLUC. For each renewable fuel pathway, GHG emissions include production and transport of the feedstock; LUC; production, distribution, blending, and end use of the renewable fuel (Scarlat and Dallemand, 2011; Endres et al., 2015).

The California Low Carbon Fuel Standard (LCFS) was set in 2007 to reduce GHG emissions from the transportation sector in California by at least 10% by 2020. The LCFS does not limit the carbon intensity of different types of fuels but imposes a cap of the carbon intensity of all fuels used for transport. The carbon intensity, calculated on a full life cycle basis, includes all direct emissions from production, transportation, and use, as well as any other effects, both direct and indirect. The GHG emissions are calculated using an emission factor to estimate the GHG emissions from LUC depending on the type of land conversion and the carbon storage of land, using the Model "Greenhouse Gases, Regulated Emissions, and Energy Use in Transportation" (CA-GREET) and the Global Trade Analysis Project (Scarlat and Dallemand, 2011).

A large number of national and international initiatives, regulations, and voluntary sustainability standards have been developed, driven by the legal sustainability requirements set in the United States and in the EU.

Sustainability certification systems for biofuels in the EU, such as International Sustainability & Carbon Certification (ISCC), RSB, REDCert, NTA 8080, 2BSvs, etc., have been developed to show compliance with the European RED requirements. These are more generic standards which cover a wide range of feedstock categories to be used for biofuels or bioliquids. Compared to more general agricultural or forest certification systems, these new schemes were required to address carbon emissions because of the EU-RED requirements for biofuels to reduce emissions compared to their fossil fuel reference. More recently, new efforts were targeting a broader certification for bio-based materials, such as the Roundtable for Sustainable Biomaterials, or ISCC certification to cover all end uses expanded to cover bioenergy (liquid biofuels as well as biomass and biogas for energy generation) and bio-based materials (biochemicals, bioplastics, etc.) (Scarlat et al., 2015; EC, 2018).

10.2.2.3 Solid and gaseous biomass

While sustainability criteria apply to liquid biofuels for transport, there are currently no mandatory criteria for solid and gaseous biomass on EU level. The European Commission report on the requirements for a separate sustainability scheme for the use of biomass other than biofuels or bioliquids (COM(2010a,b,c) 11) provided EU Member States recommendations for developing national schemes for solid and gaseous biomass used in electricity, heating, and cooling. In the absence of an EU-wide sustainability scheme, the Commission recommended that national sustainability schemes for biomass used in electricity, heating, and cooling comply with the same requirements as those laid down in the RED for biofuels and bioliquids. This minimizes the risk of diverse and possibly incompatible criteria, leading to discrimination in the use of raw materials based on their final use, barriers to trade, and limits the growth of bioenergy. Small-scale producers and users, below 1 MW capacity, would be excluded from the application of sustainability criteria.

The Commission acknowledged the sustainability concerns on biomass production in terms of protecting the biodiversity, ecosystems, and carbon stocks. Thus, similar to biofuel feedstocks, biomass for heat and power should therefore not be sourced from land converted from forest or other areas of high biodiversity or high carbon stock. The Commission recommended the differentiation of national support schemes for electricity, heating, and cooling installations, to provide incentives to achieve high-energy conversion efficiencies. The report sets out a common methodology for calculating the GHG performance of biomass, to include the conversion of biomass to electricity, heating, or cooling. However, sustainability criteria do not apply to waste, which is covered by environmental rules laid down in a separate waste legislation at national and European levels. Forestry-related sustainability framework and cross-compliance rules for agriculture ensure the biomass sustainability in Europe.

Since sustainability requirements do not apply for biomass used for the production of bio-based products and biochemicals, several importing countries of solid biomass (Belgium, Denmark, Netherlands, the United Kingdom) developed their own national sustainability requirements and certification schemes. Wood pellet certification systems such as the Green Gold Label, Laborelec, and the Sustainable Biomass Program are designed to prove sustainability of woody biomass to demonstrate compliance with the sustainability legislation and facilitate trade within the sector, mostly for wood pellets and wood chips, used in industrial, large-scale energy production. They allow the use of other schemes show compliance with the sustainability criteria set out in the forest standards (e.g., FSC, PEFC) (Scarlat and Dallemand, 2011; Endres et al., 2015).

10.2.2.4 Sustainability requirements

In the EU, biofuels had to meet a minimum requirement for GHG savings of 35% relative to fossil fuels, which increased to 50% in 2017 and 60% in 2018 for new biofuel plants. In order to encourage the development of advanced, second-generation biofuels produced from residues, nonfood cellulosic material, and lignocellulosic material, they would be double credited towards the 10% renewable energy target in transport. Besides the sustainability criteria, the EU-RED includes rules and a methodology for the calculation of the GHG emissions and provides actual and default values. The GHG emissions shall include all emissions from the extraction or cultivation of raw materials, emissions from processing, transport and distribution, and annualized emissions from carbon stock changes caused by LUC, calculated over a period of over 20 years. The European Commission provided guidelines establishing the rules for the calculation of land carbon stocks, including soil organic carbon and carbon stock in the above and below ground vegetation both for the reference and the actual land use and values for different soil types and land-use categories.

The EU-RED excludes several land categories, with recognized high biodiversity value, from being used for biofuel production: (1) primary forests and other wooded land; (2) areas designated for nature protection or for the protection of rare, threatened, or endangered ecosystems or species; and (3) highly biodiverse grassland, either natural or nonnatural. Biofuels should not be made from material from peatland and land with high carbon stock, such as (1) wetlands, (2) continuously forested areas, and (3) land covered by trees higher than 5 m and a canopy cover between 10% and 30%. For the biomass feedstock produced in the EU, the cross-compliance rules of the CAP apply, in accordance with the requirements for good agricultural and environmental conditions. The EU cross-compliance regulations refer to preservation of soil and water quality, of biological diversity, careful use of fertilizers and pesticides, and air pollution.

The EU sustainability scheme includes monitoring and reporting requirements. The EU Member States shall report on the impact of the biofuels and bioliquids on biodiversity, water resources, water quality, and soil quality; the net GHG emission reduction and changes in commodity prices and land use associated with increased use of biomass. The fuel suppliers are required to report on the compliance with the sustainability criteria and on the measures taken for soil, water, and air protection, the restoration of degraded land, and the avoidance of excessive water consumption in areas with water deficit. Although there are no criteria for social sustainability included, the European Commission must report on the biofuels impact on social aspects and on the impact on the availability of food at affordable prices. The European Commission will monitor the origin of biofuels consumed in the EU and impacts of their production in the EU and third countries, land use and LUC, commodity prices, and food security.

The Global Bioenergy Partnership (GBEP) has developed a set of 24 voluntary sustainability indicators for bioenergy (Table 10.3), for facilitating sustainable development of bioenergy in line with multilateral trade obligations (GBEP, 2011). The mentioned criteria take into account the various economic, environmental, and social aspects of modern bioenergy production. The GBEP Task Force on sustainability had the goal of establishing relevant, practical, science-based, voluntary sustainability criteria, and relevant indicators and best practices examples regarding the sustainability of bioenergy. The GBEP criteria include a set of indicators that can be interpreted according to national circumstances and include

Table 10.3 GBEP bioenergy sustainability indicators

Environmental	Social	Economic
GHG emissions, productive capacity of the land and ecosystems, air quality, water availability, use efficiency and quality, biological diversity, land-use change	Price and supply of a national food basket, access to land, water, and other natural resources, labor conditions, rural and social development, access to energy, human health and safety	Resource availability and use efficiencies in bioenergy production, conversion, distribution, and end-use; economic development; economic viability and competitiveness of bioenergy; access to technology and technological capabilities; energy security/diversification of sources and supply; energy security/infrastructure and logistics
1. Lifecycle GHG emissions	9. Allocation and tenure of land for new bioenergy production	17. Productivity
2. Soil quality	10. Price and supply of a national food basket	18. Net energy balance
3. Harvest levels of wood resources	11. Change in income	19. Gross value added
4. Emissions of non-GHG air pollutants, including air toxics	12. Jobs in the bioenergy sector	20. Change in consumption of fossil fuels and traditional use of biomass
5. Water use and efficiency	13. Change in unpaid time spent by women and children collecting biomass	21. Training and requalification of the workforce
6. Water quality	14. Bioenergy used to expand access to modern energy services	22. Energy diversity
7. Biological diversity in the landscape	15. Change in mortality and burden of disease attributable to indoor smoke	23. Infrastructure and logistics for distribution of bioenergy
8. Land use and land-use change related to bioenergy feedstock production	16. Incidence of occupational injury, illness, and fatalities	24. Capacity and flexibility of use of bioenergy

Source: From GBEP, 2011. The Global Bioenergy Partnership Sustainability Indicators for Bioenergy.

supporting information and descriptions of methodological approaches for measurement. The indicators have been tested in a number of countries at both regional and national level. The report on sustainability indicators for bioenergy includes guides how the indicator values should be interpreted to assess the sustainability of bioenergy and how to relate the contribution (positive or negative) of bioenergy to sustainable development, to that of the fossil fuels or other energy sources they substitute fossil fuels. The set of 24 GBEP bioenergy sustainability indicators is set out under the three pillars: environmental, social, and economic, with the relevant themes listed at the top of each pillar.

The standard ISO 13065:2015 on sustainability criteria for bioenergy was developed to provide a practical framework to facilitate the assessment of environmental, social, and economic aspects and to facilitate the evaluation and comparability of bioenergy production and products, supply chains, and applications. ISO 13065 provides sustainability principles, criteria, and measurable indicators to provide objective information for assessing sustainability. ISO 13065:2015 specifies principles, criteria, and indicators for the bioenergy supply chain to facilitate assessment of environmental, social, and economic aspects of sustainability. The standard ISO 13065 applies to all forms of bioenergy, all raw materials, technology, or end use and covers the whole bioenergy supply chain. Direct effects are defined in the standard as measurable environmental, social, and economic effects caused by bioenergy production that are under the direct control of the project owner. The standard, however, does not address indirect effects, as no consensus was reached on how to quantify and link to a specific economic activity.

10.2.2.5 Outlook for sustainability of bioenergy

The proposal for a new directive on the promotion of renewable energy sources ((COM(2016a,b,c,d) 767 final) in the EU includes new, reinforced EU sustainability criteria for bioenergy and extended their scope to cover biofuels but also biomass and biogas for heating and cooling and electricity generation. The sustainability criterion applying to agricultural biomass is streamlined to reduce the administrative burden; the criterion on cross compliance is removed, as already dealt with under the CAP. A new requirement is being introduced for ensuring proper carbon accounting of forest biomass used in energy generation, in line with the LULUCF sector rules.

Sustainability criteria for forest biomass aim to minimize the risk of unsustainable forest harvesting, requiring that the following minimum requirements apply to forest biomass, both domestic or imported: (1) legality of harvesting, (2) forest regeneration, (3) protection of high conservation value areas, including wetlands and peatlands, (4) minimization of harvesting impacts on soil and biodiversity, (5) harvesting does not exceed the long-term forest production capacity. In order to minimize the risk of negative impacts on forest carbon stock, forest biomass has to meet the following LULUCF requirements: Option A: The country of origin of the biomass (1) is a party of/has ratified the Paris Agreement, (2) has submitted an NDC to the UNFCCC, covering emissions and removals from agriculture, forestry, and land use (LULUCF) accounting, (3) has a national system for LULUCF reporting; or Option B: Management systems are in place at forest holding level to ensure that forest carbon stock and sink level are maintained.

Sustainability and GHG criteria apply to all biofuel/bioliquid and biogas plants with a fuel capacity equal or above 0.5 MW and to solid biomass installations with fuel capacity equal or above 20 MW. Bioenergy from waste and processing residues (e.g., saw dust, manure, and black liquor) needs to meet only the GHG saving criteria. The GHG saving performance requirement applying to biofuels is increased to 60% for plants in operation after October 2015 and to 70% for new plants (plants in operation after January 2021). A 80% GHG saving requirement is applied to biomass-based heating/cooling and electricity (plants in operation after January 1, 2021) and 85% for those plants starting operation after January 1, 2026. Electricity from biomass in large scale plants (equal or above 20 MW) must be produced through high-efficient cogeneration technology and meet the sustainability and GHG criteria.

The RED proposal establishes an EU-level obligation for fuel suppliers to provide a share of 6.8% in 2030 of low-emission and renewable fuels (including renewable electricity and advanced biofuels). In order to address iLUC issues, the maximum share of biofuels and bioliquids produced from food or feed crops shall be 7% of final consumption of energy in road and rail transport, a limit shall decrease to 3.8% in 2030. A specific, increasing sub-mandate is introduced for advanced biofuels that should reach at least 3.6% by 2030.

The new legal sustainability requirements covering all bioenergy pathways are a major step forward. Ensuring sustainability of biomass is a key issue in the wider context of a bio-based economy. Biomass for energy can be produced from various feedstock categories, which can also be used for food, feed fiber, or biomaterials production. Currently sustainability requirements have been established only regarding the use for biofuels and bioenergy. Similar commodities, with other uses, even having similar environmental, social, and GHG impacts, do not need to fulfill such requirements. Applying a double-standard policy between biomass production for bioenergy and food, feed, fiber, or biomaterials is very likely to lead to indirect displacement effects, since biomass suppliers must ensure sustainability production only for the part of biomass that will be used for bioenergy (Scarlat and Dallemand, 2011).

The experience has shown that it is unlikely that voluntary certification, targeting forest certification only, will be able to stop the production and use of nonsustainable wood and avoid LUCs and deforestation. Therefore, sustainability requirements should address sustainable biomass production and GHG emissions ant not biomass use to avoid leakage. Sustainability criteria could also include other aspects such as resource efficiency to differentiate between different biomass pathways (Scarlat et al., 2015). Only a certification scheme addressing biomass feedstock production, regardless the final use (with consistent sustainable land management criteria or regulations that apply to forestry and agricultural management practices, nature and environment protection regulations, and land use and land-use planning) could be more effective in addressing global sustainability concerns and might be able to avoid various effects, either direct or indirect. Labeling can play an important role for bio-based products, providing consumers with clear information on the characteristics and environmental performance of the products. A global initiative, with strong international participation, is needed to establish a global sustainability governance framework based on a global agreement on sustainability (Scarlat and Dallemand, 2011; Pelkmans et al., 2014).

10.2.3 GHG emissions and carbon accounting for bioenergy

10.2.3.1 Bioenergy and GHG emissions

Bioenergy can play a key role in the attempt to reduce GHG emissions from the energy sector as a part of the global effort to decarbonize the energy sector. The replacement of fossil fuels (mostly oil and coal) by low-carbon energy sources and the removal of fossil carbon emissions in the shortest possible time are essential to achieve the goal of limiting climate change. Along with this pressing need, there are legitimate concerns over the sustainability impacts and carbon emissions from certain bioenergy pathways. The main concerns relate to large-scale bioenergy expansion and the real GHG emission reduction from the use of biomass in comparison to the fossil fuels and the LUCs that could lead to additional GHG emissions, negating the emission reductions from biomass use.

Biomass is a limited resource and sustainability is the key issue to ensure that bioenergy contributes to the low-carbon economy. Bioenergy used in the EU has to be sustainable, and that it is produced in a way that does not cause deforestation or degradation of habitats or loss of biodiversity, is used efficiently, and delivers optimal GHG savings. Therefore, the promotion of the various bioenergy pathways and their prioritization should be based on the maximum GHG emission reductions that can be achieved.

To ensure a significant contribution to energy and climate goals of the EU, the RED already includes specific, mandatory threshold for GHG savings, based on a simplified methodology for their calculation. The RED II extended these sustainability criteria to solid and gaseous biomass and provides new GHG emission thresholds for all bioenergy (e.g., gaseous, liquid, and solid biomass). The methodology defined for GHG savings calculation includes all emissions, from the extraction or cultivation of raw materials, emissions from processing, transport, and distribution and emissions from carbon stock changes caused by dLUC (Agostini et al., 2014).

Biomass use for energy generation is considered carbon neutral over its life cycle because the combustion of biomass releases the same amount of CO_2 that was captured by the plant during its growth. The fundamental difference between fossil energy and bioenergy relies on the fact that fossil fuels release CO_2 that has been locked up for millions of years, while burning biomass emits carbon that is part of the biogenic carbon cycle. Thus, the use of fossil fuel increases the total amount of CO_2 into the atmosphere while bioenergy production operates inside this system (IEA, 2018). However, there are GHG emissions associated with fossil fuels used for biomass production and harvesting, processing conversion, and transport. Furthermore, the harvest of biomass may lead to a change in carbon stored above and below ground with the release of additional carbon emissions (Cherubini et al., 2009).

10.2.3.2 Life cycle assessment of GHG emissions from bioenergy

Life cycle assessment (LCA) is the tool currently used to quantify the environmental impacts of products and services, which covers all processes along the supply chain of the product or service. LCA is commonly used to quantify the GHG emissions savings of bioenergy, by comparing the bioenergy system with a reference (fossil) energy system following a life cycle approach. In the LCA analysis, all energy inputs and GHG emissions occurring along the supply chain, from biomass production (planting, harvesting etc.), transport and storage, processing, distribution, and final use, must be accounted. The utilization of by-products that can displace other materials, having GHG and energy implications, must also be considered in the analysis.

To ensure that relevant comparisons could be made, LCA should be conducted following harmonized and standard procedures. The Task Force on GHG methodologies of the GBEP has developed in 2009 (GBEP, 2009) a methodological framework for assessing GHG emissions associated with bioenergy based on a life cycle assessment (LCA). The GHG framework serves as guidance and provides a template for LCA to be used as a tool for calculating GHG emissions in a consistent manner and to enable GHG emissions to be compared on equal basis. The framework consists of a 10-step analysis procedure covering the emissions from biomass feedstock production to end use, including LUC, co-products and by-products, transport of biomass, conversion, and transport. The ISO provides two international standards on the general principles and requirements of the environmental management—LCA: ISO 14040:2006 (Principles and framework) and ISO 14044: 2006 (requirements and guidelines). The ISO standards specify the principles, requirements, and guidelines for the quantification of the carbon footprint of a product.

Several LCA models are available for GHG emission estimation, including Biograce, E3 Database in Europe, the Argonne National Laboratory GREET model (GHGs, Regulated Emissions and Energy in Transportation) in the United States and the GHGenius model in Canada. LCA requires large amounts of data on a specific product or service for assessing the complete supply chain. The wide range of results of LCA studies occurred depending on the data that are generally valid for certain regions and conditions. Several LCA databases for the GHG and energy balance of bioenergy systems are available worldwide, such as ECOINVENT, ELCD (European reference Life Cycle Database), GEMIS (Global Emission Model for Integrated Systems), CPM LCA Database or US Life Cycle Inventory Database (LCI) from NREL (Openlca, 2018, E3database, 2018, Ghenius, 2018 GREET, 2018, Biograce, 2018). Several commercial LCA software packages (e.g., SimaPro, GaBi, Umberto, openLCA, eBalance, EIME, Quantis Suitye, Team5, and REGIS) can also be used as tools for biofuels LCA (ESU Services, 2018; Ecoinvent, 2018; ELCD, 2018; Gemis, 2018; CPM, 2018; LCI, 2018).

Given the numerous methodological choices and assumptions made for an LCA, the results of GHG emissions balances differ significantly among different pathways but even for apparently similar systems. The most striking is the wide range of results reported for the same biofuel and biomass pathway by different authors. The differences occur for a multitude of reasons including differences in approach, system boundaries, reference systems, different methods of allocation of co-products, and LUCs considered. The system boundary should be defined to ensure that bioenergy and reference fossil systems provide equivalent products and services and include all processes along the value chain, energy uses, material flows, GHG emissions, and coproducts in both systems. The choice of the reference system to which the bioenergy system is compared is critical, since the estimated benefits of bioenergy can differ widely depending on the assumed energy system replaced (natural gas, oil, or coal) and the use of coproducts. In addition, the data used, the main assumptions made, and the reference systems are not always described in detail in a transparent manner.

Bioenergy systems usually require fossil energy for biomass production, transport, and conversion to bioenergy. Depending on the feedstock types, crop yield, fossil energy intensities for farming (machinery, fertilizers, irrigation, etc.), harvesting, transport, feedstock processing, conversion requirements and the types of co-products, some production chains are more desirable than others. The crucial factors appear to be the amount and type of fossil fuel used to produce, transport, and process the feedstock and the efficiency in the conversion process (Cherubini et al., 2009). Major aspects that lead to wide ranges in LCA results are input parameter values, various assumptions on crop yield, fertilizer application ranges, or energy input for producing fertilizers. Various methods are used for the allocation of by-products (e.g., Distiller's dried grains with solubles from ethanol, meals from biodiesel, fertilizers from biogas production, etc.), according to the economic value of co-products, their mass and energy content, or by system expansion. Since some key parameters used in LCA analysis vary widely between different systems and locations, and are subject to huge uncertainties, the real energy balance and GHG emissions for a specific system could be substantially different from the default values. Therefore, LCA results based on selected default values may significantly increase the risk of drawing misleading conclusions (Cherubini et al., 2009; Bird et al., 2011; Carvalho et al., 2015).

The source of the process fuel used and the level of fossil fuel input are the main factors responsible for the differences in GHG emission reductions observed between different biofuels pathways (such as sugarcane ethanol compared to corn ethanol). GHG emissions will be lower and can be close to carbon neutrality if biomass is produced with low-energy input or derived from residue streams, converted with high conversion efficiency and if the energy used is renewable, including for transport. Thus, energy and GHG balances of bioenergy pathways can be improved by using by-products resulted from within the system for fueling the processes (such as biofuels for farming, by-products for electricity, and heat production for conversion processes, etc.). Thus, the GHG emissions mostly reflect the carbon intensity of fossil fuels used in the bioenergy pathway. In this respect, energy balance (or energy ratio) can be better and a sound basis to assess and compare the performances of different bioenergy pathways.

Most studies demonstrate that the majority of bioenergy pathways have lower GHG emissions than fossil energy systems if an LUC (direct or indirect) is avoided. The Figs. 10.14 and 10.15 show the ranges for GHG emissions from different biofuels, electricity, and heat pathways. The figures show that in the case of first generation or food-based biofuels, GHG emission savings typically vary between 50% and 65% in the case of biodiesel, and 40% and 80% for bioethanol. In the case of the use of waste cooking oil or animal fats, GHG emission savings typically vary between 70% and 90%. The use of upgraded biogas in transport brings much higher benefits, typically ranging between 70% and 200%, if using manures or various residues in comparison to biogas produced from energy crops (60%–70%), due to avoided methane emissions. Some bioenergy pathways can be close to carbon neutrality. For example, net GHG emission reductions from electricity generation from various feedstocks are usually 90%–95% in comparison to fossil fuel-based electricity generation. The favorable energy balance for bioethanol from sugarcane (70%–80% GHG reduction) and advanced, lignocellulosic biofuels (80%–85% GHG reduction) are mostly due to the assumptions that they are processed using biomass residues, in comparison to biofuels from food-based crops that rely on fossil energy input both in farming, biofuels conversion, and feedstock transport. If energy supply is progressively decarbonized, the GHG emissions from the use of fossil energy in the

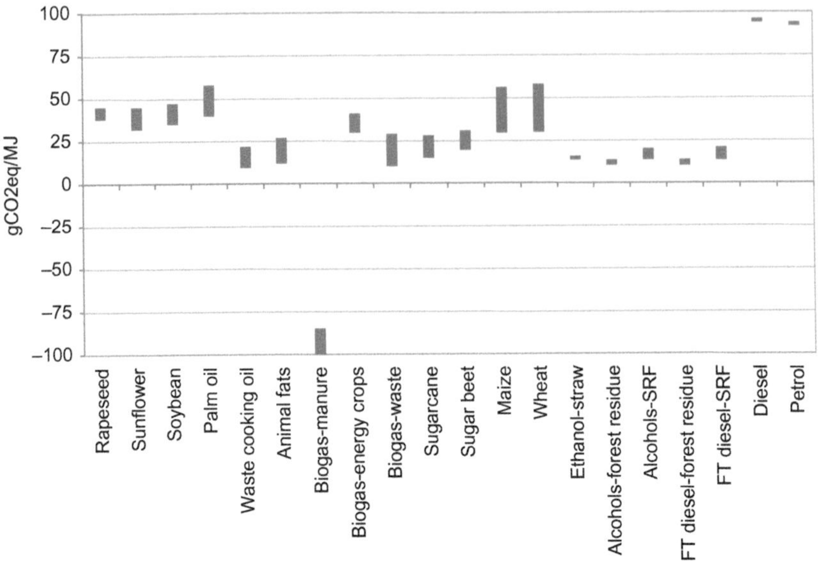

Figure 10.14 GHG emissions from different biofuels pathways. *From Cherubini F., Bird N.D., Cowie A., Jungmeier G., Schlamadinger B., Woess-Gallasch S., 2009. Energy- and GHG-based LCA of biofuel and bioenergy systems: Key issues, ranges and recommendations. Resources, Conservation and Recycling. 53: 434–447; Strengers B., et al., 2015. Greenhouse Gas Impact of Bioenergy Pathways. The Hague: PBL Netherlands Environmental Assessment Agency; Giuntoli J., Agostini A., Edwards R., Marelli L., 2015. Solid and Gaseous Bioenergy Pathways: Input Values and GHG Emissions. Joint Research Centre Report EUR 27215 EN; Edwards, R., Padella, M., Giuntoli, J., Koeble, R., O'Connell, A. Bulgheroni, C. et al., 2016. Biofuels Pathways. Input Values and GHG emissions. Database. European Commission, Joint Research Centre (JRC).*

bioenergy supply chain will be reduced and thus the GHG chain emissions from bioenergy will decrease. Since advanced biofuels are not yet commercial, LCA of GHG emissions are estimated based on projections of technical data from different stages of development of processes involved, at pilot plant, and even laboratory scale that need to be confirmed in commercial operation.

10.2.3.3 LUCs and GHG emissions

Increased demand of bioenergy is likely to lead to both dLUC and iLUC. Major concerns about the impact of biofuel production on LUC were expressed widely, based on the past experience of cropland expansion into uncultivated areas, grasslands, and forestlands due to food and feed production. LUC has been the most controversial issue, since it could lead rise to

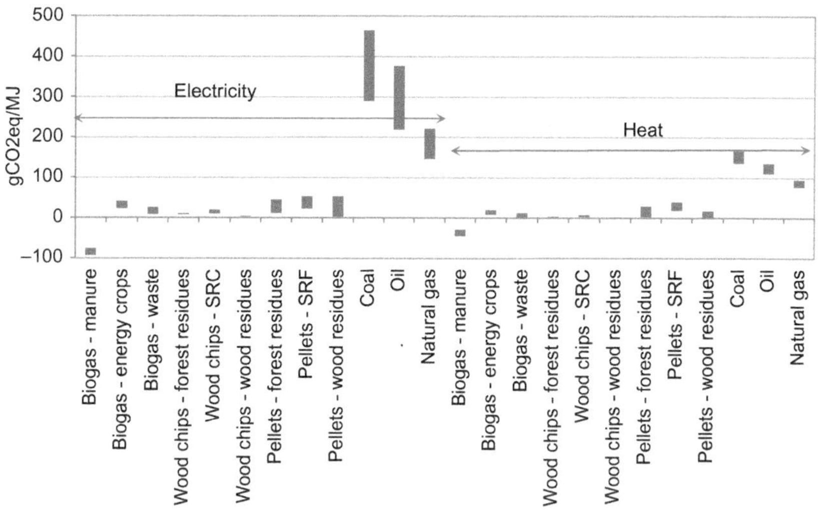

Figure 10.15 GHG emissions from different bioenergy pathways. *From Cherubini F., Bird N.D., Cowie A., Jungmeier G., Schlamadinger B., Woess-Gallasch S., 2009. Energy- and GHG-based LCA of biofuel and bioenergy systems: Key issues, ranges and recommendations. Resources, Conservation and Recycling. 53: 434–447; Strengers B., et al., 2015. Greenhouse Gas Impact of Bioenergy Pathways. The Hague: PBL Netherlands Environmental Assessment Agency; Giuntoli J., Agostini A., Edwards R., Marelli L., 2015. Solid and Gaseous Bioenergy Pathways: Input Values and GHG Emissions. Joint Research Centre Report EUR 27215 EN; Edwards, R., Padella, M., Giuntoli, J., Koeble, R., O'Connell, A. Bulgheroni, C. et al., 2016. Biofuels Pathways. Input Values and GHG emissions. Database. European Commission, Joint Research Centre (JRC).*

significant changes in above and below ground carbon stocks with potential GHG emissions. LUCs usually result in a sudden change in the carbon stock in above and belowground biomass, and a slow change in the carbon content in soil. Depending on the previous use of the land and the crop to be established, this can have a positive or a negative impact. Thus, if high soil carbon stocks land (e.g., grassland, forest) is converted into cropland, this might lead to high carbon emissions. When marginal or degraded land, with low-carbon stock, is taken into production, or when a perennial grass or a forest plantation is established on cropland, this leads to an increase in the carbon stock (Hiederer et al., 2010). Thus, depending on the land cover categories involved, this can severely affect the net GHG emissions of the bioenergy system. The high ranges of the carbon stock in different areas and land cover types, depending on soil and climate conditions, make the quantification of carbon emissions and carbon sequestration associated with LUC highly uncertain.

10.2.3.4 Direct land-use change

GHG emissions from dLUC have been included in LCA studies of biofuels recently. DLUC occurs when a new biofuel crop is established and displaces a prior crop that was cultivated on that land. Thus, a direct link can be established between biofuel production and the LUC. DLUC accounts for changes in land use associated with the direct expansion of biofuel feedstock production on cropland, the displacement of food, feed, or fiber crops, and the possible conversion of other land-use types (grasslands, forest lands, etc.) into cropland for biofuel crops. For annual bioenergy crops, or perennial grasses, the annual change in carbon stock is equal to zero, since the whole crop is harvested annually. A forest plantation requires longer time to grow after being harvested (for example, an SRF/SRC needs between 3 and 5 years until harvest) and until carbon stock reaches the initial value. The calculation of GHG emissions from LUC needs to consider soil and biomass carbon stocks to evaluate how much carbon would be released, which depends on the location and the characteristics of the land that would be converted. Rules for the calculation of land carbon stocks changes due to land conversion were incorporated into the EU legislation (Commission Decision 2010/335/EU). The methodology followed the Tier 1 approach of the Intergovernmental Panel on Climate Change (IPCC) based on the definition of default values for direct land use and soil-carbon changes of carbon stocks over a time horizon of 20 years (IPCC, 2006; Hiederer et al., 2010).

10.2.3.5 Indirect land-use change

ILUC comprises the change in land use outside a biomass production area that is induced by changing the use or production quantity of a feedstock that was produced in that area. Thus, crops previously produced in the biomass production area are being produced elsewhere to meet demand, resulting in some land being converted to agricultural land. The increased demand of biomass for bioenergy might have multiple effects: substitution of food and feed due to biofuel production; food crop price increase inducing reduced demand; crop area expansion; multiple cropping; and yield increment through agriculture intensification. As result, the total cropland area might not change, or additional land might be converted to cropland from other land uses (such as fallow land, pasture, or forest). This conversion can occur anywhere in the world, due to the complex interactions of the global agricultural commodities market (Searchinger et al., 2008; Gnansonou et al., 2008; Cherubini et al., 2009).

The iLUC and the conversion of other land to cropland would involve GHG emissions into the atmosphere through the changes in carbon storage in land. The main questions are how much land might be converted, where the conversion would occur, what LUCs would be involved and how much would be the GHG emissions that could be associated to biofuels production. In order to assess the GHG emissions associates with LUC, the crucial issue is to identify the areas where the expansion of cropland is likely to occur, and how this additional land is spatially distributed. At the beginning of the debate on ILUC effects, the magnitude of GHG emissions from of ILUC are claimed to be large enough to negate the GHG emission benefits of the biofuel supply chain, despite the high uncertainties and high inaccuracy in the estimations (Searchinger et al., 2008; Fritsche et al., 2010; Cherubini et al., 2009).

Since food crop markets are global and food production can take place in various regions, LUC estimates have to be made globally. Since iLUC is not empirically observable, the estimates are determined mostly through modeling and few studies have been conducted to find evidence of iLUC in historical data. Although dLUC and iLUC are conceptually different, models analyze both effects together and thus the ILUC factor is a result of a combination of both. Bioenergy emissions associated to LUC are modeled in several steps: biofuel impact on crop market is evaluated through global models, crop production is translated into LUC, the total GHG emissions are estimated depending on the specific types of LUC, and finally total emissions are converted to an iLUC factor. Methods to quantify LUC and associated GHG emissions include economic models (market equilibrium models): PE models (AGLINK-COSIMO, CAPRI, IFPRI-IMPACT, FAPRI-CARD, etc.) that address specific markets (e.g., agricultural markets) or computable general equilibrium models (GTAP, IFPRI-MIRAGE, LEI-TAP, etc.) that provide an economy perspective over various sectors and area covered. The quantification of GHG emissions from ILUC requires to couple economic models with biophysical models to transfer the changes in markets, trade, and production into changes in land use. Alternative approaches include statistical analyses and deterministic approaches that use simplified calculations based on a set of statistical data and assumptions on production, trade patterns, and displacement ratios of land use for biomass production (Bauen et al., 2010; Fritsche et al., 2010; Edwards et al., 2010; Moreira et al., 2012; Wicke et al., 2012).

LUC is a complex and dynamic process that is affected by many economic and policy drivers and is difficult to model, given the complexity of the economic reality. The results show high ranges, associated to the methodology used, scenario, and underlying assumptions (changes in production between regions, changes in food consumption or substitution of crops, type of land converted, etc.), data sets (spatial resolution, carbon stock, yields on new land, etc.), and coproduct allocation method. Modeling LUC requires a large amount of unobservable and unverifiable parameters, which are uncertain and poorly supported by real data. Elasticities between crop production and crop yields and land expansion are exogenous and play a critical role in the assessment when using market equilibrium models. In addition, by-products and coproducts of bioenergy production, which can be used to meet regional food, feed, and fuel demands, leading to lower LUC, are not always included or not properly accounted for in the market equilibrium models. The reductions in the estimated magnitude of iLUC-induced emissions over time are a result of improvements in the assumptions and data used in iLUC calculation, or the accounting for coproducts and substitution patterns in economic models. Despite recent improvements of the models, still large uncertainties and shortcomings exist in the current modeling efforts (Fritsche et al., 2010; Edwards et al., 2010; Wicke et al., 2012). Fig. 10.16 provides an overview of LUC-related GHG emissions as determined by different studies.

10.2.3.6 Carbon accounting and carbon debt

Bioenergy can be carbon neutral because the carbon that is released during combustion has previously been sequestered from the atmosphere and will be sequestered again as plants regrow. Bioenergy production can also lead to GHG emissions if its production and use lead to changes in carbon stocks in soils or vegetation that needs to be included in the calculation (IEA, 2017c). The internationally agreed IPCC guidelines state that emissions from the combustion of biomass can be accounted as zero in the energy sector since they are reported in the LULUCF sector at the moment and at the point of harvest, independently from the end-use of biomass. Therefore, to avoid double counting, the carbon emissions from biomass combustion are not added to the total energy sector emissions (EU Regulation No. 525/2013). As a safeguard, to avoid the loss of carbon stock in forests, the EU-RED does not allow the use of raw materials obtained from high-carbon stock land or land with high

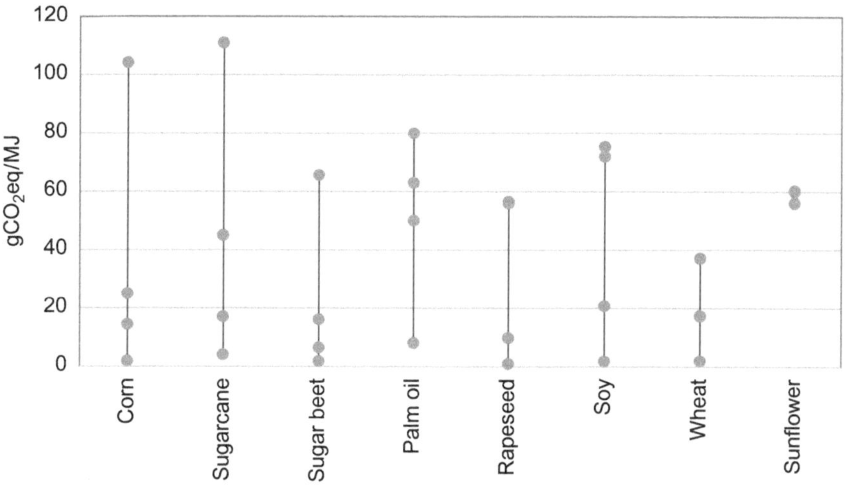

Figure 10.16 Ranges in ILUC GHG emissions from biofuels. *From Searchinger T., et al., 2008. Use of US croplands for biofuels increases greenhouse gases through emissions from land-use change. Science 319, 1238–1240; EPA, 2010. Renewable Fuel Standard Program (RFS2) Regulatory Impact Analysis. Assessment and Standards Division, Office of Transportation and Air Quality, U.S. Environmental Protection Agency; CARB, 2009. CARB Staff Report: Proposed Regulation to Implement the Low Carbon Fuel Standard. California Air Resources Board; Tipper R., Hutchison C., Brander M.A., 2009. Practical Approach for Policies to Address GHG Emissions From Indirect Land Use Change Associated With Biofuels. Edinburgh, UK: Ecometrica; Al-Riffai P., Dimaranan B., Laborde D., 2010. Global Trade and Environmental Impact Study of the EU Biofuels Mandate. Specific Contract No SI2.537.787 Implementing Framework Contract No TRADE/07/A2; Hertel, T., Golub, A.A., Jones, A.D., O'Hare M., Plevin, R.J., Kammen, D.M., 2010. Global land use and greenhouse gas emissions impacts of U.S. maize ethanol: estimating market mediated responses. Bioscience 60 (3), 223–231. doi:10.1525/bio.2010.60.3.8; E4Tech, 2010. A Causal Descriptive Approach to Modeling the GHG Emissions Associated With the Indirect Land Use Impacts of Biofuels. Final Report. A study for the UK Department for Transport, October 2010; Tyner, W.E., Taheripour, F., Zhuang, Q., Birur, D., Baldos, U., 2010. Land Use Changes and Consequent CO2 Emissions Due to U.S. Corn Ethanol Production: A Comprehensive Analysis; West Lafayette, IN: Purdue University; Laborde D., 2011. Assessing the Land Use Change Consequences of European Biofuel Policies. Washington, DC: IFPRI; Marelli, L., Ramos F., Hiederer R., Koeble R., 2011. Estimate of GHG Emissions From Global Land Use Change Scenarios. Joint Research Centre (JRC) Report EUR 24817 EN; Moreira, M., Nassar, A., Antoniazzi, L., Bachion, L.C., Harfuch, L., 2012. Direct and indirect land use change assessment. In: Poppe, M.K., Cortez, L.A.B., (Eds.), Sustainability of Sugarcane Bioenergy. Brasília, DF: Center for Strategic Studies and Management (CGEE); Elliott, J., Sharma, B., Best, N., Glotter, M., Dunn, J.B., Foster, I., et al., 2014. Spatial modeling framework to evaluate domestic biofuel-induced potential land use changes and emissions. Environ. Sci. Technol. 48 (4), 2488–2496, CARB, 2014. iLUC Analysis for the Low Carbon Fuel Standard (Update). California Air Resources Board.*

biodiversity value, such as wetlands, primary forests, and highly biodiverse grasslands.

There is a concern that the assumption of biogenic carbon neutrality is not valid since harvest of wood for bioenergy might cause a temporary decrease of the forest carbon stock, which may not be recovered during the time frame considered in the analysis, leading to a temporary increase in atmospheric CO_2 (Agostini et al., 2014). In the calculation of GHG emissions, the EU-RED methodology does not consider the change in the carbon stock on land if it does not involve LUC. The assumption of carbon neutrality is commonly accepted for annual crops, as the carbon emitted from bioenergy production will be sequestered again within a short time frame. If the whole tree is used for bioenergy, the carbon emitted from combustion spends a longer time in the atmosphere until it is recaptured through plant growth (Agostini et al., 2014). Thus, the assumption of carbon neutrality depends on the time frame considered in the analysis, with the difference that annual crops have a time frame of 1 year before it grows, while the forest biomass requires longer time until a tree reaches maturity and it is harvested.

The concept of carbon debt describes a temporary loss of carbon stock in the forest due to harvesting. If there is a time gap between emissions generated by biomass combustion and the carbon recapture, then a carbon debt can occur (Allen et al., 2016). Biogenic carbon cycle for forests can be long because forest rotations are longer and the decay of dead biomass with associated CO_2 emissions are relatively a slow process. This introduces a time component into the consideration about carbon benefits using forestry biomass. If the forest carbon stock is reduced, there may be a time delay until the savings from avoided fossil fuel emissions lead to a net reduction in atmospheric CO_2. Forest carbon stock may increase after harvesting, but at a slower rate compared to the absence of harvesting for bioenergy (Matthews et al., 2014; Agostini et al., 2014; IEA, 2017c). The payback time can be almost immediate when annual crops or crop residues that would otherwise decay rapidly are used to displace fossil energy. Similarly, the use of wood by-products from forest industry (sawdust, bark, etc.) and forestry residues (tops and branches) could achieve carbon emission reduction in the short term. If the use of biomass for energy causes a reduction in forest carbon stocks, this carbon debt can be repaid if the biomass displaces fossil energy sources. There is a large variability in the results of payback time of forest bioenergy in comparison to fossil fuel. Several studies claim that bioenergy systems might require payback times of decades when considering roundwood as feedstock for bioenergy or that result in GHG emissions higher than those from the fossil system (Searchinger et al., 2009; Cherubini et al., 2011; Mitchell et al., 2012; Pingoud et al., 2012). From an economic perspective, it is unlikely that good quality wood and whole trees can be used for energy purposes since their value is much higher in other uses than energy.

In reality, a forest usually comprises stands of different ages that are harvested each year and forest operations are coordinated to maintain or even increase the wood volume in the forest. Carbon losses through harvest in some stands are balanced by carbon gains (forest growth) in other stands, and the fluctuations in carbon stock are even out across the whole forest (IEA, 2018). Sustainable forest management, the avoidance of clear-cut and keeping harvesting below the net annual increment (NAI) are key instruments that allow net sequestration of carbon in forests, while providing wood for the manufacture of products and biomass for energy generation with no reduction in carbon stock. If biomass extraction exceeds NAI and forest rotation length is reduced, the carbon stock of the forest may decrease, and this should be accounted in the GHG emissions of bioenergy. Studies show that unmanaged forests tend to have lower growing stock than managed forests (or forests available for wood supply), due to environmental constraints or due to the fact that non-managed forests are often less productive. Unharvested forests have declining carbon stock over time because the carbon sequestration rate diminishes as forests approach maturity (IEA, 2017c). Increased bioenergy demand may lead to improvements in forest conditions through reforestation and afforestation of degraded lands, improved forest management, control and prevention of natural disturbances, etc. (Agostini et al., 2014; van der Velde et al., 2017).

10.2.3.7 Land use, land-use change and forestry

Carbon emissions associated with LUCs are included in the LULUCF accounting system but not included in the process-related emissions to avoid double counting. Under the IPCC Guidelines for National Greenhouse Gas Inventories (IPCC et al., 2006), for the preparation of national GHG inventories, CO_2 emissions from biomass combustion are not reported in the energy sector (zero rating). This avoids double counting, because it is assumed that these emissions are accounted as part of the emissions from the LULUCF sector. When biomass for bioenergy is traded internationally, a gap in the accounting can occur if the country where the biomass originates does not adequately take account of LULUCF emissions.

The LULUCF sector can contribute to climate change mitigation by reducing emissions and maintaining and enhancing sinks and carbon stocks. The use of biomass for energy generation leads to reductions of emissions under the ETS while this could lead to increased emissions or reduced carbon storage in the LULUCF sector. Carbon emissions from LULUCF are not covered by the EU 2020 GHG emission reduction targets, but complete emission inventories are being established. A recent EU legislative proposal ((COM(2016a,b,c,d) 479) provides the inclusion of GHG emissions and removals from LULUCF in the 2030 climate and energy framework. A no-debit rule requires to ensure that accounted emissions from land use are entirely compensated by an equivalent removal of CO_2 from the atmosphere through other actions in the sector such as afforestation or improving sustainable management of existing forests. Emissions of biomass used for energy will be recorded and counted towards 2030 climate commitments. Hence, after 2020, biogenic emissions from the use of forest biomass for energy will be accounted by Member States in their national LULUCF inventories and towards their 2030 commitments, while supply chain emissions occurring in the EU (cultivation, transport, etc.) will be accounted under the EU ETS and the Effort Sharing sectors.

10.2.4 Support schemes for bioenergy

10.2.4.1 Support schemes for bioenergy

A low-carbon economy with a significant share of energy from renewable sources requires major changes of the energy system. Policy support is needed to contribute to the achievement of the energy and climate targets and to ensure a competitive, sustainable, and secure energy system in a European Energy Union. Energy markets alone cannot deliver the desired level of renewables in the EU since renewables in general are not competitive with fossil fuels if environmental externalities are not considered, requiring high investments and high operation costs (e.g., bioenergy) mostly in relation to feedstock. Support schemes for renewable energy sources are a key mechanism to encourage large scale take-up and deployment of renewable energy generation, and to achieve the renewable energy targets. Support schemes remain necessary to make certain renewable energy technologies more competitive and to promote the investment in renewable energy.

The latest developments in renewables and technology improvements led to a major decrease in the investment costs (mostly for wind and solar photovoltaic) and thus to significant cost reduction of energy delivered. Bioenergy can be competitive in some cases, especially when cheap or even negative cost of biomass feedstock is available (such as waste and residues, landfill gas, and livestock manure). Therefore, support needs to be carefully designed to provide tailored support to the technology, feedstock, and plant size to avoid distortion of the energy market. Some instruments focus on investment and not energy production, such as investment subsidies and fiscal instruments. Other instruments enhance the income of the green energy producer, depending on the energy production, such as feed-in tariffs, as well as the green certificates system in the framework of renewable energy obligation (AEBIOM, 2015; CEER, 2017; RES LEGAL, 2018). Support schemes thus include:

- schemes supporting investments, such as investment grants, investment aids, loans, tax exemptions or reductions, tax refunds, and subsidies;
- schemes supporting energy production or direct price support schemes that include feed-in tariffs, feed-in premium, or renewable energy obligations and green certificates.

The major support schemes for renewables in the EU are:

- **Feed in tariff (FiT):** A price-based policy instrument whereby eligible renewable energy generators receive a fixed payment at a guaranteed level, independent of the market price, for the renewable electricity, heat, and/or biogas produced and fed into the grid.
- **Feed in premium (FiP):** A price-based policy instrument whereby renewable energy sells their renewable energy on the market and receives an additional payment on top of the market price.
- **Quota Obligation (QO):** A QO or minimum share of renewable energy is set on energy suppliers (or consumers and producers). Obliged parties can prove the fulfillment of obligation through Green Certificates (GC), which they can acquire on the market from renewable energy producers.
- **Green certificates (GC):** Power plant operators receive certificates for their green energy. The GC can be traded separately from the energy produced. Green energy producers can sell their GC to energy suppliers obliged to fulfill the quota obligation that provides an additional income on top of the market price of the energy sold.

- **Tendering or auction schemes:** A type of quantity-based policy instrument used to allocate financial support, supplied on a contractual basis, to different renewables technologies following a competitive bidding procedure to determine the support level, such as FiT or FiP.
- **Net metering:** Small-scale energy producers are allowed to consume electricity at a different time from generation. The excess electricity, which is not consumed onsite, is remunerated.
- **Investment grants:** Direct support to investment to stimulate the take-up of less mature technologies and increases the generation of renewable energy.
- **Soft loans:** A loan with an interest rate below the commercial rate and/or a longer repayment period.
- **Tax exemptions:** A taxation policy where renewable energy generators pay lower excise tax rates than conventional energy generators (CEER, 2017; AEBIOM, 2015; RES LEGAL, 2018).

Support schemes in the form of FiT are still the most widespread form of RES support throughout Europe and throughout RES technologies, although the existence of multiple schemes in parallel is a common approach. GC schemes coupled with QO are implemented in fewer countries. Investment grants are provided especially for solar PV, although they are applied to all RES in some cases, and especially for small scale households (solar, heat pumps, or biomass) (Table 10.4). In many countries new support instruments are being introduced for new RES installations while the former ones remain in place for existing operating plants. There is a large variation in the deployment of different renewables in different EU Member States, depending on resources available, technology level, energy markets, energy policies, and support provided. In terms of technology supported, solar PV, wind onshore, hydropower, and bioenergy are the most widely supported RES, while the support of wind offshore, geothermal, and concentrated solar power are less widespread. Technology-specific support schemes are provided, reflecting the differences in technology maturity, investment, and operating costs, as well as plant capacity, since the scale effect is highly important for renewable energy plants (AEBIOM, 2015; CEER, 2017; RES LEGAL, 2018).

Although there is a strong correlation between renewable energy deployment and support provided, there are significant variations between countries in terms of RES electricity support expenditure per unit of electricity produced. Fig. 10.17 shows the levels of support for electricity produced from different renewables. RES support schemes are funded

Table 10.4 Overview of the types of support schemes for bioenergy in the EU

		AT	BE	BG	CY	CZ	DE	DK	EE	EL	ES	FI	FR	HR	HU	IE	IT	LT	LU	LV	MT	NL	PL	PT	RO	SE	SI	SK	UK	CH	NO	IS
Electricity	FiT	*		*	*	*	*			*					*	*			*					*				*		*		
	FiP					*	*	*	*	*		*	*		*		*	*				*										
	GC		*											*									*		*	*						
	Subsidy		*							*		*			*		*	*	*						*		*		*			*
	Investment grants		*									*													*							
	Loans						*							*				*					*				*					
	Net metering		*		*			*		*					*		*			*		*										
	Tax regulation									*			*				*	*				*	*			*		*	*			
	Quota		*																				*		*	*			*		*	
Heat	Subsidy	*	*			*	*		*	*	*	*	*					*	*			*	*		*		*	*				
	Investment grants	*	*										*												*							
	Loans		*	*			*			*	*		*				*	*					*				*		*			
	Tax exemptions		*			*		*		*			*					*		*		*				*						
	Heat bonus/price based											*					*	*											*			
	FiP																					*										
	FiT																												*			
	Other																*															
Transport	Mandatory quota	*	*			*	*	*		*	*	*	*	*	*	*	*		*		*		*	*	*			*	*			
	Tax exemptions	*	*	*		*		*		*		*	*		*			*		*		*		*		*	*	*				
	Subsidy	*							*	*								*									*					
	GHG reduction quota																															

Source: From AEBIOM, 2015. Bioenergy 2015 Support Schemes; RES LEGAL, 2018. RES-Legal Europe Database. Available at: http://www.res-legal.eu/en/compare-support-schemes/ (accessed January 2018).

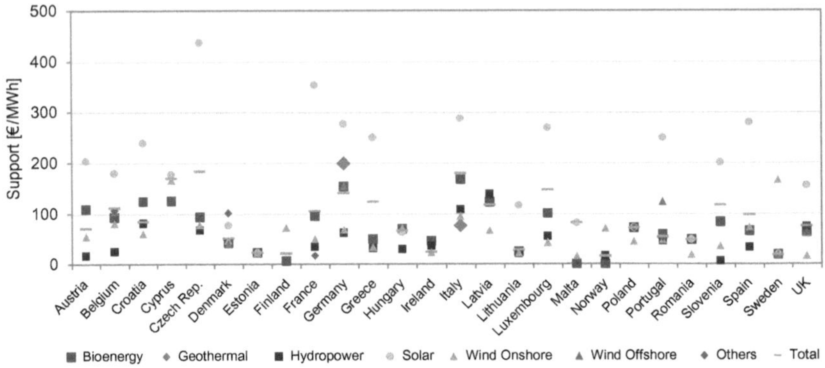

Figure 10.17 Support level for renewables in 2015, by source (€/MWh). *From CEER, 2017. Status Review of Renewable Support Schemes in Europe. Council of European Energy Regulators C16-SDE-56-0.*

either through general taxation or through nontax levies paid through the electricity bill by some or all electricity consumers. Most countries fund their RES support schemes through nontax levies and in some cases where the costs are provided through the state budget (CEER, 2017).

The FiT system proved to be highly efficient in promoting high levels of renewables, such as in Germany and Italy, for example, in PV, but incurring high costs for energy consumers. FiT may be the better choice for less mature and small-scale technologies. However, FiT is less compatible with the principles of liberalized markets, and they are not cost-effective to achieve high RES deployment. If FiT is not set to reflect the actual production costs, they could become high burden on the funding system and lead to unnecessary high energy costs. FiP system offers the advantage that is more market orientated than the FiT. Different FIP options are possible, including a fixed premium, a floating premium, and a premium with cap and floor, adapted to changing market prices to limit the price risks for plant operators. Compared to FITs, investment risks associated with FiP are higher since the renewable electricity has to be sold on the electricity market. Net metering enables the use of electricity produced onsite and thus reducing the load on the electricity network. In this case, electricity consumed onsite may be exempted from levies, charges, and taxes, which are usually part of the retail electricity price (Held et al., 2014; AEBIOM, 2015; RES LEGAL, 2018).

The QO with GC system is compatible with market principles and can contribute to the deployment of the most competitive renewable energy technologies. However, if QOs are designed in a technology-neutral way, only the most cost-effective technologies are supported, excluding certain technologies incurring higher costs. This would result in unbalanced energy systems and the market development detached from available renewable resources. Tendering or auction schemes, being based on a competitive bidding procedure, can increase the cost-effectiveness of renewables support. However, the outcome of auctions depends on the concrete design and the framework conditions: the energy market, available renewable resources, economic perspectives, and the existence of administrative and grid-related barriers. Auctions should therefore be carefully designed to match the existing framework conditions and the renewable energy deployment objectives. Auctions, like QO system, could be technology specific to better exploit the available potentials and develop an adequate energy mix (Held et al., 2014; AEBIOM, 2015; RES LEGAL, 2018).

Investment support are generally used in combination with other measures, such as FiTs or FiPs, more often for renewable heating and cooling projects compared to renewable electricity projects. Investment grants are provided to stimulate the take-up of less mature technologies and increases the generation of renewable energy. Tax incentives or exemptions are often complementary to other types of support programs and include tax incentives relating to investments (income tax deductions for the capital investment in renewable energy projects or accelerated depreciation) and production (income tax deduction for renewable electricity produced thereby reducing operational costs). Low-interest loans or soft loans are loans available at an interest rate below the market rate or longer repayment periods that help reduce investment-related costs. Soft loans have been used in combination with other support schemes, such as investment incentives to support renewable heating projects and combination with FiT systems or QOs for renewable electricity production.

10.2.4.2 Support to fossil and renewable energy

Despite the concerns related to the impact on climate change, fossil fuel energy sector still receives significant subsidies that distort energy markets, promotes inefficient use of energy, and leads to high CO_2 emissions. The value of subsidies in fossil energy consumption worldwide totaled \$325 billion in 2015 and \$260 billion in 2016, decreasing from almost \$600 billion in 2011. The subsidies to renewable energy were well below, with an estimation of \$150 billion, of which 80% are directed to the power sector, 18% to transport, and around 1% to heat. Although subsidies to fossil fuels remain well above those to renewables, the gap has narrowed substantially. The renewables subsidies are projected to increase to a peak of $240 billion in the 2030s and then decline, as it is expected that renewable energy generation becomes competitive by 2040. EU accounts for just over half of the global subsidies to renewables for power (more than $60 billion), followed by the United States ($18 billion), China (almost $17 billion), and Japan ($10.5 billion) (IEA, 2016a; IEA, 2017e).

A study commissioned by the European Commission showed that in 2012, the total value of public interventions in energy in the EU was €122 billion, including support to energy demand, support to energy savings, support to renewable energy, coal, and nuclear (Alberici et al., 2014). Remarkably, most support is given to the renewable energy technologies, particularly solar energy, although significant support is also given to nonrenewable energy (coal and nuclear). Renewable energy received a support of about €15 billion, with solar energy receiving most support (€15 billion), followed by wind (€11 billion), biomass (€8 billion), and hydro (€5 billion). There is significant support to fossil fuels (€16 billion) and nuclear (€6.6 billion). Support to energy savings is reported at a low level around €9 billion in 2012. The largest volumes of support are provided through FiTs (€2012 27 billion), followed by investment grants (€2012 13 billion) and exemptions from energy taxes (€2012 12 billion).

The analysis of cumulative historic interventions over the period 1970–2007 showed that the direct investment support resulted in cumulative interventions equaling almost €200 billion for coal, €100 billion for hydro, and €220 billion for nuclear power plants. The reported cumulative R&D expenditure in the period 1974–2007 was €87 billion for energy supply technologies, of which the nuclear sector has received around 78%, renewable energy received 12%, and fossil fuels received 10%. In addition, other historic support for coal totaled €380 billion in the period 1970–2007 and cumulative interventions totaled about €70–150 billion for renewable energy, with 40% for biomass, 25% for wind, 25% for hydropower, and 10% for solar (Alberici et al., 2014).

10.2.4.3 Latest developments in support polices for renewables

In order to address market distortions that may result from the support granted to renewable energy sources, the European Commission has issued guidance on support schemes to help governments to design and to revise support schemes. The Guidelines on State aid for environmental protection and energy should prepare the ground for achieving the objectives set in the 2030 Framework. Member States have to adapt their national support schemes for renewable energy according to these new guidelines when they set new national schemes or when they extend their existing national schemes (EC, 2014).

The new State Aid framework established by the European Commission encourages a decreasing support to renewables and further market integration. Thus, financial support should be limited to what is necessary to ensure competitiveness and deployment of renewables and avoid excessive burden on the energy market. Support schemes should be flexible and to respond to market developments and decreasing energy production costs. It is expected that in the period between 2020 and 2030, RES technologies will become competitive and, as technologies mature, support should be phased out gradually. Market instruments, such as auctioning or competitive bidding process open to all renewable electricity providers, on the basis of clear, transparent, and nondiscriminatory criteria, should be used from January 1, 2017, to avoid distortion of the competition. For the support for electricity production from RES, FiPs allocated through tenders and based on the technology neutrality principle have to replace FiTs. Given the different stages of technological development of renewable energy technologies, technology-specific tenders could be carried if EU Member States can justify the longer term potential of a new and innovative technology, the need to achieve diversification of energy supply, network constraints, and grid stability.

The aid to renewable energy can be granted as investment or operating aid. As a general rule, it is granted as a premium in addition to the market price with the exemption of RES installations with an installed electricity capacity of less than 500 kW or demonstration projects, and wind plants with an installed electricity capacity of 3 MW. For installations with an installed electricity capacity of less than 1 MWe or demonstration projects, aid may be granted without a competitive bidding process. For energy other than electricity from renewable sources, operating aid can be provided. The aid per unit of energy should not exceed the difference between the total levelized costs of energy (LCOE) of renewable energy produced and the market price of energy. Investment aid is deducted from the total investment amount, if granted, in calculating the LCOE. State aid for energy from renewable sources using waste, including waste heat, as input fuel can make a positive contribution to environmental protection, provided that it is in line with the EU legislation on waste and waste hierarchy. Investment aid in new and existing capacity for food-based biofuel will not be allowed. Investment aid to convert food-based biofuel plants into advanced biofuel plants is allowed. Investment aid to biofuels can only be granted in favor of advanced biofuels. Support cannot be given to biofuels that are subject to a supply or blending obligation, unless a Member State can demonstrate that the aid is limited to sustainable biofuels that are too expensive be commercialized with a supply or blending obligation only (EC, 2014). In the last years, major changes made to national renewable energy support schemes include new support instruments, such as tendering procedures for the determination of support levels (FiT or market premium) in several EU Member States (Germany, Italy, Spain, etc.) and certificates schemes; other countries are likely to follow this trend. FiT schemes often remain in place for smaller installations while FIP become mandatory for new larger installations.

10.2.4.4 Outlook

An appropriate policy framework and strong policy measures are needed to support the expansion of bioenergy and to move toward low-carbon energy system. Targets for emissions reduction, shares of renewable energy, and policies phasing out fossil fuels would provide a favorable investment climate in low-carbon technologies (IEA, 2017b). Appropriate and compensating mechanisms for high investment costs need to promote low-carbon energy production. The first step toward a low-carbon energy system should include the reduction or phasing out the subsidies for the production and use of fossil fuels. Support mechanisms for low-carbon energies should also consider pricing of the environmental externalities caused by fossil fuel use. Carbon pricing mechanisms that reflect the real societal cost of carbon emissions can provide an incentive for bioenergy deployment and accounting for the higher capital and operating costs of bioenergy.

To enable the adequate development of the low-carbon energy sector, support should be differentiated between technologies, feedstock, and plant size, taking into account specific circumstances, while avoiding disproportionate support, but allowing technology deployment. This enables balanced market development and adequate energy mix and ensures the deployment of technologies that are at different level of development (Held et al., 2014). Support schemes need to be predictable, stable, and to provide flexibility to adapt to changing circumstances and technology improvement. Support policies need to target the deployment of renewable electricity and heat and include mandatory targets and quotas for bioenergy while rewarding multiple socioeconomic benefits of bioenergy, in addition to carbon emission reduction.

As the share of renewable energy in the electricity system increases, new challenges arise regarding the integration of a high variable renewable share into the electricity grid. Both bioenergy and hydropower offer good opportunities for accommodating higher integration into the energy systems of variable, intermittent sources of renewable energy (such as wind and solar). Given the current higher generation costs of biomass electricity than variable renewable technologies, technology-specific support needs to take into account the bioenergy ability to offer flexible generation and receive extra credit for its ability for balancing the grid. Technology neutral measures, while useful, are unlikely to promote the deployment of the new technologies that are still in development phase and did not yet reach commercial maturity. For novel, less technically mature technologies (such as advanced biofuels) with significant longer term decarbonization potential, but higher investment risks and costs, a dedicated quota could be considered the boost initial market growth required.

10.2.5 Bioenergy for sustainable development

10.2.5.1 Bioenergy and UN sustainable development goals

Sustainable development is a concept defined as the *development that meets the needs of the present without compromising the ability of future generations to meet their own needs* by the World Commission on Environment and Development (the Brundtland Commission). Sustainable development has become a fundamental, overarching objective of EU policies especially since 1997. The EU has taken the initiative to build a sustainable low-carbon and low-input economy, to increase resource efficiency, to decrease energy consumption, to reverse the loss of biodiversity and natural resources, and to limit climate change. The United Nations Conference on Sustainable Development (Rio + 20) in Rio de Janeiro, Brazil, in 2012 led to the development of a set of sustainable development goals (SDGs), built upon the Millennium Development Goals. The United Nations General Assembly formally adopted in September 2015 the 2030 Agenda for Sustainable Development and the set of 17 SDGs with 169 associated targets (UN, 2018).

Sustainable energy is a key enabler for sustainable development. The new SDGs include a specific goal on energy to e*nsure access to affordable,* *reliable, sustainable, and modern energy for all.* This goal includes three targets and indicators for 2030: (1) ensuring universal access to modern energy services; (2) doubling the share of renewable energy in the global energy mix; and (3) doubling the global rate of improvement in energy efficiency. Although SDGs make no reference to the contribution of biomass for the food, feed, fiber, materials, and energy production, biomass can make a significant contribution to the achievement of the SDGs. Modern bioenergy, as part of the bioeconomy, is expected to increase globally driven by several SDGs and to play an important role in the future sustainable energy supply, fostering sustainable energy for all and climate goals (Fritsche et al., 2018).

Sustainable Development Goals

Goal 1: End poverty in all its forms everywhere

Goal 2: End hunger, achieve food security and improved nutrition, and promote sustainable agriculture

Goal 3: Ensure healthy lives and promote well-being for all at all ages

Goal 4: Ensure inclusive and equitable quality education and promote life-long learning opportunities for all

Goal 5: Achieve gender equality and empower all women and girls

Goal 6: Ensure availability and sustainable management of water and sanitation for all

Goal 7: Ensure access to affordable, reliable, sustainable, and modern energy for all

Goal 8: Promote sustained, inclusive, and sustainable economic growth, full and productive employment, and decent work for all

Goal 9: Build resilient infrastructure, promote inclusive and sustainable industrialization, and foster innovation

Goal 10: Reduce inequality within and among countries

Goal 11: Make cities and human settlements inclusive, safe, resilient, and sustainable

Goal 12: Ensure sustainable consumption and production patterns

Goal 13: Take urgent action to combat climate change and its impacts

Goal 14: Conserve and sustainably use the oceans, seas, and marine resources for sustainable development

Goal 15: Protect, restore, and promote sustainable use of terrestrial ecosystems, sustainably manage forests, combat desertification, and halt and reverse land degradation and halt biodiversity loss

Goal 16: Achieve peaceful and inclusive societies, rule of law, effective and capable institutions

Goal 17: Strengthen means of implementation and revitalize the global partnership for sustainable development

10.2.5.2 Benefits and risks of bioenergy

Modern bioenergy has emerged as result of the concerns related to the scarcity of fossil fuel and to energy security, the negative impacts of the use of fossil fuels, and the climate change. Bioenergy production brings significant opportunities to deliver a number of social, environmental, and economic benefits in addition to the climate and energy goals (IEA, 2016b). Bioenergy provides good opportunities for agricultural markets and has the capacity to promote sustainable development in rural communities. On the other hand, there are environmental, social, and economic concerns about the use of biomass for bioenergy. Bioenergy can have negative impacts if not developed properly. Key concerns are the real GHG emissions from some bioenergy pathways, food security, LUC and biodiversity, and increased competition for raw materials (food, feed, fiber, or materials).

The debate on the sustainability of biofuels, food versus fuel, and LUC often overlooked potential positive effects, such as sustainable rural development. The benefits and the impacts of biofuels or bioenergy production depend strongly on the specific context. Bioenergy synergies with food production, water, ecosystems, health, and welfare can produce multiple benefits, if properly planned and managed. Appropriate environmental and social safeguards need to be put into place to address certain potential negative effects. Bioenergy should be assessed based on its overall performances to provide sustainable energy and deliver wealth to local communities in addition to GHG emission reduction (Osseweijer et al., 2015; Fritsche et al., 2017).

10.2.5.3 Energy supply and energy access

Access to modern energy is as a key driving force for sustainable development. Energy poverty in a wide context is related to access and affordability in the developing world. Currently more than 1.4 billion people have no access to electricity, and the access to electricity of an additional 1 billion people is unreliable. Energy helps to provide clean water, food, sanitation, lighting, transport, and mechanical power. Biomass has been always a source of energy for heating, cooking, and recently, electricity, high temperature heat for industry and transportation fuel. Even now, more than 2.7 billion people, representing one-third of the world population, rely on the traditional use of biomass (firewood, charcoal, agricultural, and forest residues) for heating and cooking today (IEA, 2016c).

Bioenergy provides over half of all the renewable energy worldwide and about 10% of global primary energy supply, out of which traditional bioenergy makes up more than 60% (REN21, 2017; IEA, 2017a). Countries that lack energy access are very often dominated by subsistence agriculture, with poor people living in rural areas in developing countries lacking adequate access to basic infrastructure, clean water, sanitation, and electricity. The lack of access to modern energy services in developing countries results in a high dependency on traditional biomass to meet their energy needs (Diaz-Chavez et al., 2015). Currently, firewood and charcoal supply 80% of sub-Saharan Africa's energy needs. Unsustainable wood harvesting and charcoal production in sub-Saharan Africa, South America, and South Asia is a major driver for forest degradation and deforestation. Traditional biomass use for cooking and heating is inefficient and poses health risks associated with high emissions and indoor air pollution. In many cases, traditional bioenergy relies on the use of various crop residues and dried manure for cooking, in arid and semi-arid areas, that reduces soil fertility (Fritsche et al., 2017). Charcoal produced using inefficient technologies is also responsible for forest degradation, deforestation, and high GHG emissions.

Improved access to reliable and affordable energy, including the use of modern bioenergy, offers opportunities for poverty reduction and rural development, by supporting economic activities and economic growth. Higher income from various economic activities improves access to energy. Local modern bioenergy enhances energy access for energy-deprived and remote and off-grid communities. Local bioenergy production also reduces the need for imported fossil fuels, increases energy security, and provides income-generating opportunities, thus contributing to local and national economies (Diaz-Chavez et al., 2015; Fritsche et al., 2017; IEA, 2016b). The change to modern energy (such as improved cooking stoves) in low-income countries can considerably reduce the firewood and charcoal demand and thus reduce the pressure on forests, which have been harvested at unsustainable rates. In addition, modern bioenergy contributes to the improvement of air quality by reducing indoor and outdoor air pollution. Lack of energy creates difficulties in ensuring food supply, which can be mitigated through increased contribution of modern bioenergy that brings additional benefits for food supply including food industry, storage and refrigeration, and fuel for transport. Consequently, access to affordable and reliable (bio)energy is a precondition for improving food security.

10.2.5.4 Food security

The growth in biofuel production for the use in transport has led to a heated debate on their potential impact, first of all on food security and access to food. The feedstock used for the first-generation biofuel production includes a range of agricultural crops, such as rapeseed, soy, wheat, maize, sugar beet, sugarcane, and palm oil. The most obvious concerns include the risks of increased competition between food and non-food uses of biomass. This competition may put at risk local food supplies in developing countries, food security, and access to land and water resources, while bringing little benefits for local population other than additional income (Osseweijer et al., 2015; FAO, 2017; Fritsche et al., 2017). The food versus fuel controversy has started following the food and oil price increase between 2006 and 2008. It involved scientists, policy-makers, and non-governmental organizations. Increased use of agricultural commodities for food and energy purposes has indeed increased the interdependence between food, feed, and energy markets. However, despite the fact that the increased use of food crops for biofuels might have an impact on food price response, the peak in the food prices appeared to be connected to a combination of factors, including energy prices, increased food demand, unusual weather patterns, and the depreciation of the US dollar (COM(2008) 321 final).

Food security, according to the definition of the FAO, has multiple dimensions: availability, accessibility, stability, and utilization (FAO, 2008). Positive effects of bioenergy production include enhanced economic conditions of rural communities leading to increased energy and food security, food accessibility, and affordability (IEA, 2016b). The causes of food insecurity include poverty and household income, undermined local food production by subsidized imports, reduced water availability, lack of agricultural land and access to land, poor infrastructure, and conflicts. While we should ensure that the poor and vulnerable people are not excluded from the resources they depend, increased income and increased food production, especially in Low Income Countries, reduces food insecurity. The effects of biofuel production on food security differ between rural and urban populations, and between net food producing and importing countries. Thus, while biofuel production can improve the economic conditions of farmers and farmer incomes and hence improve the ability to purchase food, the urban population might suffer from the higher prices of agricultural commodities (Fritsche et al., 2017).

Bioenergy can increase food security through improved farming practices and investments leading to increased crop productivity. New infrastructure built to support the bioenergy can drive food industry development and can offer new job opportunities, thereby increasing overall food availability and accessibility. The use of agricultural and forestry residues as feedstock for bioenergy and the use of marginal, abandoned, or degraded land for feedstock production can minimize food–bioenergy competition (Fritsche et al., 2017; IRENA, 2018). Advanced, lignocellulosic biofuels might also be able to mitigate the competition between food and bioenergy, when using waste and residues (waste oil, crop residues, wood residues, etc.) or energy crops cultivated on land that is not currently used for crops (marginal, degraded land, etc.). However, the main issue related to biofuels and bioenergy appear to be the competition for land and water resources, not the final use of biomass, which still needs to be addressed.

10.2.5.5 Rural development

Bioenergy provides opportunities to promote sustainable agriculture, to improve agricultural practices, and drive sustainable rural development. Agriculture is expected to provide food and feed, but also other products, such as fiber, biomaterials, or bioenergy, while avoiding as much as possible negative impacts on soil, water, and biodiversity. Integration of bioenergy into the rural, agricultural, and forestry landscapes enables better use of available land and water resources while providing multiple benefits to ecosystem services and encourage sustainable land management. Positive effects of bioenergy production include enhanced economic security of rural communities through improved capacity for crop production and increased income (IEA, 2016b). Biomass production for energy stimulates rural development and leads to improvement of supply chain logistics and local infrastructure that are beneficial for food production. Agriculture and food industry are labor-intensive, with many job opportunities along the food supply chain. Bioenergy pathways, from biomass production to end use, provide job opportunities, including skilled labor that can be a driver of food industry development and of general industrial development in rural areas (Nogueira et al., 2015; Osseweijer et al., 2015).

The use of agricultural and forestry residues as sustainable feedstock for the production of bioenergy and bioproducts can provide an additional source of revenue in rural areas. The main concern is to maintain the organic carbon content in soils to ensure their long-term productivity. This requires strict rules on sustainable harvest rates (Monforti et al., 2015; Fritsche et al., 2017). Forests managed for multiple purposes, such as wood products, pulp and paper, and other products, provide biomass feedstock for bioenergy mainly as by-products from processing and residues from forestry operations with certain economic benefits for local communities. Transition from traditional biomass use to modern bioenergy in low-income countries can reduce the time needed to collect firewood, allowing women to dedicate to income-generating activities or giving children more time to study. Improving delivery of affordable and reliable energy services especially to rural communities, bioenergy can also contribute to the development of human and economic capacity to adapt to climate change. Bioenergy can be an integral part of waste management and can offer considerable opportunities for making economic use of biomass previously considered as waste, while mitigating impacts on land, water, and air (IEA, 2017b).

10.2.5.6 Bioenergy integration in agriculture and forestry

Bioenergy is part of a larger bioeconomy, covering agriculture, fisheries, forestry, and the production of food, feed, fiber, pulp and paper, wood, biomaterials, bio-based chemicals, and medicines, with high interactions among these markets. In this context, an increasing demand of biomass for multiple uses is expected. The production of biomass for energy can be achieved by enhancing agricultural or forest production systems while mitigating negative impacts on the landscape and local communities, and avoiding unsustainable land use through adequate land-use planning (IEA, 2016b).

Some approaches to exploit the synergies between food, feed production, and bioenergy include integrating bioenergy production into existing activities and different land uses. Multifunctional land uses provide food and feed for a growing population, as well as biomass for bio-based products and bioenergy (IEA, 2017b; IRENA, 2018). Integrating novel biomass production systems into agriculture and forestry landscapes using crop rotations, multi-cropping, multi-purpose crops, intercropping, and agroforestry approaches can meet the increased biomass demand for multiple uses. Intercropping could be a low land-use intensity option that minimizes competition with food production (Langeveld et al., 2014).

Developments in agriculture can promote the adoption of modern farming techniques, development of good management practices and improve food security and energy access. Sustainable intensification of agriculture can play a particularly important role in meeting the demand for biomass for bioenergy through yield increase, use of agricultural residues, and cultivation of energy crops on marginal or marginal and degraded land (Fritsche et al., 2017; IRENA, 2018).

Agroforestry systems, combining food and energy production, could improve the livelihoods of farmers in developing countries by ensuring local energy supply, providing simultaneously food, feed, fiber, other bio-based materials, and bioenergy feedstock. Integrated production systems can ensure high resource use efficiency and sustainable biomass production and mitigate pressures on current land use (IRENA, 2018). The deployment of modern bioenergy with higher conversion efficiency than traditional bioenergy will decrease the demand for firewood, a key source of deforestation today. This avoids deforestation, with significant impact on local communities, ecosystems, and water resources, especially in developing countries. Forest management and afforestation could increase the availability of woody biomass that has multiple uses: wood material, ecosystem services, and bioenergy. Applying sustainable forest management will enhance the health and productivity of forests and increase wood material availability for multiple uses, while providing the full range of ecosystems services, environmental protection, and social development (Nogueira et al., 2015; IRENA, 2018).

Perennial bioenergy crops, integrated into agricultural landscapes, contribute to rural, social, and economic development by using degraded land and land with low productivity. Perennial crops can enhance degraded ecosystem services, improve biodiversity, and provide opportunities to mitigate environmental impacts and water pollution impacts by acting as acting as vegetation filters and capturing nutrients in runoff from croplands. Perennial crops and vegetation strips can also be placed to reduce the risk of water and wind erosion and provide additional biodiversity benefits. Thus, energy crops, grown on poor soil, low quality land will not only provide biomass without competing for cropland but will also help to rehabilitate and reclaim marginal and degraded land, improving land productivity in the long term for future use. Due consideration must be given however to high biodiversity value land and traditional land use by local communities (Fritsche et al., 2017).

10.2.5.7 Renewable energy integration into energy systems

Sustainable and modern bioenergy is not only relevant for residential uses, small-scale local use in stand-alone applications or mini-grids, providing solutions for rural electrification. Bioenergy also provides solutions for large-scale production of electricity, high-temperature process heat for local industry and biofuels for agriculture and transport. Biomass use for heat and power production can be used as a base-load to provide continuous electricity, as well for load-balancing. Increasing shares of variable renewable in energy grids require changes in infrastructure including energy storage, transmission capacity, frequency, and power flow control systems. Bioenergy can provide flexibility of generation patterns to balance the expansion of intermittent and seasonal wind and solar resources and thus allowing the integration of higher shares of renewable energy into the electricity grid. Biogas from anaerobic digestion can be used in connection with gas storage to provide electricity at the moment when demand is higher or to compensate for the variable renewable energy production. Biogas can also be upgraded at natural gas quality and injected into the natural gas to be used at the location where energy is needed and at the moment where it is needed. Biogas can be upgraded and be used as sustainable transport fuel in natural gas vehicles. Biorefineries allow sustainable processing of biomass into a range of bio-based products and bioenergy in a concept similar to conventional oil refineries, in integrated multi-feedstock, multiproduct, multiprocess systems. Biorefineries thus entail the integration of a number of technologies and combinations of several thermochemical and biochemical processes to produce bio-based materials, biochemicals, and bioenergy.

DISCLAIMER

The views expressed are purely those of the authors and may not in any circumstances be regarded as stating an official position of the European Commission.

REFERENCES

AEBIOM, 2015. Bioenergy 2015 Support Schemes.

Agostini A., Giuntoli J., Boulamanti A., 2014. Carbon Accounting of Forest Bioenergy. Conclusions and Recommendations From a Critical Literature Review. Joint Research Centre (JRC) Report EUR 25354. EN.

Alakangas, E., Heikkinen, A., Lensu, T., Vesterinen, P., 2007. Biomass Fuel Trade in Europe. Summary Report VTTR0350807. Jyväskylä, Finland: VTT, p. 57.
Alberici A., Boeve S., van Breevoort P., Deng Y., Förster S., et al., 2014. Subsidies and Costs of EU Energy. Project number: DESNL14583.
Allen B., Paquel K., Baldock D., Bowyer C., Nanni S., Nesbit M., 2016. Ensuring the Sustainability of Bioenergy and Minimising the Risk of Carbon Debt. Institute for European Environmental Policy (IEEP) Report.
Al-Riffai P., Dimaranan B., Laborde D., 2010. Global Trade and Environmental Impact Study of the EU Biofuels Mandate. Specific Contract No SI2.537.787 Implementing Framework Contract No TRADE/07/A2.
Asikainen, A., Liiri, H., Peltola, S., Karjalainen, T., Laitila, J., 2008. Forest energy potential in Europe (EU27). Working Papers of the Finnish Forest Research Institute. Joensuu: Finnish Forest Research Institute. 69, p. 33.
Batidzirai, B., Smeets, E.M.W., Faaij, A.P.C., 2012. Harmonising bioenergy resource potentials—methodological lessons from review of state of the art bioenergy potential assessments. Renew. Sustain. Energy Rev. 16 (9), 6598–6630.
Bauen, A., Woods, J., Hailes, R., 2004. Bioelectricity Vision: Achieving 15% of Electricity From Biomass in OECD Countries by 2020. E4tech (UK) Ltd.
Bauen A., Chudziak C., Vad K., Watson P., 2010. A Causal Descriptive Approach to Modelling the GHG Emissions Associated With the Indirect Land Use Impacts of Biofuels. Lausanne, Switzerland: E4tech.
BEE, 2010. Status of Biomass Resource Assessments Version 3 Deliverable No: D 3.2.
Beringer, T., Lucht, W., Schaphoff, S., 2011. Bioenergy production potential of global biomass plantations under environmental and agricultural constraints. GCB Bioenergy 3, 299–312.
Biograce, 2018. Available at: http://www.biograce.net/.
Bird N., Cowie A., Cherubini F., Jungmeier G., 2011. Using a Life Cycle Assessment Approach to Estimate the Net Greenhouse Gas Emissions of Bioenergy. IEA Bioenergy: ExCo:2011:03.
CARB, 2009. CARB Staff Report: Proposed Regulation to Implement the Low Carbon Fuel Standard. California Air Resources Board.
CARB, 2014. iLUC Analysis for the Low Carbon Fuel Standard (Update). California Air Resources Board.
Carvalho Macedo I., Nassar A.M., Cowie A., Seabra J.E.A., Marelli L., Otto M., et al., 2015. SCOPE Bioenergy & Sustainability—Greenhouse Gas Emissions from Bioenergy, pp. 583–616.
CEER, 2017. Status Review of Renewable Support Schemes in Europe, Council of European Energy Regulators C16-SDE-56-0.
Cherubini, F., Bird, N.D., Cowie, A., Jungmeier, G., Schlamadinger, B., Woess-Gallasch, S., 2009. Energy- and GHG-based LCA of biofuel and bioenergy systems: Key issues, ranges and recommendations. Resour. Conserv. Recycl. 53, 434–447.
Cherubini, F., Strømman, A.H., Hertwich, E., 2011. Effects of boreal forest management practices on the climate impact of CO2 emissions from bioenergy. Ecol. Modell. 223 (1), 59–66.
COM, 1997. (97) 599 Final. Communication from the Commission: Energy for the Future: Renewable Sources of Energy—White Paper for a Community Strategy and Action Plan, European Commission.
COM, 2002. 27. Life Sciences and Biotechnology—A Strategy for Europe. Communication from the Commission to the European Parliament, the Council, the Economic and Social Committee and the Committee of the Regions.
COM, 2006. 105 Final. Green Paper: A European Strategy for Sustainable, Competitive and Secure Energy.

COM, 2008. 321 Final. Tackling the Challenge of Rising Food Prices. Directions for EU Action. Communication From the Commission to the European Parliament, the Council, the European Economic and Social Committee and the Committee of the Regions.

COM, 2010a. 11 Final. Report from the Commission to the Council and the European Parliament on Sustainability Requirements for the Use of Solid and Gaseous Biomass Sources in Electricity, Heating and Cooling.

COM, 2010b. 2020 Final. EUROPE 2020. A Strategy for Smart, Sustainable and Inclusive Growth. Communication from the Commission.

COM, 2010c. 639 Final. Energy 2020. A Strategy for Competitive, Sustainable and Secure Energy. Communication from the Commission to the European Parliament, the Council, the European Economic and Social Committee and the Committee of the Regions.

COM, 2011a. 885 Final. Energy Roadmap 2050 Communication from the Commission to the European Parliament, the Council, the European Economic and Social Committee and the Committee of the Regions, 2011.

COM, 2011b. 112 Final. A Roadmap for Moving to a Competitive Low Carbon Economy in 2050 Communication from the Commission to the European Parliament, the Council, the European Economic and Social Committee and the Committee of the Regions.

COM, 2012. 60 Final. Innovating for Sustainable Growth: A Bioeconomy for Europe. Communication from the Commission to the European Parliament, the Council, the European Economic and Social Committee and the Committee of the Regions.

COM, 2014a. 15 Final. A Policy Framework for Climate and Energy in the Period From 2020 to 2030. Communication from the Commission to the European Parliament, the Council, the European Economic and Social Committee and the Committee of the Regions.

COM, 2014b. 398 Final. Towards a Circular Economy: A Zero Waste Programme for Europe. Commission to the European Parliament, the Council, the European Economic and Social Committee and the Committee of the Regions.

COM, 2015. 614. Closing the Loop—An EU Action Plan for the Circular Economy. Communication from the Commission to the European Parliament, the Council, the European Economic and Social Committee and the Committee of the Regions.

COM, 2016a. 479. Proposal for a Regulation of the European Parliament and of the Council on the Inclusion of Greenhouse Gas Emissions and Removals From Land Use, Land Use Change and Forestry Into the 2030 Climate and Energy Framework and Amending Regulation No 525/2013 of the European Parliament and the Council on a Mechanism for Monitoring and Reporting Greenhouse Gas Emissions and Other Information Relevant to Climate Change.

COM, 2016b. 482 Final. Proposal for a Regulation of the European Parliament and of the Council on Binding Annual Greenhouse Gas Emission Reductions by Member States from 2021 to 2030 for a Resilient Energy Union and to Meet Commitments Under the Paris Agreement and Amending Regulation No 525/2013 of the European Parliament and the Council on a Mechanism for Monitoring and Reporting Greenhouse Gas Emissions and Other Information Relevant to Climate Change.

COM, 2016c. 767 Final. Proposal for a Directive of the European Parliament and of the Council on the Promotion of the Use of Energy from Renewable Sources (recast).

COM, 2016d. 860 Final. Clean Energy For All Europeans. Communication from the Commission to the European Parliament, the Council, the European Economic and Social Committee, the Committee of the Regions and the European Investment Bank.

COM, 2017a. 34 Final. The Role of Waste-to-Energy in the Circular Economy. Communication from the Commission to the European Parliament, the Council, the European Economic and Social Committee and the Committee of the Regions.

COM, 2017b. 713 Final. The Future of Food and Farming Communication from the Commission to the European Parliament, the Council, the European Economic and Social Committee and the Committee of the Regions.

Commission Decision, 2010. 2010/335/EU of 10 June 2010 on Guidelines for the Calculation of Land Carbon Stocks for the Purpose of Annex V to Directive 2009/28/EC.

CPM, 2018. CPM LCA Database. Available at: http://cpmdatabase.cpm.chalmers.se/AboutDatabase.htm.

de Vries, B.J.M., van Vuuren, D.P., Hoogwijk, M.M., 2007. Renewable energy sources: their global potential for the first-half of the 21st century at a global level: an integrated approach. Energy Policy 35, 2590–2610.

de Wit, M.P., Faaij, A.P.C., 2008. Biomass Resources Potential and Related Costs. Assessment of the EU-27, Switzerland, Norway and Ukraine. Report. Utrecht, The Netherlands: Utrecht University, Copernicus Institute. Available at: www.refuel.eu/fileadmin/refuel/user/docs/REFUEL_D9.pdf.

Diaz-Chavez R., Johnson F.X., Richard T.L., Chanakya H., 2015. SCOPE Bioenergy & Sustainability—Biomass Resources, Energy Access and Poverty Reduction. pp. 711–728.

Directive, 2009a. 2009/30/EC of the European Parliament and of the Council amending Directive 98/70/EC As Regards the Specification of Petrol, Diesel and Gas-Oil and Introducing a Mechanism to Monitor and Reduce Greenhouse Gas Emissions and Amending Council Direc.

Directive, 2009b. 2009/28/EC of the European Parliament and of the Council on the promotion of the Use of Energy From Renewable Sources and Amending and Subsequently Repealing. Directives 2001/77/EC and 2003/30/EC.

Directive, 2015. 2015/1513 of the European Parliament and of the Council Amending Directive 98/70/EC Relating to the Quality of Petrol and Diesel Fuels and Amending Directive 2009/28/EC on the Promotion of the Use of Energy From Renewable Sources.

Directive, 2001. 2001/77/EC of the European Parliament and of the Council of 27 September 2001 on the Promotion of Electricity Produced From Renewable Energy Sources in the Internal Electricity Market.

Directive, 2003. 2003/30/EC of the European Parliament and of the Council of 8 May 2003 on the Promotion of the Use of Biofuels or Other Renewable Fuels for Transport.

Directive, 2008. 2008/98/EC of the European Parliament and of the Council of 19 November 2008 on Waste and Repealing Certain Directives.

Doornbosck Doornbosch and Steenblik, 2007. Biofuels: is the cure worse than the disease? Background paper for the OECD Round Table on Sustainable Development, Paris, France: OECD, September 11–12, 2007. OECD-Dokument SG/SD/RT (2007)3.

E3database, 2018. Available at: http://www.e3database.com/.

E4Tech, 2010. A Causal Descriptive Approach to Modeling the GHG Emissions Associated with the Indirect Land Use Impacts of Biofuels. Final Report. A study for the UK Department for Transport, October 2010.

Elliott, J., Sharma, B., Best, N., Glotter, M., Dunn, J.B., Foster, I., et al., 2014. Spatial modeling framework to evaluate domestic biofuel-induced potential land use changes and emissions. Environ. Sci. Technol. 48 (4), 2488–2496.

Ericsson, K., Nilsson, L.J., 2006. Assessment of the potential biomass supply in Europe using a resource-focused approach. Biomass Bioenergy 30 (1), 1–15.

EC, 2014. Guidelines on State Aid for Environmental Protection and Energy 2014-2020 Communication from the Commission (2014/C 200/01).

EC, 2016. EU Reference Scenario 2016. Energy, Transport and GHG Emissions Trends to 2050. European Commission.

EC, 2018. Voluntary Schemes. European Commission. Available at: https://ec.europa.eu/energy/en/topics/ renewable-energy/biofuels/voluntary-schemes.

Ecoinvent, 2018. The Ecoinvent Database. Available at: https://www.ecoinvent.org/database/database.html.

Edwards R., Mulligan D., Marelli L., 2010. Indirect Land Use Change From Increased Biofuels Demand. Comparison of Models and Results for Marginal Biofuels Production From Different Feedstocks. Joint Research Centre (JRC) Report EUR 24485 EN.

Edwards, R., Padella, M., Giuntoli, J., Koeble, R., O'Connell, A. Bulgheroni, C. et al., 2016. Biofuels Pathways. Input Values and GHG emissions. Database. European Commission, Joint Research Centre (JRC).

EEA, 2006. How Much Bioenergy Can Europe Produce Without Harming the Environment?, European Environment Agency, Report No 7/2006 ISBN 92–9167–849-X2006.

EEA, 2007. Environmentally Compatible Bio-Energy Potential From European Forests.. European Environmental Agency, Copenhagen, p. 54.

Elbersen, B., Startisky, I., Hengeveld, G., Schelhaas, M.-J., Naeff, H., Böttcher, H., 2012. Atlas of EU Biomass Potentials. Deliverable 3.3: Spatially Detailed and Quantified Overview of EU Biomass Potential Taking Into Account the Main Criteria Determining Biomass Availability From Different Sources.

ELCD, 2018. European Reference Life Cycle Database. Available at: http://eplca.jrc.ec.europa.eu/ELCD3/.

Endres, J., Diaz-Chaves, R., Kaffka, S.R., Pelkmans, L., Seabra, J.E.A., & Walter, A., 2015. Sustainability Certification. Bioenergy & Sustainability: Bridging the Gaps. Scientific Committee on Problems of the Environment (SCOPE), Paris Cedex, pp. 660–681.

EPA, 2010. Renewable Fuel Standard Program (RFS2) Regulatory Impact Analysis. Assessment and Standards Division, Office of Transportation and Air Quality, U.S. Environmental Protection Agency.

Erb, K.H., Haberl, H., Kraussmann, F., Lauk, C., Plutzar, C., Steinberger, J.K., et al., 2009. Eating the Planet: Feeding and Fuelling the World Sustainably, Fairly and Humanely—A Scoping Study. Institute of Social Ecology and PIK Potsdam, Vienna, 2009.

ESU Services, 2018. Available at: http://esu-services.ch/software/ (accessed January 2018).

Eurostat, 2018. European Statistics. Available at: http://ec.europa.eu/eurostat.

FAO, 2008. An Introduction to the Basic Concepts of Food Security. Rome: Food and Agriculture Organization of the United Nations. Available at: http://www.fao.org/docrep/013/al936e/al936e00.pdf.

FAO, 2009. Algae-Based Biofuels: A Review of Challenges and Opportunities for Developing Countries. Food and Agriculture Organisation.

FAO, 2016. The State of World Fisheries and Aquaculture 2016. Contributing to Food Security and Nutrition for All. Rome.

FAO, 2017. The future of food and agriculture: trends and challenges.. Food and Agriculture Organization of the United Nations, Rome.

FAO, 2018. Food and Agriculture Organisation. Available at: http://www.fao.org/faostat/en/#home.

Field, C.B., Campbell, J.E., Lobell, D.B., 2008. Biomass energy: the scale of the potential resource. Trends Ecol. Evol. 23, 65–72.

Fischer, G., Schrattenholzer, L., 2001. Global bioenergy potentials through 2050. Biomass Bioenergy 20 (3), 151–159.

Fritsche U., Hennenberg K., Hunecke K., 2010. The iLUC Factor as a Means to Hedge Risks of GHG Emissions From Indirect Land Use Change. Darmstadt: Oeko Institute.
Fritsche, U.R., Berndes, G., Cowie, A., Johnson, F.X., Dale, V., Langeveld, H., et al., 2017. Energy and Land Use—Global Land Outlook Working Paper. Unpublished. Available at: https://doi.org/10.13140/rg.2.2.24905.44648.
Ganko, E., Kunikowski, G., Pisarek, M., Rutkowska Filipczak, M., Gumeniuk, A., Wróbel, A., 2008. Biomass Resources and Potential Assessment. Final Report WP 5.1 of RENEW Project, Contract no SES-CT-2003-502705. Project Final Report. Warsaw: EC BREC/IPiEO Institute for Fuels and Renergable Energy. Available at: www.renew-fuel.com.
GBEP, 2009. The Global Bioenergy Partnership Common Methodological Framework for GHG Lifecycle Analysis of Bioenergy.
GBEP, 2011. The Global Bioenergy Partnership Sustainability Indicators for Bioenergy.
GEA, 2012. Global Energy Assessment—Toward a SustainableFuture, Cambridge University Press, Cambridge, UK, and New York, NY, USA, and the International Institute for Applied Systems Analysis, Laxenburg, Austria.
GEMIS, 2018. GEMIS Database. Available at: http://iinas.org/database.html.
Ghenius, 2018. Available at: https://www.ghgenius.ca/.
Giuntoli J., Agostini A., Edwards R., Marelli L., 2015. Solid and Gaseous Bioenergy Pathways: Input Values and GHG Emissions. Joint Research Centre Report Report EUR 27215 EN.
Gnansonou E., Panichelli L., Dauriat A., Villegas J.D., 2008. Accounting for Indirect Land Use Changes in GHG Balances of Biofuels—Review of Current Approaches. Working Paper REF. 437.101, Ecole Poytechnique Federale de Lausanne, 2008.
GREET, 2018. Available at: https://greet.es.anl.gov/index.php.
Haberl, H., Beringer, T., Bhattacharya, S.C., Erb, K.-H., Hoogwijk, M., 2010. The global technical potential of bio-energy in 2050 considering sustainability constraints. Curr. Opin. Environ. Sustain. 2 (5–6), 394–403.
Held A., Ragwitz M., Gephart M., de Visser E., Klessmann C., 2014. Design Features of Support Schemes for Renewable Electricity. Project number: DESNL13116.
Hertel, T., Golub, A.A., Jones, A.D., O'Hare, M., Plevin, R.J., Kammen, D.M., 2010. Global land use and greenhouse gas emissions impacts of U.S. maize ethanol: estimating market mediated responses. Biosci 60 (3), 223–231. Available from: https://doi.org/10.1525/bio.2010.60.3.8.
Hiederer, R., F. Ramos, C. Capitani, R. Koeble, V. Blujdea, O. Gomez, et al., 2010 Biofuels: A New Methodology to Estimate GHG Emissions From Global Land Use Change. Joint Resrarch Centre Report EUR 24483 EN.
Hoogwijk, M., Faaij, A., van den Broeka, R., Berndes, G., Gielen, D., Turkenburg, W., 2003. Exploration of the ranges of the global potential of biomass for energy. Biomass Bioenergy 25, 119–133.
Hoogwijk, M., Faaij, A., Eickhout, B., 2005. Potential of biomass energy out to 2100, for four IPCC SRES land-use scenarios. Biomass Bioenergy 29, 225–257.
IEA, 2016a. World Energy Outlook.
IEA, 2016b. Mobilising Sustainable Bioenergy Supply Chains: Opportunities for Agriculture. Summary and Conclusions from the IEA Bioenergy ExCo77 Workshop.
IEA, 2016c. Medium-Term Renewable Energy Market Report 2016. Market Analysis and Forecasts to 2021. International Energy Agency.
IEA, 2017a, Energy Technology Perspectives. International Energy Agency.
IEA, 2017b. Technology Roadmap. Delivering Sustainable Bioenergy. International Bioenergy Agency.
IEA, 2017c. World Energy Outlook (WEO). International Energy Agency.

IEA, 2017d. State of Technology Review—Algae Bioenergy. International Energy Agency.
IEA, 2017e, World Energy Outlook.
IEA, January 2018. Is Energy From Woody Biomass Positive for the Climate? International Energy Agency Bioenergy.
IFEU, 2007. Nachwachsende Rohstoffe für die chemische Industrie: Optionen und Potenziale für die Zukunft. Institute for Energy and Environmental Research (IFEU), Heidelberg.
IPCC 2006. In: Eggleston H.S., Buendia L., Miwa K., Ngara T. and Tanabe K., (Eds.), IPCC Guidelines for National Greenhouse Gas Inventories, 2006, The National Greenhouse Gas Inventories Programme, The Intergovernmental Panel on Climate Change, Hayama.
IPCC, 2011. Special Report on Renewable Energy Sources and Climate Change Mitigation, Report of the Intergovernmental Panel on Climate Change(IPCC). Cambridge, United Kingdom and and New York, NY: Cambridge University Press. ISBN 978-92-9169-131-9.
IRENA, 2018. Bioenergy for Sustainable Development. International Renewable Energy Agency (IRENA), IEA Bioenergy—IEA Technology Collaboration Programme on Bioenergy, Food and Agriculture Organization of the United Nations (FAO).
ISO 13065:2015 Sustainability Criteria for Bioenergy.
Jiang, R., Nivrutti Ingle, K., Golberg, A., 2016. Macroalgae (seaweed) for liquid transportation biofuel production: what is next? Algal Res. 14, 48—57.
Judd, S., van den Broeke, L.J.P., Shurair, M., Kuti, Y., Znad, H., 2015. Algal remediation of CO_2 and nutrient discharges: a review. Water Res. 87, 356—366.
Johansson, T.B., Kelly, H., Reddy, A.K.N., Williams, R.H., 1993. A renewables-intensive global energy scenario (RIDGES). In: Johansson, T.B., et al., (Eds.), Renewable Energy: Sources for Fuels and Electricity.. Island Press, Washington, DC.
JRC, 2015. The JRC-EU-TIMES model. Bioenergy potentials for EU and neighbouring countries. Joint Research Centre, EUR 27575 EN 2017.
Junginger, M., et al., 2014. International bioenergy trade: history, status & outlook on securing sustainable bioenergy supply, demand and markets. Lecture Notes in Energy 17.
Kostova B., 2017. Current status of alternative aviation fuels. US Department of Energy. Presented at: DLA Energy Worldwide Energy Conference on April 12, 2017. http://www.wwenergyconference.com/wp-content/uploads/2017/03/FINAL-Alternative-Renewable-Fuels-Panel-Discussion-Technology-Development-and-Certification-Kostova.pdf.
Langeveld, J., et al., 2014. Analysing the effect of biofuel expansion on land use in major producing countries: evidence of increased multiple cropping". Biofuels Bioprod. Biorefining 8, 49—58.
Laborde, D., 2011. Assessing the Land Use Change Consequences of European Biofuel Policies. IFPRI, Washington, DC.
Lamers, P., Junginger, M., Hamelinck, C., Faaij, A., 2012. Developments in international solid biofuel trade—an analysis of volumes, policies, and market factors. Renew. Sustain. Energy Rev. 16 (5), 3176—3199.
LCI, 2018. The US Life Cycle Inventory (LCI) Database. http://www.nrel.gov/lci/.
López Barreiro, D., Prins, W., Ronsse, F., Brilman, W., 2013. Hydrothermal liquefac-tion (HTL) of microalgae for biofuel production: State of the art review and future prospects. Biomass Bioenergy 53, 113—127.
Lysen, E., van Egmond, S., et al., 2008. Biomass assessment—assessment of global biomass potentials and their links to food, water, biodiversity, energy demand and economy—inventory and analysis of existing studies. Netherlands Research Programme on

Scientific Assessment and Policy Analysis for Climate Change, WAB report no. 500102014.
Marelli, L., Ramos F., Hiederer R., Koeble R., 2011. Estimate of GHG emissions from global land use change scenarios. Joint Research Centre (JRC) Report EUR 24817 EN.
Matthews R., Mortimer N., Mackie E., Hatto C., Evans A., Mwabonje O., et al., 2014. Carbon impacts of using biomass in bioenergy and other sectors: forests DECC project TRN 242/08/2011 Report URN 12D/085.
Matzenberger, J., Kranzl, L., Tromborg, E., Junginger, M., Daioglou, V., Goh, C.S., et al., 2015. Future perspectives of international bioenergy trade. Renew. Sustain. Energy Rev. 43, 926–941.
Mehrabadi, A., Craggs, R., Farid, M.M., 2015. Wastewater treatment high rate algal ponds (WWT HRAP) for low-cost biofuel production. Bioresour. Technol. 184, 202–214.
Milledge, J., Heaven, S., 2013. A review of the harvesting of micro-algae for biofuel production. Rev. Environ. Sci. Biotechnol. 12, 165–178.
Milledge, J., Heaven, S., 2014. Methods of energy extraction from microalgal biomass: a review. Rev. Environ. Sci. Biotechnol. 13, 301–320.
Mitchell, S.R., Harmon, M.E., O'Connell, K.E.B., 2012. Carbon debt and carbon sequestration parity in forest bioenergy production. GCB Bioenergy 4 (6), 818–827.
Monforti, F., Lugato, E., Motola, V., Bodis, K., Scarlat, N., Dallemand, J.-F., 2015. Optimal energy use of agricultural crop residues preserving soil organic carbon stocks in Europe. Renew. Sustain. Energy Rev. 44, 519–529.
Moreira, J.R., 2006. Global biomass energy potential. Mitig. Adapt. Strat. Global Change 11, 313–342.
Moreira, M., Nassar, A., Antoniazzi, L., Bachion, L.C., Harfuch, L., 2012. Direct and indirect land use change assessment. In: Poppe, M.K., Cortez, L.A.B. (Eds.), Sustainability of Sugarcane Bioenergy. Center for Strategic Studies and Management (CGEE), Brasília, DF.
Nielsen J.B.H., Oleskowicz-Popiel P., Al Seadi T., 2007. Energy crop potentials for bioenergy in EU-27.
Nogueira, L., Leal, M.R.L.V., Fernandes, E., Chum, H., Díaz-Chavez, R., Endres, J., et al., 2015. SCOPE Bioenergy & Sustainability—Sustainable Development and Innovation. Scientific Committee on Problems of the Environment, Paris, France, pp. 184–217.
OECD, 2017. OECD-FAO Agricultural Outlook 2017-2026. http://stats.oecd.org/index.aspx?DatasetCode = HIGH_AGLINK_2017.
Openlca, 2018. http://www.openlca.org/lca-data/ (accessed January 2018).
Osseweijer, P., Watson, H.K., Johnson, F.X., Batistellad, M., Cortez, L.A.B., Lynd, L.R., et al., 2015. SCOPE Bioenergy & Sustainability—Bioenergy and Food Security. Scientific Committee on Problems of the Environment, Paris, France, pp. 91–136.
Pelkmans, L., et al., 2014. The role of sustainability requirements in international bioenergy markets. In: Junginger, M., Goh, C., Faaij, A. (Eds.), International Bioenergy Trade. Lecture Notes in Energy, 17. Springer, Dordrecht.
Pingoud, K., Ekholm, T., Savolainen, I., 2012. Global warming potential factors and warming payback time as climate indicators of forest biomass use. Mitig Adapt Strateg Glob Change 17, 369.
Popp, J., Lakner, Z., Harangi-Rákos, M., Fári, M., 2014. The effect of bioenergy expansion: food, energy, and environment. Renew. Sustain. Energy Rev. 32, 559–578.
Proskurina, S., Junginger, M., Heinimö, J., Tekinel, B., Vakkilainen, E., 2018. Global biomass trade for energy—Part 2: Production and trade streams of wood pellets, liquid biofuels, charcoal, industrial roundwood and emerging energy biomass. Biofuels, Bioprod. Bioref. Available from: https://doi.org/10.1002/bbb.1858.

Redwood, M.D., Paterson-Beedle, M., Macaskie, L.E., 2009. Integrating dark and light bio-hydrogen production strategies: towards the hydrogen economy. Rev. Environ. Sci. Biotechnol. 8, 149–185.

REN21, 2017. Renewables 2017 Global Status Report.

RES LEGAL, 2018. RES-Legal Europe Database. Available at: http://www.res-legal.eu/en/compare-support-schemes/ (accessed January 2018).

Rocca S., Agostini A., Giuntoli J., Marelli L., 2015. Biofuels from algae: technology options, energy balance and GHG emissions. Insights from a literature review. JRC report.

Rokityanskiy, D., Benítez, P.C., Kraxner, F., McCallum, I., Obersteiner, M., Rametsteiner, E., et al., 2007. Geographically explicit global modelling of land-use change, carbon sequestration, and biomass supply. Technol. Forecast. Soc. Change 74, 1057–1082.

Scarlat, N., Dallemand, J.F., 2011. Recent developments of biofuels/bioenergy sustainability certification: a global overview. Energy Policy 39 (3), 1630–1646.

Scarlat N., Dallemand J.F., 2016. Technology Development Report. Heat and Power from Biomass. Low Carbon Energy Observatory Joint Research Centre Deliverable D 2.1.

Scarlat, N., Martinov, M., Dallemand, J.F., 2010. Assessment of the availability of agricultural crop residues in the European Union: potential and limitations for bioenergy use. Waste Manage. 30 (10), 1889–1897.

Scarlat, N., Dallemand, J.F., Monforti-Ferrario, F., Nita, V., 2015. The role of biomass and bioenergy in a future bioeconomy: policies and facts. Environ. Dev. 15, 3–34.

Scarlat N., Fahl F., Dallemand J.F., 2018a. Status and opportunities for energy recovery from municipal solid waste in Europe. Waste Biomass Valor., 1–20.

Scarlat, N., Dallemand, J.F., Fahl, F., 2018b. Biogas: developments and perspectives in EuropeRenew. Energy 129, 457–472ISSN 0960-1481. Available from: https://doi.org/10.1016/j.renene.2018.03.006.

Searchinger, T., et al., 2008. Use of US croplands for biofuels increases greenhouse gases through emissions from land-use change. Science 319, 1238–1240.

Searchinger, T.D., Hamburg, S.P., Melillo, J., et al., 2009. Fixing a critical climate accounting error. Science 326 (5952), 527–528.

SEC, 2011. 1565 Energy Roadmap 2050 Commission Staff Working Paper. Impact Assessment Accompanying the Document. Communication from the Commission to the European Parliament, the Council, the European Economic and Social Committee and the Committee of the Regions.

Sharmina, M., McGlade, C., Gilbert, P., Larkin, A., 2017. Global energy scenarios and their implications for future shipped trade. Marine Policy 84, 12–21.

Siemons, R. M. Vis, van den Berg, D., Mc Chesney, I., Whiteley, M., Nikolaou, N., 2004. Bioenergy's role in the EU Energy market, a view of developments until 2020.

Sims, R., Hastings, A., Schlamadinger, B., 2006. Energy crops: current status and future prospects. Global Change Biol. 12, 2054–2076.

Slade, R., Saunders, R., Gross, R., Bauen, A., 2011. Energy From Biomass: The Size of the Global Resource. Imperial College Centre for Energy Policy and Technology and UK Energy Research Centre, London.

Smeets, E., Faaij, A., Lewandowski, I., Turkenburg, W., 2007. A bottom-up assessment and review of global bio-energy potentials to 2050. Prog. Energy Combust. Sci. 33, 56–106.

Smith, A., Brown, K., Ogilvie, S., Rushton, K., Bates, J., 2001. Waste Management Options and Climate Change: Final Report. European Commission, DG Environment, Luxembourg.

Strapasson, A., Woods, J., Chum, H., Kalas, N., Shah, N., Rosillo-Calle, F., 2017. On the global limits of bioenergy and land use for climate change mitigation. GCB Bioenergy 9, 1721–1735. Available from: https://doi.org/10.1111/gcbb.12456.

Strengers, B., et al., 2015. Greenhouse Gas Impact of Bioenergy Pathways. PBL Netherlands Environmental Assessment Agency, The Hague.
Thrän, D., Weber, M., Scheuermann, A., Fröhlich, et al., 2006. Sustainable Strategies for Biomass Use in the European Context. Institute for Energy and Environment, Leipzig, p. 387.
Thrän, D., Seidenberger, T., Zeddies, J., Offremann, R., 2010. Global biomass potentials—resources, drivers and scenario results. Energy Sustain. Dev. 14, 200–205.
Thrän, et al., 2017. Global Wood Pellet Industry and Trade Study 2017. IEA Bioenergy Task 40.
Tipper, R., Hutchison, C., Brander, M.A., 2009. Practical Approach for Policies to Address GHG Emissions From Indirect Land Use Change Associated With Biofuels. Ecometrica, Edinburgh, UK.
Tyner, W.E., Taheripour, F., Zhuang, Q., Birur, D., Baldos, U., 2010. Land Use Changes and Consequent CO2 Emissions Due to U.S. Corn Ethanol Production: A Comprehensive Analysis. Purdue University, West Lafayette, IN.
UN, 2018. United Nations Conference on Sustainable Development, Rio + 20. Available at: https://sustainabledevelopment.un.org/rio20.
UNEP, 2015. Global Waste Management Outlook. United Nations Environment Programme.
USDA, 2017. EU Biofuels Annual 2017 Foreign Agricultural Service GAIN Report Number NL7015, 2017.
van der Velde M., et al., 2017. Report on current biomass supply and technical potential with a focus on Europe. Joint Research Centre Deliverable D2.2 - Final version M24.
Vigani, M., Parisi, C., Rodriguez-Cerezo, E., Barbosa, M.J., Sijtsma, L., Ploeg, M., et al., 2015. Food and feed products from microalgae: market opportunities and challenges for the EU. Trends Food Sci. Technol. 42, 81–92.
WBGU, 2009. Future Bioenergy and Sustainable Land Use. German Advisory Council on Global Change (WBGU).
Wicke, B., van der Hilst, F., et al., 2014a. Model collaboration for the improved assessment of biomass supply, demand, and impacts. GCB Bioenergy 7 (3), 422–437.
Wicke, B., Verweij, P., van Meijl, H., van Vuuren, D.P., Faaij, P.C., 2012. Indirect land use change: review of existing models and strategies for mitigation. Biofuels 3 (1), 87–100.
Wolf, J., Bindraban, P.S., Luijten, J.C., Vleeshouwers, L.M., 2003. Exploratory study on the land area required for global food supply and the potential global production of bioenergy. Agric. Syst. 76 (3), 841–861.
Yamamoto, H., Yamaji, K., Fujino, J., 1999. Evaluation of bioenergy resources with a global land use and energy model formulated with SD technique. Appl. Energy 63, 101–113.
Yamamoto, H., Yamaji, K., Fujino, J., 2000. Scenario analysis of bioenergy resources and CO2 emissions with a global land use and energy model. Appl. Energy 66, 325–337.
Yamamoto, H., Fujino, J., Yamaji, K., 2001. Evaluation of bioenergy potential with a multi-regional global-land-use-and-energy model. Biomass Bioenergy 21, 185–203.

FURTHER READING

ECOFYS, Alberici, S., Spöttle, M., Toop, G., 2014. Assessment of sustinability standards for biojet fuel. Final report. https://www.ecofys.com/files/files/ecofys-2015-assessment-of-sustainability-standards-for-biojet-fuel.pdf.
EU Regulation No 525/2013 of the European Parliament and of the Council of 21 May 2013 on a mechanism for monitoring and reporting greenhouse gas emissions and for

reporting other information at national and Union level relevant to climate change and repealing Decision No 280/2004/EC.

Fritsche U.R., Eppler U., Fehrenbach H., Giegrich J., 2018. Linkages between the sustainable development goals (SDGs) and the GBEP sustainability indicators for bioenergy (GSI). Technical Paper for the GBEP Task Force on Sustainability.

ICAO, 2016. On board a sustainable future: 2016 environmental report. International Civil Aviation Organization. http://www.icao.int/environmental-protection/Documents/ICAO%20Environmental%20Report%20 2016.pdf

IRENA, 2017. Biofuels for Aviation: Technology Brief. Abu Dhabi: International Renewable Energy Agency. http://www.irena.org/documentdownloads/publications/irena_biofuels_for_aviation_2017.pdf.

de Jong, S., Antonissen, K., Hoefnagels, R., Lonza, L., Wang, M., Faaij, A., Junginger, M., 2017. Life-cycle analysis of greenhouse gas emissions from renewable jet fuel production. Biotechnol. Biofuels. 10, 64. Available from: https://doi.org/10.1186/s13068-017-0739-7. http://rdcu.be/IZ8l

Staples, M.D., Malina, R., Suresh, P., Hileman, J.I., Barrett, S.R.H., 2018. Aviation CO2 emissions reductions from the use of alternative jet fuels. Energy Policy 114, 342–354. Available from: https://doi.org/10.1016/j.enpol.2017.12.007

INDEX

Note: Page numbers followed by "*f*" and "*t*" refer to figures and tables, respectively.

A

aCBP. *See* Advanced consolidated bioprocessing (aCBP)
Acetobacterium woodii, 138
Acid catalysis, 128–129
Ad Hoc Working Group on the Durban Platform for Enhanced Action (ADP), 345–346
Advanced biofuels. *See* Second-generation (2G), biofuels
Advanced consolidated bioprocessing (aCBP), 411–412, 413*f*
Advanced control systems, 11–12
Advanced conversion technologies, 132–138. *See also* Conversion technologies
  biological processes, 133–135
  combination of thermochemical and biochemical technologies, 138
  thermochemical processes, 135–137
Advanced liquid biofuels, 138
AEDP. *See* Alternative Energy Development Policy (AEDP)
Afforestation, 536
Agenda for Sustainable Development, The (2030), 341–343
Agricultural/agriculture, 42–50, 232, 307, 381–382, 389, 493–494, 534
  bioenergy integration in, 535–536
  biomass, 36–37, 42–43
  by-products, 38
  certification schemes, 499
  energy crops, 42–43
  land, 36–37
  lignocellulosic crops, 48–50
  oil crops, 47–48
  policies, 182
  productivity, 305
  residues, 472–473, 535
  starch rich crops, 45–46
  sugar-producing crops, 44–45
Agro-industrial
  composition of agroindustrial biomass, 157*t*
  residues, 38, 64–70, 65*t*
  wastes, 156–157
Agroforestry, 234, 536
Agronomy, 59
AICD. *See* Country infrastructure diagnostic for Africa (AICD)
Aircraft emissions, 463
Alcohol to jet-SPK/SKA (ATJ-SPK/SKA), 416–417
Algae
  for bioenergy, 81–97
  biofilm PBRs, 89–91
  biomass, 64
  biorefinery, 95–97
  cultivation for bioenergy production, 468
  eukaryotic, 82–85
  green, 93–94
  macroalgae and bioenergy, 92–95
  microalgae cultures, 86–92
  perspectives for bioenergy
    energy options for algae cultivation, 466–467
    outlook, 468–469
  Phyla, 94–95
  prokaryotic, 82
  systems, 464–469
"Algal beds and reefs", 28–29
Algal Turf Scrubber, 90
Alkali metals, 166
Alkaline-catalyzed transesterification, 128
Alternative Energy Development Policy (AEDP), 389–390
Anaerobic digesters, 474
Anaerobic digestion, 75*t*, 77, 80
  of algae, 466–467
  of manure, 474
Annual biomass production efficiency in biosphere, 29*b*

APR. *See* Aqueous phase reforming (APR)
Aquatic biomass, 37
Aqueous phase reforming (APR), 139
Arthrospira (Spirulina) maxima, 86
Ash composition, 35
ASP. *See* US Aquatic Species Program (ASP)
ASTM d7566 standard, 409–410
ATJ-SPK/SKA. *See* Alcohol to jet-SPK/SKA (ATJ-SPK/SKA)
Auction schemes, 522
Austria, biorefineries in, 197–200
Autothermal gasification, conventional, 413–414
Aviation biofuels, perspectives for, 459–464
  market for biofuel use in aviation, 459–462
  outlook, 463–464
  potential for emissions savings in aviation, 462–463

B

B2B communications. *See* Business-to-business communications (B2B communications)
B2DS. *See* Beyond 2 Degree Scenario (B2DS)
Bagasse, 64–66, 124
Barcelona City Council's tender procedure, 335
Barley (*Hordeum vulgare* L.), 118
Base catalysis, 128–129
BCP. *See* Biorefinery Complexity Profile (BCP)
BE. *See* Bioeconomy (BE)
BECCS. *See* Bioenergy with Carbon Capture and Storage (BECCS)
BEFS approach. *See* Bioenergy and Food Security approach (BEFS approach)
β-glucosidases (βGs), 134
Beyond 2 Degree Scenario (B2DS), 448–452
Bio-based economy. *See* Knowledge-based bioeconomy
Bio-based products, 64–66, 322, 332
  B2B marketing, 333
Bio-based resources, 10
"Bio-based" label, 331
Bio-dimethyl-ether (Bio-DME), 135
Bio-oil, 467
Bio-polyester, 95
Bio-solar cell factories (BSCF), 407–409, 407*f*
Biocascading approach, 180
Biochemical process, 466
Biochemical technologies, 74
Biochemical-based technologies, biofuels from, 406–412
  aCBP, 411–412
  BSCF, 407–409
  LCFAc, 411
  LCFAls, 409–410
Biodiesel, 47–48, 128, 198–200, 199*f*, 303, 324–325, 380–381. *See also* Biofuels
  market, 480
  production, 118–121, 128*f*
  trade, 480
  from VOs, 127–129
Biodiversity, 229, 232
  loss, 7, 347–348
Bioeconomy (BE), 3, 5–6, 11, 26, 180–181, 227, 489–498
  cross-cutting nature, 326
  developments, 494–498
  energy and climate change, 490–493
  framework for developing, 325–326
  towards low-carbon economy, 489–490
  and research interests, 243–245, 245*f*
  social issues in, 246*b*
  social methods, 245–247
  sustainability
    food production, GMOs, and links to sustainable development goals, 252–253
    job creation, working conditions, and rural development, 247–252
    social considerations for assessment, 241–245
    society perception, corporate sustainability reporting, and monitoring, 255
    stakeholders, gender, skills, and training, 254–255

transitions in societal perceptions, 242–243
Bioeconomy–circular economy, 12–15
Bioenergy, 3–4, 40, 435–489. *See also* Biofuels
algae for, 81–97
application to bioenergy chains, 282–284
and biodiversity, 232
biomass resources, 435–449
carriers, 42
categories, 42–43
challenges, 348–351
and climate change mitigation, 232–234
considerations for developing countries, 312–316
development opportunities for women, 315–316
improvements in health, education, and communication, 314–315
poverty alleviation, 313–314
traditional energy forms, 312–313
consumer studies, 321–325
observations from bioenergy and biofuel consumer studies, 324–325
observations from wood pellet consumer studies, 323
contribution to rural development
access to modern energy services, 303
energy availability and supply in rural areas, 300–303
environmental benefits, 309–310
experience of "bioenergy villages", 311–312
multifunctionality in integrated production system, 304–305
multiplier effect in local economic development, 305–309
risks and requirements for sustainable and equitable development, 316–317
status of world's rural population, 298–300
crops, 43–44
deployment, 308
environmental impact for full supply chain, 284–286
factors for bioenergy market development, 320–321
B2B marketing of bio-based products, 333
European Commission market pull recommendations, 330–332
framework for developing bioeconomy, 325–326
national bioeconomy strategies, 328–330
policy context in Europe, 326–328
public procurement as market pull measure, 333–336
feedstock production, 260
in future low-carbon energy systems, 450–459
governance, 341–354
innovation
advanced renewable fuels, 418–421
emerging biofuels from biochemical-based technologies, 406–412
emerging biofuels from thermochemical-based technologies, 413–418
future trends on heat, cooling, and power technologies, 421–427
life cycle assessment of GHG emissions, 508–512
markets and trade, 476–477
modern, 4–5, 18–19, 58, 315–316
opportunities, 344–347
perspectives
of algae for, 464–469
for aviation biofuels, 459–464
policies and measures to support sustainable, 489–537
and policy context, 17–19
production, 280–281, 516–518
roadmap, 450
role of trade in bioenergy, 476–489
socioeconomic implications, 259–266
sustainability, 226–229
challenges for bioenergy, 232–234
environmental pillar, 280–296
governance of sustainable bioenergy, 235
IO methodology, 266–276

Bioenergy (*Continued*)
sustainable biomass, 229–232
way forward, 235–236
waste and residues for sustainable, 470–476
Bioenergy and Food Security approach (BEFS approach), 59–60
Bioenergy with Carbon Capture and Storage (BECCS), 233–234, 451–452
Bioethanol, 26, 44, 380–381. *See also* Biofuels
feedstock balance for, 395–396
policy target, 391–394
estimated water requirement for achieving policy mandate, 393*t*
land-GHG emissions nexus of, 394–395
water requirement for cassava and molasses bioethanol production, 392*t*
production, 130, 479
nexus management, 397–398
from straw, 210
from wood, 209
wood chips for, 191
Biofilm PBRs with mobile supports, 91
Biofuels, 26, 67, 146, 180–181, 243, 244*f*, 458
from biochemical-based technologies, 406–412
blends, 349
development, 380–381
Directive 2003/30/EC, 490–491
feedstock production, 514
4G, 138
industry, 380–381
international trade in, 479–480, 480*f*
market for biofuel use in aviation, 459–462
observations from biofuel consumer studies, 324–325
1G, 113–114, 121, 122*f*
policy, land–water–energy nexus, 388–398
feedstock balance for food and bioethanol, 395–396
government policy on biofuels promotion, 389–391
land-GHG emissions nexus of bioethanol policy target, 394–395
land–water–energy nexus of irrigation system, 396–397
nexus management for sustainable bioethanol production, 397–398
about Thailand and agriculture, 389, 390*f*
water scarcity and bioethanol policy target, 391–394
production, 95, 113–114, 503, 514
land–water–energy nexus and sustainability, 381–383, 383*f*
sustainability certification, 499–501
from thermochemical-based technologies, 413–418
3G, 138
Biofuels, conventional, 113–114
biofuels, gaseous, 39
Biogas, 191–192, 193*f*, 231, 417, 537
from anaerobic digestion, 537
production, 52, 79, 310, 473–474
upgrading, 474
Biogreen economy, 247–250
Bioliquids, 39
Biological conversion routes, 158–159
Biological material, 145
Biomass, 16–17, 26, 137, 146, 180, 229, 235, 303, 508, 531–532, 537
agricultural, 36–37, 42–43
algae, 64
aquatic, 37
assessments, 59
availability, 470
categories, 445
chemical composition and characterization
elemental composition, 30–31, 31*f*
inorganic compounds and ash composition, 35
water content and heating value, 34–35
classification, 35–40
according to biomass end use, 39–40
according to chemical composition, 35–36
according to origin, 36–38
combustion

boilers, 169
systems, 159–160
conversion process, 137
cultivation, 232
edible, 26
electricity production, 457–458
feedstocks, 478
in future
EU low-carbon energy system, 457–459
global contribution of bioenergy to energy supply, 453*f*
global low-carbon energy system, 452–454
gaseous, 501–502
novel fuels from HTL, 414–415, 415*f*
partitioning, 59–60
potentials, 442, 470
processing, 283–284
production, 501–502, 534
raw materials, 162
as renewable source of energy
nature and terminology, 25–27
photosynthesis process, 27–29
residual biomass *vs.* biomass by-product *vs.* waste, 37*b*
resources, 40–97, 435–449, 439*t*
agriculture, 42–50
algae for bioenergy, 81–97
assessing biomass potentials, 435–438
by-products, residues, and wastes, 58–81
energy production, 40*f*
European biomass potentials, 445–448, 446*f*
forestry, 50–58
global biomass potentials, 438–445, 440*f*
making use of full biomass potential, 448–449
for solid biofuels, 150–157
secondary, 26
sustainability governance, 235
technologies, 3–4
thermal energy and/or electricity generation costs with, 171–172
traditional, 58
woody, 36
Biomass Energy Europe project, 445
Biomass Future project, 445–448
Biomass power plants, modern, 148
Biomass-to-gas (BtG), 420
combined PtG and, 420
Biomethane, 137, 474
Bioplastic, 191–192
Bioreactors, 91–92
Biorefinery Complexity Profile (BCP), 203, 204*f*
Biorefinery Fact Sheets, 206, 210–216
biorefinery plant, 211–213
value chain sustainability assessment, 214–216
Biorefinery/biorefineries, 180–195, 205*f*, 398, 467, 537. *See also* Energy-driven biorefineries
advanced, 246–247
biomass for, 39–40
classification of systems, 185–189, 186*f*
complexes, 187–189
complexity index, 201–206
conventional, 189
economic and environmental aspects, 206–216
energy-driven, 190–194
integration in existing infrastructure, 196–201
motivations, 183–185
perspectives, 217–220
plant, 211–213, 212*f*, 213*t*
energy balance, 213*f*
mass balance, 213*f*
share of costs, 214*f*
share of revenues, 214*f*
product-driven, 194–195
Biorefining, 412, 462
Biosphere, biomass production efficiency in, 29*b*
BioSTEP project, 329–330
Biotechnology, 187–189
BSCF, 408
Biowaste, 38
Black liquors from pulping kraft process, 414
Blue-green algae. *See* Cyanophyceae
Bottom-up approach, 342, 436–437
Bourgogne Pellets (BP), 270–271, 285*f*

Bourgogne Pellets (BP) (*Continued*)
environmental impact of 1 GJ of miscanthus, 287–291
environmental impact of bioenergy, 284–286
environmental performances and profitability, 291–292
optimizing environmental impact of logistic chains, 286–287
Brazil, sugarcane production in, 44
Brown algae, 93–94
BSCF. *See* Bio-solar cell factories (BSCF)
BtG. *See* Biomass-to-gas (BtG)
Building block approach, 342
Business-to-business communications (B2B communications), 331
marketing of bio-based products, 333
Business-to-consumer communications, 331
Butyribacterium methylotrophicum, 138

C

C-stock. *See* Carbon stock (C-stock)
C5 and C6 sugar, 191, 192*f*, 199*f*
CAEP. *See* Committee on Aviation Environmental Protection (CAEP)
California Low Carbon Fuel Standard (LCFS), 500
Calvin cycle, 27, 86
Camelina, 121
CAP. *See* Common Agricultural Policy (CAP)
Cap-and-trade system, 497
Carbohydrates, 31–33
Carbon
accounting for bioenergy
bioenergy and GHG emissions, 507–508
and carbon debt, 516–519
direct land-use change, 514
indirect land-use change, 514–516
LCA of GHG emissions from bioenergy, 508–512
LUCs and GHG emissions, 512–513
LULUCF, 519–520
carbon-negative bioenergy, 233–234
debt, 516–519
emissions, 519–520
Carbon capture and storage technologies (CCS technologies), 450
delaying, 455
Carbon dioxide (CO_2), 113, 135–136, 160–161, 164, 468–469
emissions, 450
fuels from, 418–419
Carbon footprint (CF), 384, 387–388
assessment, 386
Carbon monoxide (CO), 135–136, 164
Carbon Offsetting and Reduction Scheme for International Aviation (CORSIA), 461
Carbon stock (C-stock), 386
Carrageenans, 93–94
Cascading use
bioenergy and bioeconomy, 16–17
principle, 471
"Case-by-case" analysis, 16–17
Cassava (*Manihot esculenta*), 118
Cassava ethanol, 391
Castor bean (*Ricinus communis* L.), 121
CBP. *See* Consolidated bioprocessing (CBP)
CCS technologies. *See* Carbon capture and storage technologies (CCS technologies)
CDS. *See* Condensed distiller's soluble (CDS)
CE. *See* Circular economy (CE)
Cell factory, 242
Cellobiohydrolases (CHBs), 134
Cellulases, 134
Cellulose, 32–33, 134
Cellulosic biomass, 134
Cellulosic ethanol production, 93–94
Cercozoa, 83*t*
Cereals, 117
crops, 45–46
straw, 130
CF. *See* Carbon footprint (CF)
Charcoal, 52, 151, 483, 531–532
Charophyta, 83*t*
CHBs. *See* Cellobiohydrolases (CHBs)
Chemical oxygen demand (COD), 67
Chi watersheds, 393–394
Chlorella, 91–92

C. pyrenoidosa, 86
C. vulgaris, 86
Chlorophyta, 83*t*
Chlorophyta Phyla, 92
Chlorophytes, 82
CHP. *See* Combined heat and power (CHP)
Circular economy (CE), 3–4, 13–14
package, 494
and waste, 470–472
Climate change, 5–7, 341–354, 490–493
agenda, 342–343
mitigation, 180, 342
carbon-negative bioenergy, 233–234
climate impacts from ILUC, 232–233
Clostridium, 412
C. carboxidivorans, 138
C. lentocellum, 412
C. thermocellum, 412
CNG. *See* Concentrated natural gas (CNG)
Co-combustion, 162–163
boilers, 169
Co-combustion, direct, 163
Co-combustion, indirect, 163
COD. *See* Chemical oxygen demand (COD)
Cogeneration. *See* Combined heat and power (CHP), generation
Colombia's biodiversity strategy as multilevel approach, 351–352
Combined heat and power (CHP), 39, 160, 166–169, 474
generation, 166
technologies, 147
Combustion, 160–162
ashes composition of certain biomass, 161*t*
biomass combustion
boilers, 169
systems, 159–160
FB, 148
Commercial scale energy-driven biorefineries, 189–190
Commercialization, 408–409
Committee on Aviation Environmental Protection (CAEP), 461
Common Agricultural Policy (CAP), 493–494
Communication, improvements in, 314–315
Concentrated natural gas (CNG), 417
Condensed distiller's soluble (CDS), 126–127
"Consequential" approach, 292–293
Consolidated bioprocessing (CBP), 411
Consumer
acceptance, 324
studies, 321
Conversion technologies, 121–129. *See also* Advanced conversion technologies
1G bioethanol, 122–127
biodiesel from VOs, 127–129
Cooling technologies, 421–427
COP13. *See* 13th meeting of the Conference of the Parties (COP13)
COP21. *See* 21st Conference of the Parties (COP21)
Copenhagen conference (2009), 348–349
Corn, 46
Corn stover, 63, 130
Corporate social responsibility, 248*t*
Corporate Sustainability Reporting (CSR), 255
CORSIA. *See* Carbon Offsetting and Reduction Scheme for International Aviation (CORSIA)
Cost–supply analyses, 437
Country infrastructure diagnostic for Africa (AICD), 368
Country-focused studies, 360
CPI. *See* Current Policy Initiatives (CPI)
Crop evapotranspiration (ET_C), 385
Crop growth models, 436–437
Crop residues, 38, 58–63, 444–445, 472
Crop yield (Y), 59–60, 283, 396–397, 436–437, 441–442, 510–511, 516
Crop yield improvement, 397–398
Crop-based biofuel production, 263–264
Crops, conventional, 42–43
Cross-flow gasifier, 165
Crypthecodinium, 91–92
Cryptophyta, 83*t*

CSR. *See* Corporate Sustainability Reporting (CSR)
Cultivation systems, 469
Current Policy Initiatives (CPI), 454
Cyanophyceae, 82

D

Dandelion Rubber and Inulin Valorization and Exploitation for Europe (DRIVE4EU), 195
Dark reactions, 27
DDGS. *See* Dried distiller's grains (DDGS)
Decarbonization scenarios, 454–455, 457
Decoupling, 13–14
Deforestation, 394–395
DEFRA. *See* Department for Environment Food & Rural Affairs (DEFRA)
Demand-driven assessments, 437
Demonstration scale energy-driven biorefineries, 190
Department for Environment Food & Rural Affairs (DEFRA), 37, 50
Di-methyl-ether (DME), 137
Diesel substitute, 409–410
Direct land-use change (dLUC), 233, 500, 514
Direct sugar conversion to hydrocarbons synthesized iso-paraffinic fuel (DSHC-SIP fuel), 409–410, 416–417
Direct sugar to hydrocarbon (DSH), 139
Diversified supply technologies, 455
dLUC. *See* Direct land-use change (dLUC)
DME. *See* Di-methyl-ether (DME)
Dolomite, 166
Dried distiller's grains (DDGS), 126–127
DRIVE4EU. *See* Dandelion Rubber and Inulin Valorization and Exploitation for Europe (DRIVE4EU)
Driver for innovation, 308–309
Drop-in aviation biofuels, 460
Drop-in biofuels, 407
Dry manure, 68
Dry milling process, 126–127
DSH. *See* Direct sugar to hydrocarbon (DSH)
DSHC-SIP fuel. *See* Direct sugar conversion to hydrocarbons synthesized iso-paraffinic fuel (DSHC-SIP fuel)

E

Earth's regenerative capacity, 14
"Eco cars", 335
"Ecolabels", 331
Ecological footprint (EF), 384, 386–387
Economy, 13, 380–381
Edible biomass, 26
Education
  educational and health services, 308
  improvements in, 314–315
EEIO analysis. *See* Environmentally Extended Input–Output analysis (EEIO analysis)
EF. *See* Ecological footprint (EF)
EGs. *See* Endoglucanases (EGs)
EIA. *See* US Energy Information Administration (EIA)
ELCD. *See* European reference Life Cycle Database (ELCD)
Electricity, 302–303
  biorefinery using grass silage and food residues, 191–192
  generation costs, 171–172, 171*t*
  and heat, 186, 191–194, 193*f*, 194*f*
  production, 422
  wood chips for, 191
Electrification of long-haul flights, 415–416
Electrons, 425
"Emergy assessment", 281
Emissions Trading Scheme (ETS), 454
Employment, direct, 306
EMS. *See* Environmental management systems (EMS)
Endoglucanases (EGs), 134
Energy, 382–383, 490–493. *See also* Bioenergy
  access, 531–533
  availability and supply in rural areas, 300–303
    basic needs, 302
    economic activities, 302

limited access to energy, in developing countries, 300–301
balance, 511
from biogas, 474
carriers, 476–477
intermediates or, 40
requiring for different uses, 302–303
conservation, 182
crops, 536
demand, 301–302
density, 478
efficiency, 451–452, 455–456
and material recovery, 181
opportunities for waste and residues for, 470
options for algae cultivation, 466–467
policies, 53
poor, 369–370
poverty, 369–370
production, 521
ratio, 511
recovery from waste, 471–472, 475
Roadmap, 454–455, 490
supply, 309, 531–533
system
models, 437
renewable energy integration into, 537
from waste, 470–471
Energy 2020–A Strategy for Competitive, Sustainable and Secure Energy, 327
Energy-driven biorefineries, 184, 188*f*, 190–194, 196–197, 200*f*
using grass silage and food residues, 191–192
using straw for FT-biofuels and methanol, 192–194
using wood chips for bioethanol, electricity, heat, and phenols, 191
ENplus, 149–150
Entrained flow gasifiers, 136
Environmental
impact of 1 GJ of miscanthus at BP facility gate and scale effect, 287–291, 288*f*
aspects of biorefineries, 206–216
benefits, 309–310
impact of bioenergy for full supply chain, 284–286
goods and services sector, 250
impact of logistic chains, optimizing, 286–287, 286*t*
performances and profitability of alternative logistic options, 291–292
pillar
advanced topics, 292–294
application to bioenergy chains, 282–284
BP, 284–292
methodological background, 280–282
potential, 436
Environmental management systems (EMS), 248*t*
Environmentally Extended Input–Output analysis (EEIO analysis), 270
Enzymatic hydrolysis-based process, 134
Enzymatic paradigms, 412
Equivalence factors, 387
Escherichia coli, 135
ET_C. *See* Crop evapotranspiration (ET_C)
Ethanol, 51, 123
fuel from sugar beet, 116
production, 118, 390–391
ETS. *See* Emissions Trading Scheme (ETS)
EU. *See* European Union (EU)
EU Reference Scenario (REF2016), 458–459
EU-RED. *See* European Union-Renewable Energy Directive (EU-RED)
Eubacterium limosu, 138
Eucalyptus, 49–50
EUEI PDF. *See* European Union Energy Initiative Partnership Dialogue Facility (EUEI PDF)
Euglenophytes, 82
Euglenozoa, 83*t*
Eukaryotic algae, 82–85, 83*t*
Eukaryotic cells, 81–82
European Commission, 325–326, 333–334, 476
Directorate General for Industry, 334–335
market pull recommendations, 330–332

European Industrial Biofuels Initiative, 132–133
European reference Life Cycle Database (ELCD), 509–510
European Renewable Ethanol Association, 118
European Union (EU), 17–18, 43, 118, 127, 445, 482–483, 489, 529
bioeconomy for, 493
bioethanol market, 479
EU-wide public procurement program, 331
European biomass potentials, 445–448, 446*f*
contribution of biomass sources, 447*f*
funded project COSMOS, 251–252
low-carbon energy system
biomass in future, 457–459, 457*f*
future, 454–457
Member States, 471–472
policy context in, 326–328
sustainability scheme, 503
waste policies and targets, 475–476
European Union Energy Initiative Partnership Dialogue Facility (EUEI PDF), 309
European Union-Renewable Energy Directive (EU-RED), 503
Eurostat, 299
Exoglucanases, 134
Extremophiles microorganisms, 408–409

F

"Fa cai", 86
Fachagentur Nachwachsende Rohstoffe (FNR), 311
False flax. *See* Camelina
FAO. *See* Food and Agriculture Organization of United Nations (FAO)
Farms scale, 422–423
domestic micro-CHP technologies, 422*t*
Fatty acids, 128–129
saturated, 33–34
FB combustion. *See* Fluidized bed combustion (FB combustion)
FB gasifiers. *See* Fixed-bed gasifiers (FB gasifiers)
FC. *See* Feature Complexity (FC)
FCI. *See* Feature Complexity Index (FCI)
FCV. *See* Fuel cell vehicles (FCV)
Feature Complexity (FC), 203
Feature Complexity Index (FCI), 203
Federation of Organic Agriculture Movements (FOAM), 11–12
Feed in premium (FiP), 521, 525
Feed in tariff (FiT), 521, 524–525
"Feed", 198–200, 199*f*
Feedstocks, 114–121, 435–436, 488, 533
advanced, 130–132
balance for food and bioethanol, 395–396
diversification, 129
oil crops, 118–121
production, 246, 372
starch-containing crops, 117–118
sugar-containing crops, 115–117
Fermenters, 91–92
Fertilizer, 191–192
Field crops, 58
Financing, 317
FiP. *See* Feed in premium (FiP)
First-generation (1G)
bioethanol, 122–127
ethanol production from starch, 126*f*
sugar and fuel bioethanol production process, 123*f*
biofuels, 113–114, 121, 122*f*
feedstocks, 42–43, 114–121
technologies, 18–19
conversion technologies, 121–129
for liquid biofuels production, 114–129
Fischer–Tropsch synthesis (FT synthesis), 135
Fischer–Tropsch-SPK/SKA (FT-SPK/SKA), 416–417
FiT. *See* Feed in tariff (FiT)
Fixed-bed gasifiers (FB gasifiers), 136, 164–165, 167*t*
Flue gases, 468–469
Fluidized bed combustion (FB combustion), 148

Fluidized bed gasifiers, 136
FNR. *See* Fachagentur Nachwachsende Rohstoffe (FNR)
FOAM. *See* Federation of Organic Agriculture Movements (FOAM)
Food
  feedstock balance for, 395–396
  production, 252–253
  residues, 191–192
  waste, 68–69
Food and Agriculture Organization of United Nations (FAO), 10–11, 37, 115, 117–118, 304, 314, 356
Food security, 229, 533–534
  bioenergy and, 355–378
    growth typologies of African countries, 369*f*
    key household types and linkages, 364*t*
    mapping of countries onto prioritization matrix, 371*t*
    prioritizing bioenergy development pathways and policies, 368–373
    rethinking food security and bioenergy linkages, 361–366, 363*f*
    viable value chains and feedstock choice, 367–368
"Food *vs.* fuel" debate, 357–361
Footprints
  and land–water–energy nexus relationships, 387–388, 388*t*
  water availability, 384
  water scarcity, 384–385, 391
Forest Stewardship Council (FSC), 11–12, 498–499
Forest(ry)
  bioenergy integration in, 535–536
  biomass, 37
  by-products/residues, 38
  carbon stock, 518
  forest and other wooded lands, 50–51
    changes in wood fuel production, 56–58
    wood fuel production, 54–56
    wood fuel types, 51–52
    wood resources in forest and other wooded lands, 52–54
  management, 536
  production, 264–265
  residues, 472–473, 535
Fossil energy, support to, 525–526
Fossil fuels, 113–114, 452, 502
Fourth generation biofuels (4G biofuels), 113–114, 138
FP7. *See* Seventh Framework Programme (FP7)
FQD. *See* Fuel Quality Directive (FQD)
Framing realistic feedstock production systems, 283
FSC. *See* Forest Stewardship Council (FSC)
FT synthesis, 136
FT synthesis. *See* Fischer–Tropsch synthesis (FT synthesis)
FT-biofuels, straw for, 192–194
FT-SPK/SKA. *See* Fischer–Tropsch-SPK/SKA (FT-SPK/SKA)
Fucoidans, 93–94
Fuel, 26
  cells, 424–427
  from CO_2, 418–419
Fuel cell vehicles (FCV), 406, 425
Fuel Quality Directive (FQD), 327, 460–461, 491–492
Fuelwood, 41–42, 52, 482–483, 485–486
  trade, 482–483
Functional unit, 281–282
Furfural, 64–66

G

Gasification, 74, 75*t*, 135, 163–166, 180, 406, 422–423
  co-combustion based on, 163
  commercialization, 423
  reactors, 136
  solar fuels from, 413–414
  syngas composition from oxidants, 164*t*
Gasifier, 163–164
GBEP. *See* Global Bioenergy Partnership (GBEP)
GBEP Sustainability Indictors for Bioenergy (GSI), 6
GC. *See* Green certificates (GC)

GCAM model, 7
GEMIS. *See* Global Emission Model for Integrated Systems (GEMIS)
Gender, skills, and training, 254–255
Genetically Modified Organisms (GMOs), 246–247, 252–253
Geographical boundaries, 389
German Bioeconomy Council, 328
German strategy, 329
GHG. *See* Greenhouse gas (GHG)
GHI. *See* Global Hunger Index (GHI)
Glaucophyta, 83*t*
Global biodiesel production, 480
Global Bioenergy Partnership (GBEP), 8, 149–150, 227, 229, 230*f*, 350, 503–505
  sustainability indicators, 504*t*
Global biomass potentials, 438–445, 440*f*
  contribution of biomass sources to, 443*f*
Global Emission Model for Integrated Systems (GEMIS), 509–510
Global Hunger Index (GHI), 370
Global landscape, 232–233
Global liquid biofuel trade, 485–486
Global solid biomass trade, 480–485
  international trade, major producers and users
    of fuelwood, 482*f*
    of roundwood, 481*f*
    of wood pellets, 484*f*
Global trade
  in bioenergy carriers, 486*t*
  statistics, 478–479
Global warming, 182
Global warming potential (GWP), 386
Global wood chip production, 484–485
GLOBIOM model, 7
Glucose polymer, 32
Glycerin, 198–200, 199*f*
Glycerol, 80
GMOs. *See* Genetically Modified Organisms (GMOs)
GO. *See* Guarantees of Origin (GO)
Gothenburg city, 335
Governance coordination, 348–351
Grain cereals, 118
Grass silage, 191–192
Green algae, 93–94
Green cane management, 131
Green certificates (GC), 521
Green fibers, 191–192, 193*f*
Green juice, 191–192, 193*f*
Green palm, 120
Greenhouse gas (GHG), 4–7, 132–133, 148, 264, 282, 320, 341
  emissions, 182, 229, 383, 437–438, 500, 512*f*, 513*f*
    bioenergy and, 507–508
    carbon accounting and carbon debt, 516–519
    direct land-use change, 514
    from food production, 381–382
    indirect land-use change, 514–516, 517*f*
    LCA, 508–512
    LUCs and, 512–513
    LULUCF, 519–520
Greenhouse gases, regulated emissions, and energy use in transportation (GREET), 500
Gross domestic production, 274–275
Gross heating value. *See* Higher heating value (HHV)
Gross impact study, 265
GSI. *See* GBEP Sustainability Indictors for Bioenergy (GSI)
Guarantees of Origin (GO), 496
GWP. *See* Global warming potential (GWP)

H

Haptophyta, 83*t*
Harvest Index (HI), 59
Health
  impact assessment, 248*t*
  improvements in, 314–315
Heat, 303
  and power biomass for, 39
  and power generation technologies, 158–170
    CHP, 166–169
    co-combustion, 162–163
    combustion, 160–162
    gasification, 163–166
    incineration, 169–170

technologies, 421–427
wood chips for, 191
HEFA. *See* Hydroprocessed esters and fatty acids (HEFA)
Hemicellulose, 32–33, 134
Herbaceous biomass, 36, 152–154
syngas composition from different oxidants, 154*t*
Heterogeneous catalysis, 129
Heterotrophic approaches, 465
Heterotrophs, 464
HHV. *See* Higher heating value (HHV)
HI. *See* Harvest Index (HI)
High energy efficiency, 454
Higher heating value (HHV), 33–35
Horizon 2020 European Commission's research program, 327
Horizontal static biofilm systems, 90
Hot spot database (HSDB), 248*t*, 251*f*
Household-level
production process, 362
savings and investment decisions, 362–363
HSDB. *See* Hot spot database (HSDB)
HTC. *See* Hydrothermic carbonization (HTC)
HTL. *See* Hydrothermal liquefaction (HTL)
Human and animal power, 312
HVO. *See* Hydrogenated vegetable oils (HVO)
HVO biodiesel. *See* Hydrotreated vegetable oil (HVO biodiesel)
Hydrogen (H_2), 135–137, 164, 418–419
"Hydrogen Mobility Europe" project, 427
Hydrogenated vegetable oils (HVO), 113–114, 129, 129*f*
Hydrological boundaries, 389
Hydroprocessed esters and fatty acids (HEFA), 129
Hydrothermal liquefaction (HTL), 96, 414, 466–467
fuels from HTL of biomass, 414–415, 415*f*
Hydrothermic carbonization (HTC), 414
Hydrotreated vegetable oil (HVO biodiesel), 43–44

I

IAMs. *See* Integrated assessment models (IAMs)
ICAO. *See* International Civil Aviation Organization (ICAO)
ICRISAT, 45
IEA. *See* International Energy Agency (IEA)
IFAD. *See* International Fund for Agricultural Development (IFAD)
IFPRI. *See* International Food Policy Research Institute (IFPRI)
IGCC. *See* Integrated gasification combined cycles (IGCC)
IGOs. *See* Intergovernmental organizations (IGOs)
ILO. *See* International Labour Office (ILO)
iLUC. *See* Indirect land use change (iLUC)
Incineration, 74, 75*t*, 169–170
Income, 306–307
iNDCs. *See* Intended Nationally Determined Contributions (iNDCs)
Indirect employment, 306
Indirect land use change (iLUC), 6–7, 132–133, 233, 473–474, 514–516, 517*f*
Industrial infrastructure, existing, 197*f*
Austria, 197–200, 198*f*
methodology, 196–197
Industrial processing techniques, 189
Industrial residues, 473
Industrial roundwood, 480–482
Industry, Research and Energy (ITRE), 496–497
Innovation fund, 497
Inorganic compounds, 35
Input–output analysis (IO analysis), 261, 263, 266–267, 269–270, 273
case study, 270–276
dynamic IO optimization model, 264
methodology, 266–276, 267*t*
table, 267, 267*t*
Insulation material, 191–192
Integrated approaches, 437–438
Integrated assessment models (IAMs), 437–438

Integrated gasification combined cycles (IGCC), 423
Integrated production systems, 536
Intended Nationally Determined Contributions (iNDCs), 341
Intercropping, 535–536
Intergovernmental organizations (IGOs), 341
Intergovernmental Panel on Climate Change (IPCC), 514
Interindustry flows, 267
Intermittent renewable electricity (IRE), 424
International bioenergy
  international agenda on bioenergy, 347–348
  market, 487–488
  trade, 477–479
International Civil Aviation Organization (ICAO), 461
International community, 341
International Energy Agency (IEA), 4–5, 10–11, 77–79, 87, 113, 300–301, 305–306, 341, 344–346, 350, 450
  Bioenergy Roadmap data, 459
  Bioenergy Task 37, 311
  Bioenergy Task 42, 185
  modeling, 452–453
International Food Policy Research Institute (IFPRI), 358
International Fund for Agricultural Development (IFAD), 300
International Labour Office (ILO), 309, 317
International Organization for Standardization (ISO), 149–150, 384
  ISO 13065 standard, 505
  ISO-WF, 384
International Renewable Energy Agency (IRENA), 10–11, 148, 305–306, 341, 345
International standards, 483–484
International Sustainability & Carbon Certification (ISCC), 500–501
International trade in biofuels, 479–480, 480*f*
Inulin, 32
Investment grants, 522
IO analysis. *See* Input–output analysis (IO analysis)
IPCC. *See* Intergovernmental Panel on Climate Change (IPCC)
IRE. *See* Intermittent renewable electricity (IRE)
IRENA. *See* International Renewable Energy Agency (IRENA)
Irrigation system
  EF, water scarcity footprint, and CF of sugarcane production, 397*t*
  for improving biofuel crop productivity, 396–397
Irrigation water, 396–397
ISCC. *See* International Sustainability & Carbon Certification (ISCC)
ISO. *See* International Organization for Standardization (ISO)
ITRE. *See* Industry, Research and Energy (ITRE)

J

Jatropha curcas, 367
Jerusalem artichoke (JA), 46
Jet biofuels, 416–417
Jet fuels, 416–417
Jobs, 305–306
Joint Research Centre (JRC), 448

K

Knowledge-based bioeconomy, 242–243, 326, 493
Kyoto Protocol, 342, 344

L

Labeling, 331, 507
Laminar PBR, 90–91
Land, 234
  degradation, 381
  direct land occupation, 387
  land-based systems, 469
  rehabilitation, 234
  resources, 438
Land use, land-use change and forestry (LULUCF), 498, 506, 519–520

Land-GHG emissions nexus
of bioethanol policy target, 394–395
net GHG emissions estimation and average GHG emission factors, 395*t*
Land-use change (LUC), 232–233, 381, 449, 512–513, 516, 531
Landfill, 77
Land–water–energy nexus of biofuels development
of biofuels policy mandate in Thailand, 388–398
emerging economies and biofuels development, 380–381
life cycle–based indicators for, 384–388
relationships of footprints and, 387–388, 388*t*
and sustainability of food and biofuels production, 381–383, 383*f*
Large-scale cultivation systems, 468–469
LCA. *See* Life cycle assessment (LCA)
LCC. *See* Life Cycle Costing (LCC)
LCFAc. *See* Long-chain fatty acids (LCFAc)
LCFAls. *See* Long-chain fatty alcohols (LCFAls)
LCFS. *See* California Low Carbon Fuel Standard (LCFS)
LCI. *See* US Life Cycle Inventory Database (LCI)
LCOE. *See* Levelized costs of energy (LCOE)
LCSA. *See* Life Cycle Sustainability Assessment (LCSA)
Lead Market Initiative, 332
Lead Market Initiative Advisory Group for Bio-based Products, 334–335
Levelized costs of energy (LCOE), 527–528
LHV. *See* Low heating value (LHV)
Life cycle analysis. *See* Life cycle assessment (LCA)
Life cycle assessment (LCA), 7, 182, 270, 281, 384, 509
of GHG emissions, 508–512
software packages, 509–510
Life Cycle Costing (LCC), 182
Life Cycle Sustainability Assessment (LCSA), 182
Life cycle–based indicators, 384–388
CF assessment, 386
EF, 386–387
relationships of footprints and land–water–energy nexus, 387–388, 388*t*
WF assessment, 384–385
Light reactions, 27
Lignins ($C_n(H_2O)_m$), 32–33, 134, 191, 414
Lignocellulosic
biofuels, 534
biomass, 36, 130, 132, 134
crops, 48–50
feedstocks, 61, 132–133
non-food biomass, 7
Lipids, 33–34, 95
Liquefied natural gas (LNG), 417
Liquid biofuels, 305–306. *See also* Biofuels
challenges, opportunities, and barriers, 139–140
first-generation technologies for production, 114–129
markets, 479
production, 52
prospective technologies, 138–139
second-generation technologies for production, 130–138
Liquid biofuels, conventional, 62–63
LNG. *See* Liquefied natural gas (LNG)
Logistics for Energy Crops biomass (LogistEC biomass), 271
Long-chain fatty acids (LCFAc), 411
Long-chain fatty alcohols (LCFAls), 409–410
biochemical and catalytic conversions, 409*f*
hydrocarbon production, 410*f*
Long-haul flights, electrification of, 415–416
Low heating value (LHV), 34–35
Low-carbon
bioenergy in future low-carbon energy systems, 450–452, 454–457
biomass in, 452–454, 457–459, 457*f*
cumulative global CO_2 reductions, 451*f*

Low-carbon (*Continued*)
economy, 489–490, 495
energy production, 528
LUC. *See* Land-use change (LUC)
LULUCF. *See* Land use, land-use change and forestry (LULUCF)

M

Macroalgae, 81–82, 464–465. *See also* Algae; Microalgae
and bioenergy, 92–95
marine, 92, 464–465
Macrocystis, 81–82
Maize, 117–118
Malnutrition, 315
Manual labor, 312
Manure, 474
Maritime biofuels
next generation of aviation and, 415–418
technological biochemical platforms, 416*f*, 417*f*
Market equilibrium models, 515
Market instruments, 527
Market Readiness Level, 202
Mass media communications, 322
MCFC. *See* Molten carbonated fuel cell (MCFC)
Mediterranean regions, 61–62
Methane (CH_4), 77, 164
Methanol
production, 137
straw for, 192–194
Methyl tertiary butyl ether (MTBE), 127
MFC. *See* Microbial fuel cells (MFC)
Microalgae, 81–82, 138–139, 464–466. *See also* Macroalgae
algae utilization, 86–87
cultivation systems, 88*f*
production systems, 87–92
algae biofilm PBRs, 89–91
fermenters, 91–92
open systems, 87–89
suspended microalgae PBRs, 89
Microalgae, harvesting, 465–466
Microbial fuel cells (MFC), 426–427, 426*f*
Microbial oils from yeast/bacterial cultivation, 418
Mild gasification process, 415
Milling process, 123–124
Minerals, 66
Miscanthus (*Miscanthus sinensis*), 7, 49, 132
Mixotrophic conditions, 465
Mixotrophy, 91–92, 464–465
mixotrophic algal growth, 465
Modernization fund, 497
Moisture content of biomass, 34
Molasses, 66
ethanol, 391
Molten carbonated fuel cell (MCFC), 425
Moorella, 138
MRIO. *See* Multiregional IO (MRIO)
MSW. *See* Municipal solid waste (MSW)
MTBE. *See* Methyl tertiary butyl ether (MTBE)
Multifunctionality of bioenergy, 304–305
Multiplier effect of bioenergy, 305–309
driver for innovation, 308–309
economic activity growth and business diversification, 307–308
income, 306–307
jobs, 305–306
reduction of rural exodus, 309
Multiregional IO (MRIO), 269
Multistage cascade, 16–17
Mun watersheds, 393–394
Municipal solid waste (MSW), 37, 70–79, 130, 146, 155, 416–417, 420–421, 461–462
approximate net calorific values of MSW fractions, 77*t*
calorific content of materials, 156*t*
characteristics of waste management strategies, 75*t*
composition of global MSW and global waste generation, 73*f*
energy recovery from MSW and industrial waste, 78*f*
generation, 154
incineration process, 170*t*
types and sources of MSW fractions, 72*t*
Municipal waste, 154–156
Municipal wastewater, 80
Myzozoa, 83*t*

N

NAI. *See* Net annual increment (NAI)
National bioeconomy strategies, 328–330
National bioenergy strategy, 349
National peculiarities, 299
National Renewable Energy Action Plan (NREAP), 62
National Renewable Energy Laboratory (NREL), 86–87
Nationally Determined Contributions (NDCs), 348–351, 450
Natural degradation of manure, 474
Natural factors, 52
Natural gas (NG), 417
Natural resources, 9
Natural rubber (NR), 194, 196*f*
NDCs. *See* Nationally Determined Contributions (NDCs)
Negative emission technologies (NETs), 233–234
Nelson's complexity index, 202
Net annual increment (NAI), 519
Net feedstock balance, 395–396, 396*t*
Net heating value. *See* Low heating value (LHV)
Net impact study, 265
Net metering, 522
Net primary production (NPP), 28–29
NETs. *See* Negative emission technologies (NETs)
Neurospora crassa, 412
Nexus assessment, 381
Nexus bioenergy–bioeconomy, 4–6
  BE and sustainable development, 6–12
  bioeconomy–circular economy, 12–15
  bioenergy and policy context, 17–19
  cascading uses, 16–17
Nexus management for sustainable bioethanol production, 397–398
"Nexus", 381
NG. *See* Natural gas (NG)
Nickel (Ni), 166
Nitrogen oxides (NOx), 33, 162
Nitrous oxide (N_2O), 283
Non-reactant alcohol, 128–129
Nonconventional unit "m3 H2Oeq", 385
Nonedible biomass, 26
Norwegian adopters of wood pellet heating, 323
Nostoc flagelliforme, 86
NPP. *See* Net primary production (NPP)
NR. *See* Natural rubber (NR)
NREAP. *See* National Renewable Energy Action Plan (NREAP)
NREL. *See* National Renewable Energy Laboratory (NREL)
Nucleic acids, 34

O

O&M costs. *See* Operation and maintenance costs (O&M costs)
Oat (*Avena sativa* L.), 118
Ochrophyta, 83*t*
OECD. *See* Organisation for Economic Co-operation and Development (OECD)
Oil
  crops, 47–48, 118–121
  extraction process, 466
  oil-containing crops, 114–121
  oil-rich biomasses, 36
Oil crisis (1973), 86–87
Oleaginous yeasts, 411
Olive oil industry, 66
Olive pomace, 66
Open Raceway Ponds (ORP), 465–466
Open systems, 87–89
Open-Bio Project, 333, 335
Operation and maintenance costs (O&M costs), 172, 172*t*
ORC. *See* Organic Ranking Cycle (ORC)
Organic matter, 31–34
  carbohydrates, 31–33
  lipids, 33–34
  nucleic acids, 34
  proteins, 33
Organic Ranking Cycle (ORC), 169
Organic wastes from animal production, 310
Organisation for Economic Co-operation and Development (OECD), 5–6, 243–245, 269, 299, 304, 341, 380
ORP. *See* Open Raceway Ponds (ORP)
Oxygen gasification, 192–194
Ozone depletion, 288–289

P

Palm (*Elaeis guineensis*), 120
  oil, 47, 120, 478–479
Parallel co-combustion, 163
Paris Agreement, 6, 341–342, 346, 349–350, 494–495
Partial equilibrium models (PE models), 448, 515
PBR. *See* Photobioreactors (PBR)
PE models. *See* Partial equilibrium models (PE models)
Pectin, 32–33
PEF. *See* Polyethylene furanoate (PEF)
PEFC. *See* Program for Endorsement of Forest Certification (PEFC)
Pelletization, 488
Pellets, 151. *See also* Bourgogne Pellets (BP); Wood pellets
  heating system investment cost, 323
  for heat production, 303
PEMFC. *See* Proton exchange membrane fuel cell (PEMFC)
Peptostreptococcus productus, 138
Perennial bioenergy crops, 536
Perennial biomass crops, 284–292
Perennial crops, 449
Perennial grasses, 49
Permanent crops, 58
Personal channels, 322
PET. *See* Polyethylene terephthalate (PET)
Phenols, wood chips for, 191
Phospholipids, 33–34
Phosphorus emissions, 287–288
Photoautotrophs, 464
Photobioreactors (PBR), 86, 408, 465
Photosynthesis, 27–29
  NPP of continental and marine ecosystems, 29*f*
  stages, 28*f*
Phototrophic approaches, 465
Photovoltaics (PV), 408
Phycobilins, 82
Phycocolloids, 93
Phyla, 82, 83*t*, 94–95
Physical–chemical conversion processes, 158–159
Physicochemical technologies, 74
Pidgeon process, 166
Pigments, 95
Plan de programmation pluriannuelle de lEnergie (PPE), 271–272
Plant biomass, 30–31
  elemental composition, 31*f*
  protein-rich biomasses from, 36
Plantae, 82
Policy context in Europe, 326–328
Pollutant effluents, 310
Polyethylene furanoate (PEF), 195
Polyethylene terephthalate (PET), 195
Polyphenols, 94
Polysaccharides, 32, 95
Populus, 49–50, 132
Poverty alleviation, 313–314
Power
  biomass and, 39
  generation technologies, 158–170
  technologies, 421–427
    farms and village scale, 422–423
    fuel cells, 424–427
    smart urban cities, 423–424
Power-to-gas (PtG), 417, 419, 419*f*
  combined BtG and, 420
Power-to-liquids (PtL), 419, 419*f*
PPE. *See* Plan de programmation pluriannuelle de lEnergie (PPE)
Private sector representatives, 11–12
Private vehicles, 301
Prochlorophyceae, 82
Product-driven biorefineries, 184–185, 194–195, 195*f*
Program for Endorsement of Forest Certification (PEFC), 498–499
Prokaryotic algae, 82
Prokaryotic cells, 81–82
Proteins, 33
  protein-rich biomasses, 36
Proton exchange membrane fuel cell (PEMFC), 425
Protons, 425
PtG. *See* Power-to-gas (PtG)
PtL. *See* Power-to-liquids (PtL)
Public procurement, 334–335
  as market pull measure for bioenergy, 333–336

Pulping kraft process, black liquors from, 414
PV. *See* Photovoltaics (PV)
Pyrolysis, 74, 75*t*, 137, 406, 467, 488
  oil, 137, 192–194, 194*f*, 488
  solar fuels obtained from gasification and, 413–414

Q

Quota Obligation (QO), 521

R

R1 criterion. *See* Recovery (R1)
R1 formula. *See* Recovery (R1)
Rape (*Brassica napus* L.), 119
Rapeseed, 47–48
RE. *See* Renewable energy (RE)
Recovered wood fuels, 51
Recovery (R1), 476
RED. *See* Renewable Electricity Directive (RED); Renewable Energy Directive (RED)
Red algae, 93–94
REDP. *See* Renewable development plan (REDP)
Reed canary grass (*Phalaris arundinaceae*), 132
Reference Technology Scenario (RTS), 450, 452
Regulatory and public procurement measures, 331
REN21. *See* Renewable Energy Policy Network for 21st Century (REN21)
Renewable bio-based resources, 180
Renewable development plan (REDP), 389–390
Renewable diesel, 129
Renewable Electricity Directive (RED), 490–492
Renewable energy (RE), 264
  integration into energy systems, 537
  support to fossil and, 525–526
  technologies, 450
Renewable Energy Directive (RED), 183, 327, 460–461
Renewable Energy Policy Network for 21st Century (REN21), 148–149
Renewable energy sources (RES), 418, 455–456, 455*f*, 456*f*, 458
Renewable fuel, 500
  advanced renewable fuel, 418–421
    from CO_2 and RES–producing power, 418–419
    combined BtG and PtG, 420
    waste-to-energy, 420–421
Renewable Fuel Act, 63
Renewable Fuel Standard (RFS), 461
  RES–producing power, 418–419
Renewable Identification Numbers (RINs), 461
Renewables, 526–528
Renewables Fuels Association, 127
RES. *See* Renewable energy sources (RES)
Residual biomass, 67
Residue to product ratio (RPR), 59–60
Residue yield (RY), 59
Residues
  agricultural, 472–473, 535
  agro-industrial, 38, 64–70, 65*t*
  crop, 38, 58–63, 444–445, 472
  food, 191–192
  forest, 472–473, 535
  industrial, 473
  primary, 472–473
  secondary, 472–473
  for sustainable bioenergy, 470–476
  wood, 482
"Resource Efficient Europe" initiative, 492–493
Resource-efficient use of biomass, 471
Resource-focused assessments, 436–437
Rest of world (RoW), 269
RFS. *See* Renewable Fuel Standard (RFS)
Rhodophyta, 83*t*
Rhodosporidium sp., 411
Rice straw, 130–131
RINs. *See* Renewable Identification Numbers (RINs)
Roundtable for Sustainable Palm Oil (RSPO), 11–12
Roundtable on Sustainable Biomaterials (RSB), 149–150

RoW. *See* Rest of world (RoW)
RPR. *See* Residue to product ratio (RPR)
RSB. *See* Roundtable on Sustainable Biomaterials (RSB)
RSPO. *See* Roundtable for Sustainable Palm Oil (RSPO)
RTS. *See* Reference Technology Scenario (RTS)
Rural development, 534–535
Rural dwellers, 301
Rural exodus reduction, 309
Rural tourism, 308

S

Saccharina (Laminaria) lattissima, 94
Saccharomyces cerevisiae, 409–410, 412
Salix, 49–50
SARPs. *See* Standard and Recommended Practices (SARPs)
Saturated fatty acids, 33–34
Savings, direct, 307
SBD. *See* Sustainable Biomass Program (SBD)
SCWG. *See* Supercritical water/steam gasification (SCWG)
SDGs. *See* Sustainable development goals (SDGs)
Seaweeds, 464–465
Second-generation (2G)
  biofuels, 62–63, 113–114, 407
  technologies, 18–19
    for liquid biofuels production, 130–138
SERI. *See* Solar Energy Research Institute (SERI)
SET-Plan. *See* Strategic Energy Technology Plan (SET-Plan)
Seventh Framework Programme (FP7), 271
Short rotation coppice (SRC), 37
SIA. *See* Social impact assessment (SIA)
Side Events, 345–346
SLCA. *See* Social life cycle assessment (SLCA)
Sludge, 80
Small to medium enterprise (SME), 334
Small-scale digesters, 303
Smart urban cities, 423–424
SME. *See* Small to medium enterprise (SME)
SOC. *See* Soil organic carbon (SOC)
Social communication processes, 322
Social impact assessment (SIA), 247, 248*t*
Social life cycle assessment (SLCA), 182, 248*t*
Social methods in bioeconomy, 245–247
  examples of assessment tools and methodologies, 248*t*
  social issues in bioeconomy, 246*b*
Society perception, 255
  bioeconomy transitions in, 242–243
Socioeconomic
  implications of bioenergy, 259–266
  issues, 241–242
SOFC. *See* Solid oxide fuel cell (SOFC)
Soft loans, 522
Soil conditions, 52
Soil organic carbon (SOC), 60
Solar Energy Research Institute (SERI), 86–87
Solar fuels, 418–419
  obtained from gasification and pyrolysis, 413–414
Solar-driven allothermal gasification process, 413–414
Solid bioenergy carriers, 42–43
Solid biofuels. *See also* Biofuels
  biomass for, 150–157
    agro-industrial wastes, 156–157
    herbaceous biomass, 152–154
    municipal waste, 154–156
    wood biomass, 150–152
  trade, 485–486
Solid biomass, 39, 145, 501–502. *See also* Biomass
  consumption, 157–158
  heat and power generation technologies, 158–170
  for solid biofuels, 150–157
  thermal energy and/or electricity generation costs, 171–172
Solid oxide fuel cell (SOFC), 425
Solid residues from anaerobic digestion, 474

Solid waste management, 475
Soybean (*Glycine max* (L.)), 119
  oil, 48, 478–479
Spatially explicit methods, 436–437
"Spillovers", 360
SRC. *See* Short rotation coppice (SRC)
Stakeholders, 254–255
Standard and Recommended Practices (SARPs), 461
Standardization, 488–489
Starch, 32, 114–121
  ethanol from, 125
  rich crops, 45–46
  starch-containing crops, 117–118
  starch-rich biomasses, 36
State Aid framework, 527
Statistical analyses of bioenergy potentials, 436–437
Stirling engines, 169
Stoichiometric ratio, 160–161
Straight vegetable oils (SVO), 418
Strategic Energy Technology Plan (SET-Plan), 492–493
Straw, 131, 153
  bioethanol from, 210
  for FT-biofuels and methanol, 192–194
Structural polysaccharides, 32, 33*f*
Sucrose, 122–123
Sugar
  conventional, 114–121
  ethanol from, 122–123
  sugar-containing crops, 115–117
Sugar beet (*Beta vulgaris* L.), 115–116, 125
Sugar-rich biomasses, 36
Sugarcane (*Saccharum officinarum* L.), 44, 115, 116*f*
Sunflower (*Helianthus annuus* L.), 120
Supercritical water/steam gasification (SCWG), 414
Suspended microalgae PBRs, 89
Sustainability, 6, 43–44, 470
  of bioeconomy, 241–258
  of bioenergy, 505–507
  certification and standards, 489, 498–507
    biofuels sustainability certification, 499–501
    solid and gaseous biomass, 501–502
  challenges for bioenergy, 232–234
    and biodiversity, 232
    and climate change mitigation, 232–234
    land, and level of consumption, 234
  of crop residues, 60
  of food and biofuels production, 381–383
  requirements, 502–505
  from Stockholm to Rio to New York, 226–229
Sustainable bioenergy, 13
  governance, 235
  policies and measures to support, 489–537
    bioenergy policy context and bioeconomy, 489–498
    GHG emissions and carbon accounting, 507–520
    sustainability certification and standards, 498–507
  support schemes for bioenergy, 520–529, 523*t*
    latest developments in support polices for renewables, 526–528
    support to fossil and renewable energy, 525–526
  for sustainable development, 529–537
    benefits and risks of bioenergy, 531
    bioenergy and UN SDGs, 529–531
    bioenergy integration in agriculture and forestry, 535–536
    energy supply and energy access, 531–533
    food security, 533–534
    renewable energy integration into energy systems, 537
    rural development, 534–535
  waste and residues for, 470–476
Sustainable biomass, 229–232
  early warnings from last century, 230*b*
  sustainability initiatives and schemes for bioenergy, 231*f*
Sustainable Biomass Program (SBD), 11–12
Sustainable development, 226

Sustainable development goals (SDGs), 6, 8, 226, 228*f*, 252–253, 253*t*, 342–343, 529–531
links to, 252–253
Sustainable energy, 529–530
Sustainable intensification of agriculture, 535–536
Sustainable potential, 436
Sustainable rural development, 234
SVO. *See* Straight vegetable oils (SVO)
Swedish consumers, 323
Swedish strategy, 329
Sweet sorghum (*Sorghum bicolor* (L.), 45, 116–117
Switchgrass (*Panicum virgatum*), 49, 132, 154
Synthesis gas (syngas), 135, 164, 192–194, 194*f*
Synthetic NG, 419
System, conventional, 207–208, 208*f*, 209*f*, 210*f*

T

Tax exemptions, 522
Technical potential, 436
Technology Readiness Level (TRL), 202
Tendering schemes, 522
Tertiary biomass, 26
Thai bioethanol system, 394–395
Thailand
land–water–energy nexus of biofuels policy mandate, 388–398
nexus management for sustainable bioethanol production, 397–398
Theoretical potential, 436
Thermal energy and/or electricity generation costs, 171–172, 171*t*
Thermochemical
biofuels from thermochemical-based technologies, 413–418
biomass conversion processes, 158–159, 159*f*
conversion, 67–68, 466
processes, 135–137, 148, 180, 187–189
Thermochemical and biochemical technology combination, 138
Third-generation biofuels (3G biofuels), 113–114
Third-generation biofuels (3G biofuels), 138
13th meeting of the Conference of the Parties (COP13), 344
Thorbecke's framework, 368–369
Torrefaction, 406, 488
Trade role in bioenergy
bioenergy markets and trade, 476–477
global solid biomass trade, 480–485
international bioenergy trade, 477–479
international trade in biofuels, 479–480, 480*f*
outlook, 485–489
Transaction matrix, 267
Transesterification process, 128
Transport biofuels, 39
Transportation biofuels, 187–189
Trash. *See* Straw
Trichosporon oleaginosus, 411
Trichosporon sp., 411
TRL. *See* Technology Readiness Level (TRL)
21st Conference of the Parties (COP21), 450
2 Degree Scenario (2DS), 448–451, 453
2030 Agenda, 8

U

Ultra-poor group, 300
UN Conference on Environment and Development (UNCED), 226
UN Food and Agriculture Organization, 347
UN sustainable development goals, 529–531
United Nations, Department of Economic and Social Affairs (UNDESA), 299
United Nations Framework Convention on Climate Change (UNFCCC), 342–343, 450, 463
United Nations Statistics Division (UNSD), 64–66, 298–299
United States cellulosic ethanol plants, 63
Unsustainable production of bioenergy, 347–348
Urbanization, 299–300
US Aquatic Species Program (ASP), 86–87

US Energy Information Administration (EIA), 79
US Environmental Protection Agency, 67–68
US Life Cycle Inventory Database (LCI), 509–510

V

Value chain
approach, 208
assessment, 211
sustainability assessment, 214–216
biorefinery with conventional reference system, 215*f*
estimating cost and revenues of biorefinery plant, 217*f*
estimating cumulated energy demand of biorefinery, 216*f*
estimating cumulated fossil energy demand, 216*f*
estimating GHG emissions, 217*f*
key characteristics of biorefinery value chain, 215*t*
Value-added products, 26
Vegetable oil (VO), 47, 113–114
biodiesel from, 127–129
"Vegetal hair". *See* "Fa cai"
Vegetal waste, 41–42
Vegetation strips, 536
Verein Deutscher Ingenieure Standard (VDI Standard), 187
Vertical static biofilm systems, 90–91
Village scale, 422–423
domestic micro-CHP technologies, 422*t*
Vinasse, 124
Voluntary pledges, framework for, 342
Voluntary sustainability schemes, 489

W

Waste
biomass, 38
biorefineries approach, 74
incineration, 74
management, 15, 471–472, 475
prevention, 471–472
for sustainable bioenergy, 470–476
agricultural and forest residues, 472–473
biogas production, 473–474
circular economy and waste, 470–472
opportunities for waste and residues for energy, 470
outlook for use of waste, 475–476
Waste Framework Directive (WFD), 475–476
Waste-to-energy (WtE), 420–421
technologies, 64, 74
on waste disposal, 421*f*
Wastewater, 80–81
Wastewater treatment (WWT), 80, 468–469
Water, 383, 449
availability footprint, 384
content and heating value of biomass, 34–35
deprivation, 385
electrolysis, 418
scarcity, 391–394
footprint, 384–385, 391
Water footprint (WF), 384
assessment, 384–385
Water scarcity index (WSI), 385
Water stress index (WSI), 384
Water vapor ($H_2O_{(s)}$), 160–161
Waxes, 33–34
WBA. *See* World Bioenergy Association (WBA)
WEC. *See* World Energy Council (WEC)
Wet milling process, 125–126
WF. *See* Water footprint (WF)
WFD. *See* Waste Framework Directive (WFD)
Wheat (*Triticum aestivum* L.), 118
production, 118
straw, 130–131
"Wine", 124
WIOD. *See* World Input–Output Database (WIOD)
Women, development opportunities for, 315–316
Wood
bioethanol from, 209
by-products and residues from wood industry, 38

Wood (*Continued*)
pellets, 483–484, 488
certification systems, 502
market, 483–484, 488
observations from wood pellet consumer studies, 323
processing, 70
residues, 482
resources in forest and wooded lands, 52–54
biomass, 36, 150–152, 152*t*
Wood chips, 484–485
for bioethanol, electricity, heat, and phenols, 191
Wood fuel
changes in wood fuel production, 56–58
direct, 51
indirect, 51
production, 54–56
types, 51–52, 51*t*
World Bioenergy Association (WBA), 42–43, 61, 149–150
World bioenergy sector, 19
World Energy Council (WEC), 146
World Input–Output Database (WIOD), 272
WSI. *See* Water scarcity index (WSI); Water stress index (WSI)
WtE. *See* Waste-to-energy (WtE)
WWT. *See* Wastewater treatment (WWT)

X

Xylitol, 64–66

Y

Yarrowia lipolytica, 411
Yeasts, 409–410
species, 411
yeast/bacterial cultivation, microbial oils from, 418

Z

Zymomonas mobilis, 135

Kolibri 360 GmbH
Ferdinand-Jühlke-Straße 7
99095 Erfurt, Deutschland
produktsicherheit@kolibri360.de